Sewall Wright
AND
Evolutionary Biology

Wright holding a baby racoon at the Provine farm in Marathon, New York, May 1983. Photograph by Doris Marie Provine.

Sewall Wright
AND
Evolutionary Biology

William B. Provine

Science and Its Conceptual Foundations
David L. Hull, editor

The University of Chicago Press
Chicago and London

WILLIAM B. PROVINE is professor in the Department of History and in the Division of Biological Sciences, section of ecology and systematics, at Cornell University. He is the author of many books, including *The Origin of Theoretical Population Genetics,* also published by the University of Chicago Press.

This book has been brought to publication with the generous assistance of the Publication Subvention Program of the National Endowment for the Humanities, an independent federal agency.

The University of Chicago Press, Chicago 60637
The University of Chicago Press, Ltd., London

Library of Congress Cataloging-in-Publication Data

Provine, William B.
  Sewall Wright and evolutionary biology.

  (Science and its conceptual foundations)
  Bibliography: p.
  Includes index.
  1. Evolution—History. 2. Genetics—History.
3. Wright, Sewall, 1889–   . 4. Geneticists—
United States—Biography. 5. Biologists—United
States—Biography. I. Title. II. Series.
QH361.P87 1986   575′.092′4 [B]   85–24651
ISBN 0–226–68474–1

To Richard C. Lewontin and Ernst Mayr

Contents

Preface

Sewall Wright's life spans almost the entire history of genetics, including the various syntheses of genetics with evolutionary theory. He was born on December 21, 1889, just seven years after the death of Charles Darwin and in the same year as the publication of Alfred Russel Wallace's *Darwinism* and Francis Galton's *Natural Inheritance,* which effectively began the modern science of biometry. He was ten when Mendelism was rediscovered and twenty in the year Thomas Hunt Morgan turned his attention to *Drosophila* and Mendelian heredity. The first of Wright's scientific papers to be published appeared in 1912.

Fifty-six years later, Wright published the first of his four-volume summary work on the process of evolution in nature. These four large volumes, published between 1968 and 1978, draw heavily upon the scientific literature from the whole century (Wright 188, 193, 198, 200; throughout this book Wright's publications are referred to by number from the list of Wright's publications found in the end matter). In the fourth volume, on variability within and among natural populations, Wright speaks about F. B. Sumner's first reports on variability in natural populations of the deer mouse *Peromyscus,* published in 1915, and also about the studies on allozyme variability in populations of *Drosophila* pioneered by R. C. Lewontin and J. L. Hubby, published beginning in the later 1960s. Wright read each of these works, separated by more than fifty years, at their times of publication.

Wright made basic contributions to transmission genetics, physiological genetics, biometry, theories of breeding, and evolutionary theory. Thus an examination of his scientific work provides one perspective upon the whole development of genetics in relation to evolutionary biology in the twentieth century. I say one perspective advisedly because the relationship of genetics and evolutionary biology is extraordinarily complex and cannot possibly be encompassed by the work of any one individual. Indeed, Wright's views on evolution have remained remarkably consistent from 1925 on, but evolutionary theory as a whole has changed considerably during the same period. Throughout this biography, I will view the development of Wright's work in relation to the larger context of the interaction of genetics and evolutionary biology.

It was when I was writing my doctoral thesis on the origins of theoretical population genetics (Provine 1971) that my biology advisor, Richard C. Lewontin, and his postdoctoral student, Joe Felsenstein, first suggested that I interview Sewall Wright, who graciously agreed. I drove to Madison in the spring of 1967 for this first meeting with him. He answered my questions fully and with amazing clarity of recall. Two years later, in the final days of writing the thesis, I again interviewed Wright in Madison for an afternoon, this time tape-recording the session. He was again most helpful and informative.

I did not see Wright again until March 1976. Stimulated by the conference on the evolutionary synthesis organized by Ernst Mayr in May and October of 1974 (Mayr and Provine 1980), to which Wright was not invited, I wanted to speak to Wright about his role during the synthesis period. I was particularly interested in talking to him about Theodosius Dobzhansky and their collaboration in the 1930s and 1940s. Dobzhansky had told me at Mayr's conference that he and Wright had corresponded frequently during those years but that he had saved neither his carbons nor Wright's letters. Wright suggested during the interview that he probably had little influence on Dobzhansky. In response to my question about surviving correspondence with Dobzhansky, however, he indicated that he probably still had some that he would bring later. The next day Wright produced a stack of several inches of correspondence with Dobzhansky.

The Wright/Dobzhansky correspondence was an extraordinary record of their collaboration, showing that Wright's assessment of his influence upon Dobzhansky was far too modest. All biologists acknowledge Dobzhansky's great influence in evolutionary biology through his book *Genetics and the Origin of Species* (1937, 1941, 1951), his Genetics of Natural Populations series (Lewontin et al. 1981), and his students. If indeed Wright had significantly influenced Dobzhansky, then Wright's importance in the synthesis period was much greater, though more indirect, than I had earlier realized. Still I had no thought of writing a biography of Wright, perhaps because Ian Shine and Sylvia Wroble were planning to write a book about Wright along the same lines as their book on Thomas Hunt Morgan (Shine and Wroble 1976).

I continued, however, to interview Wright and began to see ever more clearly that his biography would provide one way of viewing the whole synthesis of genetics and evolutionary biology in the twentieth century. Thus when by early 1978 the Shine/Wroble project had fallen through, the Wright biography seemed a natural undertaking for me.

Beginning in May 1978 I initiated with Wright a series of systematic interviews both in Madison and at my farm in Marathon, New York, where Wright came for one-week visits in September 1979, June 1980, and May 1983, interspersed by many shorter interviews. In all, we recorded about 120 hours of interviews covering all aspects of his life, publications, and correspondence; most of the tapes were professionally transcribed and then edited by the two of us. (Both the audio tapes and transcriptions are available for study at the Library of the American Philosophical Society in Philadelphia.) Wright also wrote out for me and his family a large notebook of recollections from early childhood to 1980.

The length of the interview process was significant. Some of the most important interviews came near the hundred-hour mark, long after the time I had originally thought would be the end of really productive sessions. Wright and I returned again and again to certain basic themes, for examples the tension between adaptive and nonadaptive evolution or the interpretation of his fitness surfaces, and my understanding increased not uniformly but in fits and starts. Sometimes a breakthrough would come only on the third or fourth try, or even later.

Wright placed in my hands all his scientific correspondence after 1915. This amounted to perhaps 15,000 pages, of which more than two-thirds were carbons of Wright's letters (his letters to others were generally longer than their inquiries). This collection is extraordinary for the insight it provides into the history of genetics and evolutionary biology in the twentieth century. In addition to the Dobzhansky correspondence mentioned above are major files of letters to and from R. A. Fisher and Motoo Kimura, the importance of which will be obvious in the biography. Almost every major evolutionary biologist asked for Wright's help on one quantitative problem or another. The sheer sweep of this correspondence, going from 1915 to 1983, is impressive for any scholar interested in this period. (Wright's correspondence is also available for study at the Library of the American Philosophical Society.)

Many times, in writing about Dobzhansky, A. H. Sturtevant, F. B. Sumner, William Castle, and many other scientists, I have wished they were alive to answer my many questions and to criticize my writing. It has been an extraordinary experience for me as a historian to read Wright's correspondence and published papers and, while my memory was fresh and the documents before us, to engage in many hours of intense conversation. Wright carefully read each chapter in draft (the first eleven chapters he read in two different drafts), making handwritten comments as he went, with the total going to more than one hundred pages.

I have heard some historians of science say that the objectivity of the historian necessarily suffers under such interactive circumstances, that objectivity is increased by not interviewing and certainly by not submitting drafts to the subject. There is a real danger here, but it can easily be exaggerated. The reader of this biography will see that I have little use for Wright's philosophical ideas or his famous surfaces of selective value and that I disagree with Wright's interpretation of the stability of his attitudes toward adaptive and nonadaptive evolution from the 1930s to the 1950s. We agreed to disagree before our first extended interview, and I do not think my historical assessments were seriously compromised by interacting with Wright.

The other side of the coin, the one that I insist upon, is that the interaction of the usual sources (correspondence and publications) with the interviews with Wright and his critiques of my draft chapters greatly enhanced my understanding of Wright's work in relation to modern genetics and evolutionary theory. Wright saved me from making literally hundreds of perfectly plausible or even convincing mistakes, the vast majority of which would never have been caught by the most discriminating and knowledgeable reader. This experience suggests that the literature of the history of modern science is crammed with reasonable-sounding falsehoods, many of which could have been avoided.

I am fully convinced by the experience of writing this book that unless historians of science set immediately to work seriously interviewing the many willing elder scientists, and encouraging them to critique our work, we will lose much understanding of how twentieth century science has developed. More major scientists are alive now than have lived in the history of the world, yet as a profession we historians of science concentrate heavily upon those who are dead. We should seriously reconsider this allocation of our efforts.

Acknowledgments

Many persons have been of great help to me in writing this book. Indeed, one measure of Wright's stature is the extraordinary willingness of these persons to devote so much time and energy to the improvement of a manuscript written about him and his work.

First and foremost, I wish to thank Wright himself. He devoted several hundred hours to the making of the book, and he endured my persistent questioning without complaint. Our interviews frequently went more than eight hours in a single day. I often called him on the telephone, sometimes several times a day when I was having difficulty understanding an issue. And I was greatly impressed that Wright never once chided me for disagreeing on some points with his own interpretation of the development of his thought, even though he was obviously very concerned with these disagreements.

My debt to Richard C. Lewontin and Ernst Mayr, to both of whom I have dedicated this book, is immense. Lewontin introduced me to population genetics and nurtured my education in this subject and its wider implications. Mayr has been a friendly but demanding critic for fifteen years and a dedicated correspondent. He has taught me much about evolutionary biology and its history. Lewontin and Mayr are not usually associated in the same breath as evolutionary biologists, but their joint influence is basic to my work and to this book in particular.

The encouragement of James F. Crow was crucial to my decision to undertake the Wright biography, and his support has been constant and deep. No one has as great a store of Wright stories as Crow, who also has written more about Wright than any other person.

In 1981, I circulated a few copies of the first seven chapters of the manuscript and received helpful comments from Mayr, Crow, Brian Parr, Richard Burian, Garland Allen, and Michael Wade. I rewrote those chapters extensively, in response particularly to the very detailed and thoughtful critiques of Allen and Wade. Sewall Wright's nephew, Theodore P. Wright, Jr. (son of brother Ted), provided many helpful comments on family history in chapter 1.

In the fall of 1984 a number of persons read the entire manuscript and contributed a wonderful array of critical comments. Mayr, Crow, Wade, and Malcolm Kottler wrote long and detailed critiques of every chapter (Kottler was my severest critic). Arthur Cain provided an extremely insightful and informative analysis of chapter 12. My graduate students Brian Parr and Betty Smocovitis offered substantive and stylistic suggestions for revision.

In all, the critics of the manuscript provided more than one hundred pages of commentary. I am deeply indebted to them. I am certain that the book is vastly better as a result of their thoughtfulness.

I interviewed in person or by telephone a large number of persons who helped in many ways. Among them were Motoo Kimura, Harlan Lewis, Maxime Lamotte, Thomas Park, R. A. Brink, M. R. Irwin, A. B. Chapman, Willys Silvers, Janice Spofford, Joseph J. Schwab, Giuseppe Montalenti, Franco Scudo, William Kimler, Stephen Jay Gould, John Beatty, and David Hull. Many people tried to straighten me out on the subject of Wright's concept of fitness surfaces, including Lewontin, James W. Curtsinger, Stevan Arnold, Russell Lande, Henry Schaffer, Michael Wade, Bruce Walsh, Simon Levin, Michael Turelli, and Conrad Istock. None of them is likely to agree with my assessment, but I am grateful to them for clarifying my thinking on the subject.

Whitfield Bell, Ted Carter III, Murphy Smith, and Steve Catlin of the Library of the American Philosophical Society have been extraordinarily helpful in meeting my needs with their unmatchable collection of personal papers related to the history of genetics and evolutionary biology. I also wish to thank the librarians at the Regenstein Library of the University of Chicago where the papers of Quincy Wright are preserved.

Wright, Mayr, Cain, Crow, Kimura, Lamotte and Ford all allowed me to quote from their correspondence. The following executors allowed me to quote from sources indicated: Sophie Dobzhansky Coe (papers of Theodosius Dobzhansky), J. H. Bennett (correspondence of R. A. Fisher), Thea Muller (correspondence of H. J. Muller), and William Sturtevant (correspondence of A. H. Sturtevant). Other than these sources, the correspondence quoted in this book is from the collection at the Library of the American Philosophical Society. Many thanks to all who gave me access to these personal papers.

I am grateful to the American Philosophical Society for a crucial grant that made possible the transcription of the interviews with Wright and to the National Science Foundation, Section of the History and Philosophy of Science, for grant support during most of the time I was writing the book.

The Department of History at Cornell University has been wonderfully supportive. Without well-timed help from the College of Arts and Sciences (Dean Alain Seznec, Associate Dean Geoffrey Chester, Director of Finances and Administration Jack W. Lowe), the completion of this book would have been extended by many months or even years.

I am very grateful to Jill Bonnie Malefyt and Danaya Wright for extensive typing services rendered. I also wish to thank William Kimmler for the excellent index he prepared.

Finally, I want to thank Charles and Edith Long for their constant encouragement, Doris Marie Long Provine for her love and support, and our two sons Charles and Stuart for their almost incredible understanding and patience.

1
Early Life and Education, 1889–1912

From an early age Wright was interested in his ancestors. Both sides of his family can be easily traced directly back to the early settlers of America. The family tree on his father's side goes back to the early 1500s in England, and on his mother's side one line goes back to several generations before Charlemagne in the seventh century. Mary Clark Green Wright (1833–1914), his paternal grandmother, lived with the family in Galesburg, Illinois, during the period of Wright's earliest memories. She often spoke about various ancestors. Tracing Wright's ancestry is simpler than for most because his parents were first cousins. After Wright had clarified in the early 1920s the calculation of the degree of inbreeding in sexually breeding organisms, he calculated his own inbreeding coefficient. It was 6.3%, about one hundred times the average.

Grandmother Wright talked a good deal about her father, the Reverend Beriah Green (1795–1874), who had been professor of Hebrew at Western Reserve College in Hudson, Ohio. (Founded in 1826, it became Adelbert College, now Case Western Reserve, in Cleveland in 1882.) One of his colleagues was Wright's double great-grandfather, Elizur Wright III (1804–85), professor of mathematics and natural philosophy. Both of these men were deeply opposed to slavery and joined William Lloyd Garrison's Anti-Slavery Society. Their outspoken antislavery views aroused much opposition at Western Reserve College, and both resigned their professorships. Green moved to Whitesboro, New York, where he assumed the presidency of the Oneida Institute, preached in a local church, and lived on a small farm. He became active in the underground railway, his farm serving as a way station on the route. Grandmother Wright often recalled the days of the underground railway, showing great admiration for her father and for the famous Harriet Tubman, the black woman who made many trips South to bring slaves back with her. She often stopped at Green's farm on the way to Canada. Sewall Wright's father, however, thought Beriah Green was a narrowly focused reformer and activist.

The ancestor whom Wright's father continually held up as a model for his children was Elizur Wright III (1804–85), Sewall's great-grandfather. This man's father, Elizur Wright II (1762–1845), graduated Phi Beta Kappa from Yale in 1781 and published papers on fluxions (the differential calculus) in the *Proceedings of the Connecticut Academy of Sciences*. In 1810 he moved from South Canaan, Connecticut, to Ohio and became a farmer, a profession that did not prevent him from thinking and writing about fluxions. He was one of the founders of Western Reserve College and was for many years a trustee.

His son Elizur Wright III also graduated Phi Beta Kappa from Yale, in 1826. He then became the head of a Congregational church school in Groton, Massachusetts, for two years, during which time he met Susan Clark, his future wife. Like his father, Elizur Wright III was at first very religious; when he could not secure the money to study for the ministry, he decided to earn a living by distributing religious tracts. He was very successful at this job for a year but gave it up when he received an unexpected offer of a position as professor of mathematics and natural philosophy at Western Reserve College, where his father was a trustee. He taught mathematics, chemistry, geology, atmospheric sciences, and electricity. His most consuming interest became the abolition movement, and in 1833 he resigned his professorship and moved first to New York and later to Boston as secretary of William Lloyd Garrison's American Anti-Slavery Society and editor of Garrison's newspaper.

He and Garrison had a falling out in 1840, leaving Wright without a job. He disliked the idea of returning to college teaching or farming, the two obvious choices, and decided instead to pursue a literary career. He had always written light verse, but knowing that this would produce little income, he translated and published La Fontaine's *Fables,* making his living for two years selling it in a deluxe edition. He sold it in England as well, visiting there for seven months in 1844. Robert Browning aided him, and he also met Wordsworth, Carlyle, and many of the prominent social reformers of the day. He learned about the English life insurance business, and greatly admired the city parks of London.

Returning to the United States, Elizur Wright founded a newspaper in Boston, the *Chronotype,* which provided a living and an outlet for his social ideas. He attacked slavery, liquor, social stratification, sex prejudice, capitalism, the tariff, and other well-accepted features of contemporary society. One of his favorite subjects was life insurance. In the 1840s the life insurance business in England and America was unsound—rates were too low, and when companies failed, which happened frequently, customers could not get back the cash value of their policies. Wright invented the "arithmeter," a big cylindrical slide rule, and used it to calculate life tables based on Americans' longevity; the life insurance companies immediately adopted the tables. He agitated for legislative control of the companies, soon acquiring the title "Father of Life Insurance." The Massachusetts legislature passed many of the reforms he advocated and created the office of life insurance commissioner basically for him.

He also pushed successfully for the creation of a park between Medford and Melrose, Massachusetts, eventually named Middlesex Fells. Elizur Wright III was unquestionably an extraordinary man with a wide range of abilities and interests. It can hardly be surprising that Sewall Wright's parents would hold him up as an example to be emulated by their boys. Sewall's father so admired Elizur Wright III that he wrote, with the help of his wife, a full biography, the draft of which he completed just before his death (Wright and Wright 1937).

The relations among Sewall Wright's various ancestors were not altogether harmonious. Grandmother Wright disliked strongly one particular direct ancestor from Sewall Wright's mother's side of the family: Judge Samuel Sewall (1652–1729), famous among other things for his role as one of the judges in the Salem witch trials and for his later public misgivings about that role. The probable reason for her antipathy to Samuel Sewall was that other direct ancestors, from her husband's line, were put to death by Samuel Sewall's court during the witch trials. Martha Corey, the third wife of Giles Corey (born 1612), was condemned and hanged as a witch in 1692. Giles Corey, then eighty years old, objected strenuously to the trials and was in turn charged by the court. He refused to plead either guilty or innocent, because either way he would lose the legal right to leave his property to his relatives. The court invoked the English law requiring that weights be piled upon him until he pled or was killed. Corey refused to plead and was crushed to death after three days. Not until the year 1860 did the descendants of Giles Corey and Samuel Sewall (Mary Wright, 1838–79, and Joseph Sewall, 1827–1917) marry. Some family controversy undoubtedly ensued, and Grandmother Wright, who was twenty-seven years old in 1860, was still angry about Judge Samuel Sewall's actions in the later 1690s. After reading Fiske's account of Samuel Sewall's public recantation of his role in the witch trials, Sewall Wright attempted to give him some credit in the presence of Grandmother Wright. Her comment was that the recantation had not restored the lives of the supposed witches.

Sewall Wright's parents were Philip Green Wright (1861–1934) and Elizabeth Quincy Sewall (1865–1952). Her father, Philip's aunt's husband, Joseph S. Sewall (1827–1917), was a civil engineer whose specialty was designing and constructing bridges and railroads. In 1887 Joseph Sewall was constructing a bridge over the Mississippi in St. Paul, Minnesota. One of his employees was his twenty-six-year-old nephew, Philip, who in 1884 had graduated from Tufts, the Universalist college in Medford, Massachusetts, with a civil engineering degree. After teaching for two years at Buchtel, a Universalist college in Akron, Ohio, he went to Harvard and earned an M.A. degree in economics in 1887. Then he went to work for his uncle, Joseph Sewall, in St. Paul. He and his cousin Elizabeth were married in February 1889.

The young couple settled in Melrose, Massachusetts, while Philip Wright worked for the New England Mutual Life Insurance Company of Boston. Sewall Green Wright (he later dropped the middle name) was born on December 21, 1889, and his brother (Philip) Quincy a year later on December 28, 1890. Philip Green Wright greatly enjoyed the life of the mind and in 1892 accepted a teaching job at a small Universalist college, Lombard, in Gales-

Figure 1.1. Philip Green Wright.   Figure 1.2. Elizabeth Quincy Sewall Wright.

burg, Illinois. There he taught economics, mathematics, fiscal history, astronomy, and writing. Lombard had fewer than one hundred students and
each faculty member taught many courses. Sewall and Quincy Wright would
take many of their father's courses when they later attended Lombard.

Early Life in Galesburg

When he was four or five, Sewall Wright was questioned by his family about
what he remembered from the Medford years; he was two years, nine months
old at the time of the move to Galesburg. Thinking back, Wright remembers
recalling at least three separate and distinct incidents that must have occurred
in the summer of 1893, just before the move. He does not, of course, remember these incidents directly. His parents visited the Columbian World's Fair
in 1893 in Chicago (whose midway later became the Midway of the University of Chicago, where Wright would be a professor from 1926 to 1955),
leaving him and Quincy in the care of Grandmother Wright. Both she and Sewall Wright had independent personalities, despite the great difference in
age. Sewall distinctly remembers that Grandmother Wright tried to feed him
some cereal upon which she had poured somewhat soured and separated

creamy milk. Grandmother Wright tried to force it on him, and he stubbornly refused. As she became more severe, he became increasingly stubborn.

Wright's memories of his grandmother from this summer were not, however, all unpleasant. She loved to sing old Irish and Scottish songs, such as "Bendemeer's Stream," "Sweet Afton," and "Look down, look down Rapunzel and let your golden tresses down." Sewall loved these songs. When he visited our house in June 1979, one evening he spied my wife's flute case on the shelf. He took out the flute and began to play many of the songs he first learned from Grandmother Wright. Before this, I had no inkling either that he knew such songs or that he played the flute. It turned out that playing songs on the flute had long been one of his deep private pleasures. At this time, the songs he had known since childhood obviously gave Wright much pleasure.

After the summer of 1893 Wright's memories are profuse and include family activities, playmates, geographical details, trips, and toys. From the summer of 1894 he recalls in detail the lot upon which his father had their new house built, including a particular Northern Spy apple tree that unfortunately had to be cut down to make room for the house. Especially keen are Wright's recollections about natural history. For example, from the midsummer of 1894 he recalls being much impressed by a large rookery of Great Blue Herons that he saw at his Aunt Susie's cottage on Lake Minnetonka, Minnesota.

The third and last child, Ted (Theodore Paul), was born on May 25, 1895. Sewall recalled that there was a late spring snow the preceding night and that he and Quincy had suggested the middle name of Paul for the baby, in honor of one of their friends, and their parents accepted the suggestion. The three boys naturally played and competed together.

Sewall and Quincy did not enter school at the usual time, probably because Philip and Elizabeth Wright were skeptical about the quality of the local grade school, which was on "the wrong side of the railroad tracks." Elizabeth began reading to the children when Sewall was about five. He recalls Andrew Lang's *Blue Fairy Book* as among the first to be read, followed by Lang fairy books in other colors, stories of Greek mythology, Hawthorne's *Wonder Book* and *Tanglewood Tales* (two books he received on his birthday or Christmas 1896), and Lewis Carroll's *Alice in Wonderland* and *Through the Looking Glass*. Both Sewall and Quincy were reading and writing before they were six.

They also learned some arithmetic from their mother, mostly by questioning her. Wright recalls being surprised when he discovered that Grandmother Wright multiplied by the digits of the multiplier from left to right in direct order, instead of the usual reverse order from right to left. He also found an arithmetic book in his father's library and learned much from it, including the rules for extracting square roots and cube roots from any given number.

Wright's first really intense interest in science came when he was about six or seven. He had watched his mother using the balance scales in the kitchen and became fascinated by the relationships between the weights, distances from the fulcrum, and the point of balance. His parents explained to him the principles and how to calculate moments of force. He then set to work with a thin yardstick and the weights from the kitchen scales and experimented to see if the calculated results were borne out by the rod and weights. They were, accurately enough, and Wright was very excited. What he liked most of all was that the nice relationships of elementary arithmetic could be applied unambiguously to observable physical situations.

This very first enthusiasm for science anticipated a deep characteristic of Wright's mature scientific work. From an early age he greatly enjoyed trying to see physical events or processes in the simplest possible quantitative terms. In talking to Wright about his early interest in scales and moments of force, I was reminded of Isaac Newton's work. First, Wright mentioned to me that the laws governing levers looked somewhat like Newton's laws of motion. At the same time, I drew in my own mind the similarity between what Wright was telling me about his early fascination for the principles of levers and Newton's comment about the general applicability in nature of the method that had allowed him in the *Principia* to deduce the laws of planetary motion and the actions of the tides from so simple an assumption as the inverse square law of gravitational attraction. "I wish," said Newton in the preface to the first edition of the *Principia*,

> we could derive the rest of the phenomena of Nature by the same kind of reasoning from mechanical principles, for I am induced by many reasons to suspect that they may all depend upon certain forces by which the particles of bodies, by some causes hitherto unknown, are either mutually impelled toward one another and cohere in regular figures, or are repelled and recede from one another. These forces being unknown, philosophers have hitherto attempted the search of Nature in vain; but I hope the principles here laid down will afford some light either to this or some truer method of philosophy. (Thayer 1953, 10–11)

Wright, like many other scientists from many fields, hoped with Newton that the world could be analyzed into quantitative relationships. When Wright faced a difficult biological problem, he simply thought in terms of quantitative analysis. Mendelian heredity would have a predictably strong and immediate appeal to Wright.

Through the years, the Wright family managed to preserve a striking document indicating both the level of interest and sophistication that young Sewall had developed prior to his enrollment in school. In the late summer of 1897, when Sewall was seven-and-one-half years old, he wrote a little pamphlet entitled "The Wonders of Nature." The pages are plain brown wrapping paper, sewn into a booklet of twelve pages, eleven of which contain comments by Wright and the twelfth a pasted-in diagram. The handwriting is all in capital letters (Wright did not learn script until he entered grade school in the fall of 1897). The pamphlet appears to be styled upon Reverend J. G. Wood's *Illustrated Natural History of Animals,* which Wright mentions in the text of the pamphlet. Wood was a popular British writer on natural history topics; by the early 1870s he had more than six major works in print in the United States as well as in England. His books typically interspersed figures and elaborate woodcuts liberally within the text, which was easy to read and delightfully informative (though frequently based upon hearsay). The pamphlet clearly shows that, using Wood's style as a model, Wright was observing and thinking for himself about natural history topics.

The cover page carries the title "The Wonders of Nature" and pictures the earth with the stars above. The earth has figures of trees and animals on its surface. In the center of the earth is the name S. G. Wright. The pamphlet gives a good idea of what a keen observer of nature Wright was at such an early age. In a discussion about the function of fowl gizzards, young Sewall wrote, "one night when we had company, we had chicken pie. Our Aunt Polly cut open the gizzard, and in it we found a lot of grain, and some corn."

This little notebook reveals that, at age seven-and-one-half, Wright had been reading books about natural history and he understood and appreciated many of its details. More important, he was already making his own careful observations and trying to explain what he saw. Wright had seen the constellation Lyra through the small telescope his father used to show his astronomy class such sights as the rings of Saturn, moons of Jupiter, and multiple stars. Other observations in the notebook, such as those of squashes, wrens, bees, and marmosets, Wright made on his own.

The last page of the booklet was a woodcut Wright found in a pamphlet he had obtained in Universalist Sunday school. This diagram depicts many of the best-known large dinosaurs. In our interviews, Wright distinctly recalled that he believed in evolution at the time he cut out this diagram and pasted it in his pamphlet. There thus emerges the vision of a boy who had developed a very strong interest in natural history at a very young age.

Philip Wright, however, did not encourage young Sewall's interests in natural history. Instead, he encouraged the boy to appreciate poetry. For several years on Sunday mornings Philip read poems or recited them from his extensive memory and assigned poems to each boy for recitation on these occasions. Sewall recalls particularly enjoying his father's recitations of the poems of Kipling and personally reciting verses from Keats, Wordsworth, Tennyson, Whittier, Longfellow, and others. He disliked Browning. Philip frequently took the boys on walks, reciting all the while. Sewall and Quincy, however, became more interested in the butterflies and birds than in the recitations, much, in Sewall's recollection, to his father's disgust. The walks became devoted more to natural history than to poetry. Philip's generally negative reaction to Sewall's interests in natural history continued for many years. Certainly Sewall noticed that his father approved more heartily of his brothers' activities and intellectual interests than he did of Sewall's work in biology.

Elizabeth Wright, however, had a strong interest in natural history and collected a substantial number of natural history books for her own personal library. She encouraged the boys in natural history interests by giving them, for Christmas and birthdays, books by Mabel Osgood Wright (a popular writer on natural history, but not a relative), Ernest Seton-Thompson, Paul du Chaillu, and others. Sewall's first acquaintance with the works of Darwin would later come from the books in his mother's library.

Sewall and Quincy went to public school beginning in the fall of 1897. School officials questioned them to determine the grade in which they should be placed. Sewall read the passages requested, performed all the simple arithmetical operations, and happily volunteered that he could extract square and cube roots. Unaware of the very strong anti-intellectual atmosphere among the boys in the school, on the first day he was immediately taken to the eighth grade classroom where they were studying mathematics and, standing on tiptoe to reach the board, extracted the cube root of a number put to him. He discovered that the boys were disgusted by this spectacle, and thereafter he determined to exhibit his knowledge as little as possible in front of a class. He was put in the third grade and promoted the next year into the fifth, thus spending five instead of the usual eight years in grade school.

Grade school was generally a disappointment and a bore for Wright. He learned very little that he had not earlier learned at home from his parents or on his own, or with Quincy. He and Quincy, for example, learned geography from collecting stamps before they had the subject in school. He learned some useful things in school such as writing script (not too well), diagramming sentences, memorizing the succession of presidents to McKinley, and reading assigned books. He finished the assigned readings for the whole year in a few months, and in 1980 he recalled vividly the utter boredom of "long afternoons with nothing to do but listen to the teacher trying to teach boys to read who had no interest whatever in doing this" (Wright's auto-biographical notes, 9). He learned more about spelling (imperfect spelling is frequent in "Wonders of Nature") but found the subject dull. The teacher in one grade gave a spelling test with fifty words each Friday, the results to be announced on Mondays. Usually a half-dozen or so girls spelled all correctly each test; boys rarely did. One boy managed to misspell all fifty for one test and became an instant hero among the boys.

Wright, facing the hostility to learning among his peers and being generally two years younger than his classmates, learned never to volunteer in class, but he did answer fully when called upon. He always felt much out of place in the grade school. For the first two years he and Quincy played almost exclusively with each other. Wright recalled that the Spanish-American War broke out during the first year in school and that he and Quincy replayed the battles of Santiago and Manila with their little toy boats. They explored the countryside all around Galesburg, greatly enjoying the sights and the company of each other.

Sewall was usually quiet but well behaved in school. Good behavior was not admired by his classmates, so to break his steady stream of As in deportment, he and a friend gathered large numbers of tent worms and brought them into the classroom hidden in their caps. Distributed around the classroom, the worms produced the predictable excitement, especially from the girls. Even so, Wright only got a C in deportment for the month, a decline hardly worth the effort. Charles Darwin's assessment of the value of grade school to him applies equally in the case of Wright: "The school as a means of education to me was simply a blank" (Barlow 1958, 27).

Wright's reading, of course, was not confined to the school reader. For several years after they began attending school, Elizabeth continued to read books to Sewall and Quincy, with small Ted looking on. She read several of Scott's novels, Kipling's *Jungle Book*, Dodge's *Hans Brinker and the Silver Skates*, and Edith Nesbit's *Wouldbegoods*. On their own, Sewall and Quincy read Howard Pyle's versions of the King Arthur and Robin Hood stories. Sewall read other Scott novels, most of Mark Twain, Jack London's *Call of the Wild, Sea Wolf*, several of Thackeray's novels, many of Dickens's novels, most of Stevenson's novels more than once, and many others. He also read Rawlinson's four volume translation of Herodotus. Wright has always read a great deal outside of his special academic interests, but after high school his interests turned much more to biographies and histories rather than to novels.

Wright's natural history interests were furthered in the summer of 1903 when he, Quincy, and two girls took a field natural history course with Professor Herbert V. Neal of Knox College. Neal had taken his doctorate in E. L. Mark's laboratory at Harvard, working closely with Charles Benedict Davenport and publishing jointly with him. Neal was a friend of Philip Wright. He took the children on trips observing birds, insects, and marine invertebrates. Sewall Wright recalled in particular learning about the behavior of bombardier beetles from Neal, who was a Darwinian and explained the tenets of that version of evolutionary biology. Neal later went on to a career in biology at Tufts College.

Wright's most serious natural history interests in these years were butterflies, moths, and birds. He and Quincy had large boxes with cork bottoms. One was devoted entirely to butterflies and moths, another to insects, mostly large ones collected under an outdoor electric light. He and Quincy collected bird eggs (only one from each nest) and abandoned bird nests. They carefully identified all the local birds. Beginning in 1902, Sewall began recording the first sightings of birds in the spring. The records for the spring of 1902 are in a small (3 in. x 5 in.) memorandum book sent out free as an advertisement by World's Dispensary Medical Association of Buffalo, New York, and London, England, purveyors of Dr. Pierce's "Pleasant Pellets" and other medical aids. It is a striking contrast to see in the same little booklet a careful listing of the first sighting of sixty-nine different birds, along with blatantly outrageous claims that Dr. Pierce's "Golden Medical Discovery" or "Favorite Prescription" can completely cure heart disease, kidney and liver failure, tuberculosis, blood diseases, cancer, loss of potency, homosexuality, and almost all other organic or psychological ailments. Thus opposite Wright's entry, "66. bay breasted warbler. bay on crown and breast. Date: May 6" comes the following testimony from Mrs. Maggie Spelts, who had suffered from terrible throat troubles all her life: "I have taken eight bottles of Dr. Pierce's Favorite Prescription and three of Golden Medical Discovery, four of Dr. Pierce's Extract of Smart-Weed and six vials of his Pleasant Pellets. I feel that I am cured, and feel like a different person; I look well, and can do all my work."

Many of the birds reported, such as the Virginia Rail and some of the seven different species of warblers, are either infrequently seen or difficult to distinguish, as in the six species of sparrows. Wright went long distances in making the sightings, often walking several miles before breakfast. On Saturdays he frequently walked to a woods four miles south of his home. He kept accurate records in 1902, 1903, 1906, 1907, and 1909. In the latter two years he also recorded nests. He told me in considerable detail about following a pair of Bell vireos for a long time before locating their nest in a secluded little gully overhung by bushes. He also told me about the difficulties of locating the nest of a pair of Bluegrey gnatcatchers high in a tree. He described the triumph of climbing this difficult tree, climbing out on a branch overhanging the nest, and stretching down on the bending branch to collect an egg.

Sewall Wright was always a quiet child, rather shy and not given to quick, clever conversation. Both of his brothers, however, were outgoing and more gregarious. Soon after beginning grade school, Sewall and Quincy found an older friend, Willis Rich, son of the chemistry professor at Lombard. Four years older than Sewall, Rich allowed the brothers to tag along on all sorts of excursions. He was interested in natural history and later became an expert on the Pacific fisheries. He and Sewall corresponded irregularly for about fifty years. One excursion with Rich especially stuck in Sewall's memory. On this occasion, Rich led the brothers to the city dump, composed mostly of horse manure and garbage. Rats were everywhere, their tunnels honeycombed through the dump. Older boys were there killing rats with baseball bats. Sewall was watching this from the top of the dump when the chunk he was standing on crumbled and he rolled down the side. At the bottom, he and the older boys were shocked to see that he was clutching a large rat, held so it could not bite him. The older boys told him to drop it; he did, and they killed it instantly.

Gradually, about their third year of grade school, Sewall and Quincy began to be acquainted with boys several blocks away from their home. Then in the winter of 1901 the Wright boys joined with a few others to organize a social club they named Jacks of All Trades, from a book of that title given to Quincy for Christmas. The book gave directions for constructing almost anything, from cabins and treehouses to rabbit traps and diving boards. Sewall was the first president. The club became the center of their social lives. The club had a constitution, fielded athletic teams (baseball, track, and football), sponsored numerous parties and bobsled rides in winter, picnics, and excursions for swimming or roaming in the woods. Sewall began as fullback on the football team but soon changed to quarterback because of his small size. Ted joined in and became a gifted athlete in track, basketball, and football. (While at Lombard the *Chicago Tribune* rated him the best quarterback in small colleges in Illinois.)

JOAT lasted until early 1905. Sewall has many fond memories of the club. He continued playing quarterback even after JOAT ceased to exist and has clear recollections about games he played for a team called the Mohawks in the fall of 1905. One of the favorite activities of JOAT was swimming and diving. Sewall never had any fear of the water, even before he learned how to swim. In the summer of 1900 the family went to Fish Creek on Green Bay in northern Wisconsin. Green Bay was particularly beautiful at this spot, and there Sewall learned how to swim dogpaddle style. The next summer he learned a very efficient sidestroke, and swimming and diving became a favorite activity, the sport in which he was most proficient. JOAT visited many different ponds and made a springboard and a ten-foot platform at one. The clay pits at Purington Brick Works was an exciting spot because there were ledges from which to dive. Sewall worked his way up to dives of about twenty feet vertical drop. Swimming and diving remained a favorite activity well into his eighties.

Sewall and Quincy presented a number of shows to the neighborhood during the years 1901 and 1902, usually charging one cent admission. One was a pantomime enactment from Herodotus of two Egyptian thieves; another was drawn from a B'rer Rabbit tale. Another, despite Elizabeth Wright's disapproval, came from a dime novel about Jack Harkaway and the pirates. One scene required planting a stake with a person's head on it. While practicing this scene in the attic, Sewall planted his stake with homemade head right into a knothole and into the plaster lathe of the ceiling below. The shower of plaster into the room below caused much annoyance and the show was postponed. Several of the shows involved acrobatics, including Sewall doing forward and backward flips from a springboard into a net.

The family expected the boys to work. Sewall's jobs included pumping water from the cistern to the holding tank in the attic, doing dishes, mowing the lawn, and working in the garden. He took on a paper route, making seventy-five cents a week. He and Quincy also earned money from lemonade stands. Wright recalled one occasion in 1901 when a customer came to the lemonade stand and announced that President McKinley had just been shot.

High School Days, 1902–1906

Sewall Wright entered high school in September, 1902. A high school education was not required by the state, and most of the grade school students with an antipathy to learning went to work or were married after completing the eighth grade. Most of the students who attended the public high school intended to go on to college, usually Knox College in Galesburg, but a few went to Lombard or other colleges. The quality of the teachers and students was far higher than in the grade school, with a correspondingly higher level of intellectual life.

Wright took Latin all four years. He has later said on many occasions that he was not very good in languages, but he did enjoy Latin. He read the standard Caesar's *Commentaries* in the first year, followed by Cicero's orations against Catiline the second year. Wright later said he liked to spout Cicero's orations at unexpected times. In the third year he translated much of book 1 of Virgil's *Aeneid* into rhyming pentameters, after the style of Pope's translation of the *Iliad* and the *Odyssey*. In the final year the class read many of Ovid's works. Wright translated *Atalanta* and *King Midas* into unrhymed hexameters and *Atalanta* also into a very different meter, following the style of Macaulay's *Lays of Ancient Rome*. The following is a selection of "From Atalanta's Last Race" in both versions, as preserved in one of Wright's notebooks:

> Trumpets blow loudly the signals and both of the racers dart swiftly
> Over the sandy arena, so lightly they seem not to touch it,
> You would have thought o'er the waves they could skim without
>   wetting their sandals
> Or over ripening fields of golden eared grain without sinking.
> Courage is given the youth by the shouting and cheers of spectators
> Saying "now, now is the time to hasten and make a great effort
> Hasten, Hippomenes hasten, now use all of your power.
> Banish delay and you conquer." Tis doubtful whether the hero
> Or Atalanta rejoiced more greatly on hearing this shouted.

> Then loudly blow the trumpets
>   The signal for the start
> And swiftly from the barriers
>   Both of the racers dart
> So lightly go their footsteps
>   They hardly touch the plain
> One would have thought that they could fly
> Over the sea with sandals dry
> Or o'er the tassels standing high
>   Of fields of golden grain.

> The youth is given courage
>   By shouting and by cheers
> Now is the time for effort
>   Haste are the words he hears
> Now use your strength to the utmost
>   If you would win the prize

> And hard it was to tell which one
> The maiden or Megareus' son
> Rejoiced more at the cries.

Wright also had a good course in German, in which he learned many German songs. Some of these songs he played on the flute and sang during his visit to our house in September 1979. The high school was also strong in English literature. He took courses in Shakespeare, taught by a Miss Richey, who was the daughter of the county surveyor for whom Wright worked during the summer after his second year of college. He read many other authors, including Tennyson and George Eliot (*Silas Marner* and other works). He became a member of the debating club and served as secretary during his senior year. One debate in which he took part was particularly exciting—with Springfield High School on whether United States senators should be elected by popular vote. Galesburg was assigned the affirmative and won the debate decisively.

The offerings in science were rather limited. No chemistry or zoology courses were taught. The physics course, however, was rewarding. Wright became interested in electric batteries and made a considerable number of them. He constructed an electromagnet by taking a piece of soft iron bent into a horseshoe and winding an insulated wire around it. Using one of his home-made batteries as a power source, this magnet could raise a piece of iron weighing almost nine pounds. He also became interested in the relations of pitch to the lengths of vibrating air columns and lengths of wires in musical instruments. As with the scales that had so interested him at an earlier age, Wright was delighted by the arithmetical relations found in the physical analysis of pitch production in musical instruments.

Wright took courses in algebra and geometry, enjoying the mathematical reasoning in both of them. He was fascinated in particular by the idea that curves could be represented precisely or approximately by equations. He noticed a heart-shaped curve of light on the table cloth. This observation stimulated him to work out the general formula for the family of curves in which the cardioid and limaçon are special cases. He devised ways of calculating the areas and lengths of these curves and others by averaging the mathematical functions that more or less closely approximated the curves for appropriate intervals. Only later did he discover that his averages were basically definite integrals, just not calculated by the use of infinitesimals. Certainly Wright's delight in quantitative reasoning can be seen in the enjoyment he felt in these high school courses in physics and mathematics.

If some of the seeds of his later work can be vaguely seen in high school physics and mathematics, the same cannot be said for his only biology course, a purely descriptive course in botany, largely concerned with the collection of wildflowers. Wright was indeed interested in the course. When his eyesight began to fail late in life and he could no longer identify birds at a distance, he again took up the identification of wildflowers as a relaxing pastime. At least to the level of genus, he identified every wildflower appearing on his walks around our farm in mid-June and mid-September 1979 and 1980.

Wright's intellectual interests during his high school years were, of course, not limited to the subjects taught in school. Philip Wright had studied Greek in high school and loved the language. He tried to encourage Sewall to learn Greek. Sewall did learn the alphabet and by using a dictionary was able to puzzle out some passages from a Greek Bible and Xenophon, but he did not enjoy the necessary memorization and gave up the kind of humanistic understanding that Philip wanted him to pursue.

Sewall began to use his knowledge of Greek, and the other languages he knew, in a very different kind of enterprise. He often read Chambers's *Cyclopaedia* when at a loss for other reading material, following from one article to another by the cross-referencing. On one occasion he read an article on Grimm's law, which delineated the correspondence of consonants from Latin and older Indo-Germanic languages to Low and High German. He became fascinated by the evolution of the Indo-European languages and began to keep notebooks on cognate words and grammatical forms. He classified the words under his own categories, such as numbers, relatives, animals, plants, heavenly bodies, periods of time, colors, gods, minerals, and geographic features. Philip Wright was displeased with this renewed interest in Greek and Latin because it bore little relation to his humanistic concerns, being rather like Sewall's collection of butterflies or birds' eggs. Instead of animals, Sewall was now collecting words from languages including Sanskrit, Old Persian, Greek, Latin, Old Irish and Welsh, Gothic, High and Low German, Old Scandinavian, Lithuanian, and Old Church Slavic. He literally filled a number of notebooks with these philological endeavors.

This interest in philology indicates an early and deep fascination with the evolution of patterns. How languages became transformed over time, and perhaps branched out to become several languages, was often analogized to processes of evolution in nature by late-nineteenth-century intellectuals. Wright himself made no such conscious analogy, but the evolution of languages and the evolution of biological organisms share certain features in broad outline, though not in mechanisms of change.

Perhaps not coincidentally, Wright for the first time read all the way through Darwin's *Origin of Species* (6th ed., 1872) during his last year of high school. He found *Origin* sensible and convincing and had no hesitation in accepting natural selection as the primary mechanism of evolution in nature.

During his high school years, Wright pursued many activities outside the school setting. He continued to take bicycle trips of longer and longer duration with JOAT members or with Quincy. They went to lakes and rivers for swimming and roaming. On one trip they went to Dahinda, fifteen miles away on the Spoon River. There they made "otter" slides on the black mud banks. Once he swam across the Mississippi River where it was more than a mile wide. Family vacations took the Wrights to White Bear Lake and Leach Lake in Minnesota, to many towns on the Mississippi River, to Medford and Megansett on Buzzard's Bay, both in Massachusetts. These vacations stood out prominently in Wright's mind as intensely enjoyable experiences.

Another activity that involved Philip, Sewall, and Quincy was the Asgard Press. Philip made only a small salary teaching at Lombard. As an income supplement and for literary purposes, in 1900 he bought a printing press and several fonts of type and set them up in the basement. On this press Philip printed for the college such things as catalogs or publicity materials. Of more lasting importance, he published and printed several volumes of his own poetry and some of his students' work.

Although Philip primarily taught economics, astronomy, and fiscal history, he had a high reputation among the students as a teacher of writing and literature. He attracted a talented group of students, among them Charles A. Sandburg, who became the famous Carl Sandburg. In his later years, Sandburg had nothing but the highest praise for Philip: "In English and American literature he knew more than anyone on the campus about creative writing and how to get it out of students if they had it in them." After learning about the life and works of Leonardo da Vinci, Sandburg nicknamed Philip the "Illinois Prairie Leonardo" (Crane 1975, ix). Philip published four of Sandburg's books between 1904 and 1910, all printed on the press in the basement.

Sewall and Quincy helped set the type for the first book and many of the others. Quincy was quicker and more adept with the typesetting and was therefore a greater help to Philip in the publishing venture, but Sewall printed the high school paper on the press during his high school days. The Asgard Press always occupied a warm place in the hearts of Philip, Sewall, and Quincy, as well as Sandburg. The correspondence between Philip and Quincy concerning the sale of the printing press after the Wrights had moved to Boston in 1915 shows how much both of them loved it. Quincy ended up selling it in a depressed market to a poultry farmer who needed it to print up his advertisements. The details of the collaboration of Philip Wright and Carl Sandburg, and their publications with the Asgard Press, have been well documented (Crane 1975).

In May of 1906 Sewall Wright graduated from Galesburg High School with an average of 98.35. He was fifth in his class, behind four girls.

Lombard College, 1906–1911

From the fifth grade through high school Sewall Wright had been one to two years younger than his classmates. Many of the entering undergraduates at Lombard were older than his classmates in high school, so the discrepancy between his age and that of his fellow students became greater. Moreover, he was small for his age. The combination made him feel like a small boy among fully grown men and women. He had little contact with the boys in the high school who were about the same age. The first two years at Lombard were, therefore, rather lonely. He belonged to a literary society, the Erosophian Club, but that scarcely constituted a social life. He spent much of his spare time, especially in the spring, looking for birds. The high points of these observations were spotting a flock of Cape May warblers in the trees on campus and finding a screech owl nest in an elm tree, also right on campus.

The second year of college was a little better than the first. He entered the Swan-Lawton Oratorical Contest in the spring of 1908, that year devoted to the topic "Socialism and the Individualist." The topic was perfect for Wright: he was himself already a tried and true individualist, and he was a highly inner-directed person who largely kept his own counsel. Yet at the same time he came from a family seriously interested in socialist ideas. Philip Wright had voted for William McKinley in 1896, thinking that William Jennings Bryan was a demagogue and windbag. The militarism and imperialism exhibited by the United States government during and after the Spanish-American War outraged him, however, and he voted for Bryan in 1900. Sewall and Quincy wore Bryan buttons to school in the fall of 1900, becoming the butt of many taunts. A jingle frequently heard in school was

> McKinley rides a white horse
> Bryan rides a mule
> McKinley is a wise man
> Bryan is a fool.

Philip Wright also was disturbed by the huge and growing monopolies and the apparently ever-increasing gaps between the social classes. By 1904 he was convinced that equality of opportunity could only come about through socialism. In 1904 and 1908 he voted for Eugene Debs; in the latter year, Debs polled over a million votes. Philip subscribed to a socialist weekly from Chicago, and Sewall and Quincy often read it. Sewall was generally convinced by the substantive arguments but found the dogmatic and militant tone offensive. Thus the Swan-Lawton Oratorical Contest came at an opportune moment, and Sewall much later recalled that he probably won a prize at the debate. Wright's general social reticence was apparent not only at Lombard but also at home. Philip Wright was not pleased by the direction of Sewall's intellectual interests, which certainly diverged from his own. Quincy, on the other hand, was rapidly developing intellectual interests very close to those of his father and rather like those of Elizur Wright III. Quincy was also outgoing and quick in conversations, unlike Sewall, who felt all the more individualistic and somewhat isolated under the circumstances. The following remarks, taken from auto-biographical notes written by Sewall Wright in 1978, describe his recollection of feelings at that time:

Father was obviously disappointed that I did not take to the things that interested him most—poetry, music (beyond simple tunes), economics and political science. He was often impatient with my clumsiness in carpentry and typesetting and he was sometimes sarcastic about my enthusiasms, especially that for the evolution of the Indo-European languages and to some extent that for natural history and organic evolution. I usually had little to say at meals at which conversation tended to consist of dialogues between Father and Quincy on various topics, particularly foreign affairs. Quincy was very articulate whatever topic came up. Ted was too young during the period I have in mind to have much to say on such topics but was very enthusiastic about athletics. Mother was interested in literature and art but especially in gardening. She at least encouraged my interest in natural history. . . . Quincy matched Father's interests most closely, while I diverged most.

One of the many reasons Sewall Wright worked so hard to excel in biology was to prove to his father that his biological interests were really worthy intellectual ventures.

All of the Wright men, Philip and the three boys, went on to distinguished careers. Philip left Lombard in 1912 to take a one-year position at Williams College, substituting for a professor on sabbatical leave. The next year he returned to Harvard, where he had received his M.A., to serve as an assistant to his former advisor, Professor Taussig. The following year Taussig was appointed head of the Tariff Commission, where he found a position for Philip. Later Philip did research and writing at the Brookings Institution, publishing six books on economics and the posthumous book on Elizur Wright III. Quincy Wright became a world-renowned political scientist, specializing in foreign relations. He published more books than any of the others. Ted Wright was an outstanding athlete and had a notable career as an executive. He became vice-president and director of engineering at Curtis-Wright, chairman of the Aircraft Resources Board during World War II, administrator of civil aeronautics, and finally vice-president for research (acting president for one year) at Cornell University. He was the administrator in charge of Cornell's Calspan Laboratory. While at Curtis-Wright he did important work in airplane design. At the end of his life he published a large two-volume collection of his essays and addresses.

Sewall Wright wrote out a chart in his autobiographical notes detailing the interests of the family members (see table 1.1 below), cautioning that, from lack of knowledge, he probably had not done full justice to the interests of Quincy and Ted. The chart indicates clearly Sewall's impression that he was no conversationalist, the divergence of Sewall's primary interests from his father's, and the resemblance of Quincy's interests to his father's. Sewall's lack of ease in social communication, his strongly independent personality, and his confidence in his own analysis and abilities are all characteristic of his later scientific work.

Table 1.2 Wright Family Interests

	PGW	EQSW	SW	QW	TPW
Humanities					
Novels	+++	+++	+++	+++	+
Poetry	++++	++	+	+	+
Greek & Latin	+++	+	++	0	0
History	++	++	++	++	++
Music	+++	+	++	0	++
Art	+	+++	+	+	+
Social Sciences					
Economics	++++	+	++	+++	+++
Sociology	+++	+	+	+++	+
Political sci.	+++	++	+	++++	+
Mathematics	+++	+	++++	0	++
Engineering	+++	0	+++	0	++++
Physical Sciences					
Physics	++	0	++	++	++
Chemistry	+	0	+++	+	+
Geology	+	0	+	+	+
Astronomy	+++	+	++	+	+
Biology					
Natural history	0	++	+++	+	+
Theoretical biol.	0	0	++++	0	0
Evol. of language	0	0	+++	0	0
Philosophy	++	+	++++	0	0
Activities					
Executive	++	+	0	+	++++
Gardening	+	+++	0	0	+
Printing	++++	+	0	+++	0
Carpentry	+++	0	0	+++	+
Athletics	+++	0	++	0	++++
Conversation	+++	+	0	+++	+

Notes: ++++ indicates publication or concrete public achievement; +++ implies strong interest; ++ less; + some; and 0 when there was none.

In Sewall's third year of college, Quincy entered Lombard. They both joined Sigma Nu fraternity (as would Ted later), and with Quincy's more outgoing personality to lead the way, Sewall's social life began to improve. He became editor of the Lombard college annual, *The Stroller*. The editorship required much social interaction such as commissioning articles, photographs, and so forth. In the end, however, Sewall did much of the writing himself, got his mother to draw many pictures for it, and printed it on the Asgard Press in the Wright house. As a member of the fraternity, Sewall was expected to attend the spring prom. Fraternity brothers instructed him on how to dance, and he asked an Irish girl with brilliant red hair and a lively disposition whom he had known in grade school and high school; she accepted. Sewall's later recollection: "As for the prom, I danced very poorly . . . and found that I had nothing to say."

Academic Work at Lombard, Fall 1906–Spring 1909

Upon entering Lombard, Sewall Wright wished to continue his studies in foreign languages. He took a fifth year of Latin, reading Cicero's letters among other things, but he found the professor and the course dull. He then tried a year of French. The professor was a German and uninspiring, and Wright ceased the study of French. He later read many scientific papers in French but leaned heavily upon a dictionary. Far more enjoyable was his year of German with an enthusiastic young professor. He learned to read fluently the literature assigned in the course and acquired a good start. He loved the language, but later, disillusioned by the actions of the Germans in World War I, he became less enthusiastic. He read in German, both before and after the war, some of the major genetics textbooks, monographs, and papers. When Richard Goldschmidt delivered his paper in German before giving it in English (a requirement of the German government) at the Sixth International Congress of Genetics at Ithaca, New York, in 1932, Wright was able to clearly understand the gist of his argument, though not all the details.

Philip Wright was one of his sons' most effective and interesting teachers. Sewall took many of his father's courses, including a general course on economics and one on the fiscal history of the United States. More important, however, were the quantitative courses. From his father, Sewall took solid geometry, advanced algebra (Wright later described it as "somewhat" advanced algebra, including probability theory but not determinants or matrices), trigonometry, and a year of analytic geometry and differential and integral calculus. He also took a course in surveying from his father, enabling Sewall to obtain two different jobs. Wright often wished over the years of his scientific career that he had been able to take more advanced mathematics courses in college. Philip Wright was not primarily a mathematician, but he was the only one who could teach the courses at Lombard. Sewall's evaluation was that he would have been much better off later in his career with more advanced college courses in mathematics but that he did acquire "some facility in translating questions into mathematical symbolism and solving as best I could."

This brief evaluation is very important because it points out what Wright would later find most useful about his study of mathematics. He was not primarily interested in pure mathematics but in devising ways to analyze problems quantitatively. Wright's later interests in mathematics were closely tied to particular problems about the natural world. But his college courses and personality led to a second, equally important effect: he became expert at teaching himself new mathematics. One consequence is that his mathematical analysis is frequently expressed in ways that mathematicians consider unusual. Some population geneticists would later term Wright's mathematics heuristic or involving large intuitional leaps. I think this is generally true, and in substantial part is a result of his self-teaching.

In high school no chemistry was taught. At Lombard there was no

Figure 1.3. *Left to right:* Sewall, Quincy, and Ted Wright during their student days at Lombard College.

physics, and Wright took almost all the available chemistry courses. Professor Rich, father of Willis Rich, who had earlier befriended young Sewall and Quincy Wright, was the teacher. Rich was a good practical chemist, but he was not a researcher and did not convey to his students an idea of what original research might be like. Wright took courses in general chemistry, qualitative and quantitative analysis, organic chemistry, and a year of practical ore and water analysis. The practical courses were conceptually like following cookbook recipes, and Wright became bored with them, although he had enjoyed the earlier, more conceptual ones.

By the end of his third year in college, Wright was basically majoring in chemistry, but by then he was unexcited by the thought of going forward in the field. He had taken no biology courses in college and saw no particular future in seriously pushing his natural history interests into a career of some kind. His mathematical interests were strong, but he did not think seriously about becoming a mathematician. Thus after three years of college, Sewall was unsure what he wanted to do with his life. At this crucial juncture, in the early summer of 1909, he found a job that seemed like the perfect temporary salvation from his academic dilemma.

Railroad Engineering in South Dakota

In 1908 Wright had taken a summer job as assistant to the county surveyor, during which time he learned many facets of practical surveying. The course

Figure 1.4. Base camp. Wright is second from left.

he had taken with his father was a great help as well. As the time for getting a summer job approached the following year, Elizabeth Wright wrote to one of her double second cousins, Edmund Quincy Sewall (1823–1908), a vice-president of the Chicago, Milwaukee, and St. Paul railroad system. Since 1904 he had been in charge of building the new extension from South Dakota to Puget Sound. He offered Wright the job of back flag in a party making a preliminary survey along the divide between the Grand and Moreau rivers in Standing Rock Indian Reservation in northwestern South Dakota. Thus Wright had an adventure-filled summer working in this desolate but biologically interesting area. The details of the railroad work have been recorded by Wright in the biographical notes and in the transcripts of our oral interviews.

Toward the end of the summer, Wright received an offer to remain on the crew for another nine to twelve months, with a promotion in pay as well as in position to the job of instrumentman, a more responsible and interesting job than back flag. Finding the offer attractive and having no particular inclination to return immediately to Lombard, he accepted. The surveying parties worked through the bitter winter; on one occasion Wright was shedding clothing while chopping wood, only to discover later that the temperature was −46 degrees F. Occasionally his quantitative abilities were called into play. On one occasion Wright was locating track centers on an 8 degree curve (the sharpest on the route) around a horseshoe bend of 240 degrees on the Grand River. The properly calculated track centers kept running him off the grade, to his chief's dismay. Apparently the curve had been incorrectly run in the first place, and the dirt had all been moved and could not be redone. Thus Wright and the chief spent a lot of time devising a curve (no longer an 8 de-

Figure 1.5. Railroad crew in mock defensive pose. Wright is in the pipe on the left side.

gree one) with proper spirals that would stay in the middle of the roadbed over its whole length. Wright also enjoyed the wildlife of North Dakota as well as the opportunity to observe Venus, which he could locate in the middle of the day at certain times if he followed it fairly continuously, and Jupiter, which disappeared instantly as the sun rose above the horizon.

The local Indian tribe, the Lakotas, had achieved a certain fame thirty-three years previously when under their chief Sitting Bull they had destroyed General Custer's company at Little Big Horn. The revenge of the U.S. Army, the massacre at Wounded Knee, had occurred only nineteen years earlier. The Indians caused no significant problems for the surveying crews, one family taking Wright in on a night when he had lost his way riding ahead to search for a water hole when the party was moving camp by some twenty miles.

In March Wright acquired a persistent pain in his chest and in May began to lose his appetite. His chief sent him back to Mobridge for rest and recuperation. Despite his discomfiture, Wright climbed up to the top of the freight car after dark so he could get a better view of a comet. It was Halley's Comet, May 1910. (Wright is now eagerly looking forward to seeing it again in late 1985.) During a several-week period he amused himself by working problems in a book he had brought back from a Christmas visit home: P. G. Tait's *Elementary Treatise on Quaternions* (3d ed., 1890). Earlier, Wright had looked at a little book on the subject in his father's library. He was delighted at the way quaternions provided a new method of proving geometric theorems, in many respects simpler and more elegant than the usual Cartesian approach. Most exciting was the application of quaternions to physical theory, a topic addressed in detail in Tait's book. Once he had grasped the symbolism, Wright found that, for example, Kepler's three laws of planetary motion could be seen as almost intuitively obvious consequences of Newton's inverse square law of gravitational attraction. This was just the kind of relationship of mathematics to the physical world that most excited Wright. That he so obviously enjoyed reading Tait on quaternions while living in a caboose in Mobridge, South Dakota, indicates he was about ready to turn back to the life of the mind.

He had little choice. The pain and shortness of breath were diagnosed by a company doctor as pleurisy with complete flattening of the left lung, and Wright was sent home for recuperation. By fall, his chest had completely cleared up. One consequence of this attack of pleurisy was that in 1916 Wright was rejected by New England Mutual Life Insurance Company as having had tuberculosis; he was able to obtain a policy from the New York Life Insurance Company only by being rated as age thirty-three instead of his actual age of twenty-six. Not for several more years was he put on a normal policy. At age ninety-four Wright still chuckled about his difficulties getting life insurance.

Last Year at Lombard, 1910–1911

Wright's senior year at Lombard was good in every way. The college had shrunk in size and now had fewer than seventy students total, with only seven in the graduating class. Being a senior and having been away for a year, Wright was now closer in age to his classmates and older than most of the underclassmen. He entered more fully into the social life of the college, and frequently dated a girl who was a student leader and president of her sorority. This relationship gave Wright a deep sense of well-being. His academic work took on a new sense of purpose primarily because of a new teacher at Lombard, Wilhelmine Entemann Key. In his first three years at Lombard, Wright had avoided the courses in biology because he considered the professor to be uncongenial. Now in his senior year he was taking college biology for the first time, and it changed his life.

Wilhelmine Marie Entemann (the biographical information comes primarily from Wright 180) was born in Hartford, Wisconsin, in 1872. She attended the University of Wisconsin where she worked for the zoologist E. A. Birge, famous for his research on the biology of Lake Mendota. She always greatly admired Birge and of course from her college days had a clear understanding of the appeal and rigor of original research in biology. After college she taught German in the Green Bay High School for four years, then entered the University of Chicago on a fellowship to study biology. She worked primarily with three world-renowned biologists: Charles Benedict Davenport, Charles Otis Whitman, and William Morton Wheeler. She received her doctorate (one of the first awarded to a woman at the University of Chicago) in 1901 with a thesis on variation in the wasp *Polistes*. After teaching for a year as an assistant at the University of Chicago, she spent two years (1903–05) teaching biology at New Mexico Normal University. There she met and married Francis Bruté Key, who died shortly after their marriage. She never remarried, keeping on with her career. In 1953 her stationery was still headed: Mrs. Francis Bruté Key.

After teaching at Belmont College from 1907 to 1909, she came to Lombard for the academic year 1909–10, when Sewall Wright was off in South Dakota. She always had serious research interests and sometimes little patience with naïve, inexperienced undergraduates. She could be very sarcastic. When Wright began to work with her, he was grateful to have as his biology teacher this tall, graceful, vivacious, highly intellectual woman with a lively sense of humor.

Most of the courses Mrs. Key taught were traditional. Thus in his senior year Wright took her botany course in the fall term, finding it a straightforward presentation of systematics and morphology, with generally unexciting laboratory work. The winter term, however, was wholly different and a genuine revelation for Wright. The course was on recent research in theoretical biology and focused primarily upon evolution, genetics, and cell biology. Mrs. Key conducted the class like a graduate student seminar, having the students read and prepare reports on books and original research papers by well-known scientists.

Wright had of course long been familiar with the idea of evolution by natural selection. Before his winter term course with Mrs. Key he had read Darwin's *Origin of Species* (1872), most of *The Variation of Animals and Plants Under Domestication* (1868), *Expression of the Emotions in Man and Animals* (1873), and Henry Fairfield Osborn's recently published *Age of Mammals* (1910). These books were in his mother's library. But he still had no concept of original research in biology and was still thinking vaguely that he would go into engineering, something perhaps along the lines of the engineering work he had seen first-hand while working for the railroad. He was best prepared in chemistry but had little desire to pursue that field, particularly because he did not know how original research in chemistry was conducted.

During the course Wright read the following books: Alfred Russel Wallace, *Darwinism* (1889); Francis Galton, *Natural Inheritance* (1889); G. J. Romanes, *Darwin and after Darwin,* vols. 1 and 2 (1892–95); Vernon L. Kellogg, *Darwinism To-Day* (1907); J. Arthur Thomson, *Darwinism and Human Life* (1910), and G. Archdall Reid, *The Principles of Heredity* (1905). Of these, Kellogg's *Darwinism To-Day* made the greatest impression. Kellogg presented what is even at this writing one of the two best accounts of the various competing theories of evolution and heredity that abounded in the late nineteenth century (for the other see Bowler 1983). T. H. Morgan and E. B. Wilson, according to Alexander Weinstein (personal conversation), discouraged their undergraduates at Columbia from reading this book because it presented such a confusion of theories; instead they recommended R. H. Lock's

Recent Progress in the Study of Variation, Heredity, and Evolution (1906), a book Sewall Wright did not read until graduate school. Wright was not shaken from his basically Darwinian viewpoint by reading Kellogg. Reid's book seemed mostly nonsense to him.

Of far greater significance were the original research articles read and reported on by the class, and one encyclopedia article assigned to Wright by Mrs. Key: R. C. Punnett's "Mendelism," which had just been published in volume 18 of the eleventh edition of the *Encyclopaedia Britannica*, printed by R. R. Donnelly & Sons in Chicago. Punnett's article was a full five pages, and it was an artful condensation of the most recent edition of his little book, *Mendelism* (1905). The first paragraph stated:

> Within recent years there has come to biologists a new idea of the nature of living things, a new conception of their potentialities and of their limitations; and for this we are primarily indebted to the work of Gregor Mendel. . . . Mendel's ideas have steadily gained ground, and, as the already strong body of evidence in their favour grows, they must come to exert upon biological conceptions an influence not less than those associated with the name of Darwin. (Punnett 1911, 115)

Of course included in the article was the famous "Punnett square," which demonstrates graphically the simple ratios to be expected from Mendel's theory of inheritance. Here was a theory of heredity certain to appeal to Sewall Wright—the nice arithmetical analysis, closely tied to fundamental understanding of the process of heredity. At the time Wright read Punnett's article, he had never heard of Mendelian heredity. The article instantly fired his imagination, and he began his lifelong interest in genetics.

Wright reported on another article, which appeared in the *Chicago Decennial Publications* from the University of Chicago. This was William Lawrence Tower's "Development of the Colors and Color Patterns of Coleoptera, with Observations upon the Development of Color in Other Orders of Insects" (1903). Tower later achieved a measure of infamy by his fakery in extensions of the work reported in this paper (Weinstein 1980). Wright found nothing wrong with Tower's 1903 paper, and it was in fact closely related to Wright's later work on color inheritance in mammals. While conducting his later work, Wright read papers attacking Tower's ideas and results, in particular R. A. Gortner's 1911 paper, "Studies on Melanin, IV: The Origin of the Pigment and Color Pattern in the Elytra of the Colorado Potato Beetle" (Gortner 1911).

Another article in the same *Decennial Publications* volume was Davenport's report on his early research at Cold Spring Harbor, where a year later (1904) he would assume duties as the director of the Cold Spring Harbor Station for the Study of Experimental Evolution. This paper, "The Animal Ecology of the Cold Spring Sand Spit, with Remarks on the Theory of Adaptation" (1903), reported research that Wright would join the coming summer at Cold Spring Harbor.

In a short three months, Wright had gained from his association with Mrs. Key understanding of the new field of genetics, an appreciation of what it meant to engage in original scientific research, and the idea that it might be possible for him to go into the field of experimental biology as a researcher and teacher. A major problem, of course, was that he did not have an adequate background to go directly into graduate studies in biology. The spring term, he took Mrs. Key's zoology course and dissected crayfish, grasshoppers, frogs, and the semicircular canals of a shark—but neither he nor Mrs. Key believed he would have an adequate background for graduate study upon graduation from Lombard.

Mrs. Key strongly urged Wright to go to graduate school in biology. To better prepare him for graduate school, she arranged with Davenport for Wright to go to Cold Spring Harbor for summer courses in zoology and for the opportunity to see and hear well-known researchers in action. Graduate school turned out to be an easy decision. The University of Illinois had a standing scholarship offer of $250 to the valedictorian of the graduating class of any college in Illinois. The woman who was valedictorian in Wright's class declined the offer. Wright was next in line and accepted. Thus upon graduation he gladly faced the prospect of a summer at Cold Spring Harbor and graduate school at Illinois.

Mrs. Key was delighted with Wright's progress after graduation from Lombard. They managed to see each other occasionally at various scientific meetings. She resigned from Lombard in 1912 and became a eugenics field worker for Davenport, who was then director of the Eugenics Record Office at Cold Spring Harbor. Later she did independent research and administrative work for several eugenics organizations, after the early 1930s living in Somers, Connecticut. In 1932, at the Sixth International Congress of Genetics in Ithaca, New York, she arranged a luncheon for R. A. Fisher, Sewall Wright, and herself. She had become interested in their differences and saw this as an opportunity for first-hand insight.

I found no record of correspondence between Wright and Mrs. Key until 1953, forty-two years after his year of work with her. Wright had sent her a reprint of his presidential address to the American Society of Naturalists (Wright 144). Mrs. Key responded in a letter of July 28, 1953: "At last, I exclaimed, here is one I can follow and at least partly grasp." She, like so many others, had been unable to follow his quantitative papers. She paid him a wonderful compliment on the address: "I realize you have come down to be a successor and much more to your brilliant predecessors [at the University of Chicago], Dr. Whitman and Dr. Wheeler." She told Wright she was hoping to make a trip west, despite her advanced age, and wanted to see him when going through Chicago. It was lonely growing old, she indicated. She looked "in vain for a familiar name in *Science;* not easy to outlive contemporaries." Wright invited her to stay with the Wrights on her visit to Chicago.

Mrs. Key clearly valued the renewed contact with her former student. Although plans for the trip west fell through, she and Wright continued to correspond. By early fall 1954 Wright knew he would be retiring from the University of Chicago and taking a position at the University of Wisconsin. Mrs. Key was delighted because she loved the University of Wisconsin, and cherished the memory of a sixty-year friendship with her former advisor, Professor Birge. In a letter of September 27, 1954, she indicated that she wanted his Madison address so she could visit later in the fall. On October 28 she wrote to say that she was leaving on December 6 and would see him in Madison a few days later. She was nearing eighty-three and in frail health. She never made the trip. Her health failed, and she died on January 31, 1955.

A measure of Wright's admiration for Mrs. Key is evident in his brief memoir, written in 1965 and cited above. Certainly as a direct result of her influence he was prepared to actively pursue a career in biology by the time he graduated from Lombard in the spring of 1911.

Philip Wright was not overly pleased at the prospect of an academic career for his son Sewall, especially in the field of biology and natural history. By this time Philip was fed up with the life of an academic at a small college, and he wanted something better for Sewall, who certainly inherited from his father a fear of teaching his life away at a small college. In the end Phillip escaped that fate, and Sewall himself carefully avoided it. But he also saw great potential for a career in biology that his father apparently did not see.

Cold Spring Harbor, 1911

Sewall Wright's first contact with renowned research biologists came with his visit to Cold Spring Harbor in the summer of 1911. Wilhelmine Key sent Wright to Cold Spring Harbor primarily because Charles Benedict Davenport, her former teacher at the University of Chicago, was director of the Carnegie Institution of Washington Station for Experimental Evolution at Cold Spring Harbor and director of the summer school there. Davenport had obtained his doctorate under E. L. Mark at the Museum of Comparative Zoology of Harvard University in 1892, and his early work was primarily in experimental morphology, which was mostly embryology and physiology. He soon became interested in the new science of biometry and met both Karl Pearson and Francis Galton in England in 1902. His *Statistical Methods with Special Reference to Biological Variation* appeared in 1899, with a second edition in 1904. Although initially skeptical about the wide application of Mendelian heredity, as were the English biometricians, by 1907 Davenport had become an enthusiastic Mendelian.

In the early 1900s Davenport had been keenly interested in the problems of evolution in nature. His 1903 paper, "The Animal Ecology of the Cold Spring Sand Spit, with Remarks on the Theory of Adaptation," which Wright had read in Mrs. Key's course, ended with a substantial section on evolution in nature. Here Davenport argued that gradual Darwinian evolution by natural selection was only one major mode of evolution; another equally important mode was a mutation theory of evolution in which the organism with an unusual mutation actively selected its new and more suitable environment, a variant of the "preadaptation" argument. Davenport mentioned specifically that his theory was fully compatible with and supportive of the paleontological observations of Henry Fairfield Osborn and the mutation theory of evolution of Hugo de Vries.

By the time Wright came to Cold Spring Harbor in 1911, Davenport had shifted his major research interests from the genetics of chickens, canaries, and sheep to human heredity and the newly emerged science of eugenics. He was the recognized leader of the eugenics movement in America. With the financial backing of Mrs. E. H. Harriman, Davenport founded the Eugenics Record Office at Cold Spring Harbor in 1910. In addition to directing the summer school in biology, he began in 1911 a summer course for the training of eugenics field workers.

Three major areas in which Davenport might have had a significant influence upon the young Sewall Wright, beyond the obvious stimulus to a career in biological research, are genetics, Davenport's brand of evolutionary theory, and eugenics. So far as I can detect from Wright's publications and our detailed conversations on this point, Davenport exerted no significant influence upon Wright in any of these areas. In the summer school, Wright heard Davenport lecture about the organisms of the sand spit and about his research on poultry genetics. These lectures made no strong impression upon Wright. His later theory of evolution shows no significant influence from Davenport, nor did Wright show substantial interest in the very active eugenics movement, although he was for many years listed on the letterhead as a member of the board of directors of the American Eugenics Society.

The course Wright took in the summer of 1911 was with Henry S. Pratt on marine invertebrates. The students collected their own material in the mornings, then spent the afternoons in the laboratory dissecting the specimens. Wright collected many of the organisms mentioned by Davenport in his paper on the ecology of the sand spit, including squillas, sponges, various annelids, ctenophores, sea anemones, and many different protozoa. Wright particularly remembers having worked with the fresh water Bryozoan *Pectinatella*, which was not found on the sand spit. Pratt taught a very good course on invertebrates; Wright learned a great deal from it and enjoyed it enormously.

Pratt's course was excellent background for someone like Wright, whose preparation in biology was weak. While at Cold Spring Harbor this first summer, he heard two other researchers besides Davenport lecture on genetics: J. Arthur Harris and George Harrison Shull. Harris was a biometrician who had worked with Karl Pearson and W. F. R. Weldon in England and who still in 1911 had a deep distrust of the now-growing bandwagon for Mendelism. Harris's favorite target was the pure line theory promulgated by the Danish botanist Wilhelm Johannsen; this theory fit well with Mendelism and was widely accepted in the United States. Wright thought Harris's opposition to Mendelism and pure line theory was probably wrong but found his lectures very stimulating, and he liked Harris personally. Wright had absolutely no inkling that he himself would make important contributions to quantitative genetics.

Shull's lectures were far more rewarding. Shull was a full-time researcher for the Station for Experimental Evolution. His major assignment in the years 1900 to 1910 had been to observe and analyze the breeding methods of Luther Burbank in terms of modern genetics (Glass 1980b). Beginning in 1908, Shull gravitated to maize genetics. He was one of the early pioneers, along with Edward Murray East and H. K. Hayes, of the hybrid corn technique. By the summer of 1911, Shull was strongly advocating the use of hybrid seed corn, produced from crossing highly inbred strains. In pursuit of this research, he grew at Cold Spring Harbor many diverse highly inbred strains. In Shull's reports of his research, Wright was struck by the peculiar array of characteristics that would be frozen in each inbred line. The uniformity of the plants in each line indicated a strong genetic basis for the characteristics differentiating the various lines. Wright went with Shull to see the corn plants growing and saw separate inbred strains planted in parallel rows, one strain to a row. The differences between the rows, for example that the leaves of one strain might begin to branch out six inches higher in one strain than in another, were strikingly regular all along the rows. Each of the inbred strains differed from the others by many little characteristics, which most breeders would not have thought of as being inherited; but they were. After graduate school, Wright took charge of an extended inbreeding experiment in guinea pigs. In this work he again saw the differentiation and fixation of particular combinations of characters in inbred strains. Later, in 1925, Wright would use his understanding of inbreeding in working out his distinctive theory of evolution in nature.

The whole experience at Cold Spring Harbor was good for Wright. He enjoyed the exposure to working biologists and found their work and lifestyle appealing. There were many delightful parties on the sand spit, with much singing of lab songs in front of roaring fires. Wright swam a great deal, frequently swimming across the harbor, about one-quarter mile, with Davenport's daughter Janet, an excellent swimmer. One amusing incident came when the lab had a picnic on Fire Island on the southern side of Long Island. Wright, who worked as a waiter in the dining hall, and the other waiters carried the large baskets of sandwiches, pies, and drinks. Davenport, helpful but forgetful as usual, insisted upon carrying one basket but forgot and left it behind on the pier. Mrs. Davenport sent Wright and another waiter back for the basket, which was filled with pies. By the time they got across Long Island and found the basket, the last boat to Fire Island had left, so they sat down and ate pies until indigestion set in. The rest of the pies collapsed into a mess in the hot sun. Mrs. Davenport was not amused.

Wright profited so much from the experiences at Cold Spring Harbor that he determined to return the following summer.

Graduate School at the University of Illinois, 1911–1912

Wright had saved some money from his previous engineering work, but much of it he lent to fraternity brothers at Lombard, and the money was not repaid. The last $25 of these savings evaporated after Wright was already at Urbana, when he received notice that a $25 check he had guaranteed for a visiting fraternity brother from another college had bounced. He therefore had to finance his first year of graduate school on the $250 fellowship. This he did by careful management. He took a room with another graduate student, costing him $4 per week. Food cost 35 cents a day—10 cents for breakfast, a bowl of oatmeal Wright disliked but ate because of nutritional advantages; 10 cents for lunch, a dish of baked beans or a hamburger sandwich; and 15 cents for dinner, except on Sundays when he spent an extra 5 cents for a piece of pie. This came to $2.50 a week for food. There was no tuition, only a nominal laboratory fee, and Wright made it through the year on the fellowship alone.

During this year Wright took a number of useful courses. Professor F. W. Carpenter, who had been a student at Harvard with Wright's soon-to-be advisor William Castle, taught very polished and good courses on embryology and vertebrate anatomy. Wright learned in detail the embryology of the chick, microscopic techniques, and sections. The course was conventional but well executed. He took a course in field zoology from Professor Frank Smith, an expert on the morphology and taxonomy of land and fresh water annelids. The highlight of the course was the collection and identification of many species of earthworms from the gutters of Urbana. Students in the course also collected and identified fresh water clams and catfish.

Most rewarding was Charles Zeleny's course in experimental zoology. Zeleny had worked primarily on regeneration and physiological regulation and later became well known to geneticists for his work on the Bar mutation in *Drosophila*. Wright read selected chapters from Davenport's *Experimental Morphology* (1897–99) and T. H. Morgan's *Experimental Zoology* (1907). The course spent between two and three weeks on genetics, which was the only formal coursework in genetics Wright ever took. Zeleny was not as well organized as Carpenter, but he generated much discussion among the students and Wright enjoyed the course immensely.

In addition to these usual courses, Wright took an introduction to research course with the chairman of the zoology department, Henry B. Ward, who was an expert on the parasites of fish. Ward began by giving Wright whole mounts of a half dozen or so species of trematodes, a class of parasitic flatworms (phylum Platyhelminthes), and telling Wright to find out as much as possible about their anatomies. Wright had no idea what a trematode was but quickly set to work finding out what he could about the species Ward had given him. After Wright reported his findings, Ward gave him a little bottle filled with preserved specimens of a chunky little trematode about 1 mm long, a species parasitic upon the freshwater bowfin, *Amia calva* (also known as the dogfish, mudfish, or lawyer). Ward had earlier classified the organism as *Microphallus opacus* and had made preliminary observations on its anatomy. He set Wright to work making 10 micra serial stained sections, with the idea of using these to work out in detail the digestive, reproductive, excretory, and nervous systems of the organism. Wright greatly enjoyed this research.

In the spring of 1912 a visitor to the Agriculture College of the University of Illinois changed Wright's life. This was William Ernest Castle, a former student of Davenport and E. L. Mark, who in 1912 was a professor of zoology at Harvard and working at the Bussey Institution. Castle came to lecture on his selection experiments on hooded rats and mammalian genetics. Wright found Castle's talk fascinating, and although he was enjoying the year at the University of Illinois, Wright asked Castle if it were possible to do research with him. Castle's assistant and graduate student, John Detlefsen, was graduating that spring and was scheduled to take a position at the University of Illinois. Impressed by Wright's abilities and enthusiasm, and perhaps influenced by having taught for a year (1896–97) at Knox College in Galesburg, Castle immediately agreed to take Wright on as his new assistant and graduate student.

When Wright told Professor Ward of his decision to study with Castle at Harvard the following year, Ward was pleased and suggested that Wright complete in short order a master's degree at Illinois with a thesis on the anatomy of *Microphallus opacus*. Wright wrote up his research in publishable form. Especially nice was his analysis of the flame cells of the excretory system (eight on each side) and his discovery of an unusual specimen with two ovaries rather than one. The paper was published in July 1912 (Wright 1). It was Wright's first published scientific paper. After Wright completed his doctoral thesis with Castle, it was published in a volume containing papers by Castle. The Carnegie Institution sent Ward a copy and Wright received this letter from Ward, dated November 13, 1916:

My dear Wright:
      The receipt of a copy of the recent work by Doctor Castle and yourself brings me to write you my congratulations. I have been watching your success with most genuine pleasure, and I congratulate you upon what you have done as well as send you all kinds of good wishes for the future.

                                          Very cordially yours,
                                          Henry B. Ward

The midwestern branch of the American Society of Zoologists met at Urbana during the 1911–12 school year. Wright saw and heard such figures as

Frank and Ralph Lillie, Charles Manning Child, and Horatio Hackett New-man, all of the University of Chicago. Wright was encouraged by the meeting to continue toward a career in biology, but he did not guess that in 1926 he would become a colleague of these men.

Wright left the University of Illinois in the spring of 1912 with high spir-its and expectations. He had a master's degree in hand and could look for-ward to another summer at Cold Spring Harbor, a graduate career at Harvard as Castle's assistant, and an excellent chance of employment as a research bi-ologist.

Cold Spring Harbor, Summer 1912

In his second summer at Cold Spring Harbor, Wright took a research course with Davenport on the ecology of the marsh at the head of the harbor. A freshwater stream entered at the upper end, but the water quickly became brackish in the lower end of the marsh. Wright studied the ecology of the shift from freshwater to saltwater marsh. He laid out square feet, then counted all the organisms he could find in each square foot, including spi-ders, fiddler crabs, and snails. Wright was especially interested in the transi-tion between *Melampus,* the saltwater snail, and *Succinia,* a smaller, fresh-water snail. There was a very sharp boundary between them. *Melampus* could stand only a certain amount of freshwater, and *Succinia* only a certain amount of saltwater, so even though the transition from fresh to salt water was gradual, the boundary between the two kinds of snails was distinct. Wright also worked out a definite relationship between the frequency of leeches and the frequency of snails. He wrote up all of this research but never published it. About once a week Wright discussed the progress of the re-search with Davenport, who was casual about it and not particularly encour-aging. It was obvious to Wright, as it was to Davenport's associates, that Davenport's primary interests now lay in human heredity and eugenics rather than in the ecology of Cold Spring Harbor. Despite Davenport's lack of inter-est, Wright enjoyed this opportunity for careful ecological research.

He also gained much from the visiting lecturers and from talking to the other students. One student whom Wright came to know well that summer was Alfred Henry Sturtevant, who had just received his undergraduate degree from Columbia where he was working closely with Thomas Hunt Morgan on the genetics of *Drosophila*. Morgan had just begun this work in 1910, but by the summer of 1912 it was in full swing at the fly room in Schermerhorn Hall at Columbia (Allen 1978). Wright and Sturtevant became good friends, and Wright learned about the new *Drosophila* work from him. Sturtevant had se-rious interests in taxonomy, and he spent much of the summer collecting *Drosophila* around Cold Spring Harbor. Twenty-four years later, the friend-ship would lead Wright to collaborate with Sturtevant and Dobzhansky in a study of evolution in natural populations of *Drosophila pseudoobscura*. This collaboration materialized as Dobzhansky's influential and large series,

"Genetics of Natural Populations," five papers of which were jointly authored by Wright and Dobzhansky.

Wright was eager to begin experiments in genetics, but experiments with chickens or mammals would obviously take more than the six weeks of summer school. So when summer school ended, Wright was happy to begin work in genetics at Harvard with William Castle.

2
William Ernest Castle, Edward Murray East, and the Bussey Institution of Harvard

The years at Harvard graduate school (1912–15) were highly formative for Wright. He obtained a broad education in science (with the exception of higher mathematics), learned experimental genetics primarily by working for and with William Castle, and began his own original experimental work in physiological genetics. He also read widely in the literature of genetics and met many of the prominent members of the genetics community. Wright left Harvard in 1915 as an accomplished geneticist.

The great influence the three years at Harvard had upon Wright must be emphasized. This was the last time an institutional setting, or an individual scientist, would have a major impact upon the development of Wright's scientific work. No similar fundamental influence from his colleagues or from an institution is discernible in his years at the USDA in Washington, D.C. (1915–25), at the University of Chicago (1926–54), or at the University of Wisconsin (1955–present), although each setting of course had some influence upon his work.

Wright's graduate school experience at Harvard's Bussey Institution had two major effects upon his later scientific work. First, this experience led directly to his career in physiological genetics, which occupied the bulk of his research in the forty-year period from his Harvard graduation in 1915 to his retirement from the University of Chicago at the end of 1954. He used the same experimental animal, the guinea pig, and conducted breeding experiments similar to those he had done in graduate school. Second, the Harvard experience supplied two of the four major elements that went into making his shifting balance theory of evolution in nature (Wright 201), namely the universality of gene interaction in biological organisms as well as the efficacy and limitations of direct mass selection (simple direct selection upon the whole population). Thus the graduate school experience was central to the two great facets of Wright's scientific career—physiological genetics and evolutionary theory.

Wright's later scientific work was closely tied to the influences of the institutional setting at the Bussey Institution and more specifically to its two geneticists, William Ernest Castle and Edward Murray East. Wright would likely not have pursued either of the two major prongs of his career if he had gone to graduate school at Columbia with T. H. Morgan and E. B. Wilson, or to Johns Hopkins with H. S. Jennings and Raymond Pearl, which at the time were the only other places for him to do graduate work in genetics.

The lines of continuity in Wright's scientific career can first be seen in this chapter, which may seem strange since this chapter discusses the situation at Harvard and the Bussey Institution before Wright's actual arrival. At times it may even sound like a biography of Castle. In fact, I have minimized the biographical analysis of Castle and East (fascinating though it is) and concentrated on the intellectual lines leading to Wright. Indeed, this chapter is basically a historical analysis made by systematically working backward from the bibliography in Wright's thesis; the references in this chapter are almost coextensive with those in the thesis. It must be remembered that only a decade separated Castle's first paper on Mendelism and Wright's arrival for graduate study at the Bussey Institution and that East came to the Bussey just three years before Wright. The work done by Castle and East on genetics was fresh in their own minds and was well known to all of the graduate students, most of whom read every single paper published on genetics by Castle and East. More than anything else, Wright's thesis research on physiological genetics was a direct extension of Castle's work on the inheritance of coat color and hair direction in the guinea pig, and his ideas about both artificial and natural selection were deeply shaped by his understanding of Castle's selection experiments on hooded rats.

In our interviews, Wright repeatedly voiced deep and fond recollections of the years at Harvard and the Bussey. He clearly thought these years were crucial in his development as a scientist. Most of the other graduate students at the Bussey had similar fond memories of their formative years there, as for example L. C. Dunn (Dunn 1961). After this chapter was completed I had the opportunity to read two chapters of a manuscript on the Bussey by Jack Weir, a student there in the 1930s. Weir has documented the great importance of the Bussey as a center of research and training in the field of genetics.

William Ernest Castle

Castle was born in 1867 on a farm near Alexandria, Ohio. (For biographical information on Castle, see Dunn 1965.) After graduating in 1889 from Denison University in Granville, Ohio, he taught Latin for three years at Ottawa University in Ottawa, Kansas. In 1892 he entered Harvard College, concentrating in biology and earning a second bachelor's degree. Charles B. Davenport was one of his teachers during this year. Davenport was an energetic, inspiring teacher at this stage of his career, and Castle was so impressed that he asked Davenport to keep him on as a laboratory assistant. Davenport did so, and Castle earned his advanced degrees while working as Davenport's assistant—a pattern Castle would later repeat with some of his own students, including Sewall Wright. Although formally earning his M.A. (1894) and Ph.D. (1895) with E.L. Mark, director of the Zoological Laboratory of the Museum of Comparative Zoology at Harvard, Castle actually did his research under Davenport. With Davenport as senior author, together they published in 1895 a substantive paper on the acclimatization of organisms to high temperatures. Davenport wrote Castle in 1930 that "it was always a sort of satisfaction that perhaps your first scientific paper was published jointly with me" (the paper was Castle's fourth; Davenport to Castle, December 8, 1930).

Mark attracted outstanding graduate students. Castle's fellow students included Herbert Spencer Jennings, later a prominent geneticist, and Herbert V. Neal, who later taught at Knox College in Galesburg and at Tufts in Medford, Massachusetts, and who taught natural history to young Sewall Wright. Almost all of Mark's students wrote their theses on traditional morphological or natural history topics. Castle's thesis concerned the early embryology of an ascidian, a sea squirt. Davenport, Castle, and Jennings would all turn later from experimental morphology to genetics and evolution as their primary research interests.

Castle taught for a year (1896–97) at Knox College before Davenport asked him to return to Harvard as an instructor. His last papers on morphology and natural history appeared in 1900. After a three-year gap, his papers on Mendelism and evolution began to appear in rapid succession. This dramatic career switch from the study of morphology and embryology to the study of genetics occurred with many biologists of this period, including T. H. Morgan, H. S. Jennings, Davenport, and many others, and is a phenomenon that historians of biology have yet to fully explain. Obviously the problems of heredity and evolution were known. Castle had long been interested in problems of heredity and evolution, and in his graduate student days discussed them extensively with Davenport, Jennings, Neal, and other graduate students. He much admired Weismann. A good example of Castle's early interest in heredity is evident in a letter he wrote to Davenport on August 26, 1894. Castle was at Newport, Rhode Island, collecting *Ciona* eggs with Neal for their research. While out walking, Castle and Neal came across a "curious case of polydactylism in cats"—a female cat with seven toes on each foot, and a daughter of the same cat with six toes on each foot. "Neal suggested that you might like to have the cats to breed and experiment with further," Castle said, adding, "I would secure them for myself if I had anywhere to keep them." I can find no evidence that Castle brought the cats back for experimentation, but he was obviously interested in breeding them. The initial spark inducing Castle to begin experimental work in Mendelism appears not to be the papers of the rediscoverers (Correns, de Vries, Tschermak), but William Bateson's fiery *Mendel's Principles of Heredity: A Defence* (1902) and Bateson and Saunders's *Report I to the Evolution Committee of the Royal Society of London* (1902).

Even before the rediscovery of Mendelism in 1900, Castle had been studying the possibility of changing the sex ratio in mammals by selection. For this purpose he was raising mice and guinea pigs. After reading Bateson, Castle thought that sex might be inherited according to simple Mendelian segregation and began to use his mice and guinea pigs for studies on that and the inheritance of other discontinuous characters.

During the first decade of the century Castle focused his research and writing upon the following five questions: (1) To what extent do Mendelian dominance and segregation apply to heredity in animals, mammals in particular? Is Mendelism applicable to continuous, or blending, as well as to discontinuous variations? (2) Is the purity of the germinal determinant of a character inherited in simple Mendelian fashion affected by association with different germinal determinants, as occurs in hybridization? (3) To what extent can characters inherited according to simple Mendelism be modified by selection? (4) Is the assertion of Weismann correct that germ cells and body cells are distinct and therefore acquired characters cannot be inherited? (5) What are the implications of Mendelism for conceptions of the mechanism of evolution in nature? The developing and sometimes radically changing answers that Castle and his students gave to these questions form the direct background to Sewall Wright's research in the years 1912–15. With the exception of the emphasis upon mammals, these were precisely the questions that burned in the minds of most of the early geneticists. Castle was in the mainstream of genetics research.

At the time Castle began to write about Mendelian heredity, a vigorous battle between the Mendelians, led by William Bateson, and the biometricians, led by Karl Pearson and W. F. R. Weldon, was in progress (Provine 1971, 25–89; Froggatt and Nevin 1971; Cock 1973; and Norton 1973). The biometricians argued that Mendelian heredity did not explain the observed facts of heredity and that evolution in nature proceeded gradually, by natural selection acting upon small continuous variations, in accordance with the views Darwin had often expressed. The Mendelians, on the other hand, believed that Mendelism had wide applicability and that it necessitated a view of discontinuous evolution in nature, in accordance with the mutation theory of evolution expounded by the Dutch botanist Hugo de Vries.

Castle clearly sided with the Mendelians. His first paper on Mendelism, published in January 1903, expanded in detail the suggestion of Mendel, Correns, and Strasburger that sex was inherited as a simple Mendelian determinant (Castle 1903a). His second paper, published the same month and entitled "Mendel's Law of Heredity," was a spirited exposition of Mendelism, sprinkled with the preliminary results of his breeding experiments with mice and guinea pigs (Castle 1903b). Castle suggested that continuous inheritance might have a Mendelian interpretation. Originally published in the *Proceedings* of the American Academy of Arts and Sciences, this paper later appeared in *Science* and had a very wide audience. A third paper, written with his student Grover M. Allen, appeared in April and argued that albinism was a Mendelian recessive (Castle and Allen 1903). A final major paper in this productive year appeared in November and was a frontal attack upon the biometricians, arguing that Mendelian heredity accounted for the observed facts of inheritance in mice far more accurately than either Galton's or Pearson's versions of the law of ancestral heredity (Castle 1903c). Pearson replied with a scathing and well-justified attack upon Castle's knowledge of correlation and statistics, but Castle's retort was well taken: he could predict the specific results of crosses using Mendelism, and Pearson could not using the law of ancestral heredity (Pearson 1904; Castle 1905a, 22–23).

At the same time that Castle was so vigorously promoting and defending Mendelism, his experiments were indicating that Mendelian segregation was not universal and that the "purity of the gametes" (Bateson's term) was questionable. Castle was completely open about this. Throughout his scientific career he was honest and direct about his data, interpretations, misgivings, prejudices, and changes of mind. He was also very independent minded and followed his data wherever they appeared to lead. It is not surprising, therefore, that Castle frequently and openly changed his mind about scientific questions. Even in his first year of publishing on the topic of Mendelism, and clearly arguing in its favor, he was beginning to express reservations about what soon would be major assumptions of the new science of genetics.

Castle's basic attitude was that science progressed most rapidly by the publication of bold hypotheses that all could examine with the greatest critical acumen. A bold hypothesis refuted by later research was not a failure but an advance in knowledge. Wright, with his quiet self-assurance, did not fit well with Castle's style—he preferred instead to get the hypothesis right the first time and to dispense with the retractions. The spectacle of Castle publicly retracting his hypotheses was not a model Wright would follow. Given his own personal style and his reaction to Castle, it will be no surprise to learn that Wright sometimes did not make clear where or when he had changed his mind.

In his first paper on heredity of sex Castle described what he called mosaic inheritance—that is, no dominance appears in first generation hybrids, and all individuals show an intermediate or perhaps spotting pattern. Many of these mosaic individuals, Castle said, produce mosaic gametes, and so no segregation occurs in the second and successive generations. In the inheritance of sex, hermaphroditic animals and plants exhibit this mosaic inheritance. In the paper on heredity of albinism Castle concluded, "the Mendelian doctrine of gametic purity is fully substantiated by experiments in breeding mice, guinea pigs, and rabbits" (Castle and Allen 1903, 620), but he had one reservation. He had discovered that two albino animals who shared the same determinants for albinism and looked very much alike could have, in a latent form, very different color patterns. By the time Wright was doing the research for his thesis, this reason given by Castle for doubting the purity of the gametes had a complete Mendelian interpretation, but by then Castle had much stronger evidence indicating that the gametes were not pure in many, or perhaps even any, cases.

In 1903 Castle was not alone in arguing that Mendelian segregation could not account for certain hybridizations and that the Mendelian determinants were changeable. Davenport agreed on both points, as did Thomas Hunt Morgan and others. But a decade later Davenport, Morgan, and most other prominent geneticists had adopted a firm, consistent view that Mendelism, with appropriate modifications for linkage and multiple factors, could account for almost all aspects of heredity and that the germinal determinants for Mendelian characters were pure or unchangeable except by rare mutations. Castle, however, following the results of his own research, retained and strengthened the views he held tentatively in 1903. Having come into the field of genetics defending Mendelism from the attacks of the biometricians, Castle found himself agreeing with some of Karl Pearson's most basic attacks upon Mendelism and pure line theory (see Pearson 1910). Many geneticists, especially those in the Morgan school after 1910, were not amused by Castle's heresies and argued with him frequently in print.

Castle always had many experiments under way. Although only two students officially took their degrees with Castle before 1908 (students worked officially with E. L. Mark), many in actuality did their research under Castle's direction. Castle typically farmed out many of his various experiments to students but participated actively in the interpretation of the data produced by the experiments. The results were often presented as joint publications. Castle's first major report on his breeding experiments came in February 1905, in a seventy-eight-page monograph (Castle 1905a) published by the Carnegie Institution of Washington, which began to support Castle's research in 1904 (and continued to support it until 1943). Most of the report was devoted to guinea pigs (the precursor of the colony Sewall Wright would inherit seven years later), with a few pages given to rabbits. Castle found three pairs of alternative coat characteristics inherited according to simple Mendelian rules: albinism, recessive to pigmented coat; smooth coat, recessive to rough coat; and long coat, recessive to short coat. Albinism and long coat were also recessives in rabbits (rough coat was unknown at that time in rabbits). He presented much data on the crossing of guinea pigs with different coat colors, for example agouti with agouti, black, red, and albino, or black with red and albino. Castle was unable to put the results of these crossings into a coherent Mendelian scheme; many produced quite unexpected and unpredictable results. He clearly hoped to discover such a scheme by continuation of the experiments.

Some rabbits had very long ears that drooped—the lop-eared condition. When crossed with rabbits having normal-sized ears, the hybrid offspring were all intermediate. Bred together, these hybrids produced all intermediates, with no discernible segregation. Castle suggested this was a case of non-Mendelian heredity, as in the case of the *Hieracium* hybrids studied by Mendel. (The cases are fundamentally dissimilar, geneticists later discovered, because no sexual recombination occurred to produce segregation in the *Hieracium* F_2 generation.) A year later, in a report entitled "Heredity of Hair Length in Guinea-Pigs, and Its Bearing on the Theory of Pure Gametes,"

Castle was explicit about the modification of the gametes in crossing, as deducible from his data and cytological considerations. All his experimental facts were

> in harmony with the hypothesis, for which there is strong evidence on the cytological side, that each separately heritable character is represented by a different structural element in the germ (egg or spermatazoon). In fertilization the paternal and maternal representatives of a character become more or less closely united, this union persisting through all subsequent cell-generations until the new individual forms its sexual elements. At that time the paternal and maternal representatives of a character separate from each other and pass into different cells. But the paternal and maternal representatives of a character may in the meantime have exercised on each other a considerable influence. In the case of some characters, as ear-length in rabbits, they completely blend and intermingle, so that a new character is produced strictly intermediate between the conditions found in the respective parents. In other cases the modification may be slight, as if the maternal and paternal representatives of a character had been scarcely more than approximated. Sometimes in cases of alternative inheritance no influence of the cross is observable in certain of the extracted individuals, but if any considerable number of individuals is examined, others will be found in which the cross-breeding manifests its influence. From this we conclude that gametic purity is not absolute, even in sharply alternative inheritance. (Castle and Forbes 1906, 13)

By 1906 Castle was firm in his position that gametic purity was incomplete as a general feature of the hereditary process.

At the same time that Castle was firming up his belief in the impurity of the gametes, he was beginning to change his mind about the efficacy of selection and the implications of Mendelism for evolution in nature. Shortly after the rediscovery of Mendelism, most geneticists followed the lead of William Bateson and Hugo de Vries in adopting a "mutationist" view of evolution. Natural selection and artificial breeders, according to the mutationist view, utilized as their raw material sizeable mutations, whereas under the Darwinian view selection acted upon the ubiquitous quantitative or continuous variations found in every population. The mutationist view of evolution received strong support in 1903 from Wilhelm Johannsen's pure line theory. Basically Johannsen argued that selection must be ineffective in a genetically pure line unless mutations appeared. Applied to Mendelian heredity, Johannsen's ideas logically meant that selection must be largely ineffective in attempting to change the appearance of a population homozygous for a gene controlling a sharply discontinuous observable characteristic. Most Mendelians, including Castle, adopted this view at first.

In October 1904 Davenport wrote to Castle asking him to join five or six other biologists and give an address to the American Society of Naturalists on the mutation theory of organic evolution; Castle's assignment was to treat the subtopic of animal breeding (Davenport to Castle, October 21, 1904). Castle gladly accepted, "though with some misgivings that I may be considered a very amateurish breeder" (Castle to Davenport, October 29, 1904). At the meeting, held on December 28, 1904, Castle took the standard Mendelian view. Artificial and natural selection operated upon mutations, not continuous quantitative variability: "Modification of characters by selection, when sharply alternative conditions (i.e. mutations) are *not* present in the stock, is an exceedingly difficult and slow process, and its results of questionable permanency" (Castle 1905b, 524). Castle provided an example from his own laboratory. A guinea pig with a supernumerary fourth digit on one of its hind feet appeared in Castle's stocks in 1901. Five generations of selection later, Castle had created a true-breeding stock of guinea pigs with four well-developed digits on each foot. (Sewall Wright later maintained this stock until 1954.) In his address, Castle emphasized the mutationist view: "This race was not *created* by selection, though it was *improved* by that means" (Castle 1905a, 523). Castle's talk almost certainly found a receptive audience.

But Castle was moving away from Johannsen's view of pure lines and the Mendelian purity of the gametes, both of which fit together and with the mutation theory of evolution so well and obviously. If it were true that even sharply alternative Mendelian characteristics were determined by factors that had some variability, then selection acting upon this variability might be able to significantly change the appearance of the population, without new mutations. Thus the Darwinian conception of evolution might be possible, even with Mendelian heredity.

Castle was sufficiently intrigued by the possibility that late in 1905 he set one of his students, Hansford MacCurdy, to work with selection experiments on hooded rats and guinea pigs (MacCurdy and Castle 1907). MacCurdy quickly found that the sharply alternative piebald pattern of hooded rats behaved as a Mendelian recessive to the gray color of wild rats. He crossed the hooded rats with others bearing the "Irish" pattern, which was like the hooded pattern but with much more black and lacked the recessive modifier that caused the hooded pattern when homozygous. When these hybrids were backcrossed to hooded rats, the resulting offspring exhibiting the hooded pattern had a larger black stripe down the back. To Castle and MacCurdy, the reason for this change was a corresponding alteration of the Mendelian factor determining the hooded pattern. In the selection experiment, they tried to increase and decrease the size of the dorsal stripe in the hooded rats, maintaining at the same time a control stock.

After five generations, the success of the selection experiment was beyond all earlier expectation. The plus stock had a much-increased dorsal stripe, and the minus stock a much-decreased. Neither stock showed the slightest sign of regression when selection ceased. Darwinian selection not only worked, it worked rapidly and surely. Castle immediately became a Darwinian and remained one for the rest of his life. The paper that Castle had delivered in December 1904 before the American Society of Naturalists, and published in *Science* in April 1905, was an embarrassment only a year later. Castle was an honest and forthright man. But though he retained even one-page published papers in his formal bibliography, this five-page paper from *Science* he dropped and never, to my knowledge, ever referred to it again. It does not appear in the bibliography attached to L. C. Dunn's biographical memoir of Castle written for the National Academy of Sciences (Dunn 1965).

Castle and MacCurdy published their results in May 1907. They wrote a long introduction detailing the contrast between the de Vriesian, discontinuous view of evolution and the Darwinian, gradual, and selectionist view.

> The issue between the two views is sharp and clear. According to de Vries . . . selection is not a factor in the *production* of new species, but only in their *perpetuation,* since it determines merely what species shall survive; according to the Darwinian view, new species arise through the direct agency of selection, which leads to the cumulation of fluctuating variations of a particular sort. (MacCurdy and Castle 1907, 3)

Castle and MacCurdy concluded that their experiments supported the Darwinian rather than the de Vriesian view. It is crucial to understand, however, that Castle and MacCurdy thought they were selecting a changed Mendelian factor rather than hereditary modifiers of that Mendelian factor. Soon this belief would come under serious and sustained attack from other geneticists, although in 1907 no such negative reaction occurred.

Enthusiastic about the success of the initial selection experiment, Castle in October 1907 set up an ambitious plan to continue it, using some stocks from the earlier experiment but also using far more careful and sophisticated quantitative methods for grading the extent of the hooded pattern. With several associates and assistants, including Sewall Wright, he continued this experiment until 1919. During this time some fifty thousand rats were bred and carefully measured. The experiment was well known internationally, both because of the striking success of selection and because of Castle's unpopular interpretation of his results. Wright and R. A. Fisher were greatly influenced by Castle's selection experiments on hooded rats, but they reached different conclusions about the implications of the experiments for evolution in nature (see chapters 8 and 9).

Another experiment reported by Castle's group in 1906 addressed a problem that would become important for Wright during his work at the USDA from 1915 to 1925 and also central to his shifting balance theory of evolution. This experiment was the first to utilize the organism so famous in the history of genetics. Beginning in the academic year 1901–2, Castle suggested to one of his students, F. W. Carpenter (who later taught Wright embryology at the University of Illinois in 1912), that the "little fruit fly, pomace fly, vinegar fly, wine fly, and pickled fruit fly," all names for *Drosophila am-*

pelophila, was an excellent organism for inbreeding studies. This fly later of course became known as *Drosophila melanogaster.* Charles Darwin had been keenly interested in the biological effects of inbreeding, as were most breeders of domesticated animals and plants. In the 1880s and 1890s, studies by Bos, Crampe, von Guaita, and Fabre-Domengue indicated that close inbreeding, as in brother-sister or father-daughter mating, resulted consistently in decreased fertility, lack of vigor, diminution of size, partial or complete sterility, and pathological malformations. Castle wanted to test these conclusions by using the rapidly breeding *Drosophila,* whose whole life cycle could be completed in eleven or twelve days.

In the first series (called *A*) the inbred stock appeared to follow the general rule, with much-decreased fertility compared to the control stock and nearly 20% completely sterile matings. By the winter of 1905 this stock had been through fifty-three generations of close inbreeding and exhibited throughout one-half to one-third the fecundity of the control cultures. But Castle was unsatisfied by this apparent confirmation of earlier research and in October 1903 set other students to work with two other inbred series, called *M* and *N.* Contrary to expectation, both of these series remained highly fertile during fourteen generations of inbreeding. Castle concluded that loss of fecundity could not be a necessary and inevitable consequence of close inbreeding, rather that fecundity was probably a hereditary trait. Selection experiments upon the *M* and *N* series indicated that variation in fecundity was highly hereditary. After matings between the various inbred lines were carried out, Castle even concluded that low productiveness is inherited after the manner of a Mendelian recessive character in certain of the crosses (Castle et al. 1906, 786).

After 1906 Castle bred mammals rather than *Drosophila* in his own research. After the usefulness of *Drosophila* for demonstrating Mendelian inheritance became obvious soon after T. H. Morgan's work in 1910, Castle again maintained stocks of *Drosophila,* but primarily for use in the genetics course he taught once each year. Wright would follow the same pattern through the thirty years he taught at the University of Chicago. (On how Morgan came to use *Drosophila,* see Allen 1975 and 1978, 146–47.) Castle always maintained a strong interest in inbreeding, which he believed was important both in artificial breeding and in evolution in nature.

Sewall Wright absorbed from Castle a keen interest in the consequences of inbreeding. Later he would not only devise methods for bringing the quantitative analysis of inbreeding to a wholly new level of understanding but he also used the concept of inbreeding as a central factor in devising his theory of evolution in nature.

In the years 1907–8 Castle continued serious research upon color inheritance in guinea pigs, rabbits, and mice. He began a project on the hybridization of *Cavia aperea* from Brazil (later recognized as *Cavia rufescens*) with the common guinea pig, *Cavia porcellus.* At first he could obtain no second generation from the hybrids, but in 1907 he discovered that only the male hybrids were totally sterile—the females could be backcrossed with either *C. porcellus* or *C. aperea* males. This opened up the exciting possibility of studying Mendelian inheritance between species rather than races or varieties, and Castle and his assistants immediately set to work on this research.

Castle was also conducting a selection experiment on size in guinea pigs. This was an attempt to study the effectiveness of selection upon a character which clearly did not exhibit simple Mendelian heredity, as did the hooded pattern in rats. By September 1907 Castle had bred in his experiments over 11,000 guinea pigs, 1,500 rabbits, and 4,000 rats (according to Castle's annual report to the Carnegie Institution of Washington).

Castle wished to expand his research and work more rapidly. The Harvard Corporation responded by offering to double the room available for Castle's experiments and to relieve him of formal teaching duties for one whole term each year. But before Castle could expand into the extra room, an even better possibility loomed on the horizon.

The Bussey Institution and Edward Murray East

In 1871 Harvard University established the Bussey Institution in Jamaica Plain as an undergraduate school of agriculture. By 1908, however, the Harvard Corporation and faculty decided to abolish it. Also in 1908 the physicist Wallace Clement Sabine became dean of the Graduate School of Applied Sciences at Harvard. Sabine, with Castle's strong encouragement, began to consider converting the Bussey Institution into a research facility with advanced instruction in the biological sciences related to agriculture and horticulture. Although the space problem at the Museum of Comparative Zoology was becoming acute, the inconvenience of a facility so far removed from the main campus would raise many problems. But Castle wanted the isolation because it offered the space and peace to conduct his research; he also wanted his own graduate students, instead of having the students whose work he actually guided take their degrees formally with E. L. Mark.

On February 10, 1908, Castle wrote to Davenport:

> The Bussey Institution is about to go out of business as an undergraduate school of agriculture and a proposition is under consideration to turn some of its resources (unfortunately not large) toward biological investigations, indirectly, if not directly applicable to agriculture. Dean Sabine is trying to formulate a plan under which this can be realized. He is going next week to Washington and will probably visit you also in quest of information which will aid him. He needs to know *cost of equipment and running expenses for experimental work*. The scheme may not go through, but if it does, will aid our work in heredity.
>
> Any help you can give him will be appreciated. Please consider this strictly confidential.

Davenport replied on February 12: "Congratulations on the prospect [of the Bussey Institution]. Here's hopes that it will go thru. If you see Sabine, tell him I shall be very glad to receive a visit from him." On Sabine's subsequent recommendation, the Harvard Corporation directed that the Bussey Institution become a center for research and doctoral programs in biological sciences related to agriculture.

Castle pressed for better-equipped facilities, more research time, and a balanced program in experimental genetics. Sabine complied with the last two wishes by approving the appointment of a second geneticist whose expertise was in botany, who could assume half of Castle's formal teaching load, and who would share in the direction of graduate research. Castle's hope for better facilities was aided by Harvard and, at Davenport's urging, by the Carnegie Institution of Washington, which between 1908 and 1910 doubled Castle's annual support from $500 to $1,000.

Speaking of the appointment of Edward Murray East as the botanical geneticist, Castle said in 1950: "In 1907 he published a paper entitled 'The Relation of Certain Biological Principles to Plant Breeding' [East 1907]. It was of such outstanding excellence as a discussion of mutation, Mendelism, selection and evolution that Harvard University in 1909 invited him to come to the newly organized Bussey Institution, to develop the field of plant genetics in coordination with the field of animal genetics in which I was working. My pupils and I over the succeeding years profited greatly from this association" (Dunn 1951, 61). East came with a strong recommendation from William Bateson, but probably more important was Castle's attraction to East's special qualifications as a geneticist.

Originally trained as a chemist, East began graduate work at the University of Illinois in 1900. Four years earlier, C. G. Hopkins, at the Illinois Agricultural Experiment Station, had begun selection experiments to change the oil and protein content of corn, a matter of great economic importance to farmers. East assisted Hopkins by chemically analyzing the corn samples and became interested in selection and genetics. Beginning in 1902, under Hopkins's direction, East began his own selection experiments on nitrogen content and other characteristics of potatoes. This research would be the basis of his doctoral thesis. With Castle he became a charter member of the American Breeders Association, founded in 1903. In 1905 he took a position as a plant breeder at the Connecticut Agricultural Experiment Station, where he continued research on the potato and began other experiments on the genetics of corn and tobacco. He remained there until coming to the Bussey Institution in 1909.

East began working in the field of genetics at the same time that Hugo de Vries's two-volume *Die Mutationstheorie* appeared—1901–3. De Vries had a great influence upon all geneticists but more especially the botanists. We have already seen how Castle in 1904–5 had wholeheartedly supported the de Vriesian view of selection, mutation, and evolution, only to drastically change his mind by 1907. East encountered two conflicting streams of thought in his development as a geneticist: one was the popular view of de Vries; the other came from East's own selection experiments with potatoes, the results he observed in Hopkins's experiment with corn, and his detailed knowledge of the history of potato breeding. This second stream of thought yielded a clear conclusion—that selection of fluctuating or continuous differences could and often did lead in the experimental population to highly significant changes which did not regress when selection ceased. Darwinian-type selection might therefore be a major force in both artificial selection and natural selection.

East completed his doctoral thesis in June 1907, and it was published as a bulletin of the Illinois Agricultural Experiment Station in 1908. Both the 1907 article cited by Castle and East's thesis, "A Study of the Factors Influencing the Improvement of the Potato" (East 1908), exhibit the conflict between the de Vriesian and Darwinian viewpoints in his thinking. The experimental and historical evidence was insufficient for East to dismiss the de Vriesian theory as generally applicable, but he repeatedly emphasized that his own experimental results were more consistent with the Darwinian theory. Since almost all botanical geneticists in the United States were strongly, even militantly, in the de Vriesian school, one can easily understand why Castle wished to have East join him at the Bussey. Indeed, Castle and East working together created the genetics group with the deepest commitment to the Darwinian outlook to be found in the United States in the 1910s and 1920s. Castle and East between them directed over forty doctoral theses in the Bussey between 1909 and 1936, and their Darwinian outlook would have a major influence through their students as well as through their writings. Many of the students from the Bussey went on to become major influential geneticists, among them R. A. Emerson (doctorate in 1913), E. W. Sinnott (1913), C. C. Little (1914), D. F. Jones (1918), L. C. Dunn (1920), Edgar Anderson (1922), Karl Sax (1922), R. A. Brink (1923), Paul Mangelsdorf (1925), G. D. Snell (1930), and E. R. Sears (1936). All were members of the National Academy of Sciences.

Castle and East did not, however, agree on all major issues in genetics and evolution. At the time East came to Harvard after the harvest at the Connecticut Station in 1909, he had already conducted careful experiments in crossing corn varieties with some blending characteristics. In agreement with the just-published results of H. Nilsson-Ehle at the Svalov Station in Sweden, East concluded from his research that the blending inheritance in the F_1 generation and greater variability in the F_2 was best explained by multifactorial Mendelian inheritance (East 1910). East adhered strongly to the view, held by more and more geneticists, that Mendelian determinants were not altered by hybridizations and did not undergo fluctuating variation, except by infrequent mutation. Thus Castle and East took diametrically opposed views on the genetic mechanisms underlying Darwinian selection, the reality of which neither doubted.

Figure 2.1. William Ernest Castle. Archives of the Genetics Society of America.

Figure 2.2. Edward Murray East. Archives of the Genetics Society of America.

Students in the Bussey Institution all knew that Castle and East agreed on some points and differed on others. L. C. Dunn, who succeeded Wright as Castle's research assistant in 1915, recalled a seminar given by East in which the required reading was all of Darwin's *Variation of Animals and Plants under Domestication* (1868). East emphasized the plant material but also dealt with animal breeding. Some of Castle's students in East's seminar reported to Castle

> what some of East's ideas were about the domestication of animals, and Castle immediately joined the seminar, to see that the animal material was not mishandled by his botanical colleague. They sat like cat and mouse at the two ends of the table, each one looking for a mistake on the part of the other. It's a good thing to have before students two professors who disagree. (Dunn 1961, 62)

Certainly there was general agreement among the students at the Bussey that the differences between Castle and East led not to disagreeable conflict but to an educational creative tension. Part of the reason for this happy circumstance was that Castle was very easygoing with his students and often encouraged them to disagree with him and with East. Castle did not react badly when his own students disagreed openly with him or even when they effectively disproved his work. Many of Castle's students, including particularly Sewall Wright, disagreed with Castle on substantive questions. Often Castle's students took East's side, specifically in the controversy about the changeability of the Mendelian determinants and multiple factors.

Experimental Work of Castle and His Students, 1909–1912

The years 1909 to 1912 were productive for the scientific research of Castle and his students. The selection experiment on hooded rats yielded highly significant results, and Castle further developed his arguments in favor of the interpretation of changed Mendelian factors. Working with rabbits, guinea pigs, and mice, Castle deduced the first generalized schemes for understanding most of the observed patterns of color inheritance in these mammals. Castle's student C. C. Little concentrated upon color inheritance in mice. Castle's experiments on size inheritance in rabbits were very ably expanded by his student E. C. MacDowell. Castle was able to obtain the help of John C. Phillips, an independently wealthy physician whose specialty was surgery. Phillips had been a student of Castle as well as Davenport, but he was basically a talented amateur primarily interested in birds; he became curator of birds at the Agassiz Museum. Phillips worked with Castle on the selection experiment with hooded rats and did the surgical work on the widely publicized series of experiments on ovarian transplants in guinea pigs (which I will not treat here because the burning issue in 1906–1909 of the inheritance of acquired characters had, at the Bussey Institution, died away completely by the time of Wright's entrance in 1912, largely because of the experiments by Castle and Phillips). On his own, Phillips conducted a research project on size inheritance in ducks. Finally, Castle's student John Detlefsen concentrated upon the analysis of the crosses between guinea pigs (*Cavia porcellus*) and *Cavia rufescens,* examining coat colors and other characteristics of the coat, growth, and fertility. Wright arrived at the Bussey Institution in time for firsthand acquaintance with every one of these researchers and their research projects. Most of Wright's training as a mammalian geneticist derived from his familiarity with these individuals and their work.

Selection in Hooded Rats

After the success of the Castle-MacCurdy selection experiment on hooded rats reported in 1907, Castle wished to continue the experiment for two primary reasons. He particularly wanted to see if selection could drive the plus and minus series until the average rat in each series exhibited a hooded pattern beyond the range of variability found in the control stock. This was important because de Vries, Johannsen, and many others had claimed that selection of continuous or fluctuating variability could not accomplish this result.

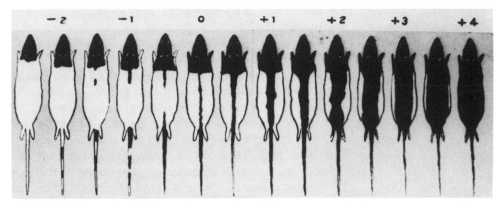

Figure 2.3. Castle's scale of gradation in the hooded pattern.

Second, Castle wanted evidence for his view that selection was changing the Mendelian factor itself. Castle understood from the beginning of the new series of selection experiments that very large numbers of rats would have to be raised to achieve convincing statistical significance and that systematic, quantifiable measurements of the extent of the hooded pattern were essential. He developed an arbitrary scale that was reproduced in nearly every genetics textbook published after 1914, when the results were first announced in detail (see figure 2.3; reproduced from Castle and Phillips 1914, plate 1).

The first two generations of the new selection experiment produced too few offspring for statistical significance, but regression appeared very strong. The efficacy of selection in the earlier Castle-MacCurdy experiment was not observed. Castle wrote in his September 1908 report to the Carnegie Institution of Washington:

> The selection experiments with rats have been continued on a considerable scale. Attempts to increase and to decrease the pigmented areas of hooded rats are constantly attended by strong regression. It is considered theoretically important to ascertain whether any permanent advance can be made by selection of fluctuating variation, such as occurs in this case.

Plainly Castle at this point entertained some doubts about the efficacy of the selection. Any such doubts were quickly and thoroughly dispelled by the obvious strong effects of selection seen in the third and later generations. By the eighth generation, measured early in 1911, selection had resulted in plus and minus series whose average pigmentation was well beyond the limits of variability found in the original or in the control stocks. Regression was greatly reduced, and no limit to the effects of selection was apparent. In his 1911 book, *Heredity in Relation to Evolution and Animal Breeding*, Castle described the results of the experiment and concluded that Darwin was right in assigning great importance to selection in evolution (p. 126). By the fall of

1913, the experiment had continued for thirteen generations and the effects of selection continued unabated.

When the first major report of this work was published by Castle and Phillips (who had assisted from the fall of 1908 until the fall of 1913) early in 1914, they were able to present the results of many subsidiary experiments. Return selection proved to take as long or longer than the original selection for extremes, indicating that selection was causing genuine hereditary change. Crosses both with wild rats and with Irish (highly black pigmented) rats were performed, with extraction of hooded rats again in the F_2 generation. The mean of the minus race, when crossed with either wild or Irish rats, regressed toward greater pigmentation as expected. The mean of the plus race, crossed with wild rats, regressed a little toward less pigmentation, also as expected.

Castle thought that crossing the plus race with Irish would increase the pigmentation, but in fact it regressed there as well, and Castle furnished this explanation for that anomalous result: "This result we can explain on the supposition that the selected plus series has accumulated *more* modifiers of the hooded pattern than the wild race contains, so that a cross with the Irish strain tends to reduce the number of modifiers in the extracted hooded individuals" (Castle and Phillips 1914, 25). None of the crosses with wild or Irish rats were carried beyond the F_2; only one extraction of the hooded pattern was attempted. The extracted hooded rats of the minus series and the plus series were still far apart in average pigmentation. As we will see later, Sewall Wright would suggest a crucial modification of the extraction procedure that would in turn cause Castle to reject his long held-view about the effects of selection upon Mendelian factors.

In the tenth generation, which occurred in the summer of 1912, two mutants (male and female) of far greater pigmentation appeared suddenly in the plus series. From these rats, by inbreeding and selection, a very highly pigmented series was founded. Castle admitted that the mutants probably had acquired a substantial modifier of the hooded pattern, but he generally, if tentatively, defended the view he had held for over ten years—that selection was changing the Mendelian factor for hooded pattern rather than accumulating modifiers of the Mendelian factor.

From 1909 onwards, Castle encountered stiff opposition to his belief that selection changed Mendelian factors. Before describing that opposition and Castle's response, examination of another set of Castle's experiments related to this problem is essential.

Size Inheritance

In 1905, when Castle published his first research on size inheritance, he concluded that size differences in general exhibited non-Mendelian inheritance— a permanent blend, with no segregation even in the F_2 generation. Between 1905 and 1909, with the help of several assistants, he continued experiments on the inheritance of ear size, weight, and skeletal dimensions in rabbits. By far the best data was on crossing a breed of very long-eared (lop-eared) rabbits with ordinary rabbits. These matings always produced offspring with ears of intermediate length. When the F_1 generation was inbred, the data indicated that the blend was permanent—no greater variability in ear length was visible in the F_2 generation than in the F_1, thus violating Mendelian expectation. The data on weight and skeletal dimensions yielded a similar conclusion. When he wrote up these data early in 1909, Castle believed that this evidence of size inheritance strongly supported his view that Mendelian factors could be changed by selection, hybridization, and perhaps other means. At this time Castle had yet to receive substantive criticism from fellow geneticists on his interpretation.

The work on size inheritance in rabbits appeared in June 1909. Castle almost immediately realized that alternative hypotheses consonant with Mendelian heredity could account for his observed results. First, he read Nilsson-Ehle's work on inheritance in wheat and oats (Nilsson-Ehle 1909). Nilsson-Ehle presented convincing evidence that two-, three-, and probably four-factor Mendelian inheritance could be discerned in what previously appeared to be simple blending inheritance. Second, Castle's new colleague, East, produced that very summer strong evidence of Mendelian multiple-factor inheritance in corn (East 1910). East strongly favored the view that Mendelian factors were stable, and after moving to the Bussey he disagreed openly with Castle's interpretation of his experiments on size inheritance and selection. Added support for East's position came from the later famous geneticist R. A. Emerson, then at the Nebraska Agricultural Experiment Station, who published a paper in 1910 showing that in gourds and summer squashes F_2 variability was much greater than F_1 variability, according to expectation under Mendelian multiple-factor inheritance (Emerson 1910).

Castle, who dealt with far fewer numbers in his experiments than the plant geneticists, had not seen greater variability in the F_2 than in the F_1 generation. The German animal geneticist Arnold Lang had a Mendelian answer to this question (Lang 1910). By constructing a model of expectation based upon an increasing number of multiple factors, he showed that, for example, with six factors affecting ear length, the original-length ears should reappear in the F_2 generation only twice in 4,096 individuals. The expected distribution was strongly concentrated in the intermediate lengths. With ten factors, the original ear length should appear twice in 16,777,216 individuals. Lang's paper was appropriately titled "Die Erblichkeitsverhältnisse der Ohrenlänge der Kaninchen nach Castle und das Problem der intermediären Vererbung und Bildung konstanter Bastard-rassen" (The inheritance of ear length in Castle's rabbits and the problem of blending inheritance and the production of constant hybrid races.) Other geneticists, including Castle's assistant John Phillips and student E. C. MacDowell, joined in the criticism of Castle's position.

Phillips conducted an experiment on size inheritance in ducks (Phillips 1912). He crossed French Rouen ducks with mallards weighing less than half as much. The F_1 offspring were uniform, but the F_2 generation exhibited a wide range of variability. The segregation expected under Mendelian inheritance was obvious. MacDowell had been given the experiment on size inheritance in rabbits after Castle published the earlier results obtained in 1909. Concentrating upon skeletal measurements and total body weight but not ear length, MacDowell produced clear evidence of greater variability in F_2 generations than in F_1 generations of the crosses between races. MacDowell also read exhaustively in all literature related to the question of size inheritance, and he came down firmly on the side of East and multiple Mendelian factors (MacDowell 1914).

Under this onslaught of criticism, Castle reconsidered and in his 1911 book apparently rejected his earlier position. After citing the evidence of Nilsson-Ehle and East, Castle concluded:

> In the light of this evidence it is clear that in maize, seemingly blending is really segregating inheritance, but with entire absence of dominance, and it seems probable that the same will be found to be true among rabbits and other mammals; failure to observe it hitherto is probably due to the fact that the factors concerned are numerous. For the greater the number of factors concerned, the more nearly will the result obtained approximate a complete and permanent blend. As the number of factors approaches infinity, the result will become identical with a permanent blend. (Castle 1911, 138–39)

Castle's apparent capitulation was short lived. On April 19, 1912, he delivered an address at the University of Illinois, and soon after it was published in *American Naturalist* (Castle 1912a). This address was the one Wright heard Castle deliver, after which he asked Castle about the possibility of studying with him at the Bussey. Castle took Wright on as his assistant then and there, little suspecting that his new assistant would a few years later suggest the experiment that would disprove the thesis Castle had presented in the lecture.

The address was entitled "The Inconstancy of Unit Characters." Castle described the controversy and then presented the summary of the data from the selection experiment with hooded rats. He stated: "The conclusion seems to me unavoidable that in this case selection has modified steadily and permanently a character unmistakably behaving as a simple Mendelian unit" (p. 356). As for size inheritance, Castle cited the work of East, Nilsson-Ehle, Phillips, and MacDowell, all suggesting a multiple-factor interpretation.

> This attractive hypothesis would account for the known facts of size inheritance fairly well, involving only the existence of multiple units which may be perfectly stable and changeless in character. Nevertheless this hypothesis has not been established beyond question. It is

quite possible that we are stretching Mendelism too far in making it cover such cases. Dominance is clearly absent and the only fact suggesting segregation is the increased variability of the second as compared with the first hybrid generation. This fact however may be accounted for on other grounds than the existence of multiple units of unvarying power. (P.361)

Between the writing of his 1911 book and April of 1912, Castle had slipped back to his earlier position. Wright recalls being much impressed by Castle's address that day; but within a year Wright would disagree fundamentally with the conclusions Castle made in the address.

After the address was published in *American Naturalist*, East decided that his appropriate response was to write a substantive, sharply worded attack upon Castle's position for the same journal. East's paper appeared that fall, when Wright was in the Bussey working with Castle (East 1912). East wrote a sophisticated paper, poking holes in the evidence of others which Castle had used to support his position and challenging Castle's interpretation of his own data. Castle's "unavoidable" conclusion that selection had modified the Mendelian unit character was to East "not only avoidable, but unnecessary" (p. 647). According to East, all the research on blending inheritance and selection, including Castle's, fit perfectly with the Mendelian view of multiple factors. The lines of disagreement were clearly drawn. A student in the midst of this controversy, observing his professors publishing papers that challenged each other, in the most prestigious journal in the field, had to choose. Wright, on this issue, chose East.

Color Inheritance

Because Castle was a critic of some cherished beliefs of Mendelians during the first two decades of genetics, it is sometimes forgotten that he was one of the great advocates of Mendelian heredity. Nowhere was Castle's admiration of the explanatory power of Mendelism greater than in the study of color inheritance, a subject to which he made a number of important contributions. A good example of Castle's belief in the power of Mendelism came in 1908 when he predicted, on the basis of Mendelian analysis, that a cross he made in guinea pigs between an agouti (the gray color of wild rodents) and a chocolate should yield definite proportions of agouti, chocolate, black, and cinnamon-agouti offspring. The exciting part of the prediction was that cinnamon-agouti guinea pigs had never been seen by fanciers anywhere, although cinnamon-agouti mice were known to occur. Sure enough, all four kinds of offspring appeared, even the unknown cinnamon-agouti, in about the predicted proportions. Castle exclaimed in the conclusion of his report to *Science*:

A moment's consideration of this case shows what a really great advance in the theory and practise of breeding has been obtained through the discovery of Mendel's law. What a puzzle this case would have presented to the biologist ten years ago! Agouti crossed with chocolate gives in the second filial generation (not in the first) four varieties, viz., agouti, chocolate, black, and cinnamon. We could only have shaken our heads and looked wise (or skeptical).

Then we had no explanation to offer for such occurrences other than the instability of color characters under domestication, the effects of inbreeding, or maternal impressions. Serious consideration would have been given to the proximity of cages containing both black and cinnamon-agouti mice.

Now we have a simple, rational explanation, which any one can put to the test. We are able to predict the production of new varieties, and to produce them.

We must not, of course, in our exuberance, conclude that the powers of the hybridizer know no limits. The result under consideration consists, after all, only in the making of new combinations of unit characters, but it is much to know that these units exist and that all conceivable combinations of them are ordinarily capable of production. This valuable knowledge we owe to the discoverer and to the rediscoverers of Mendel's law. (Castle 1908, 252)

Castle's assessment of the dramatic progress in understanding color inheritance using Mendelian analysis is wholly accurate and one measure of why the Mendelians vanquished the biometricians and all others whose theories of heredity did not allow such precise, quantitative prediction.

Color inheritance in both animals and plants was a fertile field for Mendelian analysis because many of the major color varieties were visibly distinct and controlled by single Mendelian factors. Much of the early Mendelian analysis in mammals concerned color inheritance. Lucien Cuénot in France was the first to publish a Mendelian analysis of color inheritance in mice, followed closely by the work of William Bateson and his associates (R. C. Punnett, C. C. Hurst, F. M. Durham) and that of Castle. Late in 1903 Bateson published a summary article entitled "The Present State of Knowledge of Colour-Heredity in Mice and Rats." Here Bateson outlined the fundamentals of color inheritance in mammals. He documented the existence of three independent pigments: (1) densely opaque black, (2) less opaque brown, and (3) transparent yellow. All three existed in at least two forms, one more intense, the other more dilute. Albinos exhibited a total absence of pigment. The wild coloration of mice and rats, gray or agouti, contained all three pigments in a particular geometric pattern on each hair. Full albinism was clearly a Mendelian recessive, and the dilution factor was inherited as a separate Mendelian factor. Many formerly baffling results of crossing mice began to make sense, although many others did not appear to fit the Mendelian scheme. Especially confusing were the behavior of the agouti pattern in crosses other than with albinos and the production of pied, or spotted, mice and mice with scant pigmentation. Other anomalies were frequent, despite

Bateson's optimistic assessment: "The majority of the observations are in accord with the Mendelian hypothesis in a simple form. The true solution of several subordinate problems still remains obscure" (Bateson 1903, reprint, 103).

Castle solved the basic problem of agouti coloration late in 1906. He discovered that the agouti pattern was simply a geometric distribution of the basic pure colors and that it was inherited independent of these colors. Thus the agouti/nonagouti pattern was like the previously known long-hair/short-hair pattern, also due to a Mendelian factor independent of coloration. The agouti pattern was dominant but affected only the black and brown pigments, not the yellow. This nice piece of Mendelian analysis not only cleared up many basic problems with the transmission and reappearance of the agouti pattern seen in previous crosses but also made possible the prediction of coat patterns not yet observed. The brown pigment of mice was not present in rabbits and was not generally found independent of black in guinea pigs. But in 1907 Bateson discovered and sent to Castle a pure brown (chocolate) male guinea pig. This animal differed from the wild-type agouti (homozygous for black, yellow, and agouti) by the absence of the factors for black and agouti. Castle confidently predicted that addition of the agouti factor to the factors producing the chocolate animal would produce a cinnamon-agouti, the pattern known in mice but not in guinea pigs. The fulfillment of this prediction by a cross in Castle's laboratory led to his enthusiastic report to *Science* quoted earlier.

For Castle the year 1908 was one of exciting research results. At this time he probably knew more than anyone in the world about the inheritance of coat color in mammals, and he wanted to write up his findings in a systematic and generalized manner. He wrote a summary paper that made Bateson's 1903 paper look very outdated. With the newly discovered color factors in mind, he devised simplified methods of expressing gametic and zygotic formulas. A glance at one of the formulas told immediately the genotype of the stock, and its breeding capacity was easily inferred. Written as part of Carnegie Institution of Washington Publication No. 114 (it contained also the studies of size inheritance in rabbits referred to in a previous section), *Studies of Inheritance in Rabbits* (Castle et al., 1909) was widely distributed and influential (It was also issued as one of E. L. Mark's series from the Museum of Comparative Zoology, no. 199.)

Basically Castle et al. (1) listed and explained what the authors considered to be the eight independent factors affecting coat color in rabbits, (2) gave geometric and linear models for representing these factors in a given rabbit or germ cell, (3) showed how from the gray of the wild rabbit all other color varieties of rabbits arose by loss of one or more of the eight color factors, (4) provided examples from their own experiments, and (5) suggested a simple conception of the material basis of the hereditary factors. The analysis could easily be applied to guinea pigs, mice, or rats with minor alterations. The influence of Castle's analysis can be measured by the ease with which geneticists now can read this paper and feel perfectly comfortable, with some minor modifications. It is essential to delineate Castle's conception of color inheritance in this 1909 paper because the work of Little and Detlefsen was based directly upon it and upon this basis built by Castle, Little, and Detlefsen, Sewall Wright would pursue his thesis research at Harvard. Briefly, the factors were as follows (this list comes almost verbatim from Castle's report of September 25, 1908, to President Woodward of the Carnegie Institution of Washington):

1. *C*. A general color factor necessary for the production of any pigment. It is missing only in albinos.
2. *B*. A factor for black, some substance which acting upon *C* produces black pigmentation.
3. *Br*. A factor for brown.
4. *Y*. A factor for yellow.
5. *I*. An intensity factor, which determines whether the pigmentation shall be intense (as in black and yellow), dilute (as in blue and cream), or of some intermediate degree of intensity.
6. *A*. The pattern factor which causes the ticked gray, or agouti, coat color of all wild rodents.
7. *U*. A factor for uniformity of pigmentation (in distinction from spotting with white, *S*).
8. *E*. A factor governing the extension of black and brown pigmentation but not yellow. When most restricted in distribution, the black or brown pigments are found in the eye and in the skin of the extremities only but not in the hair; when more extended they occur also in the hair generally.

Consciously attempting to follow the lead of the organic chemists, Castle arranged a geometric representation of the eight factors to indicate their interrelations in the gamete (fig. 2.4). For the zygote, each factor would be followed by the subscript 2, if homozygous. Thus, in linear presentation, the wild gray rabbit capable of producing nothing but other grays (or agoutis) was B_2 Br_2 E_2 A_2 C_2 I_2 U_2 Y_2, whereas a rabbit heterozygous for A and C would appear B_2 Br_2 E_2 A C I_2 U_2 Y_2 and could (by mere inspection of the formula) be capable of producing albinos, blacks, and grays. Castle gave the 18 different known gametic combinations and a few of the zygotic.

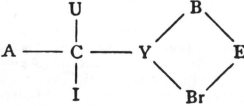

Figure 2.4. Castle's geometric representation of gametes.

Castle then showed that an all gray rabbit might conceivably have any one of 32 different zygotic formulas, of which he had produced 10 in his own experiments. Blacks, which differed from the grays only in the loss of the agouti factor, A, could have 16 different zygotic formulas, of which 5 had been found in Castle's stocks. The enormous power of Mendelian analysis, particularly in predicting the appearance of future offspring, was inescapably apparent in this paper. A breeder who was an expert on rabbits, guinea pigs, rats, or mice before 1900 could only marvel at the giant steps taken toward the understanding of color inheritance in only nine years.

Castle's scientific background was tied to a mechanist, materialist interpretation of the genetic and physiological functioning of organisms. He tried, therefore, to suggest a physiological interpretation of color inheritance. He could not be specific but said that the general color factor C was acted upon by specific substance, perhaps color enzymes (Castle et al. 1909, 46), producing specific pigments such as black, brown, or yellow. The color enzymes were somehow produced by the genetic information. The final brief section of the paper was entitled "The Material Basis of the Hereditary Factors" and is reproduced here in full:

> In what form, it may be asked, are we to suppose that the various assumed factors exist. Do they occur as so many different substances lying side by side but unmixed in every reproductive cell? To this question we may give at present no satisfactory answer.
>
> It is, however, we think, not necessary to suppose that there exist in the minute germ-cell as many complex organic substances as there are activities of the cell; neither is it necessary to suppose a different substance present for every independent factor identified. The various independent factors may have a basis no more complicated than that of so many atoms attached to a complex molecular structure. Experiment shows that the factors may be detached one by one from the organic complex. The discontinuity of their coming and going is entirely in harmony with the conception of them as components merely of complex molecular bodies. (P. 68)

Bateson, Castle, Goldschmidt, and many others, as well as the famous Garrod, were speculating about gene action even before 1910. The quote here from Castle is prophetic, not because he was unique in making such speculations but because his view would basically be verified by the revolution in molecular biology from the early 1950s on. Also, he raised here a problem that would fascinate Sewall Wright. Wright's thesis, as will become apparent in the next chapter, was about color inheritance, but consciously directed to the level of physiological genetics rather than simply to the level of transmission genetics. It was an interest central to Wright's scientific work from his graduate school days onwards and a field to which he made many influential contributions.

Castle's conceptual scheme of 1909 for color inheritance underwent a number of revisions during the three years following. The symbolization used by Castle to designate, for example, pure zygotic black was B_2. It was simpler to express this as BB, especially because the heterozygote could be expressed as Bb and the complete recessive as bb. Brown (Br) was soon found in guinea pigs to be a recessive allele of black, and the symbol Br was discarded. Also discarded was Castle's geometric representation of the gametic and zygotic formulas, probably because the geometric relation did not convey physiological interactions realistically and because the linear representation sufficed.

Some substantive problems even with fairly discontinuous coat colors and patterns remained in rabbits, mice, rats, and guinea pigs. The best known case was that of the Himalayan rabbit. This pattern was characterized by a white fur and black pigmented extremities (nose, ears, feet, and tail) and sometimes sootiness on the back. The rabbit was almost like an albino, but it certainly did not behave like one in crosses because it was clearly dominant to albinism. Yet it was also clearly recessive to dark pigmentation. In 1909 Castle tentatively suggested that the Himalayan pattern was like albinism, the recessive of the color factor C, but had to be distinct from it. He labeled it C'. Without using the term, Castle was suggesting multiple alleles (three) for the color factor, but he did not develop the suggestion.

R. C. Punnett soon presented his results on the inheritance of coat color in rabbits (Punnett 1912). He agreed with almost all Castle had said and presented in the 1909 paper but added some new experiments and observations. Most striking were his results concerning Himalayan rabbits. He discovered that there existed a Himalayan form of some full-colored varieties—black Himalayans, brown Himalayans, and agouti Himalayans; but a yellow Himalayan did not exist because the yellow pigment did not develop with the Himalayan pattern. Punnett raised the question of whether the Himalayan patterns were characterized by the absence of yellow, with a Himalayan "point factor." With this hypothesis, the only difference between the albino and the Himalayan was the Himalayan point factor. An ordinary self-colored rabbit, such as a black, would then have to possess both the yellow factor and the point factor. The F_2 offspring from crossing black with albino would be expected to produce blacks, albinos, and Himalayans. Since no Himalayans were ever seen in such crosses, Punnett rejected the hypothesis and concluded that "for the present the relations between these various forms remain obscure" (Punnett 1912, 237). Here the problem rested until Wright entered Harvard. During his graduate career the Himalayan rabbit problem was definitively solved, and Wright was able to solve the similar but more difficult problems of color inheritance in guinea pigs.

In 1911, in a trip financed by the Carnegie Institution of Washington, Castle traveled to Peru to gather wild cavies for use in breeding experiments. Three wild-caught cavies (one male, two females) about the size of domestic guinea pigs were given to Castle by a company in Ica, Peru. This Ica race was highly variable and by inbreeding experiments was shown to possess five color factors not visible in the parents. One factor was a variation called "red-eye," characterized in the genetic combinations first observed by Castle by black pigmentation, but no yellow pigment, and eyes that glowed deep red under light (instead of the normal black eyes). This red-eye variation, when crossed with the other known color variations, produced hitherto unknown color variants. This was another problem unsolved at the time Wright entered graduate school but which he solved before he left Harvard.

Spotting in guinea pigs was not a simple alternative Mendelian character, as in rabbits. Castle maintained a stock of tricolor (black, yellow, white) guinea pigs, and in 1912 he published a paper on the inheritance of tricolor in guinea pigs in relation to Galton's analysis of tricolor in Bassett hounds (Castle 1912c). "Tricolor" stock is perhaps a misnomer because the actual tricolors never bred true, and the stock consisted of tricolors, black-and-whites, and yellow-and-whites. All appeared to have the same genetic constitution as regards the known Mendelian factors: It was basically a yellow race with black and white spots. The black spots resulted from an irregular distribution of the black factor B, and the white spots resulted from distribution of the color factor C, or rather its absence. These two factors are independent in Mendelian inheritance. Castle's explanation was as follows:

> If the black factor extends over all the colored areas, the animal will be black-and-white. If the black factor falls only on areas which lack the color factor, it will produce no visible effect, and the animal will be yellow-and-white. If, finally, the black factor falls on some of the colored areas but not on all of them, those in which it falls will be black, the other yellow, and the uncolored areas of course white. Hence a tricolor will result. But the gametic composition of these tricolors will not be different from that of the black-and-whites or yellow-and-whites produced by the same race, since all alike will be characterized by irregularity in distribution of the same two factors. (Castle 1912c, 439)

Wright used the tricolor stock in his thesis research only for studying the inheritance of roughness of the hair, but he became keenly interested in the variations in patterns of color inheritance unaccounted for by Mendelian inheritance. In particular, he wished to discriminate between that part of the variability due to heredity and that part due to environmental influences. It was in connection with this problem that Wright made one of his first applications of his method of path coefficients; in Wright's thesis he gave far more attention than any other geneticist to the problems raised by the residual variation not accounted for by Mendelian inheritance.

C. C. Little and J. A. Detlefsen

Two of Castle's students whose work was of great interest to Wright when he came to Harvard were C. C. Little and J. A. Detlefsen. Little began to work

with Castle late in 1907. He had been keenly interested in dog breeding, especially color inheritance, and wanted to go on in that field. Castle persuaded him to turn to mice because he needed someone to take over the mouse stocks and because dogs took so long to breed. Little then devoted his energies in graduate school to the study of color inheritance in mice. He did not forget dogs; fifty years after entering graduate school he published a book on the inheritance of coat color in dogs (Little 1957.) Little recorded the characteristics of more than ten thousand mice between November 1907 and May 1912.

In September 1909 Castle and Little published a joint paper on pink-eyed mice. Mice and guinea pigs had similar color factors, but in mice, in addition to the usual one, there was a dilution series having pale coats with darker eyes. In this series the eye was pinkish with faint traces of black or brown pigment, if they were also present in the coat. Castle and Little conducted a mouse experiment showing that the dilution factor and the pink-eye factor segregated independently (Castle and Little 1909). This discovery added one more to the eight known Mendelian factors for color variation in mice. Little's knowledge of this series would pay a dividend three years later. When Castle went on his collecting trip to Peru in November 1911, he left Little in charge of one of his experiments with guinea pigs (most of the guinea pig stocks were under the care of John Detlefsen). The experiment involved selection for a pale black (known as blue) form, and in the stocks there appeared a very slightly pigmented animal with pink eyes. Little immediately recognized the importance of the variation, which appeared in some respects similar to the pink-eye factor in mice but which had never been seen in guinea pigs stocks (Castle 1912b). This factor proved not to be the analogue of pink eye in mice; only about fifty years later did Wright hit upon a probable genotype for this particular animal (Wright 157). The analogue of pink eye in mice did, however, turn up in the stocks Castle brought back from Peru, and Wright used this variation in his thesis research.

Castle and Little also published a paper on the well-known 1905 observation of Cuénot—that yellow mice were always heterozygous and, even though yellow was obviously a dominant (in mice) over black or brown, F_2 segregations always yielded a 2:1 ratio rather than the expected 3:1. T. H. Morgan and E. B. Wilson suggested selective fertilization as the cause of this peculiar situation, but Castle and Little showed that the missing Mendelian class of homozygous yellows was actually formed but failed to develop (Castle and Little 1910). Baur had demonstrated this situation in the plant *Antirrhinum* (Baur 1907), but Castle and Little had demonstrated the first case of homozygous lethals in animals.

Little by himself made in 1911 a significant discovery about the dilutions of yellow coat colors in mice. In a confusing set of data concerning an almost continuous series of variations from red (intense yellow) to white, Little demonstrated clearly by breeding experiments that not one but two independent dilution series were found (Little 1911). The confusion concerning the variability in the red-white series was much clarified by this discovery, and Sewall Wright was impressed by this particular piece of work.

During his graduate career, Little probably came to know more than anyone else about inheritance of coat color in mice. T. H. Morgan was anything but an amateur scientist, but when he was breeding mice for experiments on heredity during Little's graduate school days, he published his results with interpretations as he did with almost all of his experimental work. Little enjoyed poking holes in Morgan's ideas about inheritance in mice, as several of his papers attest; soon members of the Morgan school would reciprocate by attacking some of Castle's ideas. Castle's graduate students, including Wright, were well aware of these differences of interpretation.

Little's thesis covered all that was known about color inheritance in mice (Little 1913). Bateson could cover the subject in a relatively few pages in 1903, but a ninety-page monograph was required in 1913. Bateson, Cuénot, Durham, Punnett, Morgan, and many others used the mouse as their experimental mammal, and Little covered all this work and attempted to present a synthetic view along the theoretical lines Castle had developed for rabbits in 1909. Not only did Little present a comprehensive analysis covering his data and that of others but he was able, with the support of Castle, Davenport, and the Carnegie Institution of Washington, to include five pages (with twenty mice) of beautiful color plates, made from watercolors painted from life by the artist Eugene N. Fischer. Included were the primary dilution series and variations.

Another feature of the monograph was Little's obvious, but limited, concern with questions of physiological genetics. On the first page he reviewed the theories of color formation by Cuénot and Riddle. Cuénot had suggested that the general factor for color production C controlled the development of a chromogen substance, which when acted upon by enzymes controlled by the color factors produced coat colors (Cuénot 1903). Riddle, however, suggested that chromogen was widely distributed in the organism and that absence of pigmentation was caused by something other than absence of chromogen (Riddle 1909). Little's view was as follows:

> If we assume that there is only one enzyme present to act as an oxidizing agent, we must assume for it as many different degrees of activity as are required to explain the occurrence of the various colors known to mendelize (three in mice, yellow, brown, and black). If we assume that a different enzyme or group of enzymes is responsible for the production of each pigment we must suppose that in mice at least three such enzymes or groups of enzymes must exist. To determine which of these conditions occurs in mice is not a problem for the biologist, but for the chemist. The biologist must confine his attention to determining the number of distinct agencies at work in pigment formation irrespective of their chemical nature. These agencies, because of their physiological behavior, the biologist chooses to call

factors and attempts to learn what he can about their functions in the evolution of color varieties. (Little 1913, 17–18)

For purposes of discussion, Little then assumed that a different color-producing enzyme was necessary for each of the independent colors in mice: black, brown, and yellow.

Castle, who like Little was no organic chemist, surely agreed with Little's interest in the problems of pigment formation and his decision to leave any detailed analysis to the chemists. Sewall Wright, however, had a strong background in chemistry and would not be willing to let the matter rest where Little did, as will become evident in the next chapter. Little's thesis and research work was of great interest to Wright, particularly because of the similarities between coat-color inheritance in mice and guinea pigs.

John Detlefsen came as a graduate student to work with Castle in 1909. In December 1909 Castle turned over to Detlefsen the experiment begun several years earlier on the hybridization of *Cavia aperea* (by then, and from now on in this book, known as *Cavia rufescens*) with the common guinea pig, *Cavia porcellus*. In 1910, in addition to the hybridization experiment, Detlefsen was transplanting ovaries in frogs (following the example of Castle and Phillips in guinea pigs) and injecting various solutions into the reproductive glands of different kinds of animals to see if hereditary characters would be altered. Of all Castle's students, Detlefsen was the one who left open the greatest possibility of some kind of inheritance of acquired characters. In 1925 he wrote a substantial review paper on the status of the inheritance of acquired characters (Detlefsen 1925). L. C. Dunn told me that Detlefsen incurred the wrath of many geneticists with that paper because he was not more opposed to the idea of the inheritance of acquired characters.

In the crosses between *Cavia rufescens* and the guinea pig, complete sterility of the 1/2 wild and 1/4 wild males was observed, while the females were at least partially fertile. Not until the 1/8 wild generation did a partially fertile male appear. Detlefsen focused upon four basic issues: (1) inheritance of coat color and coat characters, (2) growth, (3) morphological characters, and (4) inheritance of sterility. He was very apologetic for offering more data on the Mendelian inheritance of coat color in guinea pigs because he said (accurately) that Castle had already worked out the basic scheme. His justification was finding out whether the color differences between two species of *Cavia* still obeyed Mendelian inheritance. They did, as Castle and most other geneticists suspected. One reward of this analysis was that Detlefsen discovered that the agouti pattern of the *Cavia rufescens* was different from and inherited independent of the agouti pattern in the guinea pig. This was the only novel contribution to color inheritance in guinea pigs made by Detlefsen in his thesis research.

One problem Detlefsen found baffling. In contrast to the detailed discussion of dilution factors to be found in Little's thesis, Detlefsen devoted less than one page to the subject in his. The reason was that very dilute forms arose unexpectedly on many occasions when Detlefsen expected to obtain blacks, brown, or creams. Detlefsen said "no reason for the appearance of these very dilute hybrids can be assigned. . . . No solution has yet been possible. It is possibly another of the unexpected disturbances which hybrids are prone to show, but for which was known no cause" (Detlefsen 1914, 42–43). When Sewall Wright succeeded Detlefsen, he took this problem as one that demanded a solution, and he soon had one.

Detlefsen's data on size inheritance were not very useful because the number of guinea pigs was small and F_1 and F_2 hybrid males were sterile, thus precluding the crucial inbreeding of those generations to check for increase of variability due to Mendelian segregation. The few data produced in the experiment were inconclusive, but Detlefsen seemed to agree more with MacDowell and Phillips than with Castle:

> At present we know of no adequate hypothesis, other than the Mendelian, by which to explain the uniform F_1 generation, the more variable F_2 generation, the recovery of parental types, and the tendency for certain recombinations to breed true while others split up again. There is a small number of cases of size-inheritance in which a Mendelian explanation seems well justified. It is logically defensible to resort to this explanation when possible, since it fits a large number of cases involving quantitative characters. However, it is too early to insist that size-inheritance is universally Mendelian, for the number of crucial experiments is few. (Detlefsen 1914, 43)

Thus Wright's immediate predecessor did not accept Castle's skepticism concerning the multiple-factor theory of size inheritance.

Perhaps the most interesting aspect of Detlefsen's thesis concerned whether sterility could be inherited as having a Mendelian basis. He quoted Bateson's 1913 statement: "Successful investigation of the nature even of sterility consequent on crossing, the most obscure of all genetic phenomena, may become one of the possibilities of Mendelian research" (p. 79). Because Sewall Wright suggested to Detlefsen the Mendelian analysis used by him in the published thesis, further discussion of Detlefsen's data and analysis will follow in the next chapter.

When Wright came to the Bussey, he talked with Castle, East, Phillips, MacDowell, Little, and Detlefsen and saw their experimental stocks and work firsthand. He quickly came to know each of their research projects in detail; their work and personalities form the primary background in which Wright would develop as a graduate student and working scientist.

3
Harvard, 1912–1915

In his graduate school years (1912–15) at Harvard, Sewall Wright grew from a serious science student into a sophisticated research geneticist. At Harvard, Wright demonstrated clearly the independence of mind exhibited at all stages of his life. Yet the Harvard experience came at a crucial stage in Wright's development, profoundly influencing his ideas about genetics and the direction of his later research. Early in his graduate career Wright focused upon the guinea pig as his primary research animal; his experiments with guinea pigs continue in an unbroken line from the time he entered graduate school in 1912 until his retirement from the University of Chicago in December of 1954. He continued to publish data and analysis from these experiments until 1983. The Harvard experience did not by itself determine the development of Wright's theory of evolution but did provide two of the four most essential insights that Wright would later synthesize into his shifting balance theory of evolution in nature (see chapter 9). Only the very beginnings of Wright's later accomplishments in quantitative population genetics can be discerned during the Harvard years.

During his first two years at Harvard, Wright had to take many courses required for the doctorate, so he lived in Cambridge with his aunt Elizabeth Wright. She was Philip Green Wright's sister and only eighteen-years older than Sewall. Aunt Bess, as she was called, had studied physiology at Smith College and became director of the gymnasium at Radcliffe College. She strongly supported Wright's interests in biology and did her best to make his life in Cambridge pleasant. They frequently went to plays and operas. Aunt Bess's activities at this time were somewhat limited because she was taking care of her eighty-year-old mother, Mary Green Wright, who was difficult to care for and who took a dim view of Sewall's biological interests.

On Saturdays during these two years Wright always spent the day at the Bussey Institution, where he brought up the new litters of hooded rats for Castle to grade, kept track of his own experiments with guinea pigs, and checked on the guinea pig colony. He and Castle spent much of each Saturday together and discussed genetic problems frequently. Wright was struck by Castle's remarkable patience and tolerance with views divergent from his own. These discussions were valuable in Wright's education as a geneticist.

In his third and final year of graduate school Wright lived in the graduate school dormitory at the Bussey Institution, working full time on guinea pig research and his thesis and also assisting Castle with the selection experiment with hooded rats and with Castle's genetics course. Wright found the close contact with his fellow students and with the many distinguished visitors to the Bussey very rewarding.

Introduction to Guinea Pig Genetics from John Detlefsen

In the summer of 1912, Wright was already on the eastern seaboard because of the course he took at the Station for Experimental Evolution at Cold Spring Harbor, New York. He finished that course well before the beginning of the school year at Harvard. Instead of returning home, he decided to go to Cambridge and learn what he could about the guinea pig colony he would soon inherit. John Detlefsen was there working hard to gather and organize data from the colony: having graduated in June, he was moving to take a position at the University of Illinois in the fall but wanted to publish his thesis with the added data. Wright was with Detlefsen and the guinea pigs every day for several weeks.

From Detlefsen, Wright learned all that was known of guinea pig genetics at the time. It took Wright, or any intelligent and eager student, little time to understand the results of ten years of careful research on inheritance in guinea pigs. Detlefsen had devoted much time and thought to the analysis of fertility in the ongoing crosses between *Cavia rufescens* and the common guinea pig *Cavia porcellus,* but he did not know how to get a handle on the analysis.

The first day Wright went to the Bussey, Detlefsen explained the problem and gave Wright the results so far obtained. The results through the sixth generation, using the most reliable technique for judging fertility were as follows:

Generation	% Fertile
F_1 (1/2 blood)	0
F_2 (1/4 blood)	0
F_3 (1/8 blood)	14
F_4 (1/16 blood)	33
F_5 (1/32 blood)	61
F_6 (1/64 blood)	69

Characteristically, Wright immediately set to work to think of a quantitative way of analyzing the problem. He began by assuming the possibility (typical of the multiple-factor theory at the time) that many independent Mendelian factors each contributed equally to the sterility. On this assumption, he tried by trial and error to see what number of these Mendelian factors might yield the observed results. He settled upon eight, which yielded in theory the following percentages: 0.0, 0.4, 10.0, 34.4, 60.0, 77.6. For such a quick calculation, the calculated and observed results were quite close. Detlefsen was thoroughly impressed by this way of approaching the problem and by the apparent accuracy of the theoretical calculation. Here was a quantitative theory that might indeed enhance the analysis already given in his thesis when time came for publication. Wright was surprised that Detlefsen was impressed because the idea and the calculation had seemed rather obvious to him.

Just as Wright was talking to Detlefsen about the eight-factor interpretation East happened to come by, so Detlefsen reported to East what Wright had just told him. East could be coldly critical when he disapproved of something, which he combined with a withering glare. East turned this cold look upon Wright and told him that the idea did not work. Wright responded by reproducing for East the reasoning and calculations. East was satisfied with the demonstration and, like Detlefsen, impressed by this quiet-spoken young man. East was consistently friendly to Wright after this early incident, an honor East bestowed upon few of the graduate students at Bussey.

In his published version of the thesis Detlefsen (1914) presented the close resemblance of the array predicted by the eight-factor model and the data, but he also presented other evidence indicating that the model could not be a wholly accurate interpretation. It was nevertheless an impressive attempt for someone not yet a graduate student in genetics, on his first day to visit the laboratory. After this day, and several weeks of learning guinea pig genetics from Detlefsen, Wright approached graduate school with confidence.

Formal Coursework

For the doctorate, graduate students in biology were required to take, in addition to a wide range of courses in botany and zoology, courses in geology, physics, and chemistry. Wright took geology with R. A. Daly, a very enthusiastic and knowledgeable lecturer who attracted large numbers of students. Although he had taken no physics at Lombard or the University of Illinois, Wright enrolled in an advanced physics class taught by W. C. Sabine, the same person who had been instrumental in setting up the Bussey Institution in 1908. Sabine was an authority on the acoustics of halls. He lectured once each week on some topic, required a large number of laboratory exercises, and had his assistant hand out many problems to be solved by the students. The course assistant, who had much closer contact with the students than Sabine himself, was Percy Bridgeman, later a Nobel Prize winner. Wright enjoyed all aspects of the course. At the end of the term, Bridgeman approached Wright with the news that Wright had been the only one in the class to solve correctly every single problem he had handed out. Bridgeman encouraged Wright to get out of genetics and into physics, a flattering suggestion but one that Wright rejected immediately. He wanted to be a geneticist. He did, however, continue to have a strong interest in physics long after graduate school.

Wright had a good background in chemistry and elected to take biochemistry from L. J. Henderson, then an assistant professor. Wright wanted two things out of the course he would take in chemistry—understanding of biochemistry possibly relevant to gene action (physiological genetics as opposed to physiological chemistry, the usual field in graduate schools) and a good laboratory course in biochemistry. Henderson's course unfortunately did not have a laboratory, being purely a lecture course, but Wright learned a lot from it. Henderson had just written *Fitness of the Environment* (1913; it was not yet in print when Wright took the course) and talked a great deal about "dynamic equilibrium" in the blood and in physiology generally. Wright enjoyed the course but wished he could learn more both theoretically and practically about physiological genetics. His course schedule did not allow him to take the laboratory biochemistry course given in the Harvard Medical School. These were the only courses Wright took in geology, physics, or chemistry.

Wright took cryptogamic botany with Professor Lyman. It was a good course—very thorough. The course assistant helped the students perfect their drawing abilities. Wright was not very adept at this, getting Cs on some of his drawings, but he learned much about scientific drawing by the end of the course. This was the only formal course in botany taken by Wright.

Wright had two courses from George Howard Parker, chairman of the Department of Zoology. One was on sense organs, the other on the nervous system; Wright did some original research in both. In one experiment Wright examined the behavior of catfish in relation to exposure to light. He found strong evidence that when exposed to light catfish become passive and that they are far more active in the dark. Another experiment concerned behavior of an organism when exposed to two lights of different intensities. Blowflies were known to negotiate a predictable path between the lights, keeping light intensity equal from each side. Wright decided to try earthworms, which are very sensitive to light, and discovered that their behavior was totally unpredictable and apparently random with respect to the sources of light.

Wheeler and C. T. Brues taught the entomology course. Wheeler was violently anti-Mendelian, and had nothing whatever to say about evolution in the course. Wright, who had a strong natural history bent, was nevertheless delighted by the course, which focused upon animal behavior. Wheeler took his class on a great many field trips where he would tell the students in detail about the life cycles and behavior of any insect that happened along. Students quickly discovered that Wheeler was the total naturalist—he knew everything about anything spotted by a student. He was not limited to individual behavior but also knew about the relations of insects and other animals to each other and about ecological relations in general. Wright found the course fascinating and one of the very best he ever had. Despite Wheeler's obvious objections to Wright's chosen field, he allowed Wright to try and breed two color varieties of a beetle he had collected during Wheeler's course; the experiment ended because Wright could not get beyond the F_1 generation. A course in histology taught by W. H. Rand completes the list of formal courses taken by Wright at Harvard.

Wright as Course Assistant for Castle

Notably absent from the list of formal courses taken by Wright during his graduate years are ones on genetics. Wright was never a student in a genetics course. Within the Bussey, however, researchers frequently presented their findings in seminars; Wright attended many by Castle, East, Wheeler, and by visitors from other universities or experiment stations in the United States and abroad. The graduate students constantly exchanged ideas. In addition to Detlefsen, Wright spoke most frequently to C. C. Little, E. N. Wentworth (who was at the Bussey for only a short while), Harrison Hunt (a close friend for about sixty years), Phineas Whiting, and H. D. Fish, Wright's office mate. These local seminars and interchanges were certainly more influential in Wright's development as a geneticist than the formal coursework.

Although Wright did not take Castle's undergraduate genetics course as a student, he assisted in the course and therefore learned the course content well. Castle always treated his graduate students as equals and gave them much responsibility when assisting in his genetics course. Castle lectured and his assistants held discussion sections and laboratories. Wright had four discussion sections the first time he assisted, graded all the quizzes, and gave prepared talks. On one occasion he had just read the second edition of Karl Pearson's *Grammar of Science* and talked to the students about the philosophy of science.

Castle's textbook of genetics, *Genetics and Eugenics,* published in 1916, was taken directly from the lectures Castle gave in his genetics course. The first edition of this book is therefore a good measure of what Castle expected his students to know about genetics, eugenics, and evolution; moreover, the detailed bibliography covers most of the literature that Castle recommended to his graduate students. *Genetics and Eugenics* was the most comprehensive genetics textbook available in the English language in 1916, competing very favorably with R. H. Lock's *Variation, Heredity, and Evolution,* Bateson's *Mendel's Principles of Heredity,* Punnett's *Mendelism* (all three published in England), H. E. Walter's *Genetics,* and *The Mechanism of Mendelian Heredity* by Morgan, Muller, Sturtevant, and Bridges. Castle had seven substantial chapters on ideas about evolution and heredity from Darwin to the rediscovery of Mendelism, nine chapters on Mendelian heredity, and individual chapters on sex determination, variability of Mendelian unit characters, size inheritance in relation to multiple factors and pure lines, Galton's law of ancestral heredity, and finally a chapter on inbreeding and crossbreeding. The eugenics section had only four chapters, and, according to the anonymous reviewer in the *Journal of Heredity* (8: 33), was "conservative throughout, but particularly sound in its treatment of Mendelism in man." Appended was the Royal Horticultural Society's translation of Mendel's paper on peas (Harvard University Press issued this appendix separately beginning in 1925; it has been reprinted a great many times and is still in print). Sixty-five years after assisting Castle in the genetics course, Wright recalled with clarity the issues and contents of the lectures (chapters) and was still quite familiar with all but a very few works in the extensive bibliography. Thus, although Wright was never a formal student in any genetics course, he learned well the contents of Castle's comprehensive course. The genetics textbook that Wright studied most carefully was Bateson's *Mendel's Principles of Heredity* (1913), and to satisfy the German requirement he read Erwin Baur's *Einführung in die experimentelle Vererbungslehre* (1911).

Wright as Castle's Laboratory Assistant

In the late summer of 1913 John Phillips, who had assisted Castle since 1908 on the selection experiment in hooded rats, decided to take a trip abroad. Castle then asked Wright to assist in conducting the selection experiments. Castle had vacillated a good deal about his "changed-factor" hypothesis concerning the success of the selection, and in late 1913, in writing up the results of the experiment for the period 1908–13, he did not dismiss the multiple-factor "modifier" theory of East.

Castle described East's contentions in detail. He suggested that the theory of modifiers, if true, meant that selection should reach a plateau as the stock became more homozygous for the factors affecting extent of pigmentation; but no such diminution in the power of selection had been observed. Although generally supporting the view that the Mendelian factor itself was undergoing change as a result of selection, Castle did use the theory of modifiers to explain one experimental result:

> Now it seems to us probable that what we call the unit-character for hooded pattern is itself variable; also that "modifiers" exist—that is, the extent of the hooded pattern is not controlled exclusively by a single localized portion of the germ-cell; otherwise we should be at a loss for an explanation of the peculiar results from crossing plus series hooded rats with those which are still more extensively pigmented (Irish); for by such crosses the pigmentation is rendered not *more* extensive but *less* so. This result we can explain on the supposition that the selected plus series has accumulated *more* modifiers of the hooded pattern than the wild race contains, so that a cross tends to reduce the number of modifiers in the extracted hooded individuals. No other explanation at present offers itself for this wholly unexpected but indubitable result.
>
> If a different one can be found we are quite ready to discard the hypothetical modifiers as a needless complication, contenting ourselves with the supposition that the unit character for hooded pattern is itself variable, and that for this reason racial change in either plus or minus directions may be secured at will through repeated selection. (Castle and Phillips 1914, 25)

Castle's interpretation of factor change came under immediate attack from many sides, in particular in print from A. L. and A. C. Hagedoorn

(1914) and from H. J. Muller (1914b). Castle had little respect for the Hagedoorns, who were mammalian geneticists in Holland. He had more respect for the Morgan group at Columbia but believed Morgan and his students had uncritically accepted the apparent implications of Johannsen's theory of pure lines, originally developed with beans (self-fertilization) but now applied rather freely to crossbreeding organisms. Castle hastened to answer the Hagedoorns and Muller (Castle 1914a, 1915). Muller basically argued along the same lines as East but used *Drosophila* for his evidence. Castle neatly answered Muller by asserting accurately that no experiment corresponding to his selection experiment on rats had been conducted with *Drosophila,* so Muller's support of the modifier theory had no factual basis. This controversy between Castle and the Morgan group (all members of which agreed with Muller) had effects on both camps. A. H. Sturtevant and Fernandus Payne both began substantial selection experiments in *Drosophila,* eventually obtaining results comparable to those obtained by Castle in rats (Sturtevant 1918; Payne 1918a, 1918b, 1920). Selection was effective. But Sturtevant and Payne were also able to strip away the modifiers by outcrossing and even to locate particular modifiers on specific chromosomes, a total impossibility in rats at that time. The controversy with Castle stimulated members of the Morgan school to prove their contentions experimentally. MacDowell initiated similar experiments on *Drosophila,* reaching conclusions that again supported the doctrine of multiple-factor modifiers (MacDowell 1915, 1917). For his part, Castle responded by retreating with greater determination to his hypothesis of variable Mendelian factors.

Soon after Wright came to the Bussey, East was engaged in writing his reply (East 1912) to the paper Castle had delivered at the University of Illinois in the spring of 1912 and published in *American Naturalist.* East gave a series of seminars detailing his theory of multiple factors in interpreting quantitative variability in corn. Additionally, as in his published reply to Castle, East applied his ideas to Castle's crosses between the plus and minus strains of hooded rats; East's theory fit perfectly. Wright attended these seminars and found East's explanation of Castle's results more convincing than Castle's own explanations. When Wright went to work for Castle, he tried to convince Castle that East was correct. The active criticism of Castle's interpretation, however, especially from Muller, drove Castle to hold tight to his theory.

Wright tried to deduce a way to experimentally discriminate between Castle's theory of changed factors and the theory of modifiers. He devised a crucial extension of a procedure Castle had used many times before but not carried far enough. As reported by Castle and Phillips in 1914, they had crossed both the plus and minus strains with both wild rats and "Irish" rats and extracted the hooded pattern again in the F_2 generation. With wild rats, the extracted hooded patterns had regressed toward one another but still differed significantly. Castle interpreted this result as evidence that the Mendelian factor for hooded had been changed by the selection process; but he saw no need to continue the process of extraction for more than one cycle. Wright made a simple calculation indicating that if the modifiers were independent (i.e., not all located on the same chromosome or linked), then one could not possibly expect all the accumulated modifiers to be stripped away by only one generation of outbreeding and subsequent extraction. With very many factors at least several generations would be required. Wright therefore suggested to Castle that he carry out the extraction process for more than one generation. The F_2 offspring from the hooded $\times$ wild cross exhibiting the hooded pattern would be crossed again with the wild rats, with the hooded pattern being extracted again in the second F_2 generation, and so forth. Castle thought this was a good idea and set to work on it during Wright's last year at the Bussey.

Castle and Wright wanted to use a strain of rats uncontaminated by any possible crossbreeding with rats having the hooded or Irish factors. They trapped a colony of wild rats that lived under a barn back of the Bussey building. These rats were extremely wild and difficult to handle but the crosses between the plus strain of hooded rats and the wild rats went well. Wright was still at the Bussey for the first generation offspring. When the hooded pattern was extracted in the first F_2, it had (as occurred in earlier experiments) regressed back toward the minus strain. According to Wright's prediction, Castle should have expected the second extraction of the hooded pattern in the second F_2 generation to show continued regression. This result did not occur in the second extracted F_2 hooded patterns, and indeed the regression was reversed. From original hooded rats with an average pattern of +3.73, the first F_2 offspring had an average of +3.17, and the second F_2 offspring an average of +3.34. Clearly, the regression had ceased after only one extraction. Castle took this as added evidence that selection had indeed changed the Mendelian factor for the hooded pattern, and he published the data and interpretation in the same Carnegie Institution of Washington Publication (Castle 1916) with Wright's thesis.

Still Castle came under attack. Particularly irritating was an extended critical article in *American Naturalist* of December 1916, written by Castle's former student E. C. MacDowell. Although Carnegie Publication No. 241 had appeared in September of 1916, MacDowell was unable to take the new evidence into account. Castle thus answered MacDowell in *American Naturalist* by chiding him for ignoring the new results, presenting the results by quoting at length from the Carnegie publication, and concluding:

> I have now presented the evidence which has led me to reject the hypothesis formerly held tentatively that modifying factors were largely concerned in changes produced in the hooded pattern of rats under repeated selection. This evidence seems to me to admit of only one consistent interpretation, that a single variable genetic factor was concerned in the original hooded race, that a changed condition of this same factor was produced in the minus race, and another changed condition in the plus race, and a third appeared in the mutant race. All are allelomorphs of each other, and of the non-hooded

or self condition found in wild rats, yet all tend to modify each other in crosses. The character has a high degree of genetic stability, yet is subject to continuous genetic fluctuation. (Castle 1917, 113–14)

Castle's former students had put him into a strange position. Both Mac-Dowell and Wright disagreed strongly with Castle's primary interpretation of his results in the selection experiment in hooded rats. Expecting that Castle would be proved wrong, Wright suggested a crucial experiment that instead at first appeared to favor Castle's interpretation. Criticism from MacDowell then elicited from Castle his strongest statement yet favoring his long-held view and rejecting the theory of modifiers.

But the experiment suggested by Wright was not over yet. The crosses of the minus strain with the wild rats had lagged behind those with the plus strain. The wild rats were very difficult to breed in captivity. When Castle answered MacDowell early in 1917, the results of the crosses with the minus strain still were in doubt. What happened warmed the hearts of Castle's critics. The mean grade of the minus race regressed from the control stock of −2.63 to a value of +2.55 after three extractions. One family in the crosses of the minus stock ended up with a mean value of +3.05 after three extractions, almost exactly the same as the mean value of +3.04 shown by the plus strain after three extractions. The accompanying figure 3.1 with two tables reproduced from Castle's paper tells the story (from Castle 1919a, 129). The residual heredity of the wild rats was clearly like that of the plus strain. Not only did the crosses suggested by Wright strip away the modifiers of the hooded pattern but even the variability (as measured by the standard deviation) increased and decreased in direct accord with predictions of the theory of modifiers. After seventeen years of holding open the possibility of changed Mendelian factors, and having rejected completely the competing theory of modifiers for the last two of these years, Castle now publicly recanted and joined the ranks of the Mendelians on this issue. He gave a seminar at the Bussey and published papers in the *Proceedings of the National Academy of Sciences* and in *American Naturalist* (appropriately titled "Piebald Rats and Selection: A Correction") detailing his change of view (Castle 1919a, 1919b). Castle was a scientist who trusted his own data and his own interpretations and who stuck to his views when criticized. One cannot help wondering what would have happened to Castle's views on change in Mendelian factors if Sewall Wright had not suggested the crucial experiment.

In retracting his long-held position, Castle hastened to point out the real value of his selection experiment with hooded rats. More than any other, Castle's experiment forced geneticists to give up the apparent conclusion from de Vries and Johannsen that selection of small differences could lead to no significant genetic modification. At a time when many geneticists the world over denigrated the power of selection acting upon small fluctuating differences, Castle and East beginning in 1909 at the Bussey Institution

TABLE 1

RESULTS OF CROSSING THE PLUS SELECTED RACE WITH A WILD RACE

	MEAN GRADE	STANDARD DEVIATION	NUMBER OF HOODED YOUNG
Control, uncrossed plus race, generation 10....................	+3.73	0.36	776
Once extracted hooded F_2 young...	+3.17	0.73	73
Twice extracted hooded F_2 young..	+3.34	0.50	256
Thrice extracted hooded F_2 young..	+3.04	0.64	19

TABLE 2

RESULTS OF CROSSING THE MINUS SELECTED RACE WITH A WILD RACE

	MEAN GRADE	STANDARD DEVIATION	NUMBER ON HOODED YOUNG
Control, uncrossed minus race, generation 16.................	−2.63	0.27	1,980
Once extracted hooded F_2 young...	−0.38	1.25	121
Twice extracted hooded F_2 young..	+1.01	0.92	49
Thrice extracted hooded F_2 young..	+2.55	0.66	104

Figure 3.1. Stripping away the modifiers of the hooded pattern.

trained many research geneticists, not one of whom doubted the power of selection as demonstrated by Castle with his hooded rats. Every genetics textbook displayed Castle's results. He unquestionably had a significant influence in turning geneticists toward a selectionist view of evolution in nature and in domestic populations. As will become clear in chapter 8, R. A. Fisher was one of the geneticists much influenced by Castle's selection experiments on hooded rats.

Wright did Castle another service in connection with the experiments on rats. Early in 1914 Castle had described two yellow-coated varieties of the Norway rat discovered in England, one having pink eyes and the other having dark red eyes. Both variants behaved as Mendelian recessives in crosses. Castle used the new varieties in a series of crosses with albinos, agoutis, and with each other. Some of the crosses produced rather unexpected results, and when Wright examined the entire breeding record of the new varieties he detected consistent nonrandom segregation of the two new Mendelian factors. Careful quantitative analysis of previous and continued mating revealed a strong but incomplete linkage of the two characters. Castle and Wright published their findings in *Science* in August 1915, several months before the paper by J. B. S Haldane, his sister Naomi, and A. D. Sprunt, on a case of linkage in mice, appeared in *Journal of Genetics* (Haldane, Haldane, and Sprunt 1915). To my knowledge, the paper by Castle and Wright is the first publication documenting linkage in a mammal, although cases of linkage in plants had been reported by Bateson and Punnett and in insects by Morgan and Tanaka.

Duplicate Genes in *Primula sinensis*

During his graduate school career Wright and many of his fellow students did what a graduate student in genetics could not possibly do today, or even fifty years ago. He read almost all the periodical literature dealing with genetics. One paper he read concerned the genetics of tetraploid plants in the primrose, *Primula sinensis,* reported by R. P. Gregory in the *Proceedings of the Royal Society of London* in 1914. Gregory found that in reciprocal crosses of two races of *Primula sinensis,* one cross gave normal results while the other produced an F_1 generation that was sterile with the parents and only a giant tetraploid variety in the F_2. Gregory assumed in his analysis that all four homologous chromosomes were interchangeable in synapsis, yielding certain predictions about the behavior of Mendelian characters in the tetraploid variety. Wright, although he knew little enough of the genetics of *Primula sinensis,* surmised that although the chromosomes of the two races behaved normally in the first cross, they might not in the second, perhaps because of cytoplasmic influence. It was possible that in the tetraploid only the chromosomes from one parent could pair with each other at synapsis; thus all four of any one chromosome could not pair interchangeably.

Wright said in a published note to *American Naturalist* that which of the two hypotheses was true could not be determined by Gregory's data, and he suggested an appropriate experiment. Wright never attributed much significance to this paper because H. J. Muller (1914b) had also read Gregory's paper and submitted a longer and more detailed analysis of the case to *American Naturalist*. Muller's paper appeared in August 1914, and Wright's not until October. Wright's idea had already effectively been presented by Muller two months earlier. Considering that Muller was primarily a *Drosophila* geneticist and Wright a mammalian geneticist, this instance shows how applicable the understanding of genetics in one organism was to understanding the genetics of an entirely different organism. East constantly reminded his students that the principles of the genetics of the lowly *Drosophila* or of the tobacco plant were the same as those in the "prince or potentate" (see for example East 1923, vi). A constant theme of the work of the mathematical population geneticists Fisher, Haldane, and Wright throughout their careers was that insights in understanding evolution in one organism could yield much understanding of evolution in other organisms. Sometimes they had never even seen the organisms about which they were writing. For example, Wright and Fisher wrote and argued about the rare desert plant *Oenothera organensis* over a twenty-year period, but Wright never saw one and to my knowledge neither did Fisher. (For their interchange over *Oenothera organensis* see chapter 13.)

Quantitative Genetics

Wright took no courses in mathematics while a graduate student at Harvard. Later he regretted not having taken advantage of the opportunity, but he had almost no indication as a graduate student that he would later be an expert in quantitative genetics and evolution. Wright did not lose interest in a highly quantitative approach to science during his graduate school years—his great enjoyment and superior performance in Sabine's physics course attests to this—but his primary interests were directed to color inheritance and physiological genetics. When the opportunity arose, however, Wright's basic tendency was to be quantitative where possible. A good example occurred in an episode involving Raymond Pearl, East, and the three Bussey students H. D. Fish, Phineas Whiting, and Sewall Wright.

Raymond Pearl was one of perhaps three well-trained biometricians in the United States in the decade 1905–15 (the other two being C. B. Davenport and J. Arthur Harris). From 1907 to 1918 Pearl was at the Maine Agricultural Experiment Station conducting selection and other experiments with chickens, using his quantitative skills in the analysis. He was deeply opposed to Castle's assertions about the efficacy of selection and advocated a mutationist view of evolution. Castle thought Pearl had little common sense about biological problems and was suspicious about his quantitative manipulations. Pearl and East were fast friends and corresponded frequently. East was much better prepared to appreciate Pearl's quantitative reasoning.

In connection with his breeding experiments at the Maine Station, Pearl became interested in the quantitative analysis of different systems of inbreeding. He accurately realized that the degree of inbreeding in domestic populations was crucial and that a general method of measuring the degree of inbreeding in a particular pedigree was desirable. In a long paper in the *American Naturalist* of October 1913, Pearl made an initial attempt to address the quantitative analysis of inbreeding. One of his general arguments was that inbreeding other than self-fertilization (the mathematical consequences of which Mendel himself had worked out in 1865) did not necessarily lead toward homozygosis in the population. To illustrate this argument, Pearl used the example of continued brother-sister mating. Starting with the cross of homozygotes AA and aa, Pearl demonstrated that in F_1 there were no homozygotes, in F_2 there were 50% homozygotes ($1AA:2Aa:1aa$), and in F_3 (by a more complicated calculation first done by Pearson in 1904) the expectation was $16AA:32Aa:16aa$, still 50% homozygotes. Pearl assumed the same basic calculation would continue for each generation after the F_3, so brother-sister mating did not lead to homozygosis at all, in the absence of selection. After this demonstration, Pearl felt amply justified in making the conclusion: "The automatic increase of the proportion of homozygotes which necessarily follows continued self-fertilization does not necessarily follow inbreeding of any other sort" (Pearl 1913, 614).

One does not have to be either a sophisticated statistician or a geneticist to see that Pearl's conclusion was ridiculous. (One wonders who the referee was for Pearl's paper, or if there was one.) In Castle's group, Wright's roommate, Harold D. Fish, read Pearl's paper and knew that something had to be wrong in Pearl's analysis of brother-sister mating. Fish was a very enthusias-

tic, energetic person with transitory enthusiasms for particular research problems. He was in charge of the rabbit colony (having succeeded MacDowell),
but things had become dull, and the problem with Pearl's paper appealed to
him. Fish had attended a seminar given by East and read a 1912 paper by
East and Hayes, both of which led Fish to the intuitively obvious conclusion
that inbreeding led toward homozygosis in a population. But East had not formally gone beyond the analysis of self-fertilization.

Then in 1914 a new graduate student, Phineas W. Whiting, came to
study with Wheeler in entomology. Unfortunately for this combination,
Whiting was interested in genetics, which Wheeler disliked intensely. Whiting stayed at the Bussey for only one year, but during that time he became
friends with Little, Wright, and Fish among the students. Whiting's organism
for research was the common Greenbottle fly, *Lucilia sericata,* often called
the blowfly. He raised the flies by allowing the adults to lay their eggs in de-
caying guinea pig carcasses. Wright recalls that no one but Whiting could go
within a hundred feet of the shed where he raised the flies without nearly
fainting from the terrific stench. Early in 1914 Whiting was just completing a
series of selection experiments for bristle number in his flies, using brother-
sister mating for many generations in a row. He calculated that heterozygosis
was reduced by one-eighth in going from the F_2 to the F_3, and deduced
wrongly that the remaining heterozygosis would be reduced by one-eighth in
each succeeding generation. Whiting took his calculations to East, who
agreed that homozygosis progressively increased, but not by the amount calculated by Whiting. East, however, could not at this time provide Whiting
with a satisfactory alternative analysis.

When Fish read Pearl's paper, after having already talked to both East
and Whiting about the effects of brother-sister mating, he excitedly launched
into an attempt to accurately quantify the problem because he knew Pearl had
to be wrong. Fish was not mathematically inclined, so he took a direct approach and laboriously began working out the consequences of each generation of inbreeding after the F_3 where Pearl stopped. Pearl's mistake had been
in assuming that the reasons for 50% homozygosis in the F_2 generation were
the same as for the 50% homozygosis in the F_3 generation. In the F_2 generation random mating was the same thing as brother-sister mating, but this
equivalence does not hold in later generations. In calculating the rise in homozygosis over the next few generations, Fish had the room he shared with
Wright covered with sheets of calculations. Naturally Wright became interested and for amusement worked out a formula for calculating the effects of
brother-sister mating for any number of generations. Wright had never studied matrices, but he made a table of all of the six possible kinds of matings
with two alleles ($AA \times AA$, $AA \times Aa$, $AA \times aa$, $Aa \times Aa$, $Aa \times aa$,
$aa \times aa$), including the frequencies of the six types which they produced, in
proper ratio. This was in essence a recurrence matrix which worked well for
brother-sister mating. Using Wright's procedure, Fish was able to calculate

that, for example, the least number of generations required to reduce heterozygosis to less than one-half of 1% was twenty-five.

Fish took his results to Whiting, who asked Fish to prepare a note for his paper on Greenbottle flies. Fish also took his results to East, who then independently worked out the effects of brother-sister mating using a different method. Pearl had written to East asking his opinion of the 1913 paper, and now East replied, showing Pearl wrong and saying that Fish would publish complete figures. (It seems that everyone wanted Fish to complete a publishable paper.) Pearl was indignant that Fish proposed to point out in print Pearl's mistakes. Wright recalls that Pearl came by the Bussey and wanted Fish fired. In the end, Pearl published a correction first, in January 1914, not mentioning Fish, and with the statement: "The blunder, kindly pointed out to me by Professor E. M. East, which in retrospect seems altogether too stupid even to be possible, was in the failure to recognize that after the second generation the constitution of the *family* would no longer be the same as that of the population" (Pearl 1914, 57). East called Wright to see if Wright was really interested in the problem because Fish wanted to refer in the paper to Wright's assistance. Wright truthfully told East that he was not much interested, and Fish published his note in Whiting's article (Whiting 1914) with no reference to Wright. Thus there is no published record of Wright's first venture into population genetics. The venture in no way gave Wright a hint that an important part of his life's work lay in this direction. (Most of this story can be found in Fish's note in Whiting 1914, except for the part played by Wright, which came entirely from interviews with Wright.)

A second brush with quantitative analysis early in 1914 was of far greater and general importance to Wright's development as a quantitative theorist, although here again Wright did not see the vast implications of his first tentative ideas. I am speaking about Wright's invention of the method of path coefficients, which was basically a method to quantify the causal chains in an already definite causal scheme. By late in 1913, E. C. MacDowell had completed his work on size inheritance in rabbits, and he had concluded that Castle was wrong in rejecting the idea of Mendelian multiple-factor inheritance to account for size inheritance in rabbits. Castle did not allow MacDowell's thesis to be published as a Carnegie Institution of Washington Publication without caveats of his own. Castle added both a prefatory note and an appendix. One is reminded a bit of Osiander's preface to the *De Revolutionibus* of Copernicus as Castle spoke about MacDowell's acceptance of the Mendelian multiple-factor theory as advocated by Nilsson-Ehle and East:

> Very naturally Mr. MacDowell has been strongly influenced by this idea, which seems to unify, if not to simplify, our conceptions of the method of inheritance. While not entirely sharing his views, I have tried not to bias his judgement either for or against the multiple-factor hypothesis which he adopts in this paper. But to avoid misunderstanding, I wish to say that in my own opinion the theory of the pu-

rity of the gametes has not been established, and too great
definiteness and fixity, is ascribed to Mendelian units and factors in
current descriptions of heredity; consequently, too great importance
is attached to hybridization and too little to selection, in explaining
evolution. But neither my views nor Dr. MacDowell's should bias
the judgement of the reader. We wish to place before him clearly the
results of experiments which have entailed much painstaking observation; the correct interpretation will become evident in due time.
(MacDowell 1914, 3)

Following the presentation of MacDowell's data and interpretations Castle
added an appendix reinterpreting some of MacDowell's data to fit the conception of changed Mendelian factors rather than the accumulation of modifiers.
In the course of his experiment MacDowell had indeed taken painstaking
labors to clean, measure, and preserve the bones of his rabbits, and he used
these skeletal measurements in one section of his thesis. Castle saw an additional use for MacDowell's measurements. One of Castle's positions from at
least 1905 on was that total size did not depend upon the additive effect of
Mendelian size factors for individual parts of the body such as leg, arm, or
skull. Instead, the size factors controlling the size of the organism as a whole
were general factors, affecting all parts of the body simultaneously. Castle
thought that a calculation of the relations between different bone measurements would reveal high correlations, thus indicating little independent variability for each bone. Although Castle had little mathematical facility, and
had battled with Karl Pearson, he still had much respect for standard deviations, correlation coefficients, and measurements of error. Indeed, such quantitative measurements were essential for his work. Castle also required his
students to become familiar with these quantitative concepts. Thus, as an exercise he set Sewall Wright and Harold Fish to work calculating ten of the
correlation coefficients between five different skeletal measurements: length
of skull (OM—occipital to maxilla), width of skull (Zp—posteriorly across
the zygomatic arch), and the lengths of the humerus, femur, and tibia (H, F,
and T).

Wright and Fish accomplished this exercise as Castle directed. The appendix included the complete correlation table for each of the ten correlations
and the following summary table, where σ is the standard deviation, $C.V.$ the
coefficient of variability, and r the correlation (fig. 3.2; reproduced from
MacDowell 1914):

As the table clearly indicates, Castle's original surmise appeared likely, and
he concluded:

In view of the high correlations obtaining between one skeletal dimension and another (and these agree closely with those observed in
the case of many by Pearson and others), it follows that to a large extent the factors which determine size are *general* factors affecting all

Measure-ment.	Mean.	σ	C. V.	Correlated with.	r
O. M.	7.30	0.36	49.3	Zp.	0.750
				H.	0.743
				F.	0.760
				T.	0.701
Zp.	4.00	0.17	42.5	H.	0.675
				F.	0.674
				T.	0.658
H.	6.60	0.26	39.4	F.	0.857
				T.	0.791
F.	8.28	0.31	37.4	T.	0.858
T.	9.60	0.34	35.4	....	....

Figure 3.2. Summary of correlations.

> parts of the skeleton simultaneously. When the skull is long, the legs are long and the skull is wide, and every other part varies in proportion (or within 65 to 85 per cent of the same proportion). Whatever special factors (if any) there are, which are concerned in limiting the size of particular bones, these can play only a subordinate part in determining size. The chief factors are plainly *general* factors and control the growth of the body as a whole. (MacDowell 1914, 51–52)

Castle was clearly satisfied by the exercise accomplished by Wright and Fish.

The same was not true for Wright. He had no trouble accepting the interpretation that general size factors were implicated by the calculated correlation coefficients, but the correlations were not perfect, and the rest of the variability of the bones came from somewhere, certainly in part from special size factors. The best approach would be to find a way to divide the variability observed in a given bone into two different components—that caused by general size factors and that caused by special size factors. There was, unfortunately, no available mathematical technique or biological analysis, or combination thereof, that could produce the desired quantification of the causes.

One technique, which at the time Raymond Pearl was popularizing in the United States, was that of the partial correlation coefficient, developed earlier by Karl Pearson in England. In a system of three or more variables, the idea of the partial correlation coefficient was to mathematically (not biologically) hold constant all but two of the variables and to obtain then the coefficient of correlation between them. Pearson developed a formula for calculating the partial correlation coefficient between any two of the three variables; by applying this formula serially, the partial correlation between any two of any number of variables could be calculated, but with any more than four the calculations were very laborious.

Because the partial correlation coefficient was the best available technique for what Wright wished to accomplish, he set to work with the measurements of the five bones. He calculated first all the correlations with one of the five measurements constant, then with two of the measurements constant, and finally with three of the measurements constant. When he was done, Wright found the results more suggestive than the straight correlations Castle published, but he was still unsatisfied. He wanted a way to partition the variability in accordance with biological reality, an end not realized by the partial correlation coefficient. To solve this problem, while he was still in graduate school Wright first developed his method of path coefficients. Because of the crush of his thesis research and the move to Washington in the fall of 1915, Wright did not write up these ideas until late in 1917; they were published in 1918. A more thorough discussion of the origin and application of path coefficients will follow in the next chapter.

Inheritance in Guinea Pigs

There was never any question about Wright's utilization of the guinea pig as the experimental animal for his thesis research. From their first conversation when Castle visited the University of Illinois in the spring of 1912, Wright as well as Castle understood that Wright would assume charge of the guinea pig colony when Detlefsen left and would use the colony for his original research. Compared with *Drosophila,* for example, guinea pigs had many disadvantages. They had an extended reproductive cycle and suffered from many diseases that frequently invaded the laboratory. They had a haploid chromosome number of thirty-two, as compared with four for *Drosophila melanogaster,* and the chromosomes were tiny and difficult to distinguish. (The chromosome number was not even determined until long after Wright's graduation from Harvard; estimates varied by up to ten chromosomes.) Within just a few years of the time they started work on *Drosophila,* Morgan and his students had found dozens of mutations, soon to be hundreds. At the time Wright ceased breeding guinea pigs in 1954, there were known only eleven loci with significant effects upon coat colors.

One might compare guinea pigs with mice, also a mammal whose chromosomes were difficult to analyse. When C. C. Little wrote his thesis on color inheritance in mice in 1912, he had about as many mutations affecting coat color as Wright knew in the guinea pig in 1915. But by 1979, Willys K. Silvers, a former student of Wright's, published a book on color inheritance in mice listing some 130 coat-color determinants (Silvers 1979).

There were many reasons for the lesser knowledge of inheritance in guinea pigs as compared to mice. Hundreds of geneticists, including nearly every one who worked at the Jackson Laboratory, founded by C. C. Little in Bar Harbor, Maine, worked on mice. The raising of enormous numbers of almost uniform inbred mice enabled workers there to spot any mutation and quickly utilize it. Many kinds of mutagenic agents were also used upon mice. In contrast, only Wright and two others, Heman L. Ibsen and Arnold Pictet, worked seriously on guinea pigs—and Pictet was an unreliable researcher. In the history of Mendelian genetics, research upon guinea pigs has been a backwater rather than a mainstream.

With only two possible examples of linkage as late as 1955, chromosome maps were impossible in guinea pigs. The study of evolution in nature with guinea pigs was a virtual impossibility for Wright. At first glance, through no fault of his own, Wright appears to have inherited the wrong organism for the primary focus of his research interests. In fact, this conclusion is untrue, though even Wright mused in our conversations about how his life's work might have differed had he studied *Drosophila* rather than guinea pigs. For the purpose of Wright's research in physiological genetics, he really needed no more mutations than were available in guinea pigs. There were some 3.5 million different combinations of the known genes, and Wright had enough possible combinations for a lifetime of productive research on gene interaction. So far as evolutionary studies are concerned, it is doubtful that Wright would ever have spent many years in the field, as did Theodosius Dobzhansky or F. B. Sumner. Wright always appeared to me quite happy with the choice of the guinea pig as his research organism.

Status of Guinea Pig Genetics in 1912

When Wright inherited the guinea pig colony in September 1912, the following Mendelian factors were known to occur in guinea pigs:

1. *C, the general color factor necessary for the production of color in animals.* It was dominant to albinism, represented by the recessive homozygote *cc*. The albino guinea pig was never all white, as in the albino rabbit, but had pigmented extremities, resembling the Himalayan rabbit. Castle had suggested in 1909 that the Himalayan factor in rabbits was an allele of the general color factor *C*, and Wright accepted this view from the beginning of his graduate work. No such alleles alternative to *C* were actually known in guinea pigs in 1912.

2. *A, the agouti factor.* It was dominant to its allele *a*, the recessive homozygote showing no trace of the yellow ticking characteristic of the agouti. Detlefsen had discovered what he tentatively classified as a third allele, *A'*, a darker agouti pattern found in the wild *Cavia rufescens*. The existence of Detlefsen's third allele was by no means proved in 1912.

3. *B, the factor for black.* It was fully dominant to its recessive allele in *b*, which when homozygous produced a deep rich brown, or chocolate.

4. *E, an extension factor, characterized by extended black or brown pigmentation in the fur.* In the absence of *A*, this factor produced self-colored blacks and browns. Its recessive allele *e* was characterized by the restriction of black or brown, but not yellow, pigment to the eyes; no black or brown pigment was visible in the fur except for a slight sootiness. Dark-eyed yellows were of the genotype *ee*.

5. *R, a factor for the rough or rosetted coat pattern.* The double recessive *rr* had a smooth coat.

6. *A dilution factor*. This factor had been known for many years and produced dark eyed animals with light black (blues) or brown coats; yellows were also less intense with this factor. Castle thought vaguely that dilution was recessive to intense coloration, but results of matings between animals bearing dilution with each other, with intense pigmented animals, and with albinos produced a total confusion of results. The dilutions like tortoise shell, piebald, and silvering appeared to behave so irregularly that the usual Mendelian analysis seemed impossible in 1912. Castle therefore had not named the dilution factor as a simple Mendelian factor, nor had he or Detlefsen made any serious attempt to penetrate the confusion.

7. *A factor for long hair, recessive to short hair*. Wright never worked with this factor.

What was known of Mendelian inheritance in guinea pigs left a number of problems unsolved. A question remained about whether Detlefsen had really discovered a new allele A' in the agouti series. The dominant factor for rough, R, showed a wide range of variability from animals with a single rosette to animals covered with them. Detlefsen had tentatively postulated the existence of a third allele r' over which R was incompletely dominant, but this suggestion had not been subjected to careful experimental verification. These two problems were significant, but a third one was clearly more difficult and important. This concerned the dilution factor and the wholly confusing array of coat-color patterns exhibited in crosses with animals bearing the factor. There existed no obvious handle on this last problem.

All three of these problems were part of a more general question shared in 1912 by geneticists all over the world. It had become quite obvious to any trained person that alternative Mendelian inheritance could account readily for the patterns of inheritance of visible characteristics controlled by simple Mendelian factors. Even the most vociferous opponents of Mendelism in 1912 had to grant the geneticists that much. What troubled the critics and geneticists alike was all the observed variability that existed between the easily distinguished Mendelian characters. Many Mendelians in 1912 still clung to rather narrow conceptions of dominance, thinking that, as in the case of Mendel's peas, dominance was all or nothing. Mendel's own example of incomplete dominance in the flower color of *Phaseolus vulgaris* (common princess bean), presented in the same paper as the results with peas, was hardly noticed by many Mendelians. The pressing question was whether all the variation between the obviously Mendelian characters could be encompassed within the Mendelian viewpoint. Castle in 1912 thought this variability could be put into the Mendelian view only if the Mendelian factors were themselves variable.

One approach, that taken by East, Nilsson-Ehle, and many others, was to try and explain the "graded" or "continuous" variability by hypothesizing multiple Mendelian factors, often called "modifying factors." Another complementary approach, particularly emphasized by Cuénot in France and by the Morgan school, was that of multiple alleles rather than multiple factors.

Multiple alleles were all at one locus, so only two of them, no matter how many in the series, could possibly be in any individual organism—one allele came from the mother and the other from the father. Castle's belief that Mendelian factors themselves varied was in a sense a finely graded theory of multiple alleles. In addition to the obvious (but difficult to quantify) explanation that variability was caused by environmental rather than genetic factors, multiple factors and multiple alleles were the conceptual tools with which Mendelians tried to pry apart and clarify all the variability beyond that determined by Mendelian factors with clear patterns of segregation in controlled crosses.

The three obvious problems left in guinea pig genetics in 1912 all concerned graded variation between known Mendelian factors. Sewall Wright understood immediately, from the time he first talked to John Detlefsen, that these were the problems to be solved. That Wright did indeed focus upon these problems is apparent in the very title of his thesis: "An Intensive Study of the Inheritance of Color and of Other Coat Characters in Guinea Pigs, with Especial Reference to Graded Variations."

The Guinea Pig Colony in 1912 at the Bussey Institution

Wright had very rich resources with which to pursue his research on guinea pigs. Castle had over the years obtained many stocks from the fanciers and maintained them separately, using individuals from the stocks as needed in experiments. Using Castle's names, there were the *BB* stock of very intense blacks; intense blacks and albino *BW*; the four-toe stock, developed by Castle by 1906 as described in chapter 2; the tricolor stock with spots of yellow, black, white, and all degrees of roughness in addition to smooth coats; and the sepia-and-cream and brown-eyed-cream stocks that Castle had selected for years for extreme dilution. It was in this stock that Little found the pale-coated pink-eyed guinea pig that Castle reported early in 1912.

In addition to the stocks obtained from fanciers were those that had been sent to Castle from South America or that he collected there himself. These included the *Cavia rufescens* stock used by Detlefsen in his hybridization experiment between *C. rufescens* and the common guinea pig. This stock had no control population, and Wright inherited mostly those from 1/16 to 1/64 blood, although a few from 1/8 to 1/2 blood remained. The *Cavia rufescens* stock came to Castle from Brazil in 1903. The *Cavia cutleri* stock was descended from individuals captured by Castle in Peru in 1911. Unlike the *C. rufescens* stock, these were fully fertile with guinea pigs, although smaller. The Ica stock also came from Castle's 1911 trip to Peru and were originally golden agouti in color. The Arequipa stock Castle obtained from Peruvian Indians in 1911; they contained many dilute color variations. Finally there was the Lima stock, brought to Castle by Brues, who got them near Lima, Peru, from Indians early in 1913. The *Cavia cutleri*, Ica, and Arequipa stocks were fully described by Castle in the same Carnegie Institution of Washington Publication No. 241 that contained Wright's thesis. Why Castle waited until then to publish the description of these stocks will become obvious as we understand the import of Wright's thesis research to Castle.

When Castle brought the *Cavia cutleri*, Ica, and Arequipa stocks back from Peru to the Bussey Institution in February 1912, Detlefsen was busily engaged in his thesis research on the cross between *Cavia rufescens* and the guinea pig. Castle had projects going on rats and mice, plus he had teaching duties, which left him little time to initiate new experiments utilizing his newly acquired stocks. Thus the stocks were more maintained than experimented with during the interval between February 1912 and Wright's arrival at the Bussey Institution in September 1912.

Castle's original research assignment for Wright was to continue three of the lines of research already begun by Detlefsen. These were the analysis of fertility in the *C. rufescens* × guinea pig cross, the verification of Detlefsen's proposed new agouti allele *A'*, and further analysis of intermediate roughness. But Castle did not limit Wright to these three lines of research. All of the various guinea pig stocks described above were in Wright's care, and Castle allowed him to use any of these stocks for research projects of his own choosing. Although continuing all three of the projects assigned by Castle, Wright almost immediately began to utilize the full resources of the guinea pig colony.

Inheritance of Fertility in the *Cavia rufescens* × *C. porcellus* Cross

It was unfeasible to measure the fertility of male guinea pigs by test matings because a given male might take more than a year after sexual maturity to sire offspring. Detlefsen had developed a technique for indirectly measuring fertility by examination of the sperm. If there were no motile sperm, the animal was certainly sterile. The probability of fertility increased as the percentage of motile sperm increased. Detlefsen had carried the experiment as far as the 1/64 bloods before leaving for Illinois. Wright carried the experiment on for two more generations, on the 1/128 and 1/256 bloods. The sperm counts were very laborious but significant because the pattern continued to fit closely the expectation according to the eight-factor hypothesis that Wright had suggested to Detlefsen in the summer of 1912 on his first visit to the Bussey Institution. Because the experiment had been designed and executed for many generations by Detlefsen, and because Wright had introduced no new techniques or additional interpretations, he never published his results despite the considerable time and effort devoted to gathering them.

The Agouti Series

When Detlefsen first began crossing *C. rufescens* with the guinea pig, he observed certain differences in the agouti patterns of each. In the guinea pig, the increase of yellow pigment was pronounced, the striped pattern clearly visible, and most black pigment removed from the belly, (some have black pigment at the base of the hair only), leaving it a clear light yellow (known as the agouti light-belly). *C. rufescens,* on the other hand, had less restriction of the black pigment throughout; in particular, the belly was usually ticked with black (agouti ticked-belly). These differences were observable but not striking. Detlefsen observed, as expected, that the agouti pattern derived directly from the original *C. rufescens* stock was dominant over nonagouti. Unexpectedly, however, as the agouti pattern from *C. rufescens* became placed in an animal with a greater percentage of guinea pig ancestry, the pattern frequently became noticeably darker, quite distinct from the agouti pattern usually observed in the guinea pig. Detlefsen surmised that the agouti pattern from *C. rufescens* might be different from the agouti in the guinea pig. By a series of experiments he showed that agouti light-belly was dominant over agouti ticked-belly and that the two patterns segregated independently. The problem was that the results of all of these matings could be explained equally well by hypothesizing the existence of either (1) a new independent modifying factor or (2) a new allele in the series from agouti to non-agouti. The homozygous forms would then be on each hypothesis:

	Agouti Light-Belly	Nonagouti	Agouti Ticked-Belly
Hypothesis 1	*AAa'a'*	*aaa'a'*	*aaA'A'*
Hypothesis 2	*AA*	*aa*	*A'A'*

The problem was how to experimentally discriminate between these two hypotheses.

To settle the question, Wright decided to create stocks doubly heterozygous on the first hypothesis: *AaA'a'*. These should be able to produce light-bellies, ticked-bellies, and nonagoutis when inbred. On the second hypothesis, the heterozygotes would be *AA'*, which when inbred could produce no nonagoutis, and *Aa*, which when inbred could produce no ticked-bellies. Wright therefore crossed heterozygous ticked-bellies with heterozygous light-bellies known to be free of ticked-belly from their parentage. The two hypotheses then could be represented:

	Cross	Resulting Offspring
Hypothesis 1	*aaA'a'* × *Aaa'a'* =	*AaA'a' Aaa'a aaA'a aaa'a'*
Hypothesis 2	*A'a* × *Aa* =	*AA' Aa A'a aa*

On both hypotheses, light-bellies, ticked-bellies, and nonagoutis should be produced in the ratio 2:1:1. If any light-belly from this cross could produce ticked-belly, then on hypothesis 1 it should produce nonagoutis also; on hypothesis 2 the same cross should not produce nonagoutis in addition to ticked-bellies. Carrying out the crosses with sizeable numbers, Wright found that in no case did the light-bellied offspring produce both ticked-bellies and nonagoutis. Here was decisive evidence in favor of Detlefsen's hypothesis. Wright also experimented with the agouti pattern from *Cavia cutleri,* not used by Detlefsen, finding it identical with or indistinguishable from the agouti of the domestic guinea pig. This was unsurprising because *Cavia cutleri* was generally believed to be the feral ancestor of the domestic guinea pig.

Intermediate Roughness

Back in 1905 Castle had published the view that roughness was a Mendelian dominant over smooth. All later experimentation indicated the accuracy of this initial hypothesis, but complications became apparent. In his experiments with the cross of *C. rufescens* × *C. porcellus,* Detlefsen discovered that roughness of any degree was indeed dominant over smooth coat but that the degree of roughness observed was declining as the crosses proceeded. To explain the intermediate roughness in his crosses, Detlefsen hypothesized a new and more potent smooth allele *r',* found in *C. rufescens* but not in *C. porcellus*, thus giving the roughness series three alleles: *R* (rough), *r'* (potent smooth), and *r* (smooth).

When Wright approached the problem of the inheritance of roughness, he used seven different stocks instead of only two as had Detlefsen. Roughness exhibited a wide range of variability. Wright devised a scale of five different degrees of roughness. Some stocks had all five degrees of roughness, and others only three. With this variability at his disposal, Wright made a series of ten different matings, hoping to see if any simple hypothesis could account for at least most of the observed variability. As in the case of the agouti factor, the simplest hypotheses were another allele, as proposed by Detlefsen, or an independent modifying factor. Seven of the matings produced results identical on both hypotheses; the other three were decisive. On Detlefsen's theory, a homozygous dominant rough (*RR*) could not possibly exhibit an intermediate rough coat. Wright produced individuals homozygous for rough (*RR*) that were the lowest grade of the partial roughs. The other two more complex matings also produced results inconsistent with Detlefsen's hypothesis. Wright then showed that the ten crosses were in all major features fully consistent with the hypothesis of an independent incompletely dominant modifier locus with two alleles. As in many of the apparently confusing cases analyzed by Castle and his students in the previous ten years, Wright had now provided a relatively simple yet satisfying Mendelian analysis for this problem.

The Albino Series of Alleles

When Wright took charge of the guinea pig colony in September 1912, he immediately set to work on the extensions of Detlefsen's research. At the same time he began thinking about the rich resources in variability represented in the colony. Most obvious, and seemingly impenetrable to analysis, was the wide range of color dilutions from intense to albino on an almost continuous scale. More variability of this kind appeared in the new stocks from Peru soon after Wright's arrival. Most important were a pink-eye variant in the Arequipa stock, with dilutions in the dark colors but not in the yellows, and a red-eye variant whose dilutions in the dark colors matched the previously known dilution factor, but in which yellow was completely suppressed. The pink-eye variant appeared to be homologous to the pink-eye factor in mice, and in 1914 Castle found a very similar pink-eye in rats (Castle 1914b).

How was Wright to clarify this confusing array of variability? A promising approach appeared to be the analysis of a few of the more specific variants, such as the pink-eye or the red-eye, to see where those results might lead. It occurred to Wright that the dilutions apparent in the pink-eye and red-eye might be caused either by an independent Mendelian factor or by another allele in the albino series, known in guinea pigs to contain only the dominant color factor C and the recessive factor c for albinism. There was some evidence leading in both directions. Pink-eye in mice was known to be inherited independent from albinism, and if the pink-eye in guinea pigs were similar, it too might be inherited independent of albinism. In rabbits, Castle and many of his students, including Little and MacDowell, thought the Himalayan factor was a third allele in the albino series. They had held this view since 1909.

L. C. Dunn in his Oral Memoir has stated his view (Dunn 1961) that not until A. H. Sturtevant published his paper on the Himalayan rabbit case in April 1913 did geneticists have a clear and compelling statement of a series of multiple alleles for a single locus (Sturtevant 1913). Chronologically, the publication of Sturtevant's paper fits with Wright's thought that red-eye might be an allele in the albino series. But there is no causal connection, and Dunn's view is incorrect.

Cuénot (1904, 1907) had very early proposed a series of multiple alleles in mice, and his work was well known. Castle had long thought in terms of multiple alleles (indeed, each of his "changed" Mendelian factors was a new allele), and in his selection experiment on hooded rats the mutant that appeared in the plus stock was quite clearly an allele in the hooded series. Detlefsen used the multiple allele concept twice in his research by the summer of 1912, one of which proved true and the other false as we have seen. When Wright came in 1912, everyone interested in mammalian genetics around the Bussey thought of the Himalayan factor in rabbits as an allele in the albino series. The immediate stimulus for Sturtevant's paper was Punnett's 1912 paper, "Inheritance of Coat-Colour in Rabbits." Punnett did not consider the hypothesis offered by Castle in 1909 that the Himalayan factor was an allele C' of the color factor C, and instead rejected his own hypothesis that the Himalayan pattern was the absence of yellow pigment. He ended by expressing confusion. Why Punnett did not consider or understand Castle's hypothesis, or even refer to it, is unclear to me; but it is easy to understand why Sturtevant, who was an expert on color inheritance in horses as well as in *Drosophila,* would wish to step in and clarify Punnett's obvious confusion. After all, the Morgan group had just found an eosin allele in the red-white eye-color series in *Drosophila melanogaster,* and the interpretation appeared to fit the Himalayan rabbit. Sturtevant was right in his suggestion. It did not, however, reveal anything new to Castle or his students, some of whom were a bit offended at Sturtevant's intrusion into their field. There was always some discernible tension between the Morgan and Castle camps, with Little cutting away at Morgan's ideas on mice and Muller and others working over Castle's rejection of Mendelian modifiers.

Independent of any influence from Sturtevant's paper, Wright thought it worthwhile to explore the possibility that red-eye was an allele in the albino series. Breeding tests Wright designed soon revealed that all three of the factors C, albinism, and red-eye could not be in any one individual at the same time, but any two could. C was dominant to red-eye, which in turn was dominant to albinism. Immediately one series in the great multiplicity of dilutions began to look simpler, but it was by no means solved. Encouraged by this initial success, Wright now tried a second venture: if red-eye was an allele in the albino series, could not the dilution factor known to Castle by 1905 be another allele in the series, lying in the pigmentation scale between C and red-eye? Breeding tests verified this hypothesis as well. Now the albino series contained four rather than only two alleles, and this series of dilutions began to look not only understandable but to be predictable. Wright devised a new notation to fit his discoveries: the general color factor was still C, dilution C_d, red-eye C_r, and albino C_a. C was dominant to C_d, C_d to C_r, and C_r to C_a. The albino could only have genotype C_aC_a. With ample evidence for the allelism of the four factors, but without having actually systematically carried out all the crosses (an enormous task), Wright announced this discovery in a paper sent to *American Naturalist* on Christmas Eve 1914. The paper, under the title "The Albino Series of Allelomorphs in Guinea Pigs," appeared in the *American Naturalist* of March 1915 (Wright 3). My copy of the reprint of this paper, which belonged to L. C. Dunn, is very well worn. It bears the number 26, which I think indicates that it was the twenty-sixth paper in his collection. This paper signified that order could be found in guinea pig color inheritance. Wright set to work making the crosses within the albino series, and these results, taken from his published thesis and reproduced below as figure 3.3, supported his theory in every way (from Wright 5, p.67).

By the time Wright left the Bussey in the fall of 1915, he had run through the dilution series on most of the major homozygous color patterns, as he presented in the following summary table (fig. 3.4; reproduced from Wright 5, p.67). The enormous power of Wright's albino series is obvious from inspection of the table.

Variability Unaccounted for by Mendelian Analysis

Despite Wright's dramatic success in clarifying the inheritance of color, the agouti pattern, and roughness in the guinea pig, he was careful not to mini-

Parents.	Formulæ.	Int.	Dil.	R.E.	W.	From crosses—
Intense × albino	CC × CaCa...	40	....	....	....	17a, b.
	CCd CaCa...	31	36	....	....	18b, 34, 41
	CCr CaCa...	9	....	4	....	21
	CCa CaCa...	64	....	....	71	17c, 17d, 18c, 22
Dilute × albino	CdCd × CaCa...	....	26	....	....	16a, 38a, 44
	CdCr CaCa...	....	18	24	....	19, 27
	CdCa CaCa...	....	98	....	79	16b, 16c, 19, 27, 33, 38b, 44
Red-eye × albino	CrCr × CaCa...	....	....	30	....	25
	CrCa CaCa...	....	....	6	3	23, 25
Albino × albino	CaCa × CaCa...	....	....	....	X	Long established.
Intense × red-eye	CC × $\begin{Bmatrix} CrCr \\ CrCa \end{Bmatrix}$..	9	....	....	....	20
	CCd $\begin{Bmatrix} CrCr \\ CrCa \end{Bmatrix}$..	28	31	....	....	18a, 20
	CCa $\begin{Bmatrix} CrCr \\ CrCa \end{Bmatrix}$.	25	....	20	7	18c
Dilute × red-eye	CdCd × $\begin{Bmatrix} CrCr \\ CrCa \end{Bmatrix}$..	....	15	....	....	20, 26, 43
	CdCa CrCr...	....	1	1	....	20
	CrCa...	....	13	5	8	26, 43
Red-eye × red-eye	CrCa × CrCa...	....	....	17	6	24
Intense × dilute	CC × CdCd..	14	....	....	....	28
	CC CdCa..	12	....	....	....	35
	CCd CdCd..	10	7	....	....	29, 40a
	CCd CdCa...	32	40	....	....	32, 36, 40a
	CCa CdCd..	12	10	....	....	28, 40a
	CCa CdCa...	28	22	....	15	36, 40b
Dilute × dilute	CdCd × CdCa...	....	15	....	....	30, 37, 42
	CdCa CdCa...	....	82	....	24	37, 42
Intense × intense	CCd × CCd...	57	19	....	....	31, 39
	CCd CCa...	75	35	....	....	39
	CC CC....	X	....	....	....	See rough and Lima crosses.

Figure 3.3. Summary of crosses within the albino series.

mize the variability as yet unexplained by the Mendelian analysis. In mice, rats, and rabbits, distinct Mendelian factors were known to control the appearance of white spotting or patterns of white. No such Mendelian factors were known in guinea pigs, and Wright designated the factors involved as "Σ", an assemblage of unanalyzed factors which determine white spotting." The same pair of alleles in the albino series did not yield offspring of exactly the same color when inserted into different stocks. In a section of the thesis entitled "The Relations of Imperfect Dominance, Stock, and Age to Grades of Intensity," Wright included charts of the variations in intensity of yellow and black and discussed in detail the variability left unanalyzed in his Mendelian analysis.

The problem of unanalyzed variability Wright specifically emphasized at the end of his conclusion:

Factors present.	Fur.			Eye.
	EA (agouti light-belly). EA' (agouti ticked-belly).	Eaa.	ee (A, A' or aa).	
B P C.....	Black-red agouti............	Black.........	Red............	Black.
CdCd..	Dark sepia-yellow agouti.....	Dark sepia.....	Yellow.........	Do.
CdCr...	Dark sepia-cream agouti.....	....Do........	Cream..........	Do.
CdCa...	Light sepia-cream agouti.....	Light sepia.....	....Do.........	Do.
CrCr...	Dark sepia-white agouti.....	Dark sepia.....	White (light points).	Red.
CrCa...	Light sepia-white agouti.....	Light sepia.....	....Do.........	Do.
CaCa..	White (dark points).........	White (dark points).	....Do.........	Pink.
Bpp C....	Pale sepia-red agouti........	Pale sepia......	Red............	Pink.
CdCd..	Very pale sepia-yellow agouti.	Very pale sepia.	*Yellow.........	Do.
CdCr...	Very pale sepia-cream agouti.	....Do........	*Cream..........	Do.
CdCa..	....Do....................	...Do.........	...*Do.........	Do.
CrCr...	Very pale sepia-white agouti..	....Do.	*White..........	Do.
CrCa...	....Do....................	...Do.........	...*Do.........	Do.
CaCa..	White (light points).........	White (light points).	...*Do.........	Do.
bbP C....	Brown-red agouti..........	Brown.........	Red............	Brown.
CdCd ..	Medium brown-yellow agouti.	Medium brown.	Yellow..........	Do.
CaCr...	Medium brown-cream agouti.	....Do........	Cream..........	Do.
CdCa..	Light brown-cream agouti....	Light brown....	....Do.........	Do.
CrCr...	Medium brown-white agouti..	Medium brown.	White..........	Brown-red.
CrCa...	Light brown-white agouti....	Light brown....	....Do.........	Do.
CaCa...	White (lt. br. points)........	White (lt. br. points).	....Do.........	Pink.

Figure 3.4. The dilution series.

We must recognize series of variations in which no Mendelian factors have yet been isolated. The series of white-spotted and yellow-spotted types and the series of polydactylous types are examples in guinea-pigs. Further, in all series of variations, to whatever extent analysis has been carried, there always remains some unanalyzed variation. In many cases such variations are known to be hereditary and can be assigned to the residual heredity of particular stocks. Such unanalyzed variations, however, are probably in general complicated by variation which is not hereditary, due apparently to irregularities in development. If we can measure the importance of such non-hereditary variation by the extent of irregular asymmetry met with, it is very important in white and yellow spotting, in the variations in the development of extra toes on the hind feet, and is noticeable in variations in roughness.

In the continuous series of variations several of these phenomena [those described in the thesis as a whole] have been found together. In the series from smooth to full rough we find a primary unit difference, a modifying factor, imperfect dominance in the effects of the latter, effects of residual heredity, and probably some non-heritable variation. In the series from red through yellow and cream to white we find multiple allelomorphs, imperfect dominance, and small effects due to residual heredity. In the series black through sepia to white, we find independent factors, allelomorphs which show imperfect dominance, and rather prominent effects due to residual heredity and age. . . . Thus in each case a complex of the most varied causes underlies an apparently simple continuous series of variations. (Wright 5, pp. 120–21)

If Wright did not penetrate the unanalyzed variability in his thesis, he did not despair of doing so. We will see in the next chapter that one of the first uses Wright had for his method of path coefficients was to analyze into hereditary and environmentally determined components the variability seen in the white spotting in guinea pigs. Wright's forthrightness as an individual and as a scientist is revealed by a doctoral thesis in which he elucidates a powerful new result of Mendelian analysis and at the same time emphasizes the existing limitations of the Mendelian analysis.

Physiological Genetics and Interaction Effects

Most modern biologists date the beginnings of our understanding of the physiology of gene action to the "one gene, one enzyme" hypothesis of G. W. Beadle and E. L. Tatum, working with *Neurospora crassa* in the early 1940s. Perhaps they would also include Archibald Garrod as a precursor of this idea. Garrod had proposed (Garrod 1902) that the mutant gene responsible for the symptoms of alkaptonuria no longer manufactured the enzyme that normally split the benzene ring in alkapton. Beadle himself has promoted the idea that geneticists ignored Garrod's work, resulting in a hiatus in the progress of ideas on the physiology of gene action between Garrod and the work of the late 1930s and 1940s (Beadle 1951).

Historically speaking, this is a wholly inaccurate view. There exists a large body of literature on the physiology of gene action published by biologists all over the world in the period between 1902 and the "one gene, one enzyme" hypothesis. Among the major geneticists who devoted their attention to this problem were Richard Goldschmidt, H. J. Muller, A. H. Sturtevant, Julian Huxley, E. B. Ford, Alfred Kühn, Ernst Caspari, Milislav Demerec, L. C. Dunn and, of course, Sewall Wright. Those whose influence was most strongly felt were Goldschmidt and Wright. To gain an idea of the enormous amount of thought and experimentation devoted to physiological genetics in this period one need only turn to the two books by Goldschmidt, *Physiologische Theorie der Vererbung* (1927) and *Physiological Genetics* (1938), and to Wright's long 1941 article, "The Physiology of the Gene" (Wright 107).

As mentioned in the previous chapter, there was great interest in the physiology of gene action in Castle's group, especially concerning the control exerted by Mendelian factors over the development of pigment in coat color, eyes, and skin. Castle and his students read the papers of biochemists who analyzed pigment production and discussed these issues among themselves frequently. Neither Castle nor any of his students had enough background in chemistry to do original work in this field, and as we saw, Little said the biochemistry of gene action was a problem for the chemists, not the geneticists.

Wright had the background in chemistry lacked by Castle and his other students. Even before coming to the Bussey, Wright thought the problem of the physiology of gene action to be one of the central problems of genetics, and he addressed the problem even though Mendelian factors could not be observed. Early in his Harvard stay, Wright had long talks with Little about the physiology of pigment formation and what forms genetic control might take. He read widely in the literature about the biochemistry of pigment formation, a very active field pursued by prominent biologists such as Oscar Riddle and R. A. Gortner in the United States and H. Onslow, F. Keeble, and E. F. Armstrong in England.

In 1915 Wright was stimulated by two occurrences in particular. One was a new and very substantive paper by Onslow on pigment precursors (Onslow 1915), and the other was an extended visit at the Bussey by Richard Goldschmidt, who had been stranded in the United States by the First World War. Goldschmidt, who was very chauvinistic about Germany at this time (he later fled Nazi Germany and moved to the United States), was not very popular around the Bussey. He and Wright, however, got along well, and they had extended conversations about evolution in nature, physiological genetics, and many other topics. Goldschmidt, who wrote extensively about his fellow biologists, always in print and conversation expressed the belief that he and Wright were the two geneticists who did most to investigate and stimulate interest in physiological genetics. It is no surprise, therefore, to find in Wright's published thesis a substantive section entitled "Hereditary Factors and the Physiology of Pigment," beginning with the statement: "It would be very desirable . . . to correlate color factors accurately with the variations in quality and quantity of the actual pigments and ultimately with the physiology and chemistry of pigment formation" (Wright 5, p. 67).

Wright began by stating the generally accepted view that natural melanins (pigments) were produced by the action of oxidizing enzymes, such as tyrosinases or peroxidases, on chromogens (substances capable of being chemically changed into pigments) such as tyrosin and related compounds. The two major theories concerning melanin production in animals were (1) variations in the amount of chromogen, and (2) variations in levels of oxidizing enzymes. Wright cited some evidence for the first but, following the lead of Onslow's 1915 paper, favored the second as appropriate in mammals.

Onslow's paper, "A Contribution to Our Knowledge of the Chemistry of Coat-Colours in Animals and of Dominant and Recessive Whiteness," came to the following conclusions based upon extensive experimentation: (1) pigment formation in rabbits and mice was caused by peroxidase acting upon chromogen in the presence of hydrogen peroxide; (2) recessive white was caused by lack of enzyme rather than chromogen; (3) dominant white was due to the presence of a tyrosinase or peroxidase inhibitor in the skin. To Onslow's conclusions Wright added his own strong evidence from the albino series in guinea pigs that albinism was not caused by the absence of chromogen. If that were the case, the intensity of all pigments should decrease correspondingly. With the red-eye allele C_r, no yellow developed, but the dark pigments did, and both the Himalayan rabbit and albino guinea pig had dark pigmented extremities. Wright concluded that variations in melanin production were caused in mammals by variations in the enzyme component of the color-producing mechanism.

A theory frequently expressed at this time was that the various colors were produced by simply stopping the oxidation of the chromogen at the appropriate stage. In other words, the action of the enzyme was to progressively convert the chromogen from the stage of no pigment to light pigmentation (creams, etc.) to yellow and then to the darker colors. Wright rejected this theory using genetic facts in guinea pigs. The C_r allele caused a complete absence of yellow but almost intense blacks and browns; the pink-eye factor, however, did the reverse, causing much reduced intensities of black and brown, but did not dilute yellow pigmentation at all.

Wright suggested that if Onslow were right—that the absence of tyrosinase in the skin and eye caused albinism—then the higher alleles controlling greater pigmentation were simply "quantitative variations in a factor which determines the power of producing tyrosinase. If this is so, we would expect to find that the different zygotic formulas could be arranged in a linear series with respect to their effects on pigments of all sorts" (p. 70). He accordingly presented the following table of formulas and corresponding pigmentation in the black-eye-color series, the black-fur-color series, and the yellow-fur series (fig. 3.5; reproduced from Wright 5, p.70):

Formula.	Black eye.	Black fur.	Yellow fur.
CC	Black	Black	Red.
CC_d	Do.	Do.	Do.
CC_r	Do.	Do.	Do.
CC_a	Do.	Do.	Do.
C_dC_d	Do.	Dark sepia	Yellow.
C_dC_r	Do.	Do.	Cream.
C_dC_a	Do.	Light sepia	Do.
C_rC_r	Red.	Dark sepia	White.
C_rC_a	Do.	Light sepia	Do.
C_aC_a	Pink.	White (sooty)	Do.

Figure 3.5. Color series.

Certain problems were immediately obvious to Wright from the table. The level of no pigment production was much higher in the yellow-fur series than in the other two black-color series; although the level at which pigmentation becomes evident was the same in both black series, the sequence did not agree. Thus $C_d C_a$ was distinctly lighter than $C_r C_r$ in the black-fur series but distinctly darker in the black-eye-color series.

Wright could think of three hypotheses for explaining these anomalous results, the first being a complex linkage relationship and the second assuming that the albino series C_a, C_r, C_d, and C were nonlinear quantitative variations of the same physiologically active factor. His reasons for rejecting these two hypotheses reveal a fundamental aspect of his approach to science, an aspect he shares with many other scientists in all fields. These hypotheses were, to Wright, needlessly complex and aesthetically unsatisfying. Where possible, Wright liked to find simple rather than complex or messy relationships. This is the principle of parsimony invoked frequently by scientists (for a more recent example, see Williams 1966). Wright's position was: "It seems most satisfactory to the writer to attempt to explain the results on the basis of four quantitative gradations of one factor, which determines the amount of the basic color-producing enzyme, if it is in any way possible. Let us see what assumptions must be made to do this" (Wright 5, p.71).

Wright proposed the following speculative hypothesis. The basic color-producing enzyme (I) acting upon chromogen made yellow pigment. A second enzyme (II), or more likely a substance changing the first into a different enzyme (I-II), stabilized I and, acting upon chromogen, produced the dark colors. A third enzyme (III) or substance altering the others had a modifying effect upon the dark colors but not upon yellows, as in the pink-eye factor, or the reduction of blacks to browns. Using this scheme, with added plausible assumptions about threshold levels of pigment production of the yellows and dark colors, Wright demonstrated that apparently complex changes in color patterns known to be controlled by a single Mendelian factor could be explained simply on his scheme. Even such complex patterns as the yellow and black stripes on the tiger had a simple explanation. Wright's proposal was frankly speculative, but it pointed the way toward a systematic investigation of the physiology and genetic control of pigment production.

One conclusion fundamental to Wright's speculative hypothesis, and indeed to all of his work on color inheritance, the albino series in particular, was that Mendelian factors did not act independently. Interaction of genetic factors (or rather, of their physiological products) was inevitable and had always to be taken into account at all levels of analysis of the organism. That a given Mendelian factor, such as an allele in the albino series, could have its phenotypic appearance altered by insertion into another stock with different residual heredity was an obvious sign of this interaction. Wright's thesis research must be viewed as a firm lesson in the reality, necessity, and ubiquitousness of gene interaction. All of Castle's students were keenly aware of, indeed could not escape from this lesson. I emphasize this here because all of Wright's later work in physiological genetics and population genetics was founded upon his conviction of the importance of gene interaction. To understand correctly Wright's shifting balance theory of evolution, or his theory of the evolution of dominance as opposed to R. A. Fisher's, one must clearly appreciate the centrality of the idea of gene interaction to his thinking.

Philosophy of Science

In his table of family interests (table 1.1), Wright gave himself the highest ranking in the category of philosophy. He did indeed have a keen interest in philosophy dating from his graduate school days, and he published his views briefly on many occasions before expressing them in greater detail in the period after 1950 (Wright 26, 64, 88, 104, 126). Unlike his achievements in physiological genetics and evolutionary theory, however, Wright's philosophical views had no significant influence upon others and were never respected by more than a few philosophers, Charles Hartshorne primary among them. More important, and perhaps surprisingly so, Wright's philosophical ideas had no detectable influence upon his scientific work.

Until he went to Harvard in 1912, Wright was, to the extent that he thought seriously about such issues, basically a philosophical determinist in the tradition of Laplace. But he was disturbed about the adequacy of determinist views applied to consciousness and behavior. Thus he recalls that around 1910, when Jacques Loeb, the famous mechanical materialist, and Herbert Spencer Jennings were arguing about the interpretation of animal behavior, his sympathies lay with Jennings (Wright 175, p. 116). Then in 1912 he read Bergson's *Creative Evolution*. He could not accept Bergson's ideas about the mechanisms of evolution, but he was greatly attracted by his way of dealing with mind as a fundamental aspect of the universe that was not subject to scientific analysis of the usual sort.

Probably in 1914 Wright read the crucial book that basically determined his later philosophical position, and from what I have said so far it was a most unlikely candidate. It was the second edition of Karl Pearson's *Grammar of Science* (1900). Pearson was deeply convinced that the methods of science were applicable to all phenomena, and he adamantly opposed dualistic theories of mind and matter.

> If I have put the case of science at all correctly, the reader will have recognised that modern science does much more than demand that it shall be left in undisturbed possession of what the theologian and metaphysician please to term its "legitimate field." It claims that the whole range of phenomena, mental as well as physical—the entire universe—is its field. It asserts that the scientific method is the sole gateway to the whole region of knowledge. (Pearson 1900, 24)

> Because we cannot point to the exact form of material life at which consciousness ceases, we have no more right to infer that consciousness is associated with all life, still less with all forms of matter, than we have to infer that there must always be wine mixed with water,

because so little wine can be mixed with water that we are unable to
detect its presence. (P. 58)

Wright took from *The Grammar of Science* just the opposite of what Pearson
said about dualism, panpsychism, and the limits of scientific analysis. This is
especially curious because for more than forty years, many times in print (see
Wright 144, p. 14), Wright attributed his philosophical views to Pearson and
The Grammar of Science.

The mystery here is easily clarified. What Wright found most attractive
in *The Grammar of Science* was Pearson's summary of the views of W. K.
Clifford, whose book *Common Sense of the Exact Sciences* Pearson had edited for publication after Clifford's death. Pearson was actually summarizing
Clifford's "monistic panpsychism" (as Wright later termed it) in order to refute it. Wright, however, attributed Clifford's views to Pearson.

For Clifford, and thus for Wright, science could only describe the external aspect of the world. Mind, the internal aspect, was totally beyond scientific analysis. Mind was as fundamental to the world as matter and was not
an emergent property of matter.

Wright never changed this basic view. Each time he expressed his philosophy, the response from his colleagues was generally negative. Charles
Hartshorne, a member of the philosophy department at the University of
Chicago for many years and a student of Alfred North Whitehead, was, however, favorably impressed. Wright's major statement of his philosophical
ideas came at a meeting honoring Hartshorne (Wright 175).

One of the favorite tricks in the bag of historians and philosophers of science is to show how the fundamental assumptions of a scientist determine his
or her scientific works. I tried that with Wright, but the ploy did not work.
Although Wright has since 1914 always believed that mind is a fundamental
aspect of the universe, he has also believed that science cannot touch mind.
Science provides the most reliable means for understanding the external aspect of the world. Thus Wright approached science as if the world were deterministic, but at least he had the satisfaction of knowing that there was another whole world of mind that totally escaped the science. He expressed
these views clearly in the conclusion to his presidential address to the American Society of Naturalists in 1952:

> It is the task of science, as a collective human undertaking, to
> describe from the *external* side, (on which alone agreement is possi-
> ble), such statistical regularity as there is in a world in which every
> event has a unique aspect, and to indicate where possible the limits
> of such description. It is not part of its task to make imaginative in-
> terpretations of the internal aspect of reality—what it is like, for ex-
> ample, to be a lion, an ant or ant hill, a liver cell, or a hydrogen ion.
> The only qualification is in the field of introspective psychology in
> which each human being is both observer and observed, and regular-
> ities may be established by comparing notes.

Science is thus a limited venture. It must act as if all phenomena were deterministic at least in the sense of determinable probabilities. It cannot properly explain the behavior of an amoeba as due partly to surface and other physical forces and partly to what the amoeba wants to do, without danger of something like 100 per cent duplication. It must stick to the former. It cannot introduce such principles as creative activity into its interpretation of evolution for similar reasons. The point of view indicated by a consideration of the hierarchy of physical and biological organisms, now being bridged by the concept of the gene, is one in which science deliberately accepts a rigorous limitation of its activities to the description of the external aspect of events. In carrying out this program, the scientist should not, however, deceive himself or others into thinking that he is giving an account of all of reality. The unique inner creative aspect of every event necessarily escapes him. (Wright 144, p. 17)

Under his philosophy, Wright was therefore free to proceed with his science as if evolutionary biology were a combination of deterministic and stochastic processes and nothing more.

I believe it is true that the philosophical assumptions of some scientists have very important effects upon their science. I am equally certain that many scientists proceed consciously upon absolutely ridiculous philosophical assumptions and still are outstanding scientists. In Wright's case, he believed in a philosophical view that had no discernible effect upon his science. His differences from Fisher on evolutionary theory, for example, cannot be traced to his expressed philosophical ideas. Wright's early views on philosophy therefore determined only his later views on philosophy, not his work in physiological genetics or evolutionary theory.

Philosophical views aside, Wright's three short years in graduate school were highly productive and formulative. During the next ten years in his tenure at the United States Department of Agriculture, he would extend every area of research begun at Harvard as well as develop new ones.

4
USDA and Washington, 1915–1925

In the spring of 1915 Wright was looking forward to completing his thesis, graduating, and beginning a career as a geneticist. Earlier in his graduate years Wright had planned to go to Europe for study and research after graduation, but the outbreak of the First World War dashed this possibility. In the final year of graduate school Wright was awarded a Sheldon Travelling Fellowship. He considered taking a trip to South America to collect wild cavies. The plan was that Wright would accompany William M. Mann, one of Wheeler's students, who was going to South America to collect insects for Wheeler. This trip looked probable for a while, but Mann got the chance to collect insects on a Pacific island and opted for that instead. The trip to South America fell through. This was the closest Wright ever came to studying guinea pigs in the wild. Wright was also offered a teaching position at a small college in the southern United States and a job as senior animal husbandman in animal genetics at the Animal Husbandry Division of the Bureau of Animal Industry, United States Department of Agriculture. The teaching position was unattractive to Wright because he would not have the facilities, funds, or time for his research, and he disliked the thought of unrelenting teaching responsibilities like those that had entrapped his father at Lombard. If a position at a major university with research facilities had been offered, Wright would probably have accepted it. No such offer materialized.

The job at the USDA was very specific. In 1906 the chief of the Animal Husbandry Division, George M. Rommel, had decided to extensively investigate the effects of inbreeding. The practice of inbreeding in cattle, horses, sheep, hogs, and many other economically valuable mammals was widespread but highly controversial. This same year Castle had published his study on inbreeding in *Drosophila,* and the topic was hotly debated by breeders everywhere. Indeed, inbreeding had been a controversial issue a century before Darwin's *Variation of Animals and Plants under Domestication* (1868), in which Darwin frequently discussed inbreeding. Some breeders argued that inbreeding was essential for the improvement and fixation of a herd. Others declared that inbreeding was inevitably harmful. Evidence from controlled experiments was lacking, so Rommel decided to use for this purpose the already existing stock of guinea pigs at the experimental station of the Bureau of Animal Industry at Bethesda, Maryland. Under his direction, a series of thirty-five inbred families was begun. In all but one, direct brother-sister mating was used exclusively for the duration of the experiment; one family was perpetuated by parent-offspring mating.

Data on the families were painstakingly recorded and preserved. On a visit to the USDA, Wright observed Walter J. Hall, caretaker of the animals when Wright arrived at the USDA, record the data; Wright was greatly impressed by Hall's accuracy and care. All of the data were of similar reliable quality. Recorded at birth for each individual were such factors as date, pen, number, sex, color, and weight. Coat color was drawn in a rubber stamp outline and could be done accurately. Each of the young continued to be weighed again at various intervals, and the fate of each animal recorded.

The problem in 1915 was that a great mass of this accurate data had accumulated, but it was totally unanalyzed. In addition to answering inquiries and publishing informative pamphlets, the primary requirement of the job was to quantitatively analyze the inbreeding data. Wright knew the inbreeding problem was important and found it interesting; he was also attracted by the possibility that he could continue his own research on guinea pigs. He could move samples of all of Castle's stocks to the facilities of the experiment farm in Beltsville, Maryland (the inbreeding experiment was moved there in 1911).

Castle had visited Rommel in April 1913 and gone to Beltsville to see the inbreeding experiment (Castle to Davenport, April 25, 1913). He was impressed with the work at the Animal Husbandry Division, and when Rommel wrote to ask if Castle knew a qualified candidate, he recommended Wright. Castle's view of the position was obviously favorable. East, on the contrary, was openly critical. He warned Wright that even well-conceived research projects at the USDA were ruined by federal bureaucracy (East was a strong political conservative) and that to work in Washington was to be cut off entirely from the mainstream of genetics. Wright thought East's point was well taken, but he also saw great advantages to the USDA position over the other available possibilities. He accepted the job.

Wright remained in Cambridge for the summer of 1915 to break in Castle's new assistant, L. C. Dunn. In his oral memoir, Dunn recalled that he was awed by Wright's thorough grasp of all aspects of the work at the Bussey, from feeding rations and disease prevention in the animals to the most abstract thinking. Dunn never lost this attitude toward Wright and called him (speaking of geneticists of the second generation) "pretty close to the number one man for this period of development, in this country or any other country" (Dunn 1961, 150). In September, Wright moved to Washington and the USDA.

The Isolation of Washington: Wright's Friends and Family

East was right—Washington was an isolated place for a geneticist. In Wright's early years at the USDA only one other geneticist worked there,

Guy N. Collins of the Bureau of Plant Industry. Collins was a very good corn geneticist. He had found the first case of linkage in corn and had isolated a number of genes. His work was well known and a clear counterexample to East's dictum that good genetics was impossible at the USDA. Collins was particularly interested in hybrids of corn and teosinte (wild corn), a question that later attracted the attention of many prominent geneticists including Beadle, Mangelsdorf, McClintock, and Hugh Iltis. While at the USDA, Wright took his papers to Collins for criticism before publication. Collins was very generous with his time and made detailed constructive comments on each paper Wright brought him. Collins was, for example, the only person for several years to read the first draft of Wright's big paper "Evolution in Mendelian Populations," written during the final months in Washington in 1925 but not published until 1931.

Collins was also a good friend. He frequently invited Wright to his home in Lanham, Maryland, in a neighborhood populated by project leaders in the Bureau of Plant Industry, whom Wright also met. Collins often invited Wright to luncheon meetings at Plant Industry or at the Smithsonian. World travelers who collected exotic plants attended these meetings and talked about their finds. Wright recalls the conversations as some of the most interesting he ever heard. Collins also encouraged Wright to join the Cosmos Club, where he introduced Wright to such notables as Frederick Adams Woods and Ellsworth Huntington. Also frequently there, sometimes wandering around in his slippers, was Admiral Adolphus Washington Greely, then in his midseventies, who had led an ill-fated expedition toward the North Pole in 1882. He reached the farthest north of anyone to that time but lost most of his men to cold and starvation. A rumor of cannibalism surfaced after the return of the survivors. One of Admiral Peary's associates, a Captain Bartlett, was also a member of the Cosmos Club; Wright often heard him refer to Greely as "That old pussy-footed SOB who led only one expedition and ate that!"

After a year of living by himself in rented rooms, Wright met Paul Popenoe, editor of the *Journal of Heredity*. They became friends and rented an apartment together for several years. Although he was the editor of the *Journal of Heredity*, Popenoe was not a geneticist but was keenly interested in promoting animal and plant improvement, including eugenics. The *Journal of Heredity* was formerly the *American Breeders Magazine*, official organ of the American Breeders Association, to which many practical breeders as well as geneticists like East and Castle belonged. President of the American Breeders Association in 1913 was David Fairchild, who had an important position with the exotic title of agricultural explorer in charge of foreign seed and plant introduction, Bureau of Plant Industry, USDA. Paul Popenoe had earlier worked for Fairchild's office, and his brother Wilson Popenoe continued to work there for many years. Fairchild's brother-in-law was Gilbert Grosvenor (both married daughters of Alexander Graham Bell), who had enjoyed enormous success by publishing *National Geographic Magazine*. Hoping to enjoy similar success, Fairchild encouraged the American Breeders Association (which at the urging of geneticists became the American Genetic Association) to change the name and format of its journal, becoming the *Journal of Heredity* in 1913. Fairchild then of course used the *Journal of Heredity* as a means to popularize the exotic plants his group was bringing into the country, complete with many photographs.

The *Journal of Heredity* was directed to a more popular audience than just geneticists. In addition to popular articles, however, were brief but serious articles on genetics. At Popenoe's suggestion, Wright published there an important series of eleven papers on color inheritance in mammals during 1917–18, and many later papers and reviews. When Wright first came to Washington, one of the books that came in for review was the first (1915) edition of the famous *Mechanism of Mendelian Heredity* by Morgan, Muller, Sturtevant, and Bridges. This sort of detailed genetic analysis seemed like nonsense to most breeders (and to many famous biologists, including F. B. Sumner and William Morton Wheeler), and the authors expected a superficial and negative review from the *Journal of Heredity*. Popenoe thought the book was almost useless, but Wright had a series of talks with him about it and convinced him that it represented major advances in genetics. Popenoe wrote an accurate and fair review, which Wright said evoked astonishment from the Morgan group, especially Sturtevant.

In his earlier work with Fairchild, Paul Popenoe had traveled frequently to the Middle East collecting date palms for cultivation in southern California and was largely responsible for the rise of the date industry there. Wright was friends with both Popenoes. Wilson mostly was in Central America working on the introduction of the avocado, but he stayed in the shared apartment on visits to Washington. The Popenoe brothers both ended up in the war effort, and after the war Paul founded the Institute of Family Relations. After the war, Wilson went back to the Office of Foreign Seed and Plant Introduction. Both brothers corresponded with Wright sporadically for over thirty years, and their families were friends.

Wright, along with the Popenoes and most other able-bodied young men in 1916, went to enlist; but Wright was rejected as underweight. He was not surprised because earlier in 1916 the New England Mutual Company had rejected his application for life insurance due to his history of pleurisy on the railroading expedition and a doctor's report of tuberculosis. After rejection for enlistment, Wright tried to gain weight and finally did, but the war effort was already winding down.

While in Washington Wright met the cytologist John Belling. He was brilliant but mentally unstable. Born in 1866 in England, he came to the Florida Research Station during the years 1907–15. By this time he had developed an extraordinary ability to prepare stained slides of chromosomes. In 1915 he suffered from severe depression and was hospitalized in Florida. Paul Popenoe brought him to Washington about 1918 and introduced him to Wright, but soon Popenoe had to put Belling into a mental institution in Chevy Chase, Maryland. Popenoe then left for the war, leaving Wright to see after Belling. Wright visited him frequently and remembers that the noise level was high at the hospital; Belling's favorite comment was, "Let's take a walk to get away from the other lunatics." Belling improved, and for a time he held a job caring for a large flock of chickens. When Belling was fired for odd behavior, Wright did not know what to do next.

By this time Wright's parents had moved to the Washington area. His father was working for the Tariff Commission and later the Brookings Institution. They lived in Forest Glen, Maryland, in a home formerly owned by three of his aunts on his mother's side. Wright's parents suggested that Belling come to stay with them, and they conversed on very diverse topics, from genetics to animal behavior to poetry. Wright's Aunt Hannah, who had a Ph.D. in economics, was raising bees. Belling helped her with the bees, and they got along well. Belling was also a friendly but penetrating critic for some of Wright's biological ideas during this time. Then, to Wright's astonishment, Belling and Hannah decided to marry. Belling got a job with A. F. Blakeslee at the Station for Experimental Evolution at Cold Spring Harbor. They published jointly a stunning series of papers tying together the cytology and genetics of *Datura* (jimsonweed). Unfortunately, in 1927, Belling again became depressed and had to be hospitalized. Hannah died. Belling, who by this time thought Blakeslee was exploiting him, got a job at the University of California, Berkeley, with E. B. Babcock. There he entered another very productive period until his death in 1933 (on Belling see Babcock 1933; there is also much about Belling and Aunt Hannah in the Quincy Wright papers, Regenstein Library, University of Chicago).

A more stable but no less interesting friend was Henry A. Wallace, son of Henry C. Wallace, secretary of agriculture in 1921 in the Harding administration. Henry, Jr., was managing the family business, the *Wallace Farmer,* in Des Moines, Iowa, and he came to Washington frequently to consult with his father. He had a good background in animal breeding and quantitative methods, even publishing a joint paper with the statistician G. W. Snedecor of Iowa State College in Ames, Iowa. Wallace was of course interested in Wright's quantitative theory of inbreeding and the method of path coefficients, and he frequently dropped by Wright's office. They corresponded occasionally, mostly about issues related to path coefficients. Wallace's help was decisive, as I shall discuss in the next chapter, in arranging to have Wright's long paper on corn and hog correlations published as a USDA Bulletin.

Because of the isolation from other geneticists in the Washington area, Wright attended every annual meeting of the American Association for the Advancement of Science and always read at least a short paper. At these meetings he maintained contact with genetic researchers; he roomed with Sturtevant one of the first years in Washington, for example. In 1922 the meeting was in Toronto and attracted an illustrious group of geneticists, including Bateson, Fisher, Morgan, Emerson, Castle, East, and many others. Students of course flocked to the meeting.

Figure 4.1. Wright at Cornell University, 1922. Wright is second from left on front row. Other notables in the photograph include R. A. Emerson to Wright's left, William Bateson to the left of Emerson, and Milislav Demerec behind Wright and Emerson. Archives of the Department of Plant Breeding, Cornell University.

Wright had a long discussion at the meeting with two graduate students working with Emerson at Cornell—E. G. Anderson and Milislav Demerec. Both were working in corn genetics. The three discussed the problem of variegation in corn and were considering the possibility of segregation of a highly composite unit. Wright had already at this time developed methods for analyzing a similar problem with inbreeding, and Anderson and Demerec urged him to extend the quantitative analysis to their theory. Wright decided to go with them to Ithaca, on his way back to Washington, for further discussion.

Wright worked out the required analysis for Anderson and Demerec and found that their theory did not fit the data. In the light of the later investigations of Barbara McClintock and those stimulated by her work on the causes of variegation in corn, it is not surprising that the theory of Anderson and Demerec failed. It is, however, quite certain that Demerec and Anderson were greatly impressed by Wright's ability to quantify genetic questions. Later, in 1936, when Anderson was at California Institute of Technology with Dobzhansky and Sturtevant, and Demerec was director of the Station for Experimental Evolution at Cold Spring Harbor, all would work together to get Wright to Cal Tech for planning the Genetics of Natural Populations series— a project badly in need of Wright's abilities (see chapters 10 and 11).

Bateson, who wished to see Emerson's work at Cornell, also rode back on the train with them. At breakfast at a hotel in Ithaca, Wright was explaining a ratio from his guinea pig results in which the deviation from an expected ratio was six times the probable error. Bateson was by now fed up with Wright's quantitative analysis and bitterly exploded, condemning the use of biometricianlike statistics in genetics. This outburst, so typical of Bateson, was particularly ironic in Wright's case. What Bateson could not possibly have known was that Wright had just developed his method of path coefficients precisely for the purpose of going beyond the mere correlation and regression of the biometricians to a quantitative analysis of actual causal relations. (There is a philosophical problem here that I will address in the section on path coefficients in the next chapter.)

Wright also attended the annual meetings of the Society of Animal Production, held in conjunction with the International Livestock Show, as a part of his job at the Bureau of Animal Industry. There he came to know many of the animal husbandmen at the state agricultural experiment stations all over the United States. He was well aware of the state of animal breeding. In the early years he spent much time at these meetings with John Detlefsen, his precursor at the Bussey. Detlefsen was a good animal geneticist and an inspiring teacher but utterly devoid of tact. He ridiculed research projects of others and offended nearly everyone at Illinois. Wright recalls one meeting at which Eugene Davenport, the retiring dean of the College of Agriculture at the University of Illinois and author of a widely used textbook on principles of breeding, said publicly that one of the three things he must do before retirement was fire Detlefsen. When Davenport sat down, the deans of agriculture from the universities of Pennsylvania, Iowa, and California stood up and congratulated Davenport on firing Detlefsen and stated that general agriculture had no need for young Harvard Ph.D.'s. What Wright saw on this occasion and very many others was the strong tendency of animal breeders at that time to dismiss theoretical genetics as well as quantitative theories of breeding. This situation would change drastically in the succeeding decades, in part from Wright's influence. Indeed, at the present time I would say that Wright is admired, or perhaps revered is the better word, more by animal breeders than by geneticists or evolutionists. I will in this and later chapters trace some of the changes in animal breeding theory and Wright's influence in them.

One benefit of these annual meetings was that Wright saw first hand the prize animals and discussed them with Detlefsen or colleagues from the Animal Husbandry Division. Another person with whom Wright spent considerable time at the meetings was E. N. Wentworth, geneticist and animal breeder who for a time had been a student at the Bussey during Wright's tenure there. Wentworth was a professor of animal breeding at Kansas State University in Manhattan, Kansas, before World War I. After the war he became a public relations expert for Armour and Company, using his ties with the livestock breeders. Wright was shy, and Wentworth went out of his way to introduce Wright to the breeders and anyone else of interest. One curious assessment Wright made of these meetings is that he never did understand well the basis for judging the show animals and could not predict winners with a high degree of accuracy. Wright attended one other notable meeting during his stay at the USDA. In September of 1921 the Second International Congress of Eugenics was held in New York City. The German geneticists were not invited because of postwar resentment but geneticists and eugenicists came from everywhere else. It was a huge international meeting, and Wright had the opportunity to meet many persons, such as Lucien Cuénot, a pioneer in mouse genetics and the physiology of pigment production, and to renew older acquaintances in the field of genetics. Wright read a paper on the results of the USDA experiment on inbreeding in guinea pigs and shunned the eugenical side of the meeting along with Sturtevant, Weinstein, Bridges, and Muller.

These meetings, combined with the correspondence generated by his published papers and friendships in the field of genetics, enabled Wright to escape in large part the fate East had predicted for him. The isolation, in my assessment, fit well with Wright's independent attitude of mind and in no substantive way hindered his development as a research geneticist. Indeed, Wright would be more cut off from the genetics research community as a professor at the University of Chicago than he was in the position at the USDA.

Marriage

Each year the USDA granted Wright thirty days of leave. In the summer of 1920 he spent the entire leave at the Station for Experimental Evolution at Cold Spring Harbor. At the same time Little came from the University of Maine, where he was then president, and they decided to try a joint experiment on transplantation of skin grafts in guinea pigs in an effort to learn more about the developmental genetics of coat colors. The skin grafts were awkward and rather grisly, but Wright and Little finally successfully completed one graft. MacDowell's little dog, not supposed to be in the area, promptly seized the recipient guinea pig and shook it to death. The experiment was abandoned.

The leave period was rewarding in another way. At dinner Wright always sat with his friends Phineas and Anna Rachel Whiting and a pleasant young woman named Louise Williams, who had a few years earlier taken her master's degree at Denison University with Harold Fish, Wright's former labmate at the Bussey Institution. Castle had recommended Fish for the Denison job, which Castle himself once held. Although Fish never finished his doctorate, he was a wonderful teacher, and many of his students later took advanced degrees. Louise Williams had been an instructor at Smith College since taking her degree with Fish and had come to Cold Spring Harbor with the responsibility of caring for Fish's experimental rabbits. Wright did not talk easily with strangers, but the shared background and Williams's congenial manner were conducive to friendship. Wright described the time with Williams at Cold Spring Harbor in his autobiographical notes.

> I could talk with her easily and got to like her very much. She was somewhat lame from a congenital dislocation of her left hip but this did not interfere with her walking for miles except that she was somewhat slow. She had an attractive gaiety and a good sense of humor, and was sympathetic with anyone in trouble. She and two other girls who were former students of Fish were staying at the Fish cottage. On visiting Fish the first evening I found that he had come down with a severe attack of sciatica and was about to be taken to the hospital in Huntington, L.I. One of the little Fishes began howling as his father was being put in the ambulance and I was struck by the way Louise immediately began reassuring him and soon had him in good humor.
>
> She had been helping Fish with the rabbit colony that he had brought from Denison in projects that he had started with Castle at the Bussey Institution. The animal caretaker who had cleaned the cages had just left; the first thing Louise said to me at dinner was how badly she had been disappointed in me. She had seen a strange man come and supposed it was the new animal caretaker that they expected and he turned out to be just another geneticist. I said that it was obviously up to me to remedy this and helped her with the cages the next day. We often went walking after dinner. On one such walk we collected a lot of goldenrods and cardinal flowers for the annual picnic at the sandspit which was to come the next day. We were invited together to dinner by the Director, Dr. Blakeslee. I was so much in love with her at the end of my vacation and so dubious about merely starting a correspondence that I proposed on the last evening. She demurred at first because of the shortness of our acquaintance and because I had not met her parents or she mine but finally agreed to consider us engaged. (Autobiographical notes, pp. 60–61)

During the fall Wright visited the Williams family at their home in Granville, Ohio. Louise's father was a farmer; both he and her mother traced their heritage back to mid-seventeenth-century England (his to Wales, hers to Hertfordshire). Louise came to visit Wright's parents in Maryland the same fall. They had a wonderful spring visit together at Smith. They were married in Granville on September 10, 1921, with Quincy Wright the best man, and moved into an apartment above the Cornell Restaurant on Linworth Place, S.W., near Wright's Washington office. Two years later they moved into a house about four miles closer to the experimental farm in Beltsville, some nine miles north of the USDA offices. The Wrights had two boys, Richard and Robert, during the remaining four years in Washington.

Figure 4.2. Louise Wright soon after marriage to Sewall.

Division of Responsibilities at the USDA

Wright's primary responsibility at the USDA was the continuation of the in-breeding experiments in guinea pigs and the analysis of the data collected since 1906, with an eye toward the implications for domestic animal breeding. Wright in general spent about half his time at the experimental farm in Beltsville with the guinea pigs (his stocks from Castle's laboratory included) and the other half at his office at the USDA complex in Washington. There he was responsible for answering the many general genetical questions sent to the USDA by amateur and professional breeders and for whatever special jobs were assigned to him by Rommel.

One of the first special jobs resulted from prompt and drastic action taken by Dr. J. R. Mohler, chief of the Bureau of Animal Industry. At a show of prize dairy cattle in Chicago, the dreaded hoof-and-mouth disease appeared in one of the herds. Mohler immediately ordered every exposed cow and bull destroyed and buried in quick-limed trenches. Many of these animals were very valuable, and since Mohler agreed to reimburse the owners for damages suffered, the lost cattle immediately became, in the eyes of the owners, the finest of the breed. Rommel assigned Wright the task of tracking down the pedigree of each animal so that an estimate of its value could be made realistically. This practice of tracking pedigrees became very important to Wright's research interests in 1923–26, when he was investigating in detail the history of the Shorthorn breed of cattle in an effort to elucidate how they had evolved over time.

A second job assigned to Wright as part of the war effort was to make quantitative estimates of various crops in relation to the production of animal products. In one of his files from the USDA days, I found meticulously prepared charts showing, by area of the country, production figures for winter and spring wheat, hay, peanuts, corn, oats, barley, rye, and other crops, probably obtained from the Bureau of Plant Industry. Then in the file came a series of detailed reports, complete with hand-colored charts, all prepared by Wright on such topics as "Ratio of Exports of Beef Products to Total Production of Beef and Veal in the United States," "Veal Slaughter as Percent of Total Slaughter," "Percent of Cattle Slaughtered under Federal Inspection," "Live Stock Estimates Since 1890," "The Lamb Crop," "The Pig Crop," "The Calf Crop," "Production and Consumption of Meat in the United States," "Monthly Variations in the Hog Population," and "Estimate for Hog Production in 1919." These reports seem so unrelated to any of Wright's substantive research interests that they could be attributed simply as his contribution to the war effort. This evaluation would be wrong.

With Rommel's encouragement, Wright had rented a card-sorting machine, and he was anxious to keep the machine in use to justify its rental. He thus became involved in many statistical projects with other members of the staff. After the flurry of reports described above, Wright decided that the machine was controlling his professional life too much, and he gave it up. While preparing these reports, however, Wright became interested in the relationship between corn production and hog production. He had just developed his method of path coefficients for other purposes and decided to apply it to his corn and hog data. The result was his substantial pamphlet on corn and hog correlations, using the method of path coefficients, which I will address in the next chapter.

At the Washington office, almost every letter with general questions about animal breeding appeared eventually on Wright's desk for his answer. These letters—from the sophisticated to the ridiculous—received careful, detailed responses from Wright. The topics most often addressed were the effects of inbreeding, inheritance of acquired characters, telegony, sex ratio, sex determination, freemartins, and color inheritance. For example, James Hunter of Dallas, Texas, wrote to give examples of telegony (a female "contaminated" by a previous fertilization in an earlier pregnancy) and inquire about them. Wright answered:

We have noted with interest the cases which you cite. A considerable number of such cases have been reported but the theory is now generally considered to be thoroughly discredited. The mechanism of heredity is so complex that we can seldom know enough about the parents and the more remote ancestors to predict with certainty the results of a cross, even within a pure breed. When unexpected characteristics appear, it is only too easy to refer them to telegony, the influence of maternal impressions, the inheritance of acquired characters or some such theory. The only method by which the science of breeding can be built on a secure foundation is through the acceptance of only such principles as stand the test of carefully planned experiments. A considerable number of such experiments have been made in order to test the theory of telegony but only negative results have been obtained.

Wright then referred Hunter to the pertinent literature. A Mr. Hall wrote to ask if the existence of blind fish found in caves proved the theory of the inheritance of acquired characters. To this Wright replied:

The case of the blind fish found in caves is one which has been discussed a good deal in connection with the inheritance of acquired characters. Unfortunately, as in most other apparent cases of such inheritance, there is more than one possible explanation. We know that hereditary variations are occurring from time to time, apparently by chance, i.e., without any obvious relation to the experiences of the parents. Blind variations of fish would ordinarily be at a great disadvantage, but this would not be the case with those which wandered into caves. Normal fish which wandered into caves would probably soon be attracted out. Thus the prevalence of blind fish in caves may be due merely to selection of the most suitable habitat by the normal and blind variations, respectively. It is also true, however, that chance variations are more frequently injurious than beneficial, explicable on the principle that anything done at random to a complicated mechanism is more likely to injure than improve it. Such harmful variations are generally weeded out by natural selection, but if normal fish were isolated in caves, injurious variations to unused organs such as the eyes would tend to accumulate. Thus a tendency to degeneration of any unused organ probably takes place without any real inheritance of acquired characters.

However, I would not wish to make a dogmatic statement that there is no inheritance of acquired characters in the long run. It is simply that the negative experimental evidence indicates that it is at least very unusual or very slight while the cases in nature which suggest such inheritance can, in most cases at least, be explained in other ways. Until there is thoroughly authenticated positive experimental evidence it seems best to avoid using it as an explanation. (J. K. Hall to Wright, January 11, 1924)

These two letters typify the careful, thoughtful response Wright gave to inquiries. Occasionally letters came from highly professional breeders, but mostly from those such as Hall and Hunter. Some of the inquiries were delightful. The secretary of the Illinois Vigilance Association, according to its letterhead "organized for the purpose of suppressing the traffic in women and girls and the conditions which made that traffic possible," wrote to ask the pertinent question: Was there scientific evidence to support the association's belief that benefits resulted from the restraint of the sex instinct? Wright said in reply that

> I know of no investigations on the effect of infrequent breeding of domestic or wild animals. There are generally believed to be no ill effects. A very important investigation by Orren Lloyd-Jones and F. A. Hays on the effects of too frequent breeding of rabbits will appear in the *Journal of Experimental Zoology* on April 5. Temporary sterility of the male is produced but there seem to be no effects on the young in any case.

The secretary seemed pleased with Wright's response.

> I wish to thank you most cordially for your very kind letter and for the information which you give. It is important and helpful, and I am very much obliged to you for it. If at any time you should have any additional information showing the benefits of the restraint of the sex instinct, either in animals or human beings, I shall appreciate very much hearing from you in regard to it.

One person who took advantage of Wright's detailed responses was Albert E. Wiggam, popular writer and lecturer on eugenics. Wiggam gave hundreds of lectures promoting eugenics in the late 1910s (see the Davenport-Wiggam correspondence for this period), but knew very little genetics. As he was preparing to write his popular books *The New Decalogue of Science* (published in 1923) and *The Fruit of the Family Tree* (published in 1924), Wiggam wrote to Wright for information about the genetics of human race crossing, particularly regarding color inheritance. Wright's answer was so helpful and detailed that Wiggam immediately wrote back and asked more questions, suggesting that he pay Wright for his obvious efforts and expertise. Wright was not opposed to the general idea of eugenics but was unimpressed in general by the work actually done in eugenics and absolutely refused to take any time from his research to promote eugenics. Wright replied to Wiggam: "I think that I ought to make clear that I can not go into your project to any greater extent than would come under the head of my regular work in handling correspondence on genetic questions and of course I can not receive any compensation for it from other sources than the Government." To the genetic questions, Wright replied with four additional typewritten pages (devoted mostly to elementary facts about cell division).

In addition to the letters that came to him because of his position at the USDA was his professional correspondence from other geneticists. John

Detlefsen wrote on many occasions asking Wright to classify the probable genotypes of guinea pigs or mice from skins or hair (the clipped hair from one mouse was still attached to the carbon of one reply to Detlefsen). Dunn wrote from his position at the Connecticut Agricultural Experiment Station at Storrs to ask Wright's help in elucidating a complex problem in the inheritance of color in poultry—Wright supplied a crucial hypothesis borne out by Dunn's subsequent research. Phineas Whiting wrote with an interesting query:

> I am enclosing a few hairs plucked from the scalp of our Greek professor. He seems to be a sort of sepia maltese variety and asserts that his father has the same kind of hair. There is a concentration of pigment in patches, in some of the hairs more than in others. If you will look at them with microscope you may find them interesting. I should like to know if this condition is at all common in humans and if it would be worthwhile looking it up. There are several relatives that might be investigated.

The carbon of Wright's reply was not in the file.

Wright told me many times that the ten years in Washington, in retrospect, were not as burdensome with correspondence as happened in later years. Certainly the correspondence files bear this out. I counted letters from twenty-one prominent geneticists in the 1915–25 file, but there are more than twice as many letters in the file for 1932–33 than in the one covering 1915–25; the questions are more difficult, and the answers more time consuming. The only way to really understand the totality of Wright's influence as a scientist is to read through his correspondence files. He formed the habit, at least by the time of the Washington years, of answering every letter fully and technically and reviewing every manuscript sent to him likewise. Few scientists or scholars of any kind are willing to take time away from their own research to prepare detailed answers to questions sent by students or unknown persons, even if they answer the questions of other scholars this fully. Wright often took a long time, up to a month or six weeks, to answer a complex question. But he always did answer. Wright's response to my query about why he spent so much time and care on his correspondence was simply that he felt duty bound to provide such answers. Wright exerted considerable influence and contributed much original insight through his correspondence.

Research

The ten years Wright spent in Washington were highly productive for his research. His most important contributions to genetics and evolutionary theory were all initiated during this time. His research efforts had six major thrusts: (1) color inheritance in mammals, especially the guinea pig; (2) physiological genetics, primarily focused upon color inheritance but also including developmental anomalies, hereditary resistance to disease, and transplantation; (3)

development of the method of path coefficients and its application to diverse phenomena; (4) effects of inbreeding and cross-breeding in guinea pigs and other animals; (5) Mendelian analysis of pure breeds of livestock; and (6) theory of evolution in nature. The remainder of this chapter will examine the first two of these research areas; the third through the fifth form the substance of the next chapter; and I address the sixth in chapter 9.

Color Inheritance in Mammals and Physiological Genetics

Soon after he took an apartment with Paul Popenoe, editor of the *Journal of Heredity*, Wright decided to publish there a summary article on his views of color inheritance, complete with his provisional theory of the physiological basis of pigment production. This article would be followed by others directed to color inheritance in specific animals. The summary article, entitled "Color Inheritance in Mammals" appeared in the May 1917 issue (Wright 8); the subtitle was more descriptive: "Results of Experimental Breeding Can Be Linked up with Chemical Researches on Pigments-Coat Colors of all Mammals Classified as Due to Variations in Action of Two Enzymes." Wright presented here in more detail than in his thesis a review of the biochemistry of melanin and his provisional hypothesis of the physiology of pigment production, complete with diagrams to help the reader understand the interrelations.

Throughout his career Wright would use charts, graphs, and especially diagrams to convey in the briefest way the maximum understanding of his data or ideas. Wright's early diagrams, as the ones reproduced here from his paper, were simple and easily understood. His diagrams concerning color inheritance became more complex as Wright learned more about it, but these earlier diagrams are almost clear by themselves. The first represents the basic idea that two different postulated enzymes, in different combinations and concentrations, act upon chromogen to produce different pigment colors and intensities (fig. 4.3; reproduced from Wright 8, p. 228). The second shows which colors were produced by which combinations and concentrations of the enzymes (fig. 4.4; reproduced from Wright 8, p. 230).

Following a listing and discussion of the various color factors in mammals, Wright said in conclusion that he had tried to provide a genetic and biochemical scheme "to relate the findings of the biochemist in regard to melanin pigment with the great mass of curious relations between colors which have come to light in genetic work" (p. 235). Of course, as knowledge of the actual biochemistry of melanin production proceeded in the twentieth century, Wright would have to change the details of his scheme accordingly. For comparison, one might turn to Wright's paper on genic interaction published in 1963 (Wright 170); I will trace this development chronologically in chapter 6. Wright ended by suggesting a program of research that urged geneticists to take the existing color factors and examine their effects in all combinations with each other. The resulting data should "give a very much

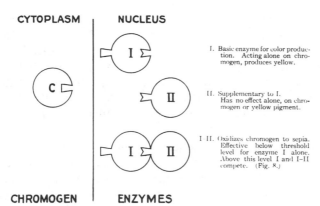

I. Basic enzyme for color production. Acting alone on chromogen, produces yellow.

II. Supplementary to I. Has no effect alone, on chromogen or yellow pigment.

I-II. Oxidizes chromogen to sepia. Effective below threshold level for enzyme I alone. Above this level I and I-II compete. (Fig. 8.)

Figure 4.3. Pigment production.

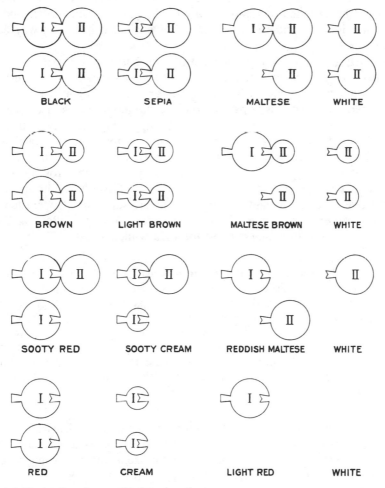

Figure 4.4. Production of coat color in mammals.

more complete understanding of the heredity of color than we have at present. By constant comparison of the deductions of such work with the findings of the biochemist, it should be possible in the end to establish a very pretty correlation of results" (Wright 8, p. 235). No one took this proposed program of research in color inheritance more seriously or pursued it more actively than Wright, who between 1912 and 1955 bred some 10,000 different combinations of the eleven known factors affecting color inheritance in guinea pigs.

The following ten papers in the series, in light of the structure provided in the first, examined color inheritance in mice, rats, rabbits, guinea pigs, cattle, horses, swine, dogs, cats, and humans (Wright 9–18). Wright's experimental expertise was limited to the first four of these, but his general scheme applied easily to all. Each paper succinctly summarized the known color factors and their interactions in each kind of animal and became the standard reference in the genetics literature. Many of the papers went further and proposed additional hypotheses to be tested or evaluated in a new way earlier hypotheses.

In "III: The Rat," Wright spent well over half of the paper making the most penetrating published criticism of Castle's interpretation of the results of his selection experiment on hooded rats. Wright had suggested the crucial final experiment, but Castle had interpreted the initial results of crossing the plus strain with the wild rats as strongly supporting his view rather than Wright's. In 1917, before experimental evidence for Wright's position was available from the cross of the minus strain with wild rats, Wright did not accept the apparent support for Castle's position given by the cross of the plus strain with wild. Instead, Wright subjected all of Castle's results so far to a quantitative assessment and found the analysis supported the existence of modifiers rather than changed Mendelian factors. Wright, following Castle, ended the paper with strong support for a Darwinian view:

> Finally, under any interpretation, Castle's selection experiment demonstrates the efficacy of Darwinian selection. It is true that one large mutation occurred with effects perhaps as large by itself as the entire plus selection series, but where such a variation gives one new level, selection has produced a continuous series of stable levels. This would give selection of small variations a more important place in evolution and animal husbandry where it is nice adjustments of one character to another or to the environment that count. (Wright 10, p. 430)

This conclusion is of interest for Wright's strong support of Darwinism, but it also clearly emphasizes the process of mass selection as in Castle's experiment, a process Wright would soon de-emphasize as the dominant process in evolution in nature at levels below that of the species.

In "VI: Cattle," Wright evaluated the theories that four, two, or one factors are necessary to explain the red-roan-white variations. The two-factor theory had been proposed by his friend Wentworth. Wright, using the Hardy-Weinberg equilibrium principle (Wright had never heard of either Hardy or Weinberg at this time) and an extension of it to multiple loci, calculated the expected proportions of each genetic formula in the population and thus the proportion of each color of the offspring in any one generation. He then compared the theoretical results with observed results and found that one factor segregating for two alleles without dominance was the hypothesis that most nearly fit. I know of no earlier attempt to use the Hardy-Weinberg equilibrium principle to discriminate between genetical hypotheses in a population; this was a technique that would later become very popular when applied to natural populations as well as to domestic.

Except for the introductory paper, the longest of the series was the last, "XI: Man." This paper is notable because variations in human eye, hair, and skin color had been intensively studied by Davenport and others, with the result being confusion rather than order. Using analogies with his albino series in guinea pigs and explicit reference to his hypothesis for the physiology of pigment production, Wright was able to suggest ways for making sense of the confusing data. He assessed the problem of color inheritance in humans to be in the same state in 1912 as color inheritance in guinea pigs. This paper, as well as many of the others in the series, became the new starting point for the analysis of color inheritance in mammals.

Because Wright directed his primary attention in the first five or six years in Washington to analysis of the inbreeding experiment in guinea pigs and pursuing the implications of his method of path coefficients, he published little of his ongoing research with the color factors in guinea pigs. But the research itself continued. In 1922–23 Wright discovered two new important color factors in the guinea pig. When he first came to Washington many families of the inbred stocks exhibited a dilution factor of the albino series of alleles. These dilutes were darker than the dark-eyed dilutes from Castle's stocks and denoted in Wright's classification as c^d, but because Castle had selected for many years for extreme dilution, Wright assumed that the dilution factor c^d also caused the dark-eyed dilutes of the Washington inbred stocks. Wright expected the differences in appearance of the two dilute forms to blend when they were crossed because of the background genetic heritage. To his surprise, the two dilute series remained wholly distinct in crosses, segregating out with their usual differences, and the breeding tests and records quickly revealed that Wright had found a new member of the albino series of alleles between the intense factor C and the dilution factor c^d. Wright named this fifth allele c^k. This new allele in the albino series was one of several such discoveries made about the same time. Castle had found a fourth allele in the albino series in rabbits, and multiple alleles of albinism had been discovered in both rats and mice.

Spurred by this success, Wright began testing all the other dilutes in the Washington stocks to see if yet another allele in the albino series could be identified. All other dilutes proved identical with c^k, except for one. This one could easily have visually passed for a member of the albino series, but in fact the breeding records and additional matings proved conclusively that an albino could transmit the intensity allele of this dilution factor to its offspring. This intensity factor could not possibly be the always dominant intense factor C (for otherwise the albino parent would not have been albino). It had to be a new independent color factor. Named F for the intense allele and f for the dilution factor, Wright found that the homozygous recessive ff exhibited some unusual characteristics. It appeared at first to intensify the dark colors at the same time that it diluted the red, which Wright explained on his hypothetical model as a release of the black-producing process from competition with the process for producing red. (By 1925 Wright discovered that the intensification of dark colors was very limited and retracted this interpretation.) Furthermore, the dilute yellow faded with age to a cream or in some cases even to white. This fading contrasted sharply with the dilutions of the albino series of alleles on yellows, whose colors were stable for the lifetime of the animal. No color factor of this fading sort had been reported in any mammal, and Wright's delight in having found it showed clearly in the published report in *American Naturalist*. Analysis of the dilution factor f assumed a prominent role over the years as Wright continually deepened his analysis of the physiological genetics of color inheritance in guinea pigs.

As soon as Wright discovered the new albino allele c^k, he set about making the rest of all possible combinations of the five alleles of the albino series. The results of these investigations were published in 1925 in a long paper (thirty-seven pages) in *Genetics* (Wright 43). Wright took the results and put them into a table showing the dilution effects of the various combinations separably upon the intensity of black pigmentation, yellow pigmentation, and eye color (fig. 4.5; reproduced from Wright 43, p.236). From the table it is obvious that a given genotypic compound can affect the dilution of black and yellow differently ($c^r c^r$ dilutes black very little but causes complete dilution of yellow into white) and that the different genotypic compounds cannot be arranged in terms of effects upon colors into a linear order of intensity. Thus homozygous $c^r c^r$ is much darker than homozygous $c^d c^d$, yet one might wish to rank c^d as being closer to C because $c^d c^d$ dilutes yellows so much less than $c^r c^r$.

In trying to work out a physiological scheme in his thesis and in the first paper of the color inheritance in mammals series (1917), Wright had thought there were two basic choices. One could suppose that the different alleles of the albino series affected the production of black and yellow in different ways or that the alleles of the series affected some more basic precursor of the color production process; but other hypotheses would then be required to explain the nonlinearity of results on black and yellow when the effects of the alleles upon the precursor should be quantitatively linear. Wright in his thesis and in 1917 opted for the second choice for two reasons. The first is historical. Castle and his students, Little in particular, had always thought of al-

	BLACK	YELLOW	EYE COLOR
$C-$	14.0	10.6	Black
$c^k c^k$	13.1	7.1	Black
$c^k c^d$	12.4	7.2	Black
$c^k c^r$	13.5	4.6	Black
$c^k c^a$	11.5	4.6	Black
$c^d c^d$	9.9	7.0	Black
$c^d c^r$	12.1	4.1	Black
$c^d c^a$	7.0	4.2	Black
$c^r c^r$	13.1	0	Dark red
$c^r c^a$	8.5	0	Light red
$c^a c^a$	0	0	Pink (no pigment)

Figure 4.5. Dilution effects.

binism as affecting just one basic process in the production of any color at all. Thus one would naturally tend to view alleles in the albino series as having their biological functions in the form of quantitative lessening of the extreme effects of albinism. Second, to propose differential effects upon black and yellow pigments to the basic physiological action of the alleles in the albino series involved a multiplication of hypotheses. Of course, explanation of the nonlinear effects of the alleles in the series upon black and yellow also required subsidiary hypotheses. To Wright, these subsidiary hypotheses were less fundamental than the more basic added hypothesis of differential effects of the alleles upon black and yellow.

By 1925 a third reason had come into play. Wright had now for ten years invested much thought and time into his provisional hypothesis of 1915 and had published it in detail on two occasions. In the absence of any compelling evidence, few scientists would voluntarily change their minds under similar circumstances. Account must also be taken of Wright's general approach to scientific research. Wright came into genetics at a time, unlike Castle, when many hypotheses could be amply justified by experimental evidence. Many geneticists (Sturtevant, for example) almost refused to publish unless they felt they had settled a question for good. William Castle had a rather different view—he tended to publish interpretations that he knew might well be disproved in the near future, but science would be advanced in the process. Wright had seen in person some of Castle's reversals and was aware of many others. Sturtevant and some others of Morgan's group unquestionably considered Castle's constant espousal of unproved hypotheses and subsequent retractions as an embarrassment to the profession. During one of Castle's public controversies (in which Castle proved the victor) involving an argument with his former teacher Davenport and fellow student Jennings, the latter wrote to the former: "I don't know of anyone that approaches him in the number of embittered controversies he has had, in which he ultimately admits he was wrong" (Jennings to Davenport, 1930). Wright simply did not wish to emulate Castle's approach to science, despite Wright's generally high estimation of Castle. Wright was far more careful than Castle in the publication of his scientific ideas and consequently had less reason to retract them. Once Wright had published an idea, he, like most scientists, was reluctant to change it unless the evidence was clear.

From the absence of linearity in the table in 1925, Wright saw that it would

> be possible to interpret albinism either as affecting a single process fundamental to any melanin production or as affecting the production of black and yellow for different reasons. If the latter be assumed, with the further assumption of polymorphic variation of the gene, the problem raised by the lack of agreement in the order of intensity in black and yellow parts of the fur ceases to exist. Under such conditions there need not be any correlation at all. (Wright 43, p. 239)

But this assumption Wright described as "rather arbitrary, . . . especially in connection with the imperfect correlation between the effects in black pigmentation in the eye and in the fur, and in view of the parallelism between the albino series in different mammals" (p. 239). He therefore assumed provisionally "that the albino series has only one primary type of effect on pigmentation and that the irregularities in the effects on different kinds of pigment on different parts of the body are due to developmental processes subsequent to the effect of the albino factors" (p. 239).On this assumption, Wright then argued that the data he had collected showed that the two genes in the zygote affected pigment production independently rather than the immediate gene products interacting together before affecting the pigmentation process. And he reiterated his previous hypothesis about threshold effects in the differential between production of black and yellow pigmentation and about the role played by competition between enzymes.

Wright was careful in the summary to distinguish clearly between "what is fact and what is speculation." It was a fact, he said, that there were at least five alleles in the albino series in guinea pigs, and the recorded data about colors exhibited by different combinations of the alleles was equally factual. His theory of the mechanisms by which genes controlled the production of pigmentation, and in particular the relations between the production of black and yellow, were frankly speculative. Over the years Wright would reluctantly abandon substantive parts of his speculative theory while retaining others.

The next obvious extension of Wright's work on color inheritance was to the effects of combinations of genetic factors beyond those in the albino series alone. He had of course produced much data on this question during the

Washington years, although the work was not published until 1927. For this paper, Wright produced various combinations of the four series C (albino series, 5 alleles), F (the new dilution series, 2 alleles), P (pink eye series, 2 alleles), and B (black and brown, the brown series, 2 alleles). These combinations were modified in terms of the distribution of brown and yellow pigments by the three additional series S (spotting or piebald, 2 alleles + ΣS, unknown modifiers), E (extension, 3 alleles + ΣE, unknown modifiers), and A (agouti, 3 alleles). Wright carefully measured, with the Milton Bradley color wheel, the percentage compositions of the coat colors in terms of white, yellow, orange, and black. This paper therefore represents the most sophisticated and precise analysis Wright had yet published about the interactions of the various color factors. The data unmistakably showed that the albino series of zygotic combinations fell into different orders of intensities when in different combinations with the other color factors. The interaction effects of gene products were fundamental to any understanding of coat color production.

The long discussion section provided the most recent summary of research on the biochemistry of melanin production. Wright's comment upon this survey of the literature is worth quoting because it represents his basic view of gene action in 1927:

> On the whole, the various lines of evidence seem to be converging toward a fairly simple interpretation of the mode of action of the major genetic color factors. The most plausible hypothesis as to the genes themselves seems to be that they are to be looked upon as chemical units not much if at all beyond the size of protein molecules arranged in a definite line or order in the chromosomes, and characterized especially by the power of somehow duplicating themselves from the building stones in their medium, following mitosis; of simultaneously separating from these duplicates under the cell condition of mitosis; and of simultaneously attracting their homologues under the conditions of the maturation prophase. As factors in development, the genes apparently behave as catalysts in not being used up themselves by the reaction for which they are responsible. In the case of color production, the effects seem to take place through the mediation of enzymes produced by the genes in the nucleus but acting upon protein decomposition products related to tyrosin in special bodies (mitochondria) in the cytoplasm of the appropriate cells. These enzymes, actually extractable from the skin of pigmented animals, are not, of course, to be identified with the genes themselves. Their production may or may not be related to the process by which the genes duplicate themselves between cell division. Much must be learned before any such hypothesis ceases to be of a highly speculative nature. (Wright 52, pp. 554-55)

To the scientist knowing little about the period before the 1940s, this statement may sound surprisingly close to an anticipation of modern molecular biology. But Wright's statements in this quote accorded closely (as Wright said

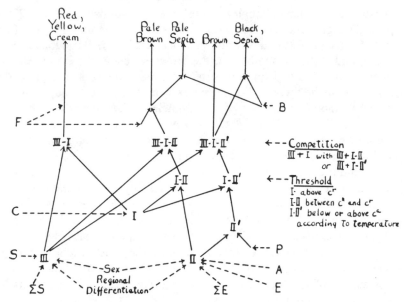

Hypothetical processes involved in melanin pigment production and points of
action of factors as deduced from the effects of factor combinations.

Figure 4.6. Melanin pigment production.

in a footnote dated after the quote was written) with those expressed by
Goldschmidt in *Physiologische Theorie der Vererbung* (1927), which Wright
had just read. The views expressed independently by Wright and
Goldschmidt also were in close accord with views expressed by Morgan,
Muller, Sturtevant, Demerec, and many other geneticists. The rise of molecular biology may have produced a revolution, but it was a revolution anticipated in broad outline by many geneticists before 1930 (see Carlson 1966,
1971; Ravin 1977). The rest of the discussion section was devoted to
Wright's efforts to interpret these new data from new combinations of color
factors (especially those involving the factor pair *Ff*) in terms of his basic
model developed in 1916–17. He found the data consistent with the model,
which he presented in his most concise diagram yet. The only real change is
that Wright now wished to identify his I, II, and III substances as probable
precursors of enzymes rather than as enzymes themselves (fig. 4.6; reproduced from Wright 52, p. 564). As Wright's ideas about the physiology of
color inheritance and color production changed over the years, the figure
changed accordingly.

One question always of considerable interest to geneticists at this time
was whether known Mendelian factors are to any extent linked in the hereditary process. This question was important not only for interpreting and predicting the results of crosses but also for constructing genetic maps of the
chromosomes, as Sturtevant had done for *Drosophila*. Cases of linkage were

far less common in rodents than in *Drosophila,* in part because fewer factors were identified in rodents and in part because rodents had up to ten times as many chromosomes as *Drosophila.* Wright took eight known factors in the guinea pig, putting all eight dominants in one homozygous stock and all eight recessives in another homozygous stock. Then he crossed the two stocks and backrossed these offspring to multiple recessives in order to examine the segregants. The experiment thus had two objectives—to see all possible combinations that could be distinguished (many had not been produced earlier) and to find any linkage.

The experiment had clear results. The distribution of the segregants indicated unmistakably that all the factors segregated randomly. Wright interpreted the results to mean that there was no sex or autosomal linkage, and he accepted recent determinations of high numbers of chromosomes in guinea pigs. The estimates available in 1927 ranged from 28 (1911) and 8 (1913) to 19 and 30 (1926). It is easy to see why Wright did little chromosome analysis. No one even knew for sure how many chromosomes the guinea pig had.

In addition to the series of experiments discussed so far about color inheritance and physiological genetics, Wright conducted, often with others, related experiments. Of the six I will briefly discuss, three are directly related to color inheritance. The first was a 1918 study in collaboration with Harrison R. Hunt, an old friend who had served with Wright as a waiter at Cold Spring Harbor back before Wright went to graduate school. They subjected guinea pig hair to chemical and microscopical examination, finding both diffuse and granular pigments. The black and brown series had only black or brown granular pigment, whereas the yellow series had granular as well as diffuse pigmentation. The distribution of the black and red pigments differed in the individual hair, and the black and red pigments responded differently to the dilution factors of the albino series (four alleles at that time). They also referred to to the chemical differences between black and red pigment found by Durham, Gortner, and Onslow. With all these differences recorded, Wright and Hunt concluded:

> At first sight it seems necessary to suppose that black and red guinea pigs differ from each other by many physiological factors to account for these many differences. But there is probably only one primary physiological difference between black and red hair because a single genetic factor is enough to effect the change. All the observed differences are, in some way, direct or indirect effects of this primary difference. (Wright 19, p. 180)

One can see in this conclusion the influence of Wright's hypothesis concerning the production of pigmentation, a relation made explicit in the paper.

From his first association with guinea pigs, Wright had found much variability left unexplained by Mendelian analysis based upon known factors. When he came to Washington, Wright immediately discovered that within the highly inbred families, by then nearly homozygous, there still existed much variability in the piebald or spotted coat pattern. He therefore supposed this variability to be mostly nonhereditary, as he determined from an analysis using path analysis, which I will describe in the next chapter.

One family exhibiting a highly variable piebald pattern also had a four-toe variation that occurred frequently. Within close families, however, three-toe × three-toe crosses produced just as many four-toed offspring as four-toe × four-toe crosses. Thus in this family the four-toe configuration also appeared to be nonhereditary. It had long been known, or at least strongly supposed by serious experimental biologists, that the frequently alleged influence of age of parents upon offspring was negligible. The view promoted by Caspar Redfield, a wealthy but ill-trained breeder, and published in *Journal of Heredity* was that older parents produced more mature offspring. Experimental biologists ridiculed Redfield. The one inbred family exhibiting both high variability of coat pattern and the extra digit seemed the perfect stock for testing the influence of parental age upon the characteristics of the offspring.

To his considerable surprise, Wright found that age of the dam was a very important factor in the amount of white in the piebald pattern and in the percentage of offspring exhibiting the four-toe characteristic. The percentage of white increased significantly with age of the dam, and the percentage of polydactyl offspring decreased significantly. Wright's surprising observations led him to make this conclusion: "The results suggest that age of dam may be more important as a factor affecting embryonic development of mammals than indicated by the small amount of previous experimental data brought to bear on this question" (Wright 48, p.559). Dunn told me that, for him, this paper was very significant for him because he had always discounted any effect of maternal age in mammals. Dunn's copy of this paper has been used so much that it is falling apart despite several different repair jobs.

The last research project directly related to color inheritance was a study of mutational mosaic coat patterns in the guinea pig, by Wright and his assistant O. N. Eaton (Wright 55). Out of some 35,000 guinea pigs, Wright had found only seven in which a somatic mutation had occurred, yielding a mosaic coat pattern. No one explanation could be applied to all seven cases, but Wright found some speculative implications for understanding color inheritance. One conclusion was that the data could be interpreted to mean that one particular stock inherited a tendency for mutations. Such cases have of course been amply verified by later research upon mice at the Jackson Laboratory and with many other organisms.

In addition to research into the physiological genetics of coat color, Wright worked with others on projects that utilized very different characteristics of the guinea pig. In 1919 Dr. Paul A. Lewis, of the Henry Phipps Institute of the University of Pennsylvania was examining resistance to tuberculosis. When he heard about the highly inbred families of guinea pigs at the USDA, he saw a chance to examine with nearly homozygous mammalian stocks possible genetic differences in susceptibility to tuberculosis. He contacted Wright, who began to send him all surplus individuals in the five remaining inbred stocks. Wright and Lewis planned to publish two papers, one providing a genetic analysis of the results with Wright as senior author and the other focused upon medical aspects with Lewis as senior author. Wright analyzed the first batch of results and presented them in a paper published early in 1921 in *American Naturalist* (Wright 25), and authored a brief analysis of the second batch of results as an abstract in 1922. Lewis continued the experiments but moved from Phipps to the Rockefeller Institute, where under very different experimental conditions the results contradicted the earlier conclusions drawn by Wright. Some effort was made to equalize the conditions at Phipps and Rockefeller. Lewis unfortunately died on an expedition studying yellow fever in 1929, whereupon all the data were sent to Wright. He wrote a very long paper, trying to take account of the variable experimental conditions, but it was basically unpublishable. A synopsis of the data and discussion finally appeared in 1977 (Wright 198, pp. 61–68).

The basic thrust of the research project is apparent in the 1921 paper. There Wright first examined the relationship of resistance to tuberculosis observed in the stocks with a number of possible variables and found no substantive correlation at all. Sex, age, size of litter, birthweight, and rate of gain were unrelated to tuberculosis resistance. There were, however, marked differences in resistance in the different inbred families. When crossed with other families, individuals from the most resistant family produced offspring even more resistant. Clearly the genetic differentiation of the inbred lines involved, among other differences such as coat colors and patterns, genetic differences in susceptibility to tuberculosis. In all the variability exhibited by differential resistance to tuberculosis, Wright estimated that perhaps 40% was caused by genetic differences.

Two other papers using different physiological processes came to similar conclusions. One, published in 1923 with Eaton's assistance, examined the determining factors behind otocephalic monsters in guinea pigs (Wright 41). Otocephaly is a deformity of varying severity affecting the face and head of the guinea pig. Those little affected have a reduced lower jaw; the higher grades range continuously from loss of mouth to entire loss of the head with only a hint of an ear. Wright devised a scale of twelve grades of otocephaly (see fig. 4.7; reproduced from Wright 41, p. 161). Of approximately 40,000 guinea pigs bred by the USDA since the beginning of the inbreeding experiments, 82 otocephalic monsters had appeared. Of these, 50 appeared in a single family, which happened to be also the most prolific of all the inbred families, and with a frequency of about 1.5%. Clearly the members of this one inbred family had a genetically determined susceptibility to otocephaly, but since the individuals within the line were genetically highly similar, no genetic explanation existed for why one member of the family exhibited otocephaly and another member did not. After largely rejecting on the basis of the data all the factors common to litter mates, condition of the mother, sea-

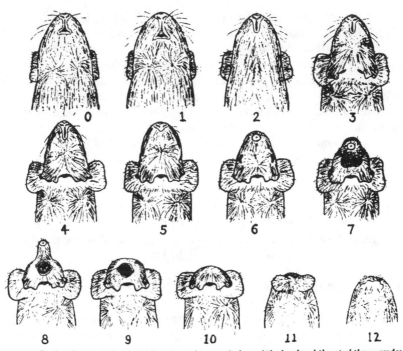

Grades of otocephaly.   Semidiagrammatic ventral views of the head and throat of the 12 grades
in comparison with the normal (o).

Figure 4.7. Grades of otocephaly.

sonal distribution, differences between early and late litters, and maternal inheritance, Wright concluded by elimination that the main factor was some
chance irregularity of development at a crucial early stage.

The final project was an extension of the one Wright and Little tried in
1921 at Cold Spring Harbor, only to have the experimental subject killed by
Macdowell's dog. The basic idea was to transplant tissues to determine the
degree of rejection by the host. Such rejection was known to be strongly dependent upon the degree of relatedness, with autotransplants having by far the
greatest rate of success. The pathologist Leo Loeb (brother of the physiologist
Jacques Loeb) had long wished to use the inbred lines of the USDA for this
purpose, and after the trial with Little failed, Wright decided to do the experiment with Loeb. The research was published in 1927 with Loeb as senior author. Loeb tried all variations of transplants—within inbred families of related and unrelated individuals, from hybrids between two families to another
such hybrid, multiple simultaneous transplantations, between inbred families,
and other combinations. The presentation of results in the paper is a deadly
bore, but very interesting conclusions emerged in harmony with other genetic
research upon the inbred families.

Within an inbred family, some variability of tissue compatibility existed,
but unrelated individuals were clearly closer in compatibility than brothers or

Figure 4.8. Sewall Wright holding a guinea pig in each hand, circa 1920.

sisters were in the absence of inbreeding. Random individuals from within an inbred line were not yet to the high level of compatibility of autotransplants, so the inbred families were not yet wholly homozygous. Loeb and Wright also found that reciprocal transplantation often led to strikingly different results. The data indicated strongly that if the donor lacked genes present in the host, or if the host had strange genes in addition to those coming from the donor, adverse reactions did not occur. The problems arose when the donor possessed genes not in the host. For Wright, the results were in harmony with his expectation that the genetic differences between inbred families should lead to significant differences in compatibility of tissues, whereas differences in compatibility of tissues between individuals within a single inbred family should be far less than those between families.

5
Path Coefficients and Animal Breeding

The method of path coefficients, invented by Wright in its first formulation during his graduate school years, provided him with a powerful quantitative tool that he would apply to many different problems. It is no exaggeration to say that, using this method, Wright revolutionized the quantitative analysis of systems of mating, a basic problem for geneticists and breeders. The method is the quantitative backbone of his work in evolutionary theory. Wright used it to calculate coefficients of inbreeding and relationship, to partition the variability in guinea pig coat patterns into components determined by heredity and environment, to analyze corn and hog correlations, and for other purposes, all during the Washington years. But when he invented the method in graduate school, Wright surely did not foresee the uses to which he would put it after 1918.

For over a year after the move to Washington, Wright had little opportunity or reason to work seriously upon quantitative analysis. Then early in 1917 Raymond Pearl again provided a stimulus. This time he published in *American Naturalist* an attack upon the familiar formula used by geneticists (a formula found in Davenport's little book on biometry) for calculating the standard deviation of a Mendelian class frequency (Pearl 1917). Pearl, as usual, was promoting a far more refined and complicated method invented by Karl Pearson. As illustration, Pearl gave a Mendelian example in which the old method was off by 40% from the new, more accurate method. But as in the earlier case of his incorrect analysis of brother-sister mating, Pearl had incorrectly applied the old method. Wright spotted Pearl's mistake and wrote a note to *American Naturalist* showing that the old method actually gave results essentially the same as the refined and complicated new one, and substantially better results than the approximation of the new method that Pearl recommended for geneticists who did not require highly precise calculations. Wright concluded: "From the nature of experimental work, great refinement in statistical treatment is often a waste of effort, and without questioning the value of Dr. Pearl's suggestion in cases in which the greatest accuracy is warranted it appears that the simple formula is still adequate for most practical purposes" (Wright 6, p. 375).

This sentence is noteworthy because one might easily conclude that Wright's views about statistical treatment must have changed substantially as he became more and more sophisticated about mathematical analysis, particularly about his method of path coefficients. The conclusion would be wrong.

Wright always avoided statistical methods he considered to be unnecessarily sophisticated or complicated in relation to the biological problems under consideration. He approached mathematics not as most mathematicians would, by investigating abstract and theoretical relations of greatest interest in themselves, but as a scientist who needed quantitative analysis for a better understanding of relations in the physical world. He studied mathematics on his own usually with a specific biological problem in mind, and he frequently altered or simplified the mathematics of others for his own particular uses. Wright unquestionably had the ability in quantitative reasoning to be a mathematician, but this was never his interest. He was always first and foremost a biologist.

We now return to the time in 1914 when Castle assigned Wright and Fish the task of computing the correlation coefficients between five different bone measurements from MacDowell's rabbit data. I have told already (in chapter 3) how Wright calculated all of the partial correlations with one, two, and three of the measurements held constant. Better than the plain correlation coefficient, the partial correlation coefficients indicated how much of the observed variability in a bone measurement was probably due to general size factors and how much to special size factors, but Wright was still dissatisfied. He saw clearly that by itself the partial correlation coefficient, like the correlation coefficient, was a mathematical quantity not tied or leading by itself to any causal interpretation of the relations under examination. Wright wanted to minimize correlational statistics and maximize the quantitative causal interpretation of the variables.

Any Mendelian in the 1910s was painfully aware that the mere correlation between parent and offspring, as admitted by Pearson, told nothing about the mechanisms of inheritance. Mendelians knew by their own work that the biometricians, using their most sophisticated statistical techniques, could never have made orderly sense of color inheritance in mammals or of many of the other problems of inheritance so strikingly solved by Mendelian analysis. The distinction between correlation coefficients and causal understanding was one drilled into the heads of geneticists. What Wright really wanted was a better way to move toward greater quantitative causal understanding than provided by correlation coefficients followed by nonquantitative attempts to fit them into a causal scheme. While still in graduate school, Wright invented the first version of the method of path coefficients as his first step toward a better quantitative causal analysis.

Soon after Wright replied to Pearl's article in 1917, Davenport published a huge seventy-six page article in *Genetics* on the inheritance of stature in humans (Davenport 1917). Davenport addressed directly the problem of general versus special size factors and quoted Castle's conclusion that special size factors played little role in comparison to general size factors with regard to the size of the body as a whole. Davenport listed the correlations calculated by Fish and Wright, noted that they were indeed rather high, but then cited much lower correlations between some human bones. Davenport also had pictures of humans from a dwarf to a South American Indian to a Nilotic Negro. He pointed out the relatively short trunk and long legs of the Negro and the relatively long trunk and short legs of the Indian. One should expect to find special size factors of greater importance than Castle thought. Indeed, Davenport's systematic observations that followed "suggest that the lengths of the different segments of stature are inherited independently as is indicated particularly by the absence of a high correlation in their variability" (Davenport 1917, 359). Here was a clear difference of opinion between Castle and Davenport.

When Wright read Davenport's paper, he realized that the additional calculations he had done with MacDowell's data, upon which Castle had based his conclusions, were pertinent to the question of general versus special size factors, and he soon wrote out his analysis and submitted the paper to *Genetics*. Entitled "On the Nature of Size Factors," the paper appeared in July 1918 (Wright 20). Wright first presented a table that included in the first column Castle's original ten correlations (calculated by Fish and Wright), all the partial correlations between bone measurements with one measurement held constant (next five columns), two measurements held constant (next ten columns), and three measurements held constant (last column). *OM* is length of skull, *Zp* is width of skull, and *H*,*T*, and *F* are the length of humerus, femur, and tibia (fig. 5.1; reproduced from Wright 20, p. 368).

From the table Wright showed that with three measurements held constant, and the consequent reduction of the influence of general size factors from an average of 75% of the variation (as measured by squared standard deviation) to less than 25%, three of the correlations (*OM* with *Zp*, *H* with *F*, and *F* with *T*) were fairly high. "These three correlations suggest the existence of growth factors which affect the size of the skull independently of the body, others which affect similarly the length of homologous long bones apart from all else, and others which affect similarly bones of the same limb" (Wright 20, p. 369). The results from having first two measurements and then one held constant, supported the same conclusion. This suggested conclusion certainly carried the analysis well beyond the simple correlation coefficients so confidently published by Castle in 1914. Wright, however, wanted more.

The next section of the paper begins with the classic understatement: "It is of interest to attempt to assign definite values to the different classes of growth factors which are indicated (by the partial correlation analysis)" (p. 370). Of course! Everyone wanted to do that, but no quantitative method for doing it existed. Wright in this case wanted to attribute the observed variability in the various bones quantitatively among the six causative hereditary factors affecting (1) general size (2) size of skull only, but all skull bones alike (3) length of leg bones only, but these alike (4) length of bones of hind limbs only, but these alike (5) length of homologous leg bones only, but these alike (6) each part independently. With this objective in mind, Wright presented the first version of his method of path coefficients.

	r	OM^R	Zp^R	H^R	F^R	T^R	$OM\text{-}Zp^R$	$OM\text{-}H^R$	$OM\text{-}F^R$	$OM\text{-}T^R$
$OM\text{-}Zp$	.750 ± .015	...	...	.502	.496	.538	...			...
H	.743 ± .016	...	.485	...	.275	.431	...	...		...
F	.760 ± .015	...	.519	.356	...	.432	...	...		...
T	.701 ± .018	...	.416	.276	.147	...	...			
$Zp\text{-}H$	.675 ± .019	.266	...		.254	.334	...		.145	.135
F	.674 ± .019	.242	...	.250	...	.282		.088		.067
T	.658 ± .020	.280	...	.275	.211	...	...	.162	.160	
$H\text{-}F$	.857 ± .009	.671	.737		.567	.649	...			.467
T	.791 ± .013	.566	.628	...	.211		.531	...	.181	...
$F\text{-}T$	.858 ± .009	.700	.746	.571			.679	.524		

	$Zp\text{-}H^R$	$Zp\text{-}F^R$	$Zp\text{-}T^R$	HF^R	HT^R	FT^R	R (three constant factors)
$OM\text{-}Zp$	...	...	...	.456	.502	.481	.448
H	...	.176	.316	...	...	.252	.172
F	.274	...	.345	...	.251	...	.224
T	.163	.051	...	.095	...	...	.022
$Zp\text{-}H$	...	...	...	...	...	.218	.119
F	...	...	...	...	.118	...	.004
T	...	...	...	.166	...	...	.136
$H\text{-}F$	...	...	.517	...	...	...	.463
T	...	.173	...	...	...	...	.161
$F\text{-}T$	.537	...	...	...	...	...	.517

Figure 5.1. Correlations between bone measurements.

A rough analysis can be made by use of the following proposition. Let X and Y be two characters whose variations are determined in part by certain causes A, B, C, etc., which act on both and in part by causes which apply to only one or the other, M and N respectively. These causes are assumed to be independent of each other. Represent by small letters, a, b, c, etc., the proportions of the variation of X determined by these causes and by a^1, b^1, c^1, etc., the proportions in the case of Y. The extent to which a cause determines the variation in an effect is measured by the proportion of the squared standard deviation of the latter for which it is responsible. This follows from the proposition that the squared standard deviations due to single causes acting alone may be combined by simple addition to find the squared standard deviation of an array in which all causes are independent of each other, i.e., $\sigma^2_{A+B+C} = \sigma^2_A + \sigma^2_B + \sigma^2_C$.

Effects	Causes					
	A	B	C	D	M	N
X	a	b	c	d	m	
Y	a^1	b^1	c^1	d^1		n^1

As a, b, etc., are the proportions of the variation of X which are determined by the various causes

$$a + b + c + d + \cdots + m = 1$$
$$a^1 + b^1 + c^1 + d^1 + \cdots + n^1 = 1$$

It is easy to demonstrate the following proposition in regard to the correlation between X and Y.

$$r_{xy} = \pm \sqrt{aa^1} \pm \sqrt{bb^1} \pm \sqrt{cc^1} \ldots$$

Where a given cause as A produces effects in the same direction in X and Y the sign of the term aa^1 is $+$. Where the effects are in opposite directions the sign is $-$. (Wright 20, pp. 371–72)

Several aspects of this first presentation of the concept of path coefficients are noteworthy. It must be clearly understood that Wright was not attempting with his method to deduce the causes of the variability of the characters X and Y; rather, he had assumed a causal scheme and was attempting by his method to quantify each cause already assumed. In this first presentation, Wright used no path diagram to represent the assumed causal scheme, as he would in later discussions of path coefficients, and he nowhere used the name path coefficients. The causes are independent, and the causal scheme is complete or, put another way, closed (the proportions of variation of character such as X or Y all add up to 1).

The proportions of variation of a character, denoted here as a, b, c as a result of causes A, B, C, are measured in terms of the proportion of the squared standard deviation. In other words, Wright was attempting to give a quantitative measure to each cause by the proportion of the squared standard deviation of the variability assigned to the cause. Any student of elementary statistics or biometry will instantly recognize the squared standard deviation as the variance and Wright's efforts in this paper as an attempt to "partition the variance." The terminology familiar to the student of statistics or biometry, and not found in Wright's paper, is of course due to Fisher. The coincidence here is extraordinary.

Wright's paper "On the Nature of Size Factors" appeared in the July 1918 issue of *Genetics*. Fisher's well-known paper, "The Correlation between Relatives on the Supposition of Mendelian Inheritance" (1918), in which Fisher first coined the term variance for the square of the standard deviation, was read to the Royal Society of Edinburgh on July 8, 1918. The similarity of statistical conception between Wright and Fisher in their attempts to quantify the causes of variability are apparent from the first paragraph of Fisher's paper:

When there are two independent causes of variability, capable of producing in an otherwise uniform population distributions with standard deviations σ_1 and σ_2, it is found that the distribution, when both causes act together, has a standard deviation $\sqrt{\sigma_1^2 + \sigma_2^2}$. It is therefore desirable in analysing the causes of variability to deal with the square of the standard deviation as the measure of variability. We shall term this quantity the Variance of the normal population to which it refers, and we may now ascribe to the constituent causes fractions or percentages of the total variance which they together produce. (Fisher 1918, 399)

Wright and Fisher had never even so much as heard of each other in the year they both published their initial analyses of the causes of variability.

After stating his new method, Wright applied it to the rabbit data. He obtained solutions to the simultaneous equations by the rather crude methods of averaging the ratios of coefficients representing proportions of the variation (as in a^1/a) and by averaging the maximum and minimum values of the components of the observed correlation, as in $\sqrt{aa^1}$. Wright put the results into a table in which the individual bones are listed on the left and the classes of six causes listed earlier are across the top (fig. 5.2; reproduced from Wright 20, p. 373). The table indicated clearly that although most of the differences in bone measurements between individuals are caused by general size factors, special size factors affected each bone length to some extent, independent of all other bone measurements. Also, some size factors affected groups of bones, (such as skull length and breadth or the three hind leg bones) independent of the rest of the body.

	General size	Skull	Legs	Hind legs	Proximal leg bones	Special	Total
OM	.746	.077	....	....	....	.177	1.0000
Zp	.620	.065	....	....	....	.315	1.0000
H	.739	....	.083	....	.030	.148	1.0000
F	.755	....	.085	.064	.030	.066	1.0000
T	.680	....	.076	.058	....	.186	1.0000

Figure 5.2. Relative importance of different classes of size factors of five bone lengths.

Wright's experience analyzing MacDowell's bone data in the period of 1914–18 convinced him completely that the array of correlation coefficients or partial correlation coefficients (and he would later add any set program of statistical analysis) could not yield by itself the pattern of causal relations among variables. What was required was an initial scheme of causal relations of the variables deduced from all available evidence from every source (including of course correlation coefficients, space-time relations, etc.), followed by the attempt to quantify the causal relations already deduced. Obviously, Wright's method could be applied more easily to some situations than others and could not be applied at all to situations in which causal relations were inscrutable for whatever reasons. It cannot be emphasized too strongly that Wright viewed path coefficients not as usual statistical facts (within confidence limits) determined by a stereotyped formula (as in correlations, regression coefficients, etc.) but instead as interpretive parameters tied to specific causal schemes.

The method of path coefficients proved to have much wider application than was apparent in Wright's first use of it or than Wright himself imagined. One major use appeared almost immediately in Wright's attempt to quantify some crucial variables connected with the inbreeding experiment with guinea pigs. Among the data meticulously kept since the beginning of the experiment were records of birth weights and weights at later times. Wright was interested in the correlation between birth weight and weight at one year for all of the twenty-four families in the experiment for fifteen generations. Since litter size was the most important factor affecting birth weight, he selected from the data only those pertaining to litters of three. Wright easily calculated the correlation between birth and year weight for the entire sample of 560 animals and the same correlation for the array of twenty-four family means (weighted, of course, by the number of individuals in each family). But of greatest interest to Wright was the average correlation within families. He therefore began the calculations, starting with the largest families. This was a tedious and boring job. After calculating the correlations for the eight largest families, Wright decided that there had to be a better way. Would it be possible, he thought, to calculate the desired correlation by finding some algebraic relation between the two correlations he knew already and the unknown one, then solving for the unknown? But there was no obvious additive or multiplicative relationship between the three correlations.

From the time in graduate school when he had first tried to interpret the partial correlation coefficients he had calculated for MacDowell's rabbit data, Wright had been in the habit of drawing diagrams of causal or statistical relations. He generally tended to think pictorially and found these diagrams a considerable aid to his thinking. In this case, he drew what later would be called a path diagram in which he represented birth weight and year weight in the whole sample as determined by the family means and the independent deviations from those means. The novel feature of this diagram in comparison to the one he used for the data on rabbit bones is that the "causes" were correlated instead of independent. Wright has recently (Wright 196, p.201) given a later reconstruction of this diagram in hindsight; I will not reproduce it here since no contemporary diagram representing his thought at that time is available. Inspection of the diagram suggested to him that the correlation between birth and year weights equaled the sum of contributions from the two connecting paths in the diagram. A simple algebraic transformation turned this equation into a form immediately recognized by Wright as having general significance.

He had discovered the additive property of Pearson's product moment $r_{xy}\sigma_x\sigma_y$, later termed covariance by Fisher—a term I will continue to use.

Variance was simply the square of the standard deviation, $\Sigma d^2/n$ where d is the deviation from the mean and n the number of deviations; the covariance is exactly similar but involving two variables. If a and b are the respective deviations from their means of two variables, then the covariance of the variables equals $\Sigma ab/n$. Since variances were known to be additive for independent variables, it was no surprise to find that covariances were additive. Crucial for Wright was that this additive property of the covariances, algebraically so simply related to correlation coefficients $r_{xy} = \text{covariance}/\sigma_x\sigma_y$, enabled him to quickly solve for the unknown average correlation within subgroups.

Wright had no rigorous proof for his deduction from his path diagram, but the additive property of covariances that he had discovered was a general truth. So he worked out a purely algebraic proof of the additive property of covariances and showed in the paper how the unknown correlation was easily calculated from this property. The results of these calculations are presented in the following table where σ_b is the standard deviation of the birth weight and σ_y that of the year weight (fig. 5.3; reproduced from Wright 7, p.535).

CONSTANTS USED IN CALCULATIONS

	σ_b^2	σ_y^2	σ_b	σ_y	$\sigma_b\sigma_y$	$r_{by}\sigma_b\sigma_y$	r_{by}
Total (560 pigs)....	130.53	14,852	11.425	121.87	1,392.4	522.15	+0.375±0.024
24 family means....	20.50	4,837	4.528	69.55	314.9	198.39	+0.630±0.083
Average family (deduced)...........	110.03	10,015	10.49	100.08	1,049.8	323.76	+0.308±0.026
Average 8 families with 297 pigs.....	108.78	8,915	10.43	94.42			+0.256±0.036

Figure 5.3. Constants used in calculations.

Wright did not, of course, publish or say anything about the path diagram that had led him to the principle of the additivity of covariances. Indeed, he considered the principle so obvious that he assumed it must be known (to someone like Pearson), but not well known: "The very simple formula discussed below has been useful to the writer and does not seem to be well known" (Wright 7, p.532). Wright himself learned from this exercise a general lesson that would soon lead to a dramatic solution to the quantitative analysis of inbreeding that Pearl, Jennings, Robbins, East, and many others had found so difficult. The general lesson was that his path diagrams allowed him in many cases to calculate the correlations between linear functions. Just how important this lesson was will become apparent later in this chapter.

By early 1920 Wright had thought a great deal more about path coefficients and had made a series of refinements of the method. He had found the coefficients of determination as earlier defined (measured by the proportion of the squared standard deviation) rather awkward in the equation

$$r_{xy} = \pm\sqrt{aa^1} \pm \sqrt{bb^1} \pm \sqrt{cc^1} \ldots$$

because of the problem of sign determination. If instead the path coefficients were defined as ratios of the standard deviations rather than the square of the standard deviation, then the $\pm$ signs and the square roots in the equation for r_{xy} above would disappear and the correlation could be expressed as (for the same variables) $r_{xy} = aa^1 + bb^1 + cc^1 \ldots$ With this assumption, the coefficients of determination for the squared standard deviation would just be the square of the path coefficients.

Once Wright had made this simplification, he set to work trying to prove rigorously what he had sensed intuitively since at least 1917—that the correlation between two variables in one of his path diagrams, in which each path between variables was a path coefficient (one path perhaps a correlation coefficient, as in the case of correlated initial causes), was equal to the sum of contributions from all connecting paths by which the two variables are connected. The contribution of each connecting pathway equaled the product of the path coefficients (including the possible correlation coefficient). Later, after he found out that the path coefficients could be treated as standardized partial regression coefficients (i.e., the deviations from the mean are measured in standard deviations rather than the original units), this proof was very simple. But in 1919–20 Wright had to put in a lot of work on the proof, which he achieved by early 1920. He then wrote a general exposition of the method of path coefficients but found his superiors at the USDA unenthusiastic about publication. This sort of quantitative analysis did not appear to them appropriate for the USDA. Thus Wright first published an example of path analysis applied to the USDA experiment on inbreeding in guinea pigs.

By 1920 Wright was still maintaining only five of the original inbred families. Each was quite clearly differentiated from the others. A keeper would have no difficulty at all knowing to which family a randomly chosen guinea pig belonged. Yet the variability within each inbred family was still considerable, especially in the spotting pattern. Even after twenty generations of brother-sister mating and a great reduction of any heterozygosis, a guinea pig with 20% of white in its coat might easily have a litter mate with up to 90% white. Wright wished to analyze this variability in coat patterns into hereditary and environmental components. The method of path coefficients was well suited for this problem. Under the title "The Relative Importance of Heredity and Environment in Determining the Piebald Pattern of Guinea Pigs," Wright published his data and analysis in 1920 in the *Proceedings of the National Academy of Science* (Wright 22). Raymond Pearl had presided at the session of the 1919 meeting of the AAAS at which Wright read the paper and suggested that Wright give the paper to him for communication to the *PNAS*.

For the analysis Wright chose one inbred family (no.35) and the control stock in which no inbreeding of even second cousins or closer had been allowed. To demonstrate the power of his new method, Wright first calculated the pertinent correlations for comparison, as shown in the following table (fig. 5.4; reproduced from Wright 22, p.327). The table indicates that the

	EXPERIMENT B (RANDOM-BRED)		FAMILY 35 (INBRED)	
	No.	Correlation	No.	Correlation
Sire-Dam.....................	105	+0.019 ± 0.066	73	+0.029 ± 0.079
Sire-Son......................	492	+0.231 ± 0.029	235	+0.013 ± 0.044
Sire-Daughter................	484	+0.194 ± 0.030	236	+0.082 ± 0.044
Dam-Son.....................	498	+0.251 ± 0.028	235	+0.042 ± 0.044
Dam-Daughter................	488	+0.165 ± 0.030	236	—0.080 ± 0.044
Average, Parent-Offspring......	1962	+0.211 ± 0.015	942	+0.014 ± 0.022
Brother-Brother..............	390	+0.219 ± 0.033	182	+0.090 ± 0.050
Brother-Sister................	437	+0.228 ± 0.031	203	+0.062 ± 0.047
Sister-Sister.................	406	+0.180 ± 0.032	194	+0.064 ± 0.048
Average, Litter Mates.........	1233	+0.214 ± 0.018	579	+0.069 ± 0.028

The matings in the correlation between sire and dam are each weighted by the number of offspring. The probable errors in this case are based merely on the number of matings. The probable errors in the other cases are based on the number of entries in the tables. Owing to repetition of individuals, they are probably somewhat too small. The correlation between litter mates in experiment B is based on 894 individuals in litters in which two or more were graded. The probable error based on this number is ±0.022 instead of ±0.018 as given in the table. Similarly there were 426 individuals in such litters in family 35, giving a probable error of ±0.031 instead of ±0.028.

Figure 5.4. Correlations between parents, etc.

matings were at random (negligible correlation between parents), that significant parent-offspring and filial correlations occurred in the control stock, and that insignificant corresponding correlations occurred in family 35. Then Wright went beyond the interpretation provided by the correlations by applying the method of path coefficients.

He assumed three primary causal factors: heredity, H; environment common to litter mates, E; and undetermined factors such as developmental irregularities, D. The path diagram he provided is reproduced below (fig. 5.5; reproduced from Wright 22, p.328). Wright explained the essentials of the method of path coefficients that he had developed but not yet published. Then he gave the first published definition of the path coefficient:

> The path coefficient, measuring the importance of a given path of influence from cause to effect, is defined as the ratio of the variability of the effect to be found when all causes are constant except the one in question, the variability of which is kept unchanged, to the total variability. Variability is measured by the standard deviation. (Wright 22, p.329)

Wright also showed with a simple path diagram how correlated as well as independent causes could be taken into account.

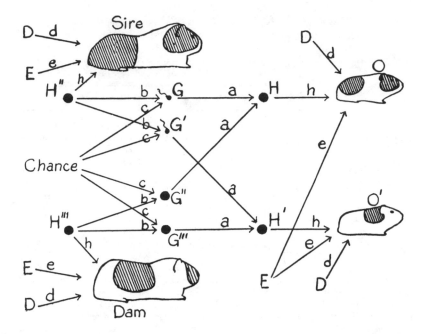

Diagram illustrating the casual relations between litter mates (O, O′) and between each of them and their parents. H, H′, H″, H,‴ represent the genetic constitutions of the four individuals, G, G′, G″, and G‴ that of four germ cells. E represents such environmental factors as are common to litter mates. D represents other factors, largely ontogenetic irregularity. The small letters stand for the various path coefficients.

Figure 5.5. Diagram illustrating the relations between two mated individuals and their progeny.

From the simultaneous equations provided by the path diagram and the closure of causation Wright calculated the path coefficients and their squares (coefficients of determination); with these latter applied to the variance measured in the two stocks, he arrived at this table (fig. 5.6; reproduced from Wright 22, p.332):

	CONTROL (B)	IMBRED (35)
Heredity, $\sigma^2_{O.H}$	0.271	0.010
Tangible environment, $\sigma^2_{O.E}$	0.002	0.020
Development, $\sigma^2_{O.D}$	0.370	0.334
Total, σ^2_O	0.643	0.364

Figure 5.6. Heredity and environment.

Only the variance due to heredity should be diminished in the inbred family, and it should be very low, as indicated by the table. The variance caused by tangible environment (E) and development (D) should be substantively equal in the two stocks. In other words, the total variance of 0.372 (and $\sigma = 0.610$) for E and D would be expected to be observed in the inbred stock. Indeed, the corresponding figure was a variance of 0.364 (and $\sigma = 0.603$). The correspondence was, in Wright's words, "an even better check than could reasonably be expected on the accuracy of the assumptions on which the degrees of determination were calculated" (Wright 22, p.332). In this procedure Wright had considered only the additive component of the variance caused by heredity; in later calculations based on similar data he also took into account the rather small components of genetic variance contributed by the effects of dominance and sex differences.

Wright had demonstrated clearly in this case that the method of path coefficients, together with a plausible causal scheme, added much more to the quantification of the causal analysis than did a mere set of correlations. Wright submitted his first general presentation of the method of path coefficients to the editor at the Bureau of Animal Industry, but the editor rejected it. Collins then induced the editor of *Journal of Agricultural Research* (he was also the assistant Chief of the Bureau of Plant Industry) to accept the paper, which appeared in January 1921 under the title "Correlation and Causation." The introduction stated succinctly Wright's justification for introducing the method.

> The ideal method of science is the study of the direct influence of one condition on another in experiments in which all other possible causes of variation are eliminated. Unfortunately, causes of variation often seem to be beyond control. In the biological sciences, especially, one often has to deal with group characteristics or conditions which are correlated because of a complex of interacting, uncontrollable, and often obscure causes. The degree of correlation between two variables can be calculated by the well-known methods, but when it is found it gives merely the resultant of all connecting paths of influence.
>
> The present paper is an attempt to present a method of measuring the direct influence along each separate path in such a system and thus of finding the degree to which variation of a given effect is determined by each particular cause. The method depends on the combination of knowledge of the degrees of correlation among the variables in a system with such knowledge as may be possessed of the causal relations. In cases in which the causal relations are uncertain the method can be used to find the logical consequences of any particular hypothesis in regard to them. (Wright 24, p.557)

Wright here, as he had from the beginning, thought of path coefficients as a way to quantify already-supposed causal relations. It is important to understand this point because many persons would interpret the method as a way of deducing causation; but Wright meant it to be a method of quantifying already deduced pathways of causation.

In this paper Wright developed and generalized the method of path coefficients. He showed that with systems of independent causes, the correlation between each independent cause and the effect was equal to the path coefficient, and in a connected chain of causes all acting in a linear fashion, the degree of determination of the effect must be a product of the component degrees of determination. In other words, the path coefficient for the entire chain of causes was the product of all the path coefficients in the chain. He showed how, in a general way, one must take into account causes that act nonadditively, as in the cases of causes with multiplicative effects or of correlated causes. Nonlinear relations between causes defeated the possibility of calculating the path coefficient of a chain of causes by the product of component path coefficients, and Wright avoided this situation. Later, he found it much easier to approximate nonlinear functions with linear functions than to develop the highly complex mathematical apparatus to cope with nonlinearity. There was a substantial section on systems of correlated causes, which could easily be taken into account by the method.

Wright derived the general formula expressing correlation in terms of path coefficients. With equations of this sort, and equations based on the idea that the sum of the degrees of determination (squares of the path coefficients) must equal unity (closure of causation), a set of simultaneous equations resulted and could be solved if the number of unknowns was sufficiently small. Finally, Wright demonstrated that for some cases the determination (square of the path coefficient) could be expressed in terms of the correlation, and he showed the close relationship between path coefficients and multiple regression coefficients. (After standardization they are the same, but Wright had not heard of that technique when he wrote the paper.) He then applied the method of path analysis to the factors determining birth weight in guinea pigs and factors affecting the rate of transpiration in plants.

Because Wright focused this paper toward correlation and causation, he did not develop the purely mathematical applications of path analysis, as he had done but not published in the 1917 paper on the additivity of covariances. His next application of path coefficients concerned their use in deducing the correlation between linear functions, in this case the linear relations between relatives on the assumption of Mendelian heredity.

Theory of Inbreeding

As Wright wrestled during the years 1915–20 with the problem of how best to analyze the great mass of data from the USDA experiment on inbreeding in guinea pigs, he came to realize that previous attempts to quantify the effects of inbreeding were inadequate. He therefore decided to work on the theory of inbreeding himself. Many geneticists had worked on theories of inbreeding in the decade 1910–20, including Fish, Pearl, Jennings, Robbins, Wentworth and Remick, and others. All of these attempts to quantitatively analyze inbreeding had proceeded upon a basically similar method of attack. First one would write out the genotypes of the zygotes or gametes, and then see how their frequencies changed by several, or perhaps more, generations of breeding under the system of inbreeding. By inspection, one hoped to see in the data a recurrence relation that could be represented by a formula. Using the formula, the genetic constitution of the population at any later generation could be calculated. This method worked on very simple systems of inbreeding such as brother-sister mating, as shown by the work of the geneticists mentioned above. There were, however, major problems with the method. Complicated systems of inbreeding or outbreeding could not be carried through even a few generations without enormous labor, and when this task was accomplished, what remained was a very confusing array of data from which by inspection one had to deduce the recurrence relation leading to the formula. Those who developed the formulas in the years 1910-20 were the first to admit the limitations of their approach.

As discussed above, the method of path coefficients best applies to variables or relations that are definite in causal lines and nicely linear. The mechanism of unlinked autosomal Mendelian inheritance was an ideal target for path coefficients because of the perfectly definite and linear relations between parent and offspring and between relatives generally. When Wright saw the applicability of path coefficients to the quantitative analysis of systems of mating, he took little time in making dramatic strides hardly dreamed of by those who had been working on inbreeding. In short order Wright wrote a series of five papers with the general title "Systems of Mating" and submitted them at one time to *Genetics* in October 1920. They were published as a sixty-seven-page block in March 1921 (Wright 27–31).

Beginning with the Hardy-Weinberg equilibrium, Wright investigated the effects of various systems of matings upon the composition of the population. The results, of course, varied—under some systems of mating the population would reach a new equilibrium and under others no equilibrium would be reached before full homozygosis.

Application of the method of path coefficients to systems of mating was comparatively simple. Wright first determined the path coefficient from gamete to zygote, $a = \sqrt{1/2(1 + f)}$, then that from zygote to gamete, $b = \sqrt{(1 + f')/2}$, and the percentage of homozygosis, $p = (1 - f)/2$. In these equations, f represents the correlation between uniting gametes, f', f'', f''', and so forth represent the same quantity in generations preceding that one. Under random breeding, f was always zero. The quantity f was of the utmost importance because it was directly related to the percentage of homozygosis and therefore an obvious measure of the degree of inbreeding. It could be used to measure inbreeding in a whole population as well as in an individual. Beginning with these equations, Wright developed a set of general formulas for populations under inbreeding, at equilibrium, and under random breeding, and presented these in a table (fig. 5.7; reproduced from Wright 27, p. 121). In addition, Wright showed how to take dominance into account, but he later corrected this procedure (Wright 64, p. 115). With all of

General formulae.

CONSANGUINE MATING	EQUILIBRIUM	RANDOM MATING
$h^2 + d^2 + e^2 = 1$	$h^2 + d^2 + e^2 = 1$	$h^2 + d^2 + e^2 = 1$
$m = \varphi(a'b'm')$	$m = \text{constant}$	$m = 0$
$g = f'$	$g = f' = f = \dfrac{m}{2-m}$	$g = f' = f = 0$
$b^2 = \tfrac{1}{2}(1+f')$	$b^2 = \dfrac{1}{2-m}$	$b^2 = \tfrac{1}{2}$
$f = b^2 m$	$f = \dfrac{m}{2-m}$	$f = 0$
$a^2 = \dfrac{1}{2(1+f)}$	$a^2 = \tfrac{1}{4}(2-m)$	$a^2 = \tfrac{1}{2}$
$ab = \tfrac{1}{2}\sqrt{\dfrac{1+f'}{1+f}}$	$ab = \tfrac{1}{2}$	$ab = \tfrac{1}{2}$
$p = \tfrac{1}{2}(1-f)$	$p = \dfrac{1-m}{2-m}$	$p = \tfrac{1}{2}$
$h^2 = \dfrac{2h_0^2(1-p)}{h_0^2(1-2p)+1}$	$h^2 = \dfrac{2h_0^2}{2-m(1-h_0^2)}$	$h^2 = h^2$
$r_{pp} = mh'^2$	$r_{pp} = mh^2$	$r_{pp} = 0$
$r_{po} = abhh'(1+m)$	$r_{po} = \tfrac{1}{2}h^2(1+m)$	$r_{po} = \tfrac{1}{2}h^2$
$r_{oo} = 2a^2b^2h^2(1+m)+e^2$	$r_{oo} = \tfrac{1}{2}h^2(1+m)+e^2$	$r_{oo} = \tfrac{1}{2}h^2+e^2$

Figure 5.7. General fomulas for inbreeding.

these formulas he could express the essential data for any generation in terms of those for the preceding generation and was prepared to analyze almost any system of mating.

Illustrating the power of these formulas, Wright calculated (ever so simply compared to the earlier approach) formulas to give the state of a population after any number of generations of self-fertilization, or brother-sister or parent-offspring matings. The results agreed with those reached earlier by the more cumbersome methods. Then Wright set to work on more complex systems of matings such as double first cousins, quadruple second cousins, octuple third cousins, half-brothers, and -sisters, first cousins, half–first cousins, and second cousins. In each case Wright drew a path diagram of the mating system, then derived by application of the formulas in his table a recurrence formula for the relationship between the successive generations, and from this he could easily deduce the corresponding recurrence relation for percentage of heterozygosis. As applied to the consequences of Mendelian heredity, the method of path coefficients was stunningly successful.

Having clarified the quantitative analysis of systems of inbreeding, Wright turned to the two most important techniques, other than inbreeding, used by breeders to alter domestic animals and plants: assortative mating based upon phenotypic resemblance and selection. Assortative mating based on phenotype means simply the breeding of those animals or plants that exhibit similar characteristics (without consideration of genetic constitution).

Wright developed general formulas applying to assortative mating, which by use of negative correlations between parents could be applied equally to disassortative mating. Assortative mating based upon resemblance was found to lead to a very different final composition of a population than did inbreeding.

Under inbreeding, the population would split up into many eventually homozygous lines, the number depending upon the initial genetic diversity; all genetic combinations (in the absence of selection) would tend equally toward fixation. With perfect assortative mating, however, the population would tend to split into two extreme homozygous types. With less than perfect assortative mating, an equilibrium might well be reached before attainment of homozygosity. Most important for breeders, Wright evaluated the consequences of combining assortative mating and inbreeding. The effects upon the population were somewhat less than strictly additive but were much greater than would occur with either system of mating alone; in particular, the rate at which complete homozygosis is approached was much increased.

One might have thought that Wright, coming as he did from direct experience with Castle's selection experiments upon hooded rats, would think of mass selection as an obviously powerful force in changing the genetic constitution of populations. Such is not quite the case. Wright pointed out, for example, that selection for heterozygotes led to balanced polymorphisms rather than fixation, no matter how long continued. This important concept was also pointed out by Fisher a year later (Fisher 1922). The fundamental effect of selection was to reduce the genetic variability in the population; thus as selection continued, the proportion of environmentally produced variability should increase and the efficacy of selection upon visible characters correspondingly reduced. On the other hand, only selection could permanently alter the genotypic constitution of the population were it to revert at any time to random breeding. Selection was by far the most effective when accompanied by inbreeding and assortative mating. "Straight selection unaccompanied by close breeding is thus not an effective method for permanently fixing characteristics" (Wright 31, p. 176). Wright was not happy with his quantitative analysis of selection in this paper and much revised the analysis in his 1931 paper on evolution in Mendelian populations (Wright 64).

The wide range of issues addressed in the fifth and last paper of the series, "General Considerations," touched upon most of the factors considered important by both breeders and evolutionists. The summary is worth quoting in full.

> It will be seen that all of the systems of mating have their advantages and disadvantages. Close inbreeding automatically brings about fixation of type and prepotency. Intermediate types are fixed as readily as extremes. It is the only method of bringing to light hereditary differences in characters which are determined largely by factors other than heredity. On the other hand, close inbreeding is likely to lead toward reduced fertility, size and vigor.

Matings between relatives more remote than first cousins have little significance as inbreeding, except in so far as there is continued breeding within a population of small size.

Assortative mating and selection can lead to fixation of extreme types only and are not very efficient in this respect. Selection, however, is the only means of permanently changing the relative proportions of the various genes present in the original stock. It is an essential adjunct of the other systems as means of improvement.

Assortative mating leads to the greatest diversification of the population as a whole, and thus is practically always accompanied by selection either in nature or in live-stock breeding. Under conditions such that all progeny are to be saved for breeding, this diversification of the population is a disadvantage. Disassortative mating is the method which best holds the whole population together, pending the fixation of the average type by close inbreeding. (Wright 31, pp. 177–78)

It is important to notice in this conclusion that Wright emphasizes breeding structure as a basis for understanding both artificial and natural selection. Only in combination with certain kinds of breeding structure was selection a powerful force in changing the genetic constitution of the population toward adaptation in nature or desired goals in artificial breeding.

The application of the method of path coefficients to systems of mating gave Wright a basic approach for analyzing both artificial and natural selection. Much of the quantitative analysis of inbreeding could be applied to Wright's vision of populations in nature. On the first page of the systems of mating series, Wright had cited the work of F. B. Sumner on variation in natural populations of the deer mouse *Peromyscus*, and in the conclusion he mentioned selection in nature and in livestock breeding in the same breath. From very early in his career, Wright saw evolution in nature as deeply related to what he knew of evolution in domestic populations.

Criticism by Henry Niles

Wright's method of path coefficients did not gain immediate acceptance or attention, even after the application to the previously insoluble problems of systems of mating. The first substantial criticism appeared in *Genetics,* where Wright had published the "Systems of Mating" series. The critic was Henry E. Niles, from Pearl's Department of Biometry and Vital Statistics, School of Hygiene and Public Health, Johns Hopkins University. Niles was a thoroughgoing Humean familiar with modern correlational analysis; he naturally found even Wright's title, "Correlation and Causation," offensive. For Niles, as for Karl Pearson (who stated the belief in all editions of his *Grammar of Science*), causation was nothing more than correlation. Thus to contrast them was nonsense.

> "Causation" has been popularly used to express the condition of association, when applied to natural phenomena. There is no philosophical basis for giving it a wider meaning than partial or absolute association. In no case has it been proved that there is an inherent necessity in the laws of nature. Causation is correlation. . . . Perfect correlation, *when based upon sufficient experience,* is causation in the scientific sense. (Niles 1922, 259, 261)

In addition to this philosophical point, Niles had three other objections to the method of path coefficients. He attributed to Wright the objectionable "assumption that a correct system of the action of the variables upon each other can be set up from *a priori* knowledge" and also pointed out "the necessity of breaking off the chain of causes at some comparatively near finite point" (pp. 262–63). Finally, Niles stated that the method when applied to simple cases led to nonsensical results.

Wright wrote a substantial reply, also published in *Genetics*. He said that he had "never made the preposterous claim that the theory of path coefficients provides a general formula for the deduction of causal relations. . . . The *combination* of knowledge of correlations with knowledge of causal relations, to obtain certain results, is a different thing from the *deduction* of causal relations from correlations" (Wright 37, p. 240). Wright insisted upon the distinction between correlation and causation, but on a pragmatic rather than philosophical basis. All geneticists were acutely aware that the measured correlation between relatives, while important, was not the same thing as understanding Mendelian inheritance as a "causal" system. Wright's experience as a biologist totally convinced him that at the working level of biological research the distinction between correlation and causation was not only valid but essential for progress in genetics.

Niles, in his reply (also published in *Genetics*), was clearly upset by Wright's refusal to meet head-on his philosophical objection: "The combination of knowledge of correlations with knowledge of causal relations means, to me, merely a combination of knowledge of correlations with knowledge of other correlations. When the true nature of causation is grasped it can not mean more than this" (Niles 1923, 259). Niles was indeed raising a substantive philosophical issue that has been debated seriously by philosophers for centuries, and the debate continues at the present. Basically, Wright replied that the philosophical distinction was irrelevant at the level at which he conducted research. Niles and Wright were not really speaking to each other on this question.

Wright acknowledged as valid Niles's statement that the chain of causes had to be broken off at a relatively near finite point, but that, he said, was true of all scientific research and the objection carried little weight. As for Niles's claim that the method did not work, Wright showed that Niles simply misapplied it. Because Niles had not understood how to apply path coefficients, Wright provided a brief but clear exposition of the method, leaving out the technicalities of the 1921 paper. This is probably the clearest brief exposition of the method of path coefficients ever published by Wright. Niles, given the opportunity to append a rebuttal to Wright's article, argued that Wright had not answered his philosophical objection to Wright's distinction between correlation and causation. Niles concluded that he could "not believe that the method of path coefficients is of the least value. It seems to be based upon a complete misapprehension of the nature of causation in the scientific sense" (Niles 1923, 260). Wright, who was certain in his own mind that correlation and causation were different in biological research, declined the offer from *Genetics* to reply yet again to Niles.

Coefficients of Inbreeding and Relationship

Animal and plant breeders had long desired an accurate quantitative measure of inbreeding. Not until the rediscovery of Mendelism and its obvious regularities was such a measure possible. Raymond Pearl was among the first to devise a measure of degree of inbreeding; he chose to base the measure upon the ratio of the actual number of ancestors for a generation to the maximum possible number. When Wright began his work on the theory of inbreeding, he was unsatisfied with Pearl's measure of inbreeding because it was not a measure of homozygosity, as Pearl had himself pointed out. For example, Wright worked out the comparison between individuals produced by continuous mating of first cousins and those produced by crossing two different lines in each of which only brother-sister mating had been allowed. Under Pearl's scheme, the individuals from each mating system had the same coefficient of inbreeding; yet in first cousin mating, the individual, after many generations, tended to be nearly homozygous, while in the other case the individual was highly heterozygous. Wright thought that a more accurate measure of inbreeding should be tied directly to the system of mating actually followed in the pedigree.

In applying his method of path analysis to systems of mating, Wright had used as a basic variable the correlation between uniting gametes denoted by f (the notation later shifted to F). In "Coefficients of Inbreeding and Relationship" published in *American Naturalist* in 1922, Wright explained why he thought f was a good measure of inbreeding:

> There are two classes of effects which are ascribed to inbreeding: First, a decline in all elements of vigor, as weight, fertility, vitality, etc., and second, an increase in uniformity within the inbred stock, correlated with which is an increase in prepotency in outside crosses. Both of these kinds of effects have ample experimental support as average (not necessarily unavoidable) consequences of inbreeding. The best explanation of the decrease in vigor is dependent on the view that Mendelian factors unfavorable to vigor in any respect are more frequently recessive than dominant, a situation which is the

logical consequence of the two propositions that mutations are more likely to injure than improve the complex adjustments within an organism and that injurious dominant mutations will be relatively promptly weeded out, leaving the recessive ones to accumulate especially if they happen to be linked with favorable dominant factors. On this view it may readily be shown that the decrease in vigor on starting inbreeding in a previously random-bred stock should be directly proportional to the increase in the percentage of homozygosis [this is true only if the initial gene frequency $p = \frac{1}{2}$]. Numerous experiments with plants and lower animals are in harmony with this view. Extensive experiments with guinea pigs, conducted by the Bureau of Animal Industry are in close quantitative agreement. As for the other effects of inbreeding, fixation of characters and increased prepotency, these are of course in direct proportion to the percentage of homozygosis. Thus, if we can calculate the percentage of homozygosis which would follow on the average from a given system of mating, we can at once form the most natural coefficient of inbreeding. (Wright 32, pp. 331–332)

The coefficient of correlation between uniting egg and sperm, $f = 1 - 2p$ (where p is the proportion of heterozygosis), was to Wright this natural coefficient of inbreeding.

Using path coefficients as he had in the "Systems of Mating" series, Wright derived a formula for the calculation of the inbreeding coefficient:

$$f = \sum \left(\frac{1}{2}\right)^{n+n'+1} (1 + f_a),$$

where f_a was the coefficient of inbreeding of the common ancestor on a path between the uniting gametes, and n and n' are the number of generations from the sire and dam respectively to the common ancestor. The summation signified that the coefficient f was the summation of all the connecting paths through common ancestors (in short, a straightforward use of the method of path coefficients). Wright then applied the formula to two different pedigrees drawn from the herd books of Shorthorn cattle, one of which had an inbreeding coefficient of .14 (as compared to .25 from only one generation of brother-sister mating) and a coefficient of .46 in the other (as compared to .50 for three generations of brother-sister mating). The second pedigree was obviously far more inbred than the first. The coefficient could be calculated for any actual pedigree, no matter how irregular, and could easily be compared to that expected from any regular system of mating. It is important to understand that the formula did not give an absolute figure for the coefficient of inbreeding but instead measured the correlation between uniting gametes relative to the correlation of uniting gametes in the foundation stock. Thus random-bred descendants of a highly inbred foundation stock would have an inbreeding coefficient of $f = 0$; the relativity of the coefficient is obvious.

There remained in 1922 two substantive problems with Wright's proposed inbreeding coefficient. First was that in extensive pedigrees, calculation of the precise inbreeding coefficient was a tedious job. Wright soon embarked on a project of calculating the history of inbreeding in the Shorthorn cattle (to be described later in this chapter) and saw the great need for an approximate method of calculating the cumulative inbreeding coefficient. Together with his assistant Hugh C. McPhee, Wright published in 1925 such a method, based upon tabulation of random samples of the pedigrees of the sire and dam (Wright 45). The approximation was not only quite close to the exact calculation but, by increased effort with larger and larger samples, could be brought as close as desired to the exact method. The second problem was that a high f might, in a very numerous breed, mean only that individuals had been mated with close relatives rather than that the whole breed was inbred; under this circumstance, random mating would soon restore full genetic diversity to the whole breed. To take account of this possibility Wright devised another inbreeding coefficient (he later named it F_{ST}) in which he took individual sire and dam lines, matched them randomly, and calculated the inbreeding coefficient for the hypothetical progeny. By 1925 Wright had not only devised a more natural and accurate inbreeding coefficient than any before it but had also provided an approximate method of calculation and an alternative coefficient to portray more accurately the degree of inbreeding in an entire population.

Wright's coefficient of inbreeding was very influential. Prominent theorists of animal breeding (the young Jay L. Lush and L. J. Cole in particular) found Wright's analysis appealing. In the ten years after Wright's initial work of 1922–25, the theory of animal breeding was much changed by the introduction of Wright's inbreeding coefficient and its associated versions (see chapter 9). Wright himself used the inbreeding coefficient as a basic variable in his quantitative analysis of population genetics, so the concept was influential in his ideas of evolution in nature as well as in artificial breeding.

Corn and Hog Correlations

Late in the 1910s, in connection with the war effort, Wright was on a committee allocating pork production in various states depending upon available corn. The committee considered many factors of both corn and hog production in their recommendations. Wright became interested in the obviously complex relations between pork production and prices and the fluctuations in the corn crop. After developing the method of path coefficients, Wright saw the possibility of applying it to the data he had collected. The project was very complex. After considerable effort, Wright wrote up the results as a substantial manuscript and submitted it for publication as a department bulletin of the Bureau of Animal Industry. Publication was rejected because officials at the Bureau of Agricultural Economics considered this intrusion from a member of Animal Husbandry to be highly inappropriate and because officials in the Animal Husbandry Division were also not eager to publish the paper. After publication was refused, Wright happened to show the paper to Henry A. Wallace, son of the secretary of agriculture, mentioned earlier as having a strong interest in quantitative analysis and breeding. Wallace, using familial influence, intervened to have the manuscript accepted as a USDA bulletin, and it was published in January 1925 (Wright 42).

The method of path coefficients was the key to the corn-hog analysis, and Wallace was keenly interested in the method. Some of the correspondence between Wright and Wallace on path coefficients survives, and it illuminates the thinking of both men. Wallace kept trying to see path coefficients in terms of well-known statistical concepts, including partial correlation and multiple regression. Wright kept trying to explain how and why path coefficients were different from the usual statistical concepts. In one letter in particular Wright gave what is to my mind the clearest presentation in a brief space of the reasoning behind the method of path coefficients. The letter is worth quoting in full for this reason.

> In reply to your recent letter, I may say that my corn-hog paper has been accepted and is to be published as a Department Bulletin. I appreciate very much your interest in the matter.
>
> In regard to path coefficients, there is no mathematical difference from a coefficient of partial regression provided that the two variables directly involved are measured in terms of their standard deviations. Thus for Yule's $b_{12.34} \ldots$ I would write $p_{1.2}\, \sigma_1/\sigma_2$ and accompany by a diagram in which the other variables in the system were indicated. The point of view, however, is rather different.
>
> Assume that we have a complex system of variables in which each one is represented as a linear function of others, as indicated by arrows in such a diagram as below [fig. 5.8]:

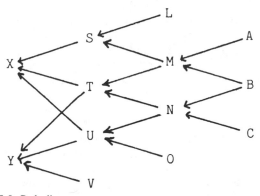

Figure 5.8. Path diagram.

> The problem in which I was interested was to find a coefficient which would measure the influence along each path and would be related in a simple way to the correlations. If I had recognized in the

beginning that the coefficient which I reached was the same as the partial regression coefficient for variables measured in terms of standard deviations, I would not perhaps have ventured to add a new name. However, I think that a short term is almost a necessity in dealing with systems in the way which I proposed. With the emphasis on actual partial regression and partial correlation, the simple way in which correlations could be analyzed in terms of what I have called path coefficients does not seem to have been noticed, i.e, the proposition that the product of the path coefficients along each path by which two variables are connected measures the contribution of that path to the correlation in such a way that the sum of all such correlations exactly equals the correlation.

Consistent results can be obtained in whatever way the variables are arranged in such a system and the method furnishes a convenient way of calculating multiple correlations and regressions. In proposing the method, however, I had primarily in mind the analysis of causal relations in cases in which direct experiment is impossible. My main point of view on this and statistical methods in general may be put as follows:

We find that everything in nature varies more or less. The first problem of statistics is to describe this variation. This involves the determination of averages, measurements of scatter, skewness, etc., and the methods of fitting frequency curves. Next we find that certain things tend to vary more or less together. The measurement of this tendency is the problem of correlation. Closely related is the problem of estimating the most probable value of one for a known value of the other regression. Where several things are varying together more or less, the degree to which one varies concomitantly with the best linear function of all of the others and the estimation of the most probable value of one for known values of the others, constitute the problems of multiple correlation and regression.

In all of this no attention is paid to the actual course of events. We know, however, that two things which vary in correlation with each other must have some physical relation. There is always either a continuous succession of events leading from one to the other or such successions of events leading up to both from more remote variables. The physicist, of course, jumps right over the question of measuring correlation to the study of causal relations, i.e., the actual sequences of events in nature. The biologist, economists, etc., have had to stop in most cases at the mere measurement of degrees of correlation, because of the complexity and uncontrolability of the factors with which they have to deal. However, we are not always entirely ignorant of the course of events in these sciences and it seemed to me important to have a method of combining such knowledge as we have of this sort with the more accurate knowledge which we have of correlation. The result is a diagrammatic representation of the course of events, in which each line of influence is measured by a coefficient which gives the portion of the variation of the dependent variable for which the causal factor is directly responsible. This coefficient is identical with the partial regression coefficient if the standard deviations are the units of measurement. The method of dealing with a complex network of relations involving perhaps hypothetical factors, is however rather different from the usual method of dealing with multiple regression.

In dealing with systems which are not intended to represent causal relations, the use of path coefficients merely gives a convenient method of calculating multiple regression and correlation. The only interpretation which can be put on the path coefficients is that of partial regression coefficients of a somewhat abstract sort. As you suggest, the idea of partial correlation is more easily grasped in such cases. The apparent simplicity of the idea of partial correlation, however, may be dangerous I think as likely to lead to an interpretation in terms of causal relation which is not justified. It goes without saying, however, that there are hundreds of cases in which multiple correlation and regression, and partial correlation (carefully interpreted) can be used for one in which there is sufficient knowledge of causal relations to make this use of path coefficients profitable. (Wright to Wallace, March 15, 1924)

Despite this letter, when Wallace published the pamphlet "Correlation and Machine Calculation" early in 1925, he continued to think of path coefficients primarily as a way to calculate multiple regression coefficients. Wright sent him a long letter on February 19, 1925, objecting to this very limited characterization of the method of path coefficients. Wright's letters in general, although not the particular one to Wallace quoted above, were often hard for those who received them to understand. Thus the ideas Wright tried to convey were sometimes misunderstood.

The paper on corn and hog correlations (Wright 42) still appears impressive to present-day econometricians. The corn variables used by Wright were corn acreage, corn yield per acre, corn crop (product of acreage and yield per acre), and average farm price of corn. The hog variables were the size of slaughter at large markets, average weight, pork production, average market price, and an average farm price. Wright considered separately the data from the eastern (primarily Atlantic seaboard) and western (primarily corn belt) sections, and he also analyzed separately the data from the western pack (size of wholesale slaughter) by winter and summer packing seasons. After presenting data on each variable and taking statistical account of long-period trends, Wright calculated the correlations between the variables, correlating most of them with each other over five years (one ahead plus three behind the year under consideration) instead of for only one year. This yielded a total of 42 variables, from which, by the use of card punching, sorting, and calculating machines, Wright calculated 510 separate correlations. Now Wright was faced with the seemingly hopeless task of trying to understand this enormous array of correlations in a quantitative causal scheme.

Here was where the method of path coefficients entered. Because Wright
had already calculated all the correlation coefficients, he could construct required path coefficients. He took the major variables and constructed a causal
scheme, which, when outfitted with path coefficients, yielded the observed
correlations. After going through a process of basically educated trial and error, Wright found a single causal scheme with the four major variables (corn
price, summer hog price, winter hog price, and hog breeding) and only fourteen paths of influence that gave "fairly satisfactory prediction of observed
correlations" (fig. 5.9; reproduced from Wright 42, p. 47). Indeed, in the
many pages of graphs and figures depicting observed correlations, Wright
had entered predicted and observed values, and they were clearly quite close.

The immediate question any interested observer might ask is whether
other causal schemes might predict the observed correlations as well or perhaps better. Speaking of the assumed system of causal relations, Wright
asked, "Is it the only hypothetical system which could do this?" (Wright 42,
p. 54). He answered that he had tried a great many other causal schemes, including some persuasive and intuitively satisfying, but found none of them
approximately as successful.

> The successful system is based on the analysis of factors which give
> the maximum percentage determination of each variable according to
> Pearson's method of multiple correlation, rather than on precon-
> ceived ideas of what the relations ought to be. From this analysis and
> the numerous tests of other systems, the writer feels confident that no
> other equally simple system can be found which differs substantially
> from the present and gives even approximately so good a fit to the
> observed correlations. Doubtless slight improvements can be effected
> by small modification in the value of path coefficients or in the addi-
> tion of minor paths. A slightly different interpretation could
> doubtless be given to some of the paths here adopted. The actions
> and reactions among the variables are necessarily too complex to be
> represented more than roughly by any simple diagram. (P. 54)

This is the most complex and extensive set of observed correlations to which
Wright ever applied the method of path coefficients; the outcome was better
than he had expected.

Limitations of Path Coefficients

During his Washington years Wright refined and demonstrated the power of
his method of path coefficients. As Wright himself stated, the method had
certain limitations that should be understood. Even the possibility of assigning quantitative values to specified causal pathways was based upon a series
of assumptions, some of which are troublesome. Deducing a causal scheme is
frequently difficult, yet a causal scheme was a basic prerequisite for the

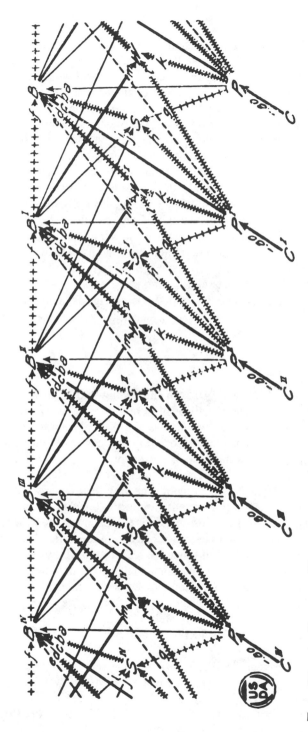

FIG. 27.—A diagram illustrating the system of interactions between corn crop (C), corn price (P), the summer price of hogs (S), the winter price of hogs (W), and the amount of hog breeding (B) in successive years, which has been found to be most successful in explaining the observed correlations. The most negative paths of influence are represented by plain arrows, the positive, by crosshatched arrows. The most important paths ($c = -.85$, $d = +.65$, $m = -.65$, $p_{pc} = -.80$) are represented by heavy lines, the least important ones ($e = -.15$, $f = +.10$, $g = +.15$) by broken lines, and the paths of intermediate importance ($a = -.45$, $b = +.35$, $h = +.50$, $i = -.40$, $j = -.40$, $k = +.45$, $l = +.25$) by light lines

Figure 5.9. Corn and hog correlations.

151

working of Wright's method. Logically, many causal schemes can possibly lead to the same observed correlations. Without added data or assumptions, path analysis provides no means of discrimination between the causal schemes. Another problem is the assumption of linear relations. Many processes in biology are known to be nonlinear, and unless they can be closely approximated by linear functions, often a difficult task, path analysis is not applicable. Also, any variable in the path diagram had to mean the same thing in all relationships, a dubious assumption in many cases.

After Wright published the method of path analysis, some researchers hoped that it was really a method for deducing causal relations. Wright repeatedly denied this, yet the hope did not die easily. He constantly reminded readers that a prior causal scheme was required and could not be derived by the method itself. The method was best applied to very definite causal relations, as in the case of Mendelian inheritance, for the analysis of systems of mating. But here the aim was not to derive the path coefficients but to use them to deduce the correlation between uniting gametes.

One thing is certain. Despite all limitations, Wright's method of path coefficients was basic to many of his own scientific achievements and to those of many others in diverse fields.

Principles of Livestock Breeding

Wright's work between 1920 and 1925 on inbreeding and crossbreeding in guinea pigs and on the Mendelian analysis of pure breeds of livestock greatly influenced the development of his own ideas about not only animal breeding but also evolution in nature. The close association of ideas of evolution in domestic and natural populations had strong precedent in the work of Charles Darwin. In *On the Origin of Species* Darwin had no direct evidence whatever about the action of natural selection in nature; his only examples were, as he clearly stated, imaginary. Darwin's presentation of the concept of natural selection is convincing because he gives abundant evidence for the efficacy of artificial selection and then analogizes natural selection to artificial selection. Darwin does this exercise so skillfully that the reader (especially the modern sympathetic one) has the feeling that Darwin actually gives much evidence of natural selection in action; but he does not. For Darwin, unlike Alfred Russel Wallace (just compare their 1858 Linnean Society papers), the conceptions of artificial and natural selection were intimately associated. Darwin of course wrote a great deal about artificial selection, particularly in *Variation of Animals and Plants under Domestication* (1868).

During Wright's ten years in Washington he was deeply immersed in the field of animal breeding—his major official responsibility at the USDA. During the second five years Wright wrote extensively upon the subject. In historical perspective it is possible to see clearly the development of Wright's ideas of animal breeding and how he applied what he had learned about it to evolution in nature. Darwin's ideas about evolution in nature gained from their association with and derivation from evolution in domestic populations; the same can be said for Wright.

There was, however, a world of difference between the analytic concepts available to Darwin and Wright. Darwin had his provisional hypothesis of pangenesis to account for the diversity of hereditary phenomena, and Wright had twenty years of modern Mendelian genetics. Just compare Darwin's analysis of color inheritance (Darwin 1868, chapter 27) with Wright's. Understandably, Wright was able to penetrate more deeply into the analysis of the effects of inbreeding and crossbreeding and derived a different view of the dominant processes involved than did Darwin. And Wright's view of evolution in nature, based upon his experience with evolution in domestic populations, thus also emphasized different aspects of the evolutionary process than did Darwin's.

Wright's ideas about animal breeding developed considerably during the period 1920–25. Aside from a brief note entitled "Scientific Principles Applied to Breeding," which Wright published in *Breeders Gazette* in 1919 (Wright 21), his first publication in the field of animal breeding came in the form of a USDA bulletin in December 1920 with the title *Principles of Livestock Breeding* (Wright 23). This bulletin was notable because Wright characteristically declined to pursue the usual approach taken in nontechnical USDA bulletins of presenting simple, straightforward directives with a few even simpler explanations. A bulletin directed toward practical animal breeders would especially be expected to be totally nontechnical. While Wright's approach was not really technical, after a brief statement about the evolution of animal breeding Wright began a basic course on the biology of reproduction and heredity, which lasted for over thirty pages. He began with cell theory and advanced to Mendelian heredity, including discussions of blending and alternative inheritance, chromosomes, linkage, and sex determination. Wright obviously wished to provide the biological background necessary for the reader to understand why the suggestions on animal breeding that followed actually worked. After this bulletin was published, Wright's superior, George M. Rommel, rewrote it in simpler form, leaving out all the background detail so the farmers could read and understand it.

Wright's bulletin actually appears very understandable to an average untrained mind, so far as I can see. Certainly the sections on actual breeding practice were understandable to almost anyone who could read or be read to and who had experience with livestock. Livestock improvement, according to Wright, meant achieving uniformity and conformity to the desired type. The means for moving toward these goals were straight mass selection, inbreeding accompanied by selection, use of prepotent males (largely homozygous for the desired qualities, hopefully mostly with dominant factors), and crossbreeding followed by inbreeding and selection.

The crucial passage concerning the relation of selection to inbreeding is important because it is here that one can see the greatest development in Wright's thinking during the following few years.

> Consistent selection toward the desired type is sometimes all that is necessary to fix a characteristic. Unfortunately, experiments have shown that what appear to be the same characteristics in two animals often depend on wholly different combinations of hereditary factors. A good example has been given in another connection in the case of two strains of light-eyed, yellow rats, each of which bred true by itself, but which produced nothing but black-eyed rats when crossed with each other. Thus progress by straight selection may be wholly upset at any time by an unfortunate cross of this kind. The whole breed must be lifted up at once if there is to be success by selection alone. Careful selection with breeding confined within a single herd or a few related herds, on the other hand, only requires that this small group be lifted up once. Once success has been obtained such a herd or group of herds becomes a powerful source of breed improvement by supplying prepotent sires. Practical experience agrees with theory in the principle that the only systematic method of fixing heredity, and so bringing out such prepotency as is in the stock, is Bakewell's old method of close breeding accompanied by careful selection. (Wright 23, p. 37)

I gain the definite impression from this passage that Wright was thinking in terms of Castle's selection experiment on hooded rats. As long as selection was confined to a relatively isolated group protected from outcrossing, the straight selection of those animals exhibiting the desired characters was the quickest means of changing the group and fixing the characters. The wider spread of the achieved desired characters was then accomplished by the use of prepotent sires. In 1922, after writing up fully for the first time the data from the inbreeding experiment on guinea pigs, Wright would show less confidence in the effects of straight selection as the best means of rapid improvement and place greater emphasis upon the role of close inbreeding as a component of the process of improvement.

Inbreeding and Crossbreeding in Guinea Pigs

Wright had been at the USDA for almost seven years when his first substantive analysis of the inbreeding experiments on guinea pigs appeared in two "professional paper" USDA bulletins in November and December 1922. The two bulletins were actually one long report divided into three sections: decline in vigor following inbreeding, differentiation among inbred families, and crosses between the inbred families (Wright 33–35). This initial report covered primarily the years 1906 to 1915, with some data from the period 1916–19. Most of the data from the years 1916–24 was not published until 1929, in a technical bulletin by Wright assisted by Eaton (Wright 55). The later data corroborated the conclusions published in 1922, and therefore I will not treat it separately.

By 1922 about 34,000 guinea pigs had been recorded in the experiments, with 25,000 in the inbred families, 4,000 in the control stock, and 5,000 in crosses between inbred families. Some clear results emerged from the data. Brother-sister mating for over twenty generations produced no obvious degeneration, as in bodily deformities. There was, however, a significant average decline in all measured elements of vigor: mortality at birth and before weaning, weight, frequency and size of litters, and resistance to tuberculosis. Most noticeable was the conspicuous differentiation among the twenty-three inbred families. Families differed in color patterns, number of toes, and elements of vigor. The first bulletin contained plates exhibiting the five families still in the experiment (p. 16); each family was very distinctive. One might never even suspect that so much variability as exhibited by all twenty-three families (each family having a distinct array of characteristics) could possibly be hidden away in the control stock. Yet intense inbreeding had brought all this hidden variability into plain sight.

Although each family exhibited distinct characteristics, much variability within each family remained. As described earlier, Wright had applied the method of path coefficients to the question of how much of this within-family variability was genetic and how much environmentally determined, and as expected he found very little was genetic. Finally, Wright found that two generations of crosses between the inbred families yielded offspring much superior to both parental stocks in all elements of vigor.

Wright interpreted all these results as being in harmony with the application of Mendelian theory to the problems of inbreeding and outbreeding, as exemplified particularly in *Inbreeding and Outbreeding* (1919) by East and his student Donald F. Jones:

> Analysis of the various crosses indicates that the results are all the direct or indirect consequence of the Mendelian mechanism of heredity. The fundamental effect of inbreeding is the automatic increase in homozygosis in all respects. An average decline in vigor is the consequence of the observed fact that recessive factors, more extensively brought into expression by an increase in homozygosis, are more likely to be deleterious than are their dominant allelomorphs. The differentiation among the families is due to the chance fixation of different combinations of the factors present in the original heterozygous stock. Crossing results in improvement because each family in general supplies some dominant factors lacking in the others. Dominance or even imperfect dominance in each unit character is built up into a pronounced improvement over both parent stocks in the complex characters actually observed. (Wright 35, pp. 48–49)

Wright could easily have ended the report after this conclusion relating the inbreeding experiment to the mainstream of Mendelian genetics. But the motivation behind Rommel's original establishment of the experiment was to discover the relation of inbreeding to livestock improvement. By the end of

1922 Wright had reached what was in his mind a much clearer view of this relationship than he expressed in the *Principles of Livestock Breeding* two years earlier.

To anyone familiar with Wright's shifting balance theory of evolution, this conclusion to the inbreeding experiment on guinea pigs will begin to sound familiar.

> It is believed that the results point the way to an important application of inbreeding to the improvement of livestock. Nearly all of the characteristics dealt with here, like most of those of economic importance with livestock, are of a kind which is determined only to a slight extent by heredity in the individual. About 70 per cent of the individual variation in resistance to tuberculosis and over 90 per cent of that in the rate of gain, and size of litter is determined by external conditions. Progress by ordinary selection of individuals would thus be very slow or nil. A single unfortunate selection of a sire, good as an individual, but inferior in heredity, is likely at any time to undo all past progress. On the other hand, by starting a large number of inbred lines, important hereditary differences in these respects are brought clearly to light and fixed. Crosses among these lines ought to give a full recovery of whatever vigor has been lost by inbreeding, and particular crosses may be safely expected to show a combination of desired characters distinctly superior to the original stock. Thus a crossbred stock can be developed which can be maintained at a higher level than the original stock, a level which could not have been reached by selection alone. Further improvement is to be sought in a repetition of the process—the isolation of new inbred strains from the improved crossbred stock, followed ultimately by crossing and selection of the best crosses for the foundation of the new stock. (Wright 35, p. 49)

Here was significantly less reliance upon direct selection than seen in Wright's views of 1920 and a greater emphasis upon close inbreeding as a means of bringing to light the variability, in complete interacting systems of genes, upon which selection could then act. He also discussed the selective diffusion from superior herds by use of superior males as the final step of improvement of a breed. Wright had seen the ability of close inbreeding to produce genetically distinct inbred lines in the corn plants shown to him by G. H. Shull at Cold Spring Harbor before Wright went to Harvard. Wright recalled Shull's work at the time as harmonizing perfectly with the results from inbreeding guinea pigs.

Wright closed the report by asserting that the method he advocated had in fact been used, perhaps unconsciously, by the pioneer breeders Bakewell, the Collings, Bates, Cruickshank, and others in the development of most of the modern recognized breeds of livestock. Given this assertion, it is not surprising that Wright, armed with his newly invented methods for measuring inbreeding coefficients in pedigrees and in populations, almost immediately turned to the herd books for a systematic analysis of the actual breeding practices in the foundation of pure breeds of livestock.

Mendelian Analysis of Pure Breeds of Livestock

Long-term accurate breeding records for pure breeds of livestock were generally nonexistent in the 1920s. One exception was the case of Shorthorn cattle. Intense efforts on the part of some breeders in the 1780s to improve the breed were accompanied by the keeping of accurate herd books. Thus Wright had pedigrees for Shorthorns continuously from the 1920s back to the 1780s, and from these records and other information about the early breeders he hoped to reconstruct in the light of modern genetics how the Shorthorns had actually evolved. The results of Wright's analysis were published in four parts, two in late 1923, one in 1925, and the last in 1926; on the last two Wright collaborated with his associate H. C. McPhee, who took Wright's vacant position when Wright left for the University of Chicago in 1925 (Wright 38, 39, 44, 49).

Wright first explained the foundation of Shorthorns in the hands of the Colling brothers in the 1780s and in particular discussed their famous bull Favourite. This early foundation stock was highly inbred and then selected for the most desirable prepotent animals, Favourite being the best known. Wright next turned to the Duchess family of Shorthorns bred by Thomas Bates. Bates began his herd by purchasing one cow, Duchess, from the Colling herd. From this cow he derived over the next forty years sixty-three cows and forty-five males, being thus a relatively unprolific family. They were, however, highly admired animals, commanding very high prices. (Twenty-four years after Bates's death in 1849 one Duchess cow sold for $40,600 in the United States.) Wright wished to analyze the breeding methods followed by Bates. Using the previously described methods for computing inbreeding coefficients, Wright discovered that for forty years Bates had maintained a nearly constant 40% degree of inbreeding in the herd, and a degree of relatedness to Favourite which began at 69% and dropped gradually to 57% (in other words, all of Bates's animals were more closely related to Favourite than a parent is to offspring or a brother to sister). On several occasions Bates introduced into the herd animals from outside the Duchess family, but all were closely related to Favourite and thus did not greatly lower inbreeding coefficients in the family.

Basically, Bates took a member of the highly inbred foundation stock and then maintained a degree of inbreeding great enough to yield substantial uniformity yet at the same time produce sufficient combinations upon which selection could be practiced. The key, according to Wright, was to sustain just the right degree of plasticity.

If Bates had not maintained close relationship between the animals which he mated, the relatively high degree of inbreeding, and close relationship to one animal (Favourite) his material would probably have been too plastic. The simultaneous variation in all characters would have been more than he could have contended with. If on the other hand he had bred wholly within his herd and between full brother and sister as far as possible, his material would soon not have been plastic enough to mold into shape. Undesirable characters, moreover, would almost certainly have become ineradicably fixed. (Wright 39, p. 416)

Putting together what he had learned from the analysis of Shorthorn breeding, inbreeding and crossbreeding in guinea pigs, and Mendelian heredity, Wright suggested the following general procedures in animal breeding. Clearly he had moved well beyond the widespread view that simple straight selection within a breed was the best means to successful breeding.

The first step in any case should be selection of a vigorous foundation, approaching as closely as possible to the desired type. This was the step taken by Collings. . . .

With such a foundation stock, one might practice the most intensive inbreeding in a large number of distinct lines, knowing that most lines would inevitably deteriorate greatly, but trusting that a few would be found in which desirable qualities would become fixed, and in which the deterioration in any vital respect would be so slight that they could be maintained successfully. By crossing such lines which have withstood this acid test of inbreeding, one might reasonably hope to recover more than the original vigor and retain those characters which had been fixed. Such a method is especially indicated where the characters are of a kind determined so slightly by heredity that genetic differences can be recognized only on comparing lines which have been kept distinct and free from outside blood. This method, an alteration of intensive inbreeding with selection and crossbreeding of the few successful lines must naturally be done on a large scale and with the undertaking of considerable risk. It is a method adapted rather to experiment stations than to private individuals. It is, however, an important method and has some parallel in the general history of the breeds. Many of the early breeders practiced close inbreeding. Only a few like the Collings were notably successful. The strains were crossed to found the present pure breeds.

For the individual breeder, however, theory as well as practice indicates that the most reliable method is the maintenance of a steady level in closeness of breeding coupled with persistent selection toward the desired type, the requisite closeness of breeding depending, naturally, on the heterogeneity of the foundation animals and the breeder's skill as a judge of livestock. (Pp. 417–18)

Transformation of the entire breed depended primarily upon the exportation of sires from the foundation herd.

By the time Wright published the above in December 1923, he had definitely arrived at the basic view of evolution in domestic populations that in 1925 he would extend to evolution in nature. Like Darwin, Wright saw evolution in nature as deeply similar to evolution in domestic populations.

The later two papers in the series extended the analysis of the history of breeding in Shorthorns but did not change the basic view presented in 1923 of evolution in domestic populations. The third paper in the series used the approximate method of calculating coefficients of inbreeding and relationship from livestock pedigrees discussed earlier and also developed a second measure of inbreeding in the whole population (later termed F_{ST}), which is derived by calculating the inbreeding coefficients of offspring from hypothetical matings of randomly chosen animals in the populations (Wright 44). The figure indicated that there had been little active additional inbreeding since the foundation period except for periods in the first and third quarters of the nineteenth century. The whole Shorthorn breed by 1925 was so inbred compared to the foundation stock that the degree of relationship between random animals was about 40%. In the fourth paper, McPhee showed that in being separated from other Shorthorns the dairy Shorthorns were so little inbred that they constituted (in terms of degree of relationship) basically a random sample of the Shorthorn breed. In other words, dairy Shorthorns were formed entirely by selection and not to any significant extent by inbreeding (Wright 49).

Concluding Remarks on the Washington Years

The Washington years were highly productive for Wright, not just in terms of the number of papers published but primarily in terms of the invention of major ideas that he would pursue further and refine in the years to come. Wright himself has said on many occasions that during the ten years at Washington he had, at least in beginning form, most of his major ideas in the fields of genetics and evolution. The one major piece of Wright's thinking not dealt with in these two chapters on the Washington years concerns the origins of his conception of evolution in nature. Before leaving Washington in December 1925, Wright had in typescript a long essay on evolution in nature; he had even given it to Collins for his critical reading and opinion. This version of the essay unfortunately no longer survives, the only available version being that published in *Genetics* in 1931 as "Evolution in Mendelian Populations." Because Wright's ideas of evolution in nature are so important and influential, I thought it best to treat their development in later chapters. Even with the evolutionary theory left aside, Wright's ten years in Washington were remarkable for productivity and inventiveness.

6
The University of Chicago and Physiological Genetics, 1926–1955

During the ten years at the USDA, Wright had encountered some difficulties in pursuing his own research interests. The editorial office of the Bureau of Animal Industry had disapproved of both the major paper on path coefficients ("Correlation and Causation") and the paper on corn and hog correlations. But the basic situation for research had actually improved after 1921 with the appointment of E. D. Ball, a hemiptera specialist, as director of research at the USDA. Ball was active in promoting original research at the USDA; he brought in persons with advanced training and provided funds for visiting research experts. With his doctorate from Harvard, Wright was in high favor with Ball. On one occasion Rommel had recommended a salary figure for Wright, and Ball took the unusual step of raising it considerably. Any difficulties within the Bureau of Animal Industry thus paled in comparison to the general support given to Wright's work.

A measure of Wright's attitude toward the USDA position can be seen in his refusal of two academic job offers between 1920 and 1925. The first came from Little, who was then president of the University of Maine. Little was trying to build up the field of genetics there and tried to hire Wright as well as Charles Metz. Wright and Metz discovered by talking to John Gowen and Karl Sax, both already at Maine, that Little, by trying hard to promote genetics, had engendered so much opposition that he was in danger of being forced out of the presidency. Wright and Metz both refused the offer. A second offer came to Wright from the University of Texas. The position was attractive, but Wright did not wish to go so far south. In later years Wright often said he did the University of Texas a great favor by refusing the offer. The position was then offered to H. J. Muller, who began a top-level program of genetics research that has continued to the present (although Muller himself left in 1932). In late 1924 the Wrights were sufficiently certain of a long-term continuation of their stay in the Washington area that they purchased a lot in University Park, just beyond Hyattsville, Maryland. Sewall and Louise had great fun making three-dimensional models of the house they would build. They were all set to go to a contractor when the offer of a professorship came from the University of Chicago.

The first hint Wright had of the position at the University of Chicago came in a letter from William Castle, dated November 20, 1924.

Dr. [Frank R.] Lillie of Chicago University is looking for a geneticist on his staff to replace [Albert William] Bellamy recently resigned. He wants a "good man" and can offer $2500–$5000 salary according to qualifications of the nominee. I should like to recommend you, if you have any inclinations toward a University connection, for I believe this is the most promising University post which has been offered for many years. Please let me know what your attitude is. I don't want to recommend anyone who will not consider the place seriously. Also please consider this information confidential.

Wright responded three days later:

I was much interested in hearing of the position at the University of Chicago and I would certainly consider it seriously if the opportunities for research and the salary would justify the change. The conditions for research here were much improved under Dr. [L. J.] Cole's administration and I believe will continue so. We have a laboratory well fitted out with microscopes, microtomes, etc. We have had tabulating and sorting machines in addition to calculating machines for statistical work for some time. The most serious effect of a change however would probably be in the interruption to the problems which I have under way. Guinea pigs are of course rather slow and expensive material. It takes years to work out a problem or even to get ready to start on it properly. I have 8-factor recessive and 8-factor dominant strains which took a good many years to develop.
The five inbred families, with behavior and characteristics analyzed since 1906 are also ideal for certain experiments. I have your four-toe stock in a pure condition entirely descended from three animals which you sent, and have a great deal of F_1, F_2, F_3 and backcross data from crosses with four of the inbred families, each giving characteristic results. If I left the Department, I would probably prefer to take up some other animal. As to salary, I am now getting $4000. In three of the last four years I have received $300 for teaching in the Department graduate school. I taught general genetics in 1921, animal breeding in the first half of last year, advanced statistical methods in the second half, and general genetics again this year.

Wright heard nothing more about the University of Chicago position until he received on February 2 a letter from Frank R. Lillie, chairman of the Department of Zoology, inviting him to accept "an associate professorship in charge of the subject of Genetics," at a salary of $4,000. Wright, Lillie said, "would be free to devote himself very largely, if not exclusively, to the development of the subject of Genetics within the department but would at the same time be expected to take an interest in the policies of the department and be ready to share in the work of the department as needed." The University of Chicago was on the quarter system, and Wright could be expected to teach a full load of two courses (five hours per week per course) for three of the four quarters in a calendar year.

The University of Chicago offer came as a surprise and raised serious issues for Wright, as he wrote to his old friend from Galesburg, Willis Rich, then working for the Bureau of Fisheries of the Department of Commerce at Stanford University.

> The offer has thrown me very much up in the air. In many ways it is just the sort of position I would like. The Zoology department there is a strong one, and I could direct my research in the line in which I am most interested, which is theoretical rather than practical. I think that I would like to try linking genetics more with development, in which, of course, Chicago is especially strong. On the other hand the salary is no more than I am getting here, not counting the extra I get for teaching. I also have never thought of Chicago as a desirable place to live. I hate to think of living anywhere where I am not within easy walking distance of the country. This applies especially to raising a family. Finally, not having much teaching experience, I am not sure how much teaching I could carry and still have time and energy for research. (Wright to Rich, February 7, 1925)

A day after this letter, Wright wrote to Lillie expressing his interest in the position but also the concerns in the letters to Castle and Rich. Lillie immediately wrote back to reassure Wright that research facilities were excellent and would be further tailored to meet his special needs. Members of the departments of physiology and mathematics stood ready to provide stimulus and help.

> As regards time and opportunity for research I can only say that it would defeat our own main purpose in calling you if you were to be seriously hampered by routine teaching. I suppose that during the first year, teaching problems would require an abnormal proportion of your time. This would be true I suppose of any transfer to an academic post. You know how productive some men are, e.g. [Charles Manning] Child and [Carl Richard] Moore, who have normal teaching responsibilities. . . .
>
> We feel that Genetics is a very important subject in the Department and for the University. You would have a free hand there, except for such elementary teaching in the subject as [Horatio Hackett] Newman does for instance. If you have faith in the subject and in your relation to a University post, I think you could count on promotion as surely as anyone in the University.

Wright found Lillie's letter reassuring. He had long thought of himself as a theoretically oriented researcher somewhat out of place in the Bureau of Animal Industry, but it was true that his research had the strong support of Ball, Collins, and some others at the USDA. The greatest attraction of the

University of Chicago position was the possibility of association with a department of zoology renowned for its work in developmental biology—Wright's hope, as it had been since graduate school at Harvard, was to connect genetics with development. With the existing ties and advantages in Washington, the decision was difficult.

While at the USDA, Wright had always been acutely aware of the issue raised by East in 1915—that government bureaucrats were subject to dismissal and replacement on political grounds and that the administrators in charge of the research at the USDA could easily ruin it in short order. Just as he was considering the University of Chicago offer in February 1925, Wright got word that President Coolidge, in the interests of economy, had fired Ball as director of research at the USDA. Unhappy as he and Louise were at the prospect of living in Chicago, Wright wrote on March 1 to Lillie accepting the University of Chicago offer. Lillie raised the salary offer, allowed Wright to continue at the USDA until January 1926, and guaranteed 120 pens made to order for housing guinea pigs. Wright decided to join the department.

Louise Wright was saddened to relinquish the beautiful lot and the planned house but heartily approved the move into an academic position. The Wrights sold their house and rented a small new house in Mt. Rainier, Maryland, giving them the needed flexibility for moving to Chicago. The rest of the time in Washington was spent trying to salvage everything possible from the USDA research projects and preparing for a strenuous first year of full-time teaching.

Wright had exactly ten months from the time he accepted the University of Chicago position as associate professor until he was due to assume his duties there on January 1, 1926. During these ten months Wright had to make some serious decisions about the future course of his research. He had already devoted thirteen years of intensive research to guinea pigs and had a large number of experiments under way with them; yet as Wright indicated in the letter to Castle quoted above, he thought he would probably want to change his experimental animal if he went to the University of Chicago. Wright clearly had in mind orienting his research more directly toward theoretical questions, a course he considered appropriate for a major research university. The problem was that Wright already had underway at the USDA more long-term experiments than could possibly be completed in a brief ten months. It was impossible to begin new experimental programs with guinea pigs before leaving the USDA.

Wright responded to this situation in two ways. First, he arranged with E. W. Sheets, chief of the Animal Husbandry Division, and J. R. Mohler, chief of the Bureau of Animal Industry, that he would continue working after January 1, 1926, with the USDA as "collaborator in animal genetics in the Animal Husbandry Division." This appointment would enable Wright to finish up the experimental work already underway at the USDA but which would not be completed when Wright left for Chicago. It also enabled Wright to request that samples of stocks crucial for future experimentation be shipped to Chicago so he could work with them in the facilities Lillie was having prepared for him. Late in November 1925, when Wright could see how the various projects were shaping up, he wrote a memorandum to Sheets outlining the projects he intended to pursue with stocks at the USDA (mostly they consisted of the analysis of already obtained data) and listing the stocks Wright wished to take with him to Chicago. The two major projects to be continued at the USDA were the comparative study of the five remaining inbred families and a study of the interrelations between such characters as weight, fertility, and vitality. Those Wright wished to pursue at Chicago were studies of (1) the factors determining otocephaly, (2) the factors determining polydactyly, (3) the linkage relations between known Mendelian factors, and (4) the color factors.

Few of these projects, however, could be accomplished rapidly. Wright wished to make an overt move toward the kind of theoretical research he expected to pursue (and would be expected by his academic colleagues to pursue). His second response to the situation at the USDA during his last ten months was therefore to turn his serious attention to the theory of evolution in nature. In 1925 geneticists had scarcely begun to actually study evolution in natural populations; they understandably worked in laboratories mining the obvious niches revealed by the development of genetics in the first quarter of the century. The only major studies of which Wright was aware in 1925 were those on natural populations of *Lymantria* (the gypsy moth) by Goldschmidt and those on natural populations of *Peromyscus* (the deer mouse) by Sumner. But Goldschmidt did very little Mendelian analysis in his studies of *Lymantria*, and until 1925 Sumner had vociferously objected to the attempt (by Castle and his students and Morgan's group) to interpret the results from *Peromyscus* according to Mendelian genetics. Indeed, Sumner was until 1925 a leading neo-Lamarckian who opposed a Darwinian interpretation of evolution in nature. Wright himself had made no genetic analysis of any natural population. Thus his attempt in 1925 to construct a theory of evolution in nature was in the deepest sense a theoretical venture.

Basically, as I will explain in detail in chapter 8, Wright took what he knew of genetic inheritance and genetic change over time in the populations he had really studied, particularly guinea pigs and Shorthorn cattle, and applied this knowledge to evolution in nature. Before Wright left the USDA, this project had become a manuscript of about one hundred pages and he had given it to Guy Collins for his criticism. Several features of this manuscript worried Wright. It contained a long introductory section on artificial breeding and selection, thus representing accurately the actual train of Wright's thinking, but the manuscript was way beyond the length preferred by most journals. Another problem was that Wright was dissatisfied with his quantitative analysis of the influence of dominance.

According to a letter from Lillie in October 1925, after Wright had visited with him at Woods Hole in July, Lillie was under the impression that Wright was writing a book rather than a long article. Whatever the case, when Wright moved to Chicago he was not prepared to submit the manuscript for publication. Teaching responsibilities and the writing up of the research left over from the USDA inundated Wright, and he did not get back to the evolution manuscript for several years. After an interesting interaction with Fisher about the evolution of dominance and revisions and deletions, Wright finally submitted the manuscript to *Genetics* on January 20, 1930, over five years after he first wrote it. The paper appeared in March 1931 under the title "Evolution in Mendelian Populations" (Wright 64) and immediately became a highly influential work.

After ten months of intense activity in preparation for the move to Chicago, and with a new baby in the house (son Rob was born on June 5, 1925), the Wrights finally packed all their belongings into a moving van and headed for Chicago.

Life in Chicago

The arrival in Chicago in early January 1926 was hardly auspicious. A large mirror and some other belongings were smashed in transit. Because they had no time to locate a house in the Chicago area, the Wrights considered themselves lucky to have access to a small apartment vacated midyear by a boyhood friend from Galesburg, Jerome Fisher, son of president L. B. Fisher of Lombard College. The apartment was dark and jammed with the furniture formerly spread through a whole house. To top it off, Louise and both boys came down with chicken pox.

In 1927 the Wrights moved into a rented house, and in 1928 they bought a house at 5762 Harper Avenue in which they lived until moving to Madison in 1955. While the house was not very large, it had enough room for a ping pong table and workshop in the basement. It had a relatively large back yard with a cottonwood and a mulberry tree and a smaller front yard with a linden tree. The house was near the Illinois Central railroad tracks but otherwise was situated on a quiet street. Many University of Chicago faculty members lived on the same street or nearby, including Wright's best friend in the zoology department, Benjamin H. Willier, John Wilson (of the Oriental Institute), and Ralph Gerard (physiology). The living situation was far more congenial than the Wrights had imagined before moving to Chicago.

With the heavy teaching schedule Wright assumed and with the love of the country all the Wrights shared, vacations were more a necessity than a luxury. Every year from 1925 to 1955 the Wrights took a vacation. Both Sewall and Louise greatly enjoyed driving, and they often drove long distances during the vacations; they also frequently drove to annual meetings of the many societies to which Wright belonged. In the first year Wright taught from January until August, and the family was eager for a vacation during

August and early September. There was no extra money, but the Wrights bought a used Model T Ford for $150 and decided to spend a month or so at Woods Hole, going from there to visit Sewall's parents in Forest Glen, Maryland, and then to see Louise's parents in Granville, Ohio.

Wright greatly enjoyed the visit to Woods Hole. Most of the biologists who came there every summer had already returned to their institutions, but the few who remained were most congenial. The Wrights enjoyed their circuit so much that they repeated it the next year. One of those years, Sturtevant remained at Woods Hole through September. Wright had always greatly enjoyed and profited by talking at length with Sturtevant, who obviously felt the same way. Wright considered Sturtevant to be the member of the Morgan group with the greatest sophistication about evolution and quantitative analysis. Very enjoyable contacts with biologists were always a feature of the Woods Hole visits.

The Wrights' daughter Betty was born on May 9, 1929, at Chicago Lying-in Hospital on the far side of Washington Park, to the west of Hyde Park. All of the children were generally healthy, except on one occasion in 1930 when the oldest boy Dick had whooping cough and encephalitis and went into a coma for many days. It was the sort of case now easily treated with antibiotics but very life threatening at that time. Sewall was at the hospital when Dick woke from the coma; he seemed fine and remarked that the trees outside his window had grown leaves overnight. He of course didn't realize that he had been in a coma for days. In the early 1930s, both boys spent much of the summers at a camp run by Wright's Aunt Polly (sister of his mother) on a lake in northern Minnesota. Family life for the first ten years in Chicago was rewarding and satisfying.

The Department of Zoology

The Department of Zoology at the University of Chicago was three years younger than Wright himself, having been founded in 1892, the first year of the University of Chicago. President William Rainey Harper wanted particularly to found a research university with greatest emphasis upon graduate teaching. The first chairman of the Department of Zoology was Charles Otis Whitman, an embryologist who worked with E. S. Morse, a protégé of Louis Agassiz at Harvard's Museum of Comparative Zoology. Whitman took his graduate degree with K. G. F. R. Leuckart at the University of Leipzig. This education in the Germanic scholarly research tradition was basic to Whitman's chairmanship of the fledgling department. Before coming to the University of Chicago Whitman had helped found and was the first director of the Marine Biological Laboratory at Woods Hole, was a founder and the first editor of *Journal of Morphology*, started with William Morton Wheeler the *Biological Bulletin*, and had been chairman of the zoology department at Clark

University. After the first ten years as chairman of the department at Chicago Whitman stated in his decennial report to Harper: "From its foundation the Department of Zoology has emphasized the research side of its work. The results of this are evidenced, not only by the list of original contributions to Zoology by members of the Department, but also by the national scientific organizations and enterprises with which its members have been connected" (cited by Newman 1948, 216). Whitman's stamp upon the department was still very evident when Wright arrived in 1926, even though Whitman had died in 1910.

Among the appointees to the department in its first ten years were Jacques Loeb (1892), William Morton Wheeler (1892), Frank R. Lillie (1893; left in 1894 and returned in 1900), Charles Manning Child (1896), Charles Benedict Davenport (1899), and William Lawrence Tower (1901). Of these Wright knew Wheeler from the Bussey Institution and Davenport from the time of the summer after his senior year in college (1911). Lillie and Child were both still active in the department when Wright came in 1926. The graduate fellows were also prominent, including William A. Locy, Samuel J. Holmes, Michael F. Guyer, Horatio Hackett Newman (later a member of the department), William J. Moenkhaus, and Wilhelmine Entemann. Wilhelmine ("Minnie" at the University of Chicago) Entemann was of course, after marriage, Wright's biology professor at Lombard during his senior year there and the person who opened to him the joys of research in biology.

The list of publications by members of the department was enormous. Whitman had no interest at all in undergraduate instruction and encouraged members of the department to avoid it; they did. Graduate classes were small. Faculty spent most of their time on research. When Frank R. Lillie returned to the department in 1900, Whitman himself was deeply immersed in his research on heredity and evolution in pigeons. Whitman turned over most details of running the department to Lillie, making only the important decisions; he likewise made Lillie his assistant at the Marine Biological Laboratory, and Lillie succeeded him in both the chairmanship and directorship. Lillie instituted democracy in the department, but otherwise, in his own words, ran the department "by consistent adherence to Whitman's principles." This is not entirely true because when Lillie became official chairman in 1910, undergraduate teaching in the department had reached an embarrassingly low ebb and something had to be done. Lillie hired Horatio Hackett Newman (a former student who had taken his doctorate with Lillie in 1905) away from his job as chairman of the Department of Zoology at the University of Texas with the understanding that Newman would energize undergraduate teaching. According to Newman, "within two or three years the undergraduate program was brought to a level comparable with that of other scientific departments and was regarded as satisfactory by the general administration, which had previously been critical of the department's participation in undergraduate education" (Newman 1948, 226; I have drawn heavily from Newman's memoir in this section.)

Newman's effusive praise of undergraduate education, found in his memoir of the department, accurately reflects his own personal interests. He organized one of the most popular undergraduate science courses ever taught at the University of Chicago, Evolution, Genetics, and Eugenics. The book he developed for that course, *Readings in Evolution, Genetics, and Eugenics* (1921), went through several editions and was still kept in print by the University of Chicago Press during my undergraduate days there (1958–62). The course and the book were very influential. (For example, the renowned Chinese geneticist C. C. Tan told me this book made him an evolutionist when he read it in a small fundamentalist missionary college in China.) But it would be unfair to say that many members of the Department of Zoology shared Newman's enthusiasm for undergraduate teaching, especially for students not concentrating in the sciences. Most department faculty, including Wright, preferred to teach graduate students and to concentrate on research and publication.

The great strength of the department in 1925 lay in embryology. Whitman, Lillie, Child, and later appointees Carl Richard Moore (1916) and Benjamin Harrison Willier (1919) were all embryologists of one kind or another. There was only one ecologist, Warder Clyde Allee, who had taken his degree at Chicago and returned as a faculty member in 1921, concentrating on animal social behavior. (Later, of course, both Alfred Emerson and Thomas Park would join the Department, making it very strong in ecology.)

Genetics had never fared well in the Department of Zoology. Davenport had started well but soon left to become director of the Carnegie Institution of Washington Station for Experimental Evolution at Cold Spring Harbor, New York. W. L. Tower then inherited the position as geneticist in the department and for a time between about 1905 and 1912 gained a wide international reputation for his work on inheritance and evolution in the potato beetle *Leptinotarsa*. He firmly believed in the inheritance of acquired characters, a belief apparently borne out by his research. Then between 1912 and 1915 it gradually became clear that Tower's ideas about color and pattern inheritance made little genetic or physiological sense, and his beetles were discovered to have had their coat colors and patterns manually altered. Tower resigned his tenured associate professorship in disgrace in 1915 and disappeared from the world of science (some historians point to Tower's case as a precursor to that of Paul Kammerer).

The department members felt burned by Tower, but they had little use for modern genetics in any case. Genetics to them was a side issue to the fundamental problem of development. The department did not even hire another geneticist until 1919. This geneticist, Albert William Bellamy, left after four years. There was little enthusiasm for genetics left in the department when Wright was hired. Newman taught elementary genetics but did no genetics research. The department turned to Wright for two primary reasons: a geneticist was needed because a fine department of zoology could scarcely be without one, and Wright was the most prominent geneticist in the United States both interested in and sophisticated about the physiological basis of gene action. Wright undoubtedly pleased Lillie when he said that he hoped in his research at Chicago to elucidate the relationships of genes to development.

Wright soon found that most members of the department were indifferent or antagonistic to genetics. Lillie seemed enthusiastic only to the extent of giving Wright all of genetics as his field within the department. Most skeptical about Mendelian heredity was Child, who was about twenty-years older than Wright. They had many long discussions about genetics and became good friends, but Child was intellectually utterly hostile to Mendelian genetics and the chromosome theory. He thought that the few known connections between Mendelian genetics and development were too simplistic to possibly account for the very complex phenomena of the control of development. At times Wright would seem to convince him, but Child would soon slip back into his former skepticism. If Wright felt isolated from research geneticists at the USDA, and he often did, then the Department of Zoology at the University of Chicago was an even more isolated position. At least Wright had Collins at the USDA to criticize his manuscripts and make suggestions for revisions; at Chicago, Wright had no one among the faculty. Spurred in part by an experience with Fisher (that I will describe in chapter 8), Wright quit giving his manuscripts to anyone before publishing them. The department at Chicago was becoming an anomaly among modern departments of zoology because of its clear skepticism about genetics, differing little from the attitude of Whitman in the first years of the century. By 1925 there were many centers of genetics research in the United States.

To the molecular biologist of today, the skepticism of the embryologists at the University of Chicago in the 1920s concerning the relation between Mendelian genetics and development seems perhaps anachronistic or even wrong headed. It is historically more accurate to say that Lillie and Child were amply justified in their general view that Mendelian genetics had very little indeed to offer the developmental biologist. Wright's explicit aim of relating genetics to development in his work at the University of Chicago would prove almost as impossible as Child predicted.

Despite the skepticism about genetics expressed by department members, Wright soon established friendly relations with all members of the department. He became a special friend of Willier; the two families lived close together and also developed a warm friendship. In 1933, when Willier took a job as chairman of the Department of Zoology at the University of Rochester, the loss of the Williers was strongly felt by the Wrights. The Wrights visited the Williers on many occasions in Rochester and later in Baltimore, when Willier became chairman of the department at Johns Hopkins.

In the spring of 1926, soon after his arrival in Chicago, Child and Willier invited Wright to go with them on Saturday trips to the forest preserves around Chicago and to the Indiana Dunes at the southern tip of Lake Michigan. These trips were the equivalent of the "walks into the country" so dear to Wright, and he continued such trips (with changing associates) for the entire time (1926–55) that he was in Chicago. These trips were ideal for relaxation and discussion. They also were of some scientific importance. In the spring of 1927 Wright and Willier decided that they should know more about natural populations, especially taxonomy and variability. They began collections—Willier of dragonflies and damselflies (Odonata) and Wright of spiders. Wright immediately felt again the joys of identification and collecting that he had felt so strongly as a boy, but this time there was a greater sense of importance because he had already written but not published the long theoretical manuscript on evolution in nature.

Using Comstock's *Spider-Book* for identifications, between 1927 and 1929 Wright collected 101 different species from seventeen different genera. Wright found it useful to see first hand what kinds of differences taxonomists used to distinguish between species of the same genus as compared with those they used to distinguish different genera. Wright could see no particular usefulness in nature of the kinds of differences that distinguished species of the same genus. This observation confirmed his belief in the nonadaptive quality of species differences that was emphasized strongly by many prominent taxonomists of the day (I will develop this point in chapter 9). His enthusiasm for collecting spiders began rather obviously to take time from writing up his experimental results and keeping up with correspondence, so in 1929 he ceased collecting. Only after retirement from the University of Chicago did Wright again take up another taxonomic pursuit, this time wildflower identification.

Faculty social life was centered at the Quadrangle Club (the faculty club). Wright joined and made many friends; he greatly enjoyed the conversations there. With the friendships Wright developed both inside and outside the department soon after arriving in Chicago, and after his burgeoning national and international reputation after 1931, he was not discouraged by his isolation from research geneticists in the University of Chicago community. Indeed, Wright was very much a solitary thinker and liked quietly to work out problems by himself. The setting in the Department of Zoology generally had very positive effects upon Wright's research.

Teaching

Before coming to the University of Chicago Wright's teaching experience included only the work as Castle's assistant and several courses given to small groups of graduate staff members during the last three years at the USDA. He was apprehensive about assuming the full teaching load in the Department of Zoology, amounting to six courses a year, or an average of two for each of three quarters. Wright had good reason to be apprehensive about this teaching load. From the beginning, he prepared very carefully for each class, writing out detailed notes on cards. This in itself took a huge amount of time. He was also a shy man who provided no banter in class and often lectured formally to even small classes. Undergraduates who had no major interest in biology found this routine unappealing, and Wright felt uncomfortable teaching them. He most liked to teach graduate or postgraduate students who were taking his courses because they wanted to hear what Wright had to say. Teaching occupied a large proportion of Wright's time during the first ten years at Chicago, especially the first five or six. He had far less time for his own research than had been the case at the USDA.

The first quarter (Winter 1926) he was asked to give only one course, Fundamental Genetics (Zoology 310), which he continued to teach throughout his stay at the University of Chicago. Wright knew ahead of time that he would be teaching the course, and he wanted it to be a laboratory as well as lecture course. Guinea pigs were useless as experimental animals in a course lasting ten weeks, so while still at the USDA Wright obtained some basic *Drosophila melanogaster* stocks from the Morgan laboratory at Columbia University and became familiar with their use in demonstrating basic principles of genetics. The students were led through two locus crosses with and without interaction effects, multiple alleles, placing an unknown mutation with the proper linkage group (chromosome), and three-point mapping experiments (eosin, miniature, and Bar on the X-chromosome). Wright found that even those students who had difficulty grasping genetic theory usually obtained clear and understandable laboratory results. He prepared detailed lectures that he delivered to the class of seven students.

Drosophila was, compared to guinea pigs, such easy material to work with that one cannot help wondering if Wright was strongly tempted to change his experimental organism. Yet from the beginning of the Chicago years, it quickly became clear that guinea pigs would remain the focus of his research. Some of his graduate students (six of twenty-two who did their doctoral theses with Wright) chose *Drosophila* as their experimental organism. Wright always had *Drosophila* on hand for teaching and research.

To keep to the average of two courses per quarter, Wright gave three in the spring quarter of 1926. Two were undergraduate courses—Elementary Zoology (Zoology 103) and a section of Newman's Evolution, Genetics, and Eugenics (Zoology 105). In Elementary Zoology Wright prepared lectures but had an experienced assistant who took care of the laboratory exercises. Wright taught this course only one other time (Fall 1928). Newman's course always attracted many students, the course generally having three or four sections of thirty to fifty students. There was no laboratory, the course being designed for students in the humanities and social sciences who were required to take a science course. Wright lectured to this class and gave the students weekly ten-minute quizzes. He gave this course fifteen times between 1926 and 1932, when Newman's course was expanded into a whole general education program for which the College of the University of Chicago later became and still remains justly famous.

Most of the students in Zoology 105 seemed wholly uninterested to Wright and he found teaching the course unsatisfactory. He thought the students were unwilling to work and argumentative. He later recalled one student as particularly obnoxious. This student regularly avoided the assigned readings, but when Wright gave him Ds or Fs on the weekly quizzes he would come charging up after class to argue for a raise in the grade, which Wright never granted. The student's name was Saul Alinsky, who turned out to be one of the greatest social organizers in twentieth-century America. Alinsky returned to the Hyde Park area to help found the Woodlawn Organization. Clearly Newman did not elicit the same response from students in the course that Wright did. Part of the problem was that Wright's personality and interests were not well suited to undergraduate teaching of this kind. Students probably found his lectures too technical for their interests.

The third course he gave that spring quarter was Biometry (Zoology 311) primarily a graduate course that became a basic element of his teaching pattern until 1955. This was a highly respected course not only among graduate students but also among postgraduates and faculty from other institutions as well as from the University of Chicago. Much later, a good deal of the material that Wright developed over the years in this course appeared in the first volume of *Evolution and the Genetics of Populations* (Wright 188).

In the summer quarter of 1928 Wright introduced a new course dear to his research interests, Physiological Genetics (Zoology 312). Although this course reflected Wright's long-term research interests as well as the connections Wright hoped to see forged between genetics, physiology, and development, members of the zoology faculty were skeptical. Lillie and Child were openly skeptical at the time Wright proposed the course to the faculty. Lillie, Child, and most of the others attributed the characters of organisms to physiological processes (such as induction fields, organizers, etc.) whose workings were little illuminated by the science of genetics. Wright, like Goldschmidt and many other geneticists, did not deny the physiological mechanisms but maintained that gene action, mediated by enzymes and affected by environmental circumstances, determined the course of the physiological processes. The faculty allowed him to proceed. In this course Wright devoted three lectures to Garrod's work on errors of metabolism and dealt with gene replication (he thought the genetic material was protein at this time) as well as color inheritance, sex determination, and other issues.

The last major course Wright introduced in these years was Evolution (Zoology 313), which Wright first taught in the spring of 1931 (his long paper on evolution had just appeared in the March 1931 issue of *Genetics*). Fundamental Genetics (310), Biometry (311), Physiological Genetics (312),

Figure 6.1. Sewall Wright at the University of Chicago, 1928.

and Evolution (313) became after 1932 (with only two exceptions) the only courses Wright continued to teach at the University of Chicago. All four attracted unusually thoughtful students and auditors.

In 1928, 1929, and 1931 Wright taught all four quarters because the family was short of money. In these years vacations were brief and the writing up of research became more difficult. The entire teaching schedule for the years 1926–35 appears in the following chart:

Year	Winter	Spring	Summer	Fall
1926	310	103, 105, 311	105, 310	
1927	105, 310	105, 311	105, 310	
1928	105, 310	105, 311	105, 312	103, 310
1929	312	105	105, 310	310
1930	312	105, 311		310
1931	312	105, 313	105, 310	310
1932	312	105, 311	105	
1933	310	312, 313		310
1934	312, 314	313	310, 313	
1935	310	311, 313		310

Notes: Zoology 103, Elementary Zoology; 105, Evolution, Genetics, and Eugenics; 303, Introductory Biology; 310, Fundamental Genetics; 311, Biometry; 312, Physiological Genetics; 313, Evolution; 314, General Statistics.

By the time Wright came to Chicago he had already begun to gain wide recognition as a quantitative thinker, primarily because of the method of path coefficients and his analysis of systems of breeding, including the inbreeding coefficient. He was also, of course, well known as a mammalian geneticist. Postgraduate students began coming to him almost immediately. Among those in the early years were Jay L. Lush and A. B. Chapman in animal breeding, Edgar Anderson in botanical genetics, and Sheldon C. Reed and Donald Charles in mammalian genetics. Each was brought into contact with Wright by old ties—Castle sent Anderson and Reed, Cole sent Lush and Chapman, and Dunn sent Charles. All of these persons learned a great deal from Wright. Typical is the following letter from Edgar Anderson, in 1932 on the staff of the Arnold Arboretum at Harvard, after his quarter working with Wright and auditing his Biometry course:

As I take up again some of the problems I was working with before I went to Chicago I begin to realize how much I learned in my short stay with you. My primary purpose in going there was to bridge the gap between my fragmentary mathematics and the useful statistical tools which I knew about but did not understand. That objective was more than realized. The sight of prf^{-1} [inverse probability function] or the mystical χ, or "e" with a mystical exponent now greet me on the page with a happy flash of recognition. So far I have more than kept up to my goal of an hour a day on statistical work. The radius vector method is yielding interesting and consistent results. I still feel guilty when I think of how much farther along much of your work would be if you had not had to answer my thousand and one questions this winter. I learned not only the statistical methods I went for, but I have a much firmer grasp on many points in physiological genetics. Best of all, it was really great good fun, and served as a kind of recreation. (Anderson to Wright, March 7, 1933)

A student of any level who asked Wright a question received a detailed answer, sometimes not on the spot but always within a few days when Wright had a chance to look and think further about it.

Wider Recognition of Wright's Research

In the first ten years at the University of Chicago, Wright's work rapidly gained the attention of biologists the world over. In part this growing recognition was the natural consequence of the work Wright had done at Harvard and at the USDA; much of the research begun at the USDA was published only after Wright's arrival in Chicago. Probably more important was the big 1931 paper, "Evolution in Mendelian Populations" (Wright 64), and the brief 1932 version of it, "The Roles of Mutation, Inbreeding, Crossbreeding, and Selection in Evolution" (Wright 70).

No biologist in the United States had ventured to quantify and synthesize at the same time the basic variables of evolution in nature into a compelling view of the entire process. No other biologist in the United States, indeed, was even close to such a quantitative synthesis. Wright's achievement in the paper was viewed by other geneticists as a tour de force. He was soon invited to be a member of the American Philosophical Society (1932) and the National Academy of Sciences (1934) and was elected vice-president (1933) and president (1934) of the Genetics Society of America, the most prestigious association of geneticists in the world.

What most biologists saw as unique in Wright's research was his deep understanding of biological processes combined with the ability to conceptualize complex biological questions in quantitative terms. Because of his previous quantitative work, his growing reputation was not just as a quantitative evolutionist but as a person who could bring quantitative analysis to bear upon almost any biological problem. Most biologists trained before 1930 had only rudimentary training in mathematics or statistical analysis. Yet many of them encountered situations in their research or data that cried out for quantitative analysis.

Editors began to ask Wright to join editorial boards so they could use him as a reader for manuscripts with any substantive mathematical content. Biologists began sending their manuscripts to Wright before submission for publication, asking for his criticism; some of those persons knew that Wright would be the most probable reader once the manuscript was submitted for publication. During the first ten years in Chicago Wright was on the editorial boards of *Journal of Heredity, Genetics, Journal of the American Statistical Association,* and the monograph series in experimental biology published by Harvard University Press. Editors of other journals sent manuscripts to Wright for evaluation, though he was not on the editorial boards. Often the quantitative methods used in these manuscripts were unfamiliar to Wright, who then had to work out the problem by his own methods. This cost him much time, but he undoubtedly learned something in return. In our conversations Wright said "I am not sure that I did the people who sent me manuscripts much of a favor. Most of them would probably have done better to try harder at thinking through their problems for themselves." But I am certain that the recipients of Wright's analyses were grateful, though they frequently had difficulty following his quantitative analyses.

Professional organizations began to ask Wright to participate in administrative matters. The Genetics Society of America asked him to serve in 1931 on the nominating committee for officers, and in 1932 on the program committee planning the Sixth International Congress of Genetics at Ithaca, New York, before electing him successively as vice-president and president. The National Research Council asked him to become a member of the Committee on Animal Breeding, and he became a member of the board of directors of the American Statistical Association (both 1931). He was elected to membership on the Section Committee of Section F (Zoological Sciences) of the American Association for the Advancement of Science (1933). This was a particularly nice invitation because it was issued by Henry B. Ward, Wright's major professor during his year at the University of Illinois. Wright accepted all of these administrative duties, but found by 1935 (after a brief stint as acting chairman of the Department of Zoology at the University of Chicago) that he could not perform them all. Thus he began to selectively refuse additional requests.

During the first ten years at Chicago, Wright received three requests that he write a monograph or book. Castle and Harvard University Press wanted a monograph on the guinea pig, similar to the one Castle had written on rabbits (Castle 1930). W. A. Goddijn, editor of *Bibliographia Genetica* in Leiden, Holland, asked in 1928 for a monograph on statistical genetics. And Gregory Pincus in 1934 suggested a monograph along the lines of the 1931 paper "Evolution in Mendelian Populations" for Harvard University Press.

How overburdened Wright was during these years can be seen in his response to these attractive requests. He agreed to write the first two. He had, however, not only a heavy teaching load (all four quarters in 1928, 1929, and 1931) but also the growing administrative responsibilities described above. In addition Wright had a burgeoning volume of correspondence that took another substantial chunk of his time. This combination of responsibilities gave him very little time for writing up even the data in his experiments, much less the requested monographs.

Wright really wanted to do the monograph for Goddijn and *Bibliographia Genetica*. The request for a monograph on statistical genetics had originally come in 1929 from Goddijn to Muller. Knowing that he was not the best qualified person to write such a work, Muller turned to Wright and to Alexander Weinstein, another student of Morgan's and an old friend of Muller's. Weinstein had a good background in physics and mathematics and was also an expert on linkage and crossing over; he agreed to write the sections on linkage, coincidence, and Mendelian ratios. Wright was left with the statistical consequences of selection, inbreeding, assortative mating, and similar issues, and this was much to his liking. Wright wrote to Muller in August 1928 that he hoped to work on the monograph in the fall when he planned to have no courses to prepare. But circumstances forced him to teach that fall and continuously until the summer quarter of 1930. Beginning in the fall of 1930 he again taught continuously until the fall of 1932. Weinstein had a reputation for very careful and meticulous work, but he was slow to publish and a poor correspondent. So Goddijn heard nothing from Wright or Weinstein for over two-and-a-half years.

On March 6, 1931, Goddijn wrote to Wright reminding him of the arrangements made in 1929 and adding:

> We have heard nothing from you, neither from Dr. Weinstein. We kindly beg you to inform us if the work has proceeded to a stage that the manuscripts may be expected in the near future. We think a review on these subjects is badly wanted for the series of monographs in *Bibliographia Genetica*. We shall be very much obliged by an early answer in order to make arrangements for the next volume.

Wright replied on March 30.

> I regret that I have been so slow in preparing my review of the field of statistical genetics. The task of correlating the mathematical results of various investigators (reached from widely different viewpoints and usually with little reference to each other) has suggested new problems on which it seemed very desirable to reach conclusions before publishing a summary. I have recently published some of my conclusions (*Genetics*, March 1931) and have nearly finished a manuscript which should fill some two hundred pages of *Bibliographia Genetica*. I hope to submit this in the near future.

The manuscript of which Wright spoke undoubtedly contained the essentials of path coefficients; the contents of the series on systems of mating, inbreeding coefficients, and applications; the big 1931 paper; and analysis of dominance. Basically Wright wanted to prepare for publication approximately the subject matter treated in great detail in the first two volumes of *Evolution and the Genetics of Populations*.

The undertaking was in fact enormous, as Wright realized more clearly as he tried to write a publishable manuscript while teaching continuously. J. P. Lotsy, coeditor with Goddijn of *Bibliographia Genetica,* died in 1932 and was replaced by the Dutch geneticists Tine Tammes and M. J. Sirks. All three editors then joined in February 1932 in expressing to Wright their hope that he would send the manuscript for publication. Wright wrote back:

> In regard to my review of statistical genetics, I appreciate that your patience must be very nearly exhausted. It has been necessary for me

to teach continuously for the last year and a half and I have not been able to obtain the leisure time which I find necessary for work in this field. I have, however, by no means given up the project.

Indeed he did not give up the project, but it would not appear in print for another thirty-six years. Beginning in the early 1930s, Wright produced more data from his guinea pig experiments than he had time to analyze and publish. He published many papers on evolutionary theory and statistical genetics, but no monograph or book during the Chicago years. After retirement from Chicago in 1955, it would take Wright five more years to complete the analysis and publication of the guinea pig data collected before 1955; only after 1960 did he even begin concentrated work on his treatise on statistical genetics and evolution. The book on guinea pigs Wright never found time to write either, although Castle sent frequent reminders. By 1934 the situation was clear, and Wright flatly refused the request by Pincus.

If wider recognition of Wright's work had its burdens, it also brought cheering rewards from his colleagues. When Wright was elected to the National Academy of Sciences, congratulations poured in from such biologists as Henry Fairfield Osborn ("I took great pleasure in voting for your admission to the National Academy"), Sturtevant ("My only regret is that I was not at Washington to vote for you"), Muller, Castle, Charles R. Stockard ("This is an honor that you have long deserved and very fully won"), and many others. L. H. Snyder, a well-known geneticist at Ohio State University, had anticipated Wright's success by writing him in December 1931: "I have just moved into our new laboratories of genetics, which are nicely equipped and arranged. I am anxious to have on the walls pictures of outstanding geneticists as an inspiration to students." Snyder obtained a large photograph of Wright that he sent for Wright's signature and return. The supply of reprints of Wright's papers quickly became exhausted, and Wright had to turn down many requests. A 1935 telegram from Bronson Price in the American Embassy in Moscow is illustrative:

> LIFTING OF BRIEFCASE GREATLY EMBARRASSES ME FOR LOSING YOUR 1921 MONOGRAPH SYSTEMS OF MATING MATTER SERIOUS BECAUSE BELONGED TO SEREBROVSKY BORROWED FROM IGNITIEV AND WAS ONLY KNOWN MOSCOW COPY IF YOU COULD PROVIDE ANOTHER PLEASE SEND

Wright managed to dig up one more copy and send it as requested.

Correspondence

Wright's primary means of communication with other biologists was through correspondence. This was a necessary consequence of being located at the University of Chicago. He did attend national meetings, and of course spoke frequently with his graduate students and occasional postdoctoral students and visitors, but he mostly corresponded with other geneticists. As Wright's reputation grew, so did his correspondence. Beginning in 1931 he saved approximately 500 pages of professional correspondence each year; of these, perhaps two-thirds were carbons of his replies, the other third being the incoming letters. For the historian, Wright's correspondence files are a rich resource. The correspondence helps reconstruct of Wright's own scientific work, sheds light upon Wright's views about questions on which he did not publish, and perhaps most important, shows clearly how Wright aided and influenced other scientists. As working historians are well aware, such correspondence files do not by any means solve all the pertinent problems left unanswered in the published record; they do, however, provide another useful dimension in the reconstruction of a person's life and work. This is particularly true in Wright's case because he was always a very straightforward person who spoke his mind clearly and frankly and because he was forced by circumstances to communicate by correspondence.

Wright generally kept only correspondence relating to scientific or administrative questions. The great bulk of the correspondence shares certain general characteristics. Between about 1920 and 1932 Wright became known as an individual who really knew how to quantify complex biological questions and who was willing to share his knowledge. Thus the characteristic letter asks Wright to answer a quantitative question, to evaluate a quantitative paper, or to present a quantitative paper at a conference. After 1925, working with Collins, Wright did not (except on one occasion with Fisher) ask for others to evaluate his papers or help him with quantitative problems; he instead had a steady stream of requests to which he was asked to react.

The smaller part of the correspondence can be divided into several categories: (1) correspondence with old personal friends such as Paul and Wilson Popenoe, Harrison Hunt, or David Fairchild; (2) correspondence with practical breeders; and (3) administrative correspondence pertaining mostly to the Genetics Society of America. Wright treated amateur animal breeders just as he did the most prestigious scientist, answering their questions fully. One particularly interesting letter came from G. L. Haverstock of Haverstock's Racoon Ranch in Blakeslee, Ohio. (According to the letterhead, their motto was "Ask our customers.") Haverstock had attended the Chicago World's Fair in 1933, where Wright had exhibited color variations in guinea pigs. Haverstock had several interesting color variations and, his imagination fired by the wonderful array of coat and eye colors in Wright's guinea pigs, wanted to obtain arrays of new colors in raccoons. His first letter was ungrammatical and unclear about pedigrees, but Wright wrote a substantial reply explaining that he could not make accurate conclusions without more complete data, suggested crosses for obtaining additional color varieties, and told Haverstock that he would be interested in the results. Haverstock wrote again seven months later with detailed pedigrees, color charts, data from further crosses and more questions. Wright analyzed this information as far as he could and suggested still further matings, ending with the statement, "I shall hope to hear from you further" (Wright to Haverstock, June 13, 1934). This is the last letter on the subject left in the files.

Unlike his teachers Castle and, particularly, East, and unlike many geneticists of his own generation, Wright never published on eugenics. He steered clear of the subject. In his correspondence, however, Wright's attitude toward eugenics emerges clearly. As an idea for benefiting mankind, Wright had no theoretical objection to eugenics. His hesitation came from his belief that human heredity was very complicated and little understood, giving little scientific basis for a eugenics movement at that time. He also disagreed with many of the genetical ideas expressed by leading eugenicists. A number of people tried to interest Wright in eugenics. David Fairchild, on a visit to Ceylon, wrote to Wright in 1926 from the Queen's Hotel in Kandy describing the disastrous political and administrative situation caused in his opinion by the half-caste children of former Englishmen:

> The half-caste has created an intolerable social situation here in my opinion and either miscegenation is a horrible mistake socially or else we are headed towards a general mixture and mongrelization. The most intelligent Englishman I have met here said, "You Americans are the only people who see straight in these matters." I firmly believe it is your duty to see this gigantic problem firsthand sometime. It would prove a tremendous stimulus. Success to you dear Wright.

Wright had no fear of the biological consequences of human race crossing comparable to that expressed frequently by Davenport because Wright had seen firsthand that Castle's crosses between widely different races of rabbits produced no obvious physical disharmonies.

In May 1931 Paul Popenoe had written a draft of a review of H. S. Jennings's new book, *The Biological Basis of Human Nature*. With R. H. Johnson, Popenoe had written in 1918 the most successful, pure eugenics textbook ever to be published in the United States, *Applied Eugenics*. The Macmillan Company had reprinted it several times; in 1931 Popenoe and Johnson were revising it for a second edition. Popenoe was a strong advocate of eugenics and gave a basically favorable review of Jennings's book, while being critical about some genetical points. Not being a geneticist, he wanted Wright to criticize the manuscript before sending it to *Journal of Heredity*. In the last few pages of the manuscript Popenoe spoke of some genes as valuable and some as less valuable or detrimental. At the end of a long letter helping Popenoe with the genetical points, Wright said he believed

> that genes should not be spoken of as valuable without defining what they are valuable for. Probably most people could agree on a moderate program of negative eugenics because antisocial traits are, within

limits, agreed upon, but positive eugenics seems to require as its first step, the setting up of an ideal of society to aim at, and this is just what people do not agree upon. In the South before the war, an ideal eugenic program would doubtless have been one that tended to develop certain admirable individual qualities in a relatively small white population, eliminated troublesome poor whites and equally troublesome intelligent and aggressive negroes, and developed a large population of docile good natured negroes. For an agricultural state not based on slavery, a program which tended toward a fairly uniform distribution of certain other admirable traits would be indicated. An industrial civilization is a more complex organism and requires development of many diverse types. I suspect that the political and other choices which direct the type of civilization more or less automatically tend to direct the genetic characters of the population, though perhaps with considerable lag. A positive eugenic program, it seems, then, must be directed toward correcting a maladjustment which is perceived between the genetic properties of the population and a social organization which is accepted (this might mean an average lowering of intelligence, though probably not in our present civilization, as you point out) or efforts toward changing the social organization so as to fit types of individuals which seem admirable but have insufficient scope in the existing society.

Your papers should have a very good effect in stimulating a more self critical attitude toward eugenics on the part of biologists writing on the subject. It has been easy enough for those of us who have been working with guinea pigs or corn or paramecia to see the shortcomings of many eugenists in the field of pure genetics, but not so easy to recognize our own shortcomings in applying our findings in lower organisms to the very special case of man. (Wright to Popenoe, May 13, 1931.)

The *Scientific American* of January 1932 contained "Is Eugenics Scientific," written by T. Schwann Harding. The basic argument was that livestock breeding was beset with many difficulties (Harding here cited Wright) and that improvement of humans by breeding was even far more difficult. Harding obviously opposed eugenics. Johnson, at that time secretary-treasurer of the American Eugenics Society, was disturbed by Harding's letter and wrote to Wright:

In this article you are quoted, and I am afraid readers might get the impression thereby, that you share the writer's views. As a matter of fact the article is very badly done and is very easily blown up. I am very much in hopes that you will write a counter article, exposing its fallacies, for *Scientific American*.

It would come with more force from you than from any of the others of us. Please let me know if you will not do this. (Johnson to Wright, January 2, 1932)

Wright replied:

> On reading the article by T. Schwann Harding to which you called
> my attention, it seemed to me that his presentation of the difficulties
> in livestock improvement and the much greater difficulties in im-
> proving "the quality and intelligence of the human race" was essen-
> tially correct. One might present the same points with considerable
> difference in the rhetoric and one need not adopt such a defeatist atti-
> tude in the face of difficulties. I do not feel that I am the right person
> to answer him, if an answer is necessary. Anything which I wrote
> would sound too much like his article to be effective, although my
> attitude differs considerably from his. (Wright to Johnson, January
> 13, 1932)

Clearly Johnson did not get the reply he wanted from Wright, who basically
did not wish to become involved with the eugenics movement, even though
he was listed on the letterhead of Johnson's letter as being one of the 108
members of the advisory council for the American Eugenics Society. Wright
was never active in the society in any way.

In 1935 when it began to be known in the United States that Nazi race
theories and eugenical doctrines were being applied to Jewish scientists in
Germany, Dunn drafted a statement condemning the Nazi actions and called
for support of the displaced scientists, many of whom were coming to the
United States. Dunn proposed to have the statement read but not acted or
voted upon at the annual meeting of the Genetics Society of America. Dunn
specifically did not want the society as a body to endorse the statement,
"since we shouldn't interfere as a Society in the affairs of another." Milislav
Demerec, then secretary-treasurer of the GSA, sent Dunn's statement to
members of the executive committee asking if it was appropriate to place it
on the agenda of the GSA meeting. Wright was a member of the executive
committee as past president (1934) of the GSA, and he replied to Demerec: "I
heartily approve of the proposed statement in regard to the unfair treatment of
many German geneticists by the Nazi government. Personally I would be
glad to see this published as an expression of the Genetics Society of America" (Wright to Demerec, December 14, 1935). This reply indicated a dislike
of Nazi eugenical thinking, and, because of Wright's willingness to have the
GSA endorse the statement, gives a measure of his strong feelings about the
importance of Dunn's statement.

Scientific Correspondence

Even before publication of Wright's "Evolution in Mendelian Populations" in
Genetics in March 1931, he had a steady stream of requests to answer quantitative questions or to read and evaluate quantitative papers for individuals and
editors. The stream increased dramatically in 1931 and remained at a very
high level well into Wright's eighties; even in his nineties he still received

several requests each month. A list of a few of these incoming requests illustrates what a resource Wright was for his colleagues. Muller wrote on many occasions to ask for help, from devising formulas for the rate of occurrence of lethals in populations to asking Wright to take on his graduate students for short periods of time in order to teach them quantitative methods. John Belling asked Wright to explain some of J. B. S. Haldane's inferences about the interrelation of the expected frequency of chiasmata and interference with the chance formation of chiasmata. P. E. Johnston of the Department of Farm Organization and Management at the University of Illinois wanted Wright to figure out the best method of judging the relative importance of various independent variables upon farm earnings. Demerec asked Wright to develop formulas for predicting the consequences of a complex gene hypothesis he had invented for variegation in corn. C. V. Green and John J. Bittner of the Jackson Lab at Bar Harbor, Maine, wrote to ask Wright about the best methods for determining probable errors in their respective experimental data. John W. MacArthur of the University of Toronto wanted Wright to check his theory of size differentiation in a set of chromosomes. Chester Bliss, on his way to England to study with Fisher in the fall of 1933, wanted Wright's criticisms of his new biometrical ideas before facing Fisher. The list goes on and on.

In each case Wright received a relatively short letter and replied with a long and quantitatively detailed one, often complete with diagrams, charts, or tables. Usually Wright could help the inquirer achieve his aims, but in some cases he could only show that the data or methods were insufficient to solve the problem at hand. Thus Myron Gordon sent Wright a mass of data, gathered at a thirty-year interval, concerning two populations of the same species of fish. He hoped to find evidence of the action of natural selection but did not know how to analyze the data, so wrote to Wright:

> From my observations, (the data) seem not to have changed much. What do you think the data indicate? I know the data are few but do they mean anything to you statistically?
>
> If you think the data warrant further elaboration and comment, I should be pleased to have your suggestions. Indeed, if the data appeal to you and you wish to work them out in your own manner, I should be happy to have you share in its publication. (Gordon to Wright, November 28, 1933)

Wright was able to reply with a very brief letter, saying that he had "come back to your very interesting data a couple of times, but I do not believe that the numbers warrant anything more than a comparison of the totals for the two collections" and suggesting that Gordon might want to test Fisher's concept of a balanced polymorphism from heterozygote advantage in such a stable population.

The effect of this correspondence upon Wright's scholarly work in the areas of physiological genetics, animal breeding, path coefficients, and evolutionary theory was substantial. Time he would otherwise have spent writing up his guinea pig research he spent thinking about and writing replies to the incoming correspondence. Wright reasoned that answering a scientific question was an all or nothing matter. Either he had to say that he did not have time to answer, and he really disliked doing this, or he had to go into the question thoroughly. He thought anything in between was worse than useless and probably misleading. Much of Wright's influence as a geneticist came through his correspondence.

Method of Path Coefficients

During his first ten years at Chicago, Wright's method of path coefficients became known not only among statisticians and animal breeders but also among some in other fields like psychology and economics. For example, Barbara L. Burks was working in 1927 at Stanford with the famous educational psychologist Louis Terman on a project attempting to determine the relative contribution of nature and nurture to mental characteristics. In analyzing their data, they had encountered the same problems with partial correlations as Wright had found in attempting to assess the roles of heredity and environment in guinea pigs. Terman and Burks were searching for a better quantitative technique when a professor of food research (who had probably read Wright's "Corn and Hog Correlations" pamphlet) referred them to Wright's method of path coefficients. Burks immediately wrote to Wright:

> It appears to Dr. Terman and myself that in your path coefficient method lies a very hopeful approach to some of the problems which are confronting psychologists, and especially to the field of mental heredity. We have long been in need of a method which would enable us to make a beginning toward untangling the relative contributions of nature and nurture to mental traits, and we have in mind to apply your path coefficient technique to our recently collected data upon mental test scores of parents and their children together with various environmental measures (cultural status of home, etc.). (Burks to Wright, January 5, 1927)

Burks knew the partial correlation technique well, having published two brief papers on it. But she quite clearly had not understood the difficulties of applying path coefficients to her data.

Wright wrote back to her a typical five-page single-spaced detailed letter. He derived the basic formula for path coefficients, now using standardized coefficients, then showed how the method could be applied in many situations. He tried to make clear again his conception of correlation and causation.

As you have so clearly brought out, correlations, whether primary or partial, give no indications by themselves as to the nature of their relations. Practically, everyone who works with correlation, however, follows his presentation of the correlations with an attempt at interpretation, at least by verbal reasoning if he does not use partial correlations in the delusion that these give an automatic method of grinding out interpretations. The path coefficient method also is intended . . . to be a method for putting the reasoning into mathematical form. The basis of the reasoning remains something outside the correlation data. (Wright to Burks, January 26, 1927)

Burks replied with an appreciative long letter. As suggested by Wright, she had read Niles's paper in *Genetics* and Wright's reply (Wright 37), and said, "Apart from the real illumination that I received from your paper, I enjoyed them both immensely—Niles's as an example of audacious miscomprehension, and yours as an example of restrained but relentless capital punishment." She included a series of theoretical experiments with path coefficients and a list of queries related to them and to applications to her data and added:

We are particularly anxious to apply the method to our foster children study, as it seems to provide the only possible clear-cut way to compare the results from the foster and control groups. We have also recommended the method to several educational psychologists in different parts of the country who are engaged in pieces of research concerning the relative contributions of nature and nurture to intelligence and achievement test scores. (Burks to Wright, February 11, 1927)

Wright's response to this letter must have dampened Burk's optimism. "I have done some work on the relative importance of heredity and environment with respect to characters of the guinea pig but have never given much consideration to the human case which is of course enormously more difficult," Wright began his letter, and then delineated "troublesome problems" with even his guinea pig analysis and some of the greater problems for the human data. Wright did not mean to dissuade Burks from using path coefficients, but wanted her to be aware of their difficulties. When published the next year, Burks's paper purported to use path analysis, but in actuality she had misunderstood and used multiple regression analysis without a clear causal scheme (Burks 1928).

Wright was surprised by this and decided to write a substantive paper using path coefficients to analyze Burks's published data. He prepared a thirty-one-page manuscript entitled "Heredity and Environment in Relation to Human Intelligence" and submitted it to *Journal of Heredity*. Robert C. Cook, the editor, returned it on July 17, 1931, so Wright could recheck the formulas before the paper was set in type. At this point Wright had second thoughts and decided not to publish the paper. The analysis of Burks's data using path coefficients appeared in outline form only in a short paper entitled "Statistical Methods in Biology" that Wright delivered at the annual meeting of the American Statistical Association (Wright 62). Muller heard that Wright was reanalyzing Burks's data and wrote in April 1932 to see if Wright had published the analysis. Wright sent him the short paper, saying: "I have not, however, thought it worth while to publish the complete analysis." A more detailed analysis of Burks's data did not appear until the first volume of Wright's *Evolution and the Genetics of Populations* in 1968.

Henry Schultz, the econometrician, was in the Department of Economics at the University of Chicago during the 1930s and was keenly interested in new statistical techniques useful in economic analysis. In the early 1930s Schultz occasionally accompanied the group of biologists that frequently treked in the Indiana Dunes. Wright and Schultz talked a lot about path coefficients, and Schultz read Wright's "Corn and Hog Correlations." In the spring of 1932 Schultz was giving a course entitled Correlation and Curve-Fitting in the economics department and referred frequently to the method of path coefficients, suggesting in the course of his lectures that "since the approach forces some formulation of the causal relationships before anything else can be done, it might be wise for those who planned to use statistics as more than a mechanical tool to get a more thorough grasp of the method" (Stanley I. Posner to Wright, March 3, 1932). Accordingly, Wright was asked to address the class in a special session. Wright had long been dissatisfied with his major statement of the method of path coefficients made in 1921 because some of his derivations were cumbersome and because he had presented only a small part of the range of application of the method (many of which he worked out only after 1921). Thus Wright was able to present a more lucid and compelling treatment of his method to Schultz's class (and of course to his own biometry class) than was available in his own published papers. Schultz, who was active in the American Statistical Association and on the editorial board, encouraged Wright to submit a major statement of the method of path coefficients to *Annals of Mathematical Statistics*.

Wright wrote the paper and sent it to Schultz on October 12, 1932. The paper did not appear in print for a long time, and Wright finally received this letter from H. C. Carver, the editor:

> Your manuscript on "The Method of Path Coefficients" was received by the Annals from Dr. Schultz and created a most favorable impression on the reviewer for the Annals and its editor. As soon as it was definitely decided that the Annals was to be continued, your paper was allocated to the June, 1934 issue.
>
> When the various papers were gathered together for the June number, it developed after a short search, that your paper had been either mislaid or mis-filed, and another paper was substituted for yours, which in turn was then assigned to the September issue. Since that time I have been conducting a search at spare times for your paper, but as yet have not found it. As soon as the current issue is

mailed out, I am going to tear through my house and office from one end to the other—it must be found. I fear that it went astray during the nine months that I spent in New York, for during that interval all manuscripts passed back and forth between Ann Arbor and New York by mail, and all the filing here was done by students.

I greatly regret this most embarrassing situation, and shall do my best to correct the situation. (Carver to Wright, June 20, 1934)

To this letter in Wright's files was attached a telegram from Carver dated two days later with the message "MANUSCRIPT LOCATED WILL APPEAR SEPTEMBER ISSUE I APOLOGIZE FOR THE DELAY."

The paper occupied fifty-four printed pages and included, in addition to the derivation and analysis of the basic formulas, discussions of degree of determination, correlation between linear functions, statistical effects of inbreeding, multiple regression, partial correlation, quantitative evaluation of causal relations, birth weight of guinea pigs, transpiration of plants, relative importance of heredity and environment, human intelligence, analysis of size factors, use of partial correlation in interpretation, difficulties in causal analysis, corn and hog correlations, elasticities of supply and demand (some of the data supplied by Schultz), and finally, tests of significance. This would be the most comprehensive paper on path coefficients published by Wright, the standard reference for a statement of the method and its range of application.

The Harvard Offer

In the mid–1930s genetics research at Harvard University was sure to change. Castle had already passed the usual retirement age of 65 and East was rapidly approaching it. The Bussey Institution was way out in Jamaica Plain, and the building of the new Biological Laboratories on the campus rendered the fate of the Bussey uncertain. Certainly it would not remain much longer as the center of genetics research. Castle, East, president James B. Conant, who was much interested in science research and education, and A. C. Redfield, director of the Biological Laboratories, all knew that a new leader in the genetics field was a necessity if the field were to continue to flourish at Harvard. East had just been engaged in a serious correspondence with Wright about Fisher's theory of evolution, and all of the principals easily agreed upon Wright as the logical person. Accordingly, in a letter dated February 25, 1935, President Conant offered Wright a professorship with an excellent salary and research budget. Conant's letter was followed right away by letters from Redfield, offering spacious laboratory and office facilities in the new building, Alfred S. Romer, and East. The letter from East was especially gratifying to Wright:

I suppose by this time you have heard that you were unanimously recommended by the division of biology for a professorship at Harvard. I wish personally that you may feel disposed to accept it.

There is no one with whom I had rather be associated during my remaining years than you. And I think that there is a great chance here. Castle and I worked under a handicap, two handicaps in fact—Mr. Lowell's lack of sympathy, and bickerings in the division. Mr. Conant is highly sympathetic with the retirement of Jeffrey and Wheeler and the death of Sargent. The entire division works in great harmony.

I think that you would have excellent conditions of work, and could build up genetics at Harvard in a wonderful way. This would give me a lot of joy. My life has been wrapped up in the work here, though accomplishment has been small. I have struggled with ill health all my life—infected gall bladder right now—and this may serve as a little alibi. But nevertheless I want to see genetics go ahead and you are the man that could push it ahead.

I hope that you will decide to come.

Wright was surprised by the Harvard offer. He was basically very happy with the Chicago position, his family was well settled there, and no other university position, except perhaps that at Harvard, would have interested him. The Harvard offer was indeed attractive. The higher salary, the greatly lessened teaching load, the research support, the extraordinary space for office and laboratory, and the lure of his alma mater all led Wright to explore the possibility carefully. He made lists of pros and cons.

The Harvard offer looked good. Wright brought the situation to Lillie, who was by then dean of the biology division. Lillie in turn went to President Hutchins, encouraging him to meet the Harvard offer in every way. Hutchins met Harvard's salary offer, suggested Wright cut his teaching load in half to an average of three one-quarter courses per year, exceeded by $1,000 Harvard's offer of research support, and offered better space and facilities for the guinea pig colony. From the Department of Zoology came the following letter:

We appreciate the fitness of Harvard's selection of you as a successor to Professor Castle. We trust you will accept their invitation to come to Harvard only in case you feel that you can do better scientific work and have a wider influence with that as your base. We have no wish to influence your decision unduly but if the University of Chicago can furnish you with approximately equal opportunities, we hope you will decide to remain with us.

Very cordially yours,

The letter was signed personally by all members of the department.

Louise Wright thought the schools for the Wright children would be worse and more expensive. Wright would have to leave his many good friends at Chicago and doubted that he would acquire as many new friends at Harvard. The stress of leaving, moving, and arranging for his graduate students and guinea pigs was bothersome. He decided to remain at the University of Chicago.

The University of Chicago, 1935–1955

The years at the University of Chicago between the Harvard offer of 1935 and Wright's retirement in December 1954 were highly rewarding. He continued his intense work in physiological genetics, becoming a major spokesman, along with Richard Goldschmidt, for that field. He also consolidated his position as the preeminent quantitative evolutionary theorist in the United States.

Wright loved the social life at the University of Chicago. Centered in the central meeting place for the faculty, the Quadrangle Club, social life at the University of Chicago consisted primarily of high-level conversation at the lunch or dinner table; but it also included shows, billiards, bridge, and other light social activities. One of Wright's favorite activities was the monthly meeting of the X Club, an exclusive association of leading University of Chicago scientists, who delivered papers and then discussed them intensely. Meeting with thirty or so distinguished scientists from all fields was always stimulating for Wright. Only men were in the X Club. The wives had a social club that met on the same evenings. When it was founded, Louise Wright suggested naming it the Two-X Club but the other women decided instead to call it the "Excess Club." I think both names were appropriate, the wives on these nights being, in the minds of the men, clearly excess baggage of the two-X kind.

During these years Wright had a relaxed formal teaching schedule, not exceeding three one-quarter courses per year. He alternated between only four courses: Fundamental Genetics, Physiological Genetics, Evolution, and Biometry. Work with graduate students, however, intensified. He directed eighteen doctoral theses between 1935 and 1954, and in addition he had eight postdoctoral students, as the following list indicates (year of doctorate and thesis topic also indicated):

1. John Paul Scott. 1935. The embryology of the guinea pig (Scott 1937a,1937b,1938).

2. Herman B. Chase. 1938. Studies on the tricolor pattern of the guinea pig (Chase 1939a,1939b).

3. Gertrude Heidenthal. 1938. A colorimetric study of genic effect on guinea pig coat color (Heidenthal 1939).

4. Joseph Jackson Schwab. 1938. A study of the effects of a random group of genes on shape of spermatheca in *Drosophila melanogaster* (Schwab 1940).

5. Elizabeth Shull Russell. 1939. A quantitative study of genic effects on guinea pig coat color (E. S. Russell 1939).

6. William Lawson Russell. 1939. Investigation of the physiological genetics of hair and skin color in the guinea pig by means of the dopa reaction (W. L. Russell 1939).

Figure 6.2. Department of Zoology faculty, circa 1945. *Top row:* Park, Strandskov, Moore, DuShane, Domm; *bottom row:* Allee, Wright, Weiss, Emerson, Schmidt.

7. Benson Ginsburg. 1943. The effects of the major genes controlling coat color in the guinea pig on the dopa oxidase activity of skin extracts (Ginsburg 1944).

8. Allen Fox. 1948. Immunogenetic studies of *Drosophila melanogaster* (Fox 1949a,1949b).

9. Liane Brauch Russell. 1949. X-ray induced developmental abnormalities in the mouse and their use in the analysis of embryological pattern (L. B. Russell 1950).

10. Frederick Carl Bock. 1950. White spotting in the guinea pig due to a gene (star) which alters hair distribution (unpublished).

11. Henry M. Wallbrunn. 1951. Genetics of the Siamese fighting fish (unpublished).

12. J. L. Henry Burkhelder. 1954. White locus position effects in *Drosophila melanogaster* (unpublished).

13. George L. Wolff. 1954. The effects of environmental temperature on coat color in diverse genotypes of the guinea pig (Wolff 1954, 1955).

14. Sally Lyman Allen. 1954. Linkage relations of the genes histocompatibility brachyury and kinky tail in the mouse as determined by tumor transplantation (S. L. Allen 1955).

15. Herman M. Slatis. 1954. Position effects of the brown locus in *Drosophila melanogaster* (Slatis 1955).

16. Willys K. Silvers. 1954. Pigment cell migration following transplantation (Silvers 1956). Silvers did his thesis jointly with Wright's former student Elizabeth S. Russell, who was at the Jackson Laboratory in Bar Harbor.

17. Peter Buri. 1956. Gene frequency in small populations of mutant *Drosophila* (Buri 1956).

18. Janice B. Spofford. 1955. The relation between expressivity and selection against eyeless in *Drosophila melanogaster* (Spofford 1956).

The postdoctoral students were Alan Robertson, Bruce Griffing, Sheldon Reed, Donald Charles, and Morris Foster.

Most of Wright's graduate students worked in physiological genetics; only one, Peter Buri, conducted research related to population genetics. This may come as a surprise to those who know Wright only from his work in evolutionary biology, but, as I have emphasized repeatedly in this biography, Wright considered himself during the years at the University of Chicago as primarily an experimental physiological geneticist. Wright never began at Chicago a school of theoretical population genetics comparable to Dobzhansky's school of experimental population genetics at Columbia University or E. B. Ford's school of ecological genetics at Oxford University. Wright's impact upon evolutionary biology might have been different if he had produced a dozen or so quantitative population geneticists as graduate students.

Wright's increasing prominence in both physiological genetics and evolutionary theory brought him rewards as well as greater responsibilities. Recognition of his scientific work came in many ways, from invited lectures and papers to prizes, honorary degrees, and the presidency of major scientific societies. Despite the lightened teaching load, ever-increasing demands upon Wright's time from graduate students and especially correspondence left him, if anything, less time for his research than he had enjoyed in the first decade at the University of Chicago. He was asked to review a great many papers on quantitative genetics by the editors of *Genetics, American Naturalist, Evolution, Journal of Experimental Zoology,* and other journals. The result was that Wright was perpetually juggling his research time between evolutionary theory and physiological genetics. He desperately tried to find the time to write his projected book on evolution, and planned to write it on many occasions, but the necessary freedom from his many responsibilities never materialized during the Chicago years. Similarly, it took so much time to produce the data from the guinea pig colony that Wright was unable to properly write it up in papers for publication. In the end, Wright found the time for these projects only after retirement from the University of Chicago and by being very productive in advanced age when most people are either dead or resting on their laurels.

The first acceptable opportunity the Wrights had to spend an academic term away from the University of Chicago came in the fall of 1936, when Sturtevant and Dobzhansky invited him to spend the year at the California Institute of Technology. Feeling he could spare only the fall quarter, Wright drove with the entire family for a glorious trip through Iowa, South Dakota,

Wyoming, Montana, Idaho, and Washington before turning south through Oregon to the California coast. Dobzhansky left almost immediately to give the Jesup Lectures at Columbia, an invitation that had not yet arrived when he had extended the invitation to Wright. Consequently, Wright spend most of his time with Sturtevant and Morgan, a most enjoyable experience. Wright went on many trips to the mountains with Sturtevant, who was collecting *Drosophila pseudoobscura* to determine the ratio of the sex ratio gene to its normal allele. The Wrights also had a delightful visit with Paul and Betty Popenoe, Paul having been Wright's roommate during his first years in Washington, D.C. Then they drove back following a southerly route through Arizona, New Mexico, Texas, Louisiana, and up through Mississippi, Tennessee, and Kentucky before returning to Illinois. This was a wonderful several months for the whole family. The basis was laid for Wright's participation in Dobzhansky's "Genetics of Natural Populations" series, Wright's most important collaborative project of his entire career (the topic of chapters 10 and 11; see also Provine 1981).

For several years beforehand, the Wrights had hoped to attend the Seventh International Congress of Genetics scheduled for 1937 in Moscow. The Russian genetics community, however, was in a turmoil, and the congress was canceled for that year and rescheduled for late August 1939 in Edinburgh. Sewall and Louise decided to take most of the summer and drive their Model A Ford around France and England before going to the congress. They spent three weeks driving all over France, including a very enjoyable day at the laboratory and home of Boris Ephrussi, the leading physiological geneticist in France. In London, Julian Huxley, then director of the London Zoo, took them on a tour of the zoo and out to lunch afterward. At that time, Huxley was much impressed by Wright's idea of random genetic drift, which he prominently praised in *Evolution: The Modern Synthesis* (1942) and in many other publications (but see chapter 12 for the end of this interaction between Wright and Huxley).

The Wrights drove to Coventry, the homeplace of the Sewalls, and visited other ancestral homes on their way to Edinburgh for the international congress, scheduled for August 23–30. Wright gave his paper on the interactions of color factors in the guinea pig on the morning of August 24, but on that very evening the approaching war began to be felt. The British government advised Britishers in Germany to come home, and the German government reciprocated. On the night of August 24, N. W. Timoféeff-Ressovsky of the Kaiser Wilhelm Institut invited Wright to his room to meet the German delegation. Wright came for a thoughtful visit, but the next morning the entire German delegation had disappeared, along with many other delegates from central Europe. On August 25 only about 150 of the more than 600 geneticists remained, and those only because they had no means of transportation home.

The congress limped along, and the Wrights stayed, although they worried about passage home. Some members of the congress booked passage on the *Athenia,* which the Germans sank, costing the lives of F. W. Tinney, of the University of Wisconsin, and his wife. Punnett, who edited the *Proceedings* of the congress, summarized the situation well: "For a day and a half the congress was able to immerse itself in its own enjoyable affairs. It even danced. But on the evening of the 24th its serenity was shattered. War, that outmoded futility of irrational immaturity, the antithesis of everything we presented, was about to overwhelm us" (Punnett 1941, 5). The Wrights were unable to secure passage until September 10, a full week after Britain had declared war upon Germany. But except for picking up a lifeboat full of seamen from a torpedoed freighter, the voyage home was actually uneventful. Even the outbreak of the war would not dim the intense enjoyment Sewall and Louise felt on this trip, and they both wanted to return for a longer visit, a hope finally realized ten years later.

The warmth and support Wright received from the zoology department at the University of Chicago was especially apparent upon his return in the fall of 1939, when his colleagues threw a gala birthday party for him, his fiftieth. Recognition came in other ways. In 1942 the University of Rochester awarded Wright his first of nine honorary doctorates, to be followed by ones from Yale (1949), Harvard (1951), and Michigan State (1955). He was elected president of the Genetics Society of America in 1934, the American Society of Zoologists in 1944, the American Society of Naturalists in 1952, and the Society for the Study of Evolution in 1955. In 1947 the National Academy of Sciences awarded him the Daniel Giraud Elliott Award, and Oxford University the Weldon Memorial Medal; in 1950 the American Philosophical Society awarded him the Lewis Prize for the best paper given at the society in the previous year (Wright 132).

At the instigation of a very active genetics group, including Wright's old friends William Castle and Richard Goldschmidt, the University of California at Berkeley invited Wright in the spring of 1943 to give a series of six Hitchcock Lectures on his evolutionary theory. The lectures were a success, and Wright hoped in the months following to write these lectures up in book form, a hope that J. T. McNeill, editor of the University of Chicago Press, extracted from Wright in the form of a tentative promise. But quiet writing time was insufficient, partly because Wright, to help in the war effort, joined the Manhattan Project to calculate the dangers of radiation. He also did statistical studies of the effects of temperature and humidity on mustard gas burns. The Hitchcock Lectures consequently did not appear as a book.

During his first twenty years at the University of Chicago Wright avoided administrative duties. During the war, however, Robert Maynard Hutchins, president of the university, stepped up his efforts to mold the University of Chicago into his long-held ideal "that the purpose of the University is to procure a moral, intellectual, and spiritual revolution throughout the world" (Hutchins 1944, 7). To this end, Hutchins made six specific proposals:

1. Abolition of all academic rank.
2. Full-time service of all faculty members, with all outside earnings turned over to the university.
3. Compensation of faculty members strictly on the basis of need.
4. Abolition of the University Senate (then composed of only about one hundred full professors), to be replaced by a Senate representing all faculty members.
5. Creation, by the University Senate, of an Institute of Liberal Studies devoted solely to liberal, or general, education rather than professional or technical education.
6. The placing of initiative in educational policy in the hands of the president, who, however, would serve at the pleasure of the Senate and the Board of Trustees.

These proposals engendered immediate opposition from many faculty in the graduate divisions, who were already suspicious of Hutchins's emphasis upon the College. Wright, who had little use for either Hutchins or Mortimer J. Adler, Hutchins's intellectual soul mate, was active in the opposition. Of the six proposals, Hutchins was able to realize two—all outside earnings of faculty had to be turned over to the university, and a new Senate was established (the latter with universitywide support). Wright was elected a member of the new Senate and became very involved in its activities. He fought hard against Hutchins. This was his only significant attempt to participate in the political affairs of the University.

One of the Hutchins's proposals had a practical effect on Wright. In November 1951 Columbia University invited Wright to give the prestigious Jesup Lectures, given previously by such luminaries as Dobzhansky, Mayr, Stebbins, and Simpson. The deciding factor in Wright's declination of the invitation (aside from the real press of other professional obligations) was the policy that required the $1,000 stipend to be turned over to the University of Chicago.

When the University of Edinburgh offered Wright a Fulbright professorship for the academic year 1949–50, he accepted enthusiastically. He planned to give a short series of lectures at Edinburgh and to spend the year writing his book on evolution. But when his presence became known he was invited to give lectures at Edinburgh and all over England, including Cambridge University, Oxford University, University of London (the Galton Lecture), University of Durham, and elsewhere. Some of the lectures, especially that at Oxford, stirred opposition from Fisher, Ford, and their followers. Everywhere he found large audiences. It was a marvelous year for both Sewall and Louise, who especially loved to travel. The book on evolution, however, had to be set aside.

When the Wrights returned from Scotland in October 1950 they found that their neighborhood in Hyde Park had deteriorated dramatically. Poor blacks had moved around the periphery of Hyde Park during the mid- to late 1940s, and the crime rate had risen more than tenfold during their absence of only a year. Muggings and rapes of university employees had become common. Faculty members who had reveled in the Hyde Park community were now afraid to walk at night, and many moved to faraway suburbs. Combined with Hutchins's firm policy of forcing faculty members out of the university at retirement, Wright began early thinking about the possibility of retiring to a different university community. Thus although Lawrence Kimpton replaced some of Hutchins's policies beginning in 1953, Wright nevertheless had his mind pretty well set on moving from Chicago. The offer from the University of Wisconsin as L. J. Cole Professor of Genetics was the perfect attraction.

Guinea Pig Research, 1925–1935

The Department of Zoology at the University of Chicago had from its inception been skeptical about genetic explanations for embryological development and in particular saw little importance in Mendelian heredity. Lillie and Child found Wright an attractive geneticist because he explicitly wanted to tie genetics to development. In 1921, before Wright or the members of the department had any idea that Wright might join the Department, he had reviewed Child's *The Origin and Development of the Nervous System* for *Journal of Heredity*. In this book Child had attempted a completely mechanistic interpretation of embryological development using his concept of metabolic gradients; he did not at any point bring in genetics, primarily because he saw development as tied to the cytoplasm and genetics as tied only to the nucleus. Wright had no major quarrel with Child's physiological conceptions, but he objected strongly to the omission of genetics: "From the standpoint of the geneticist it is to be regretted that the author has made no attempt to bring the facts of genetics into relation with his theory" (Wright 26, p. 74).

After explaining that geneticists in 1921 knew a great deal about the inheritance of unit factors and their location on chromosomes, Wright questioned Child's dismissal of the chromosome theory of inheritance as an extreme and impossible preformist view and argued for a more comprehensive view of development that included genetics.

> The genetical and cytological conception of the cell, as an association of independent organisms, living in a relatively large, less specialized mass of protoplasm and controlling the behavior of the whole in response of course to external stimuli, is not at all incompatible with a simple mechanism of heredity and a simply physiological conception of development such as that offered by Child. It is, of course, impossible to assume that the genetic basis of the differ-

ences among the millions of species of animals and plants, and the millions of individuals within each species can be very simple. It must be remembered, however, that a given cell complex has not developed out of nothing in the course of a few weeks, as the individual seems to do. It is the result of millions of years of uninterrupted slow change. The problem of heredity is merely to explain the lack of interruption in this history, i.e. the persistence of the complicated cell organization through cell division and fertilization. . . .

As to individual development, there seems to be no incompatibility with Child's explanation of its course as the behavior pattern of a particular kind of cell in relation to the metabolic gradient determined ultimately by the environment. The hypothesis that the activities of the different unit factors vary at different rates according to the position of the cell in the gradient pattern of the developing organism and to the specialization which the cell has already undergone as a result of its past history, thus determining the details of the "organismic pattern," in reaction of course with the environment, seems a necessary supplement to Child's highly suggestive hypothesis. (Wright 26, pp. 74–75)

With Wright's strong emphasis upon the relation of genetics to development, it is understandable that Lillie tried to attract him to the University of Chicago and one reason that Wright in turn found the offer attractive.

The problem, although perhaps not so obvious in 1921, was the extreme difficulty in going beyond the general principles of the connections between genetics and development as stated by Wright in his review of Child's book. While at the University of Chicago, Wright pursued vigorously his research into the physiological genetics of guinea pigs, including those concerned with color inheritance, polydactylism, abnormal development (otocephaly), and hair patterns. Such research could not, however, reveal more than the larger outlines of the mechanistic action of genes, relating genetics and development.

Not until 1934 at the second Cold Spring Harbor Symposium on Quantitative Biology did Wright return again to a published statement on the general problem. This was also the year in which Morgan published his summary statement *Embryology and Genetics*. Morgan had been interested, since he initiated research into the heredity of *Drosophila* in 1910, in connecting embryology (his original field, and that to which he returned in the mid-1920s) with Mendelian genetics. But Morgan's book epitomized the problem; it was literally a book about embryology and genetics, not embryology in relation to genetics. Wright, after reviewing his own research on guinea pigs, provided this summary statement about genes and development at the Cold Spring Harbor Symposium on Quantitative Biology for 1934:

It seems to me that the view toward which we are tending is that the specificity in gene action is always a chemical specificity, probably the production of enzymes which guide metabolic processes along particular channels. A given array of genes thus determines the production of a particular kind of protoplasm with particular properties—such, for example, as that of responding to surface forces by formation of a special sort of semipermeable membrane, and that of responding to trivial asymmetries in the play of external stimuli by polarization, with consequent orderly quantitative gradients in all physiologic processes. Different genes may now be called into play at different points in this simple pattern, either through the local formation of their specific substrates for action, or by activation of a mutational nature. In either case the pattern becomes more complex and quantitatively differentiated. Successive interactions of differentiated regions and the calling into play of additional genes may lead to any degree of complexity of pattern in the organism as a largely self-contained system. The array of genes, assembled in the course of evolution, must of course be one which determines a highly self-regulatory system of reactions.

On this view the genes are highly specific chemically, and thus called into play only under very specific conditions; but their morphological effects, if any, rest on quantitative influences of immediate or remote products on growth gradients, which are resultants of all that has gone on before in the organism. (Wright 82, p. 143)

Wright's conception here has a basically modern ring and of course fits perfectly with recent molecular biology as well as with his guinea pig research in the 1930s. Two observations, however, are necessary. First, Wright could pose no detailed mechanistic basis for his view, and none would exist before the rise of molecular biology in the 1950s. Second, Wright was by no means alone in proposing a view of this sort and was aware of its shortcomings. Indeed, Wright concluded his 1934 paper with the revealing statement:

There is little that is novel in what I have said. I find that I have added little if anything in principle, to what I said in a paper written 13 years ago, in which I attempted to correlate genetics with Professor Child's views of development. More or less similar views have been expressed by other geneticists almost since the rediscovery of Mendelian heredity. Goldschmidt, of course, has been urging for years that gene action is thorough control of developmental rates. Closely allied is the idea of balance among the effects of multiple factors, most definitely formulated by Bridges. Huxley has discussed the genetic control of growth gradients. I suspect that most of you have felt that I have been uttering platitudes.

Apparently Wright's last sentence was thought incorrect by the audience, which initiated a vigorous discussion involving Ross Harrison, MacDowell, Demerec, MacCurdy, Riddle, Davenport, Gowen, and others.

What is important to understand here is that Wright's ongoing research in the physiological genetics of guinea pigs, which both he and the University of Chicago embryologists hoped might shed light upon the fundamental connections of embryology and genetics, did not achieve that aim. Fundamental progress in this very complex field had to wait first for the biochemical genetics of lower organisms (such as Beadle and Tatum with *Neurospora* or Avery with bacteria) and then for an understanding of the molecular structure of the hereditary material.

This is not to say that Wright's work in physiological genetics was unimportant. He, probably more than any other mammalian geneticist at that time (although Dunn was moving strongly in that direction), was making clear the effects of gene action in higher animals and particularly was demonstrating the reality and detail of gene interaction and environment-gene interaction. Next to Goldschmidt, whose work was focused upon the moth *Lymantria,* Wright was the leading researcher in physiological genetics. When Wright was on the program committee of the International Congress of Genetics held at Ithaca, New York, in August of 1932, he constantly urged East, head of the committee, to arrange a major session on physiological genetics; East acceded to this request. In retrospect, it is generally remembered only that Wright gave his famous evolution paper at the congress. But he also organized a comprehensive exhibit of guinea pig materials at the congress to demonstrate his research in physiological genetics.

Although Wright had written his long paper on evolution in the months preceding his departure for Chicago from the USDA in 1925, he fully expected his major research interest at the University of Chicago to be the field of physiological genetics. He also had no firm intention to continue working with guinea pigs as his experimental animal. As pressures upon his time increased from a heavy teaching load, manuscript evaluation, and correspondence, especially after 1931, he settled into the research program in physiological genetics using his familiar guinea pig materials. As mentioned in the previous chapter, most of the papers Wright wrote on guinea pig genetics between 1925 and 1930 were based upon research completed or under way at the USDA before he left for the University of Chicago. Not until 1934 did he begin to publish in detail the results of research begun after the move to Chicago. Wright's work in physiological genetics in the years 1925–37 concerned five different projects: (1) he continued and expanded the work on otocephaly begun at the USDA; (2) he conducted a series of experiments on the inheritance of polydactylism in guinea pigs, a subject originally investigated by Castle in 1906; (3) he began new investigations of the inheritance of hair direction, culminating in 1949 with a series of three major papers; (4) he continued research on color inheritance, particularly the inheritance of the spotted pattern; and finally, (5) he interpreted the phenomenon of dominance in terms of physiological genetics. This last topic, which he also treated in an evolutionary perspective, will be analyzed in the next chapter along with the development of Wright's evolutionary ideas.

As described in the previous chapter, otocephaly is rare in most domestic and wild guinea pigs, occurring at a frequency of less than 0.05%. One strain that arose in one of the inbred families at the USDA produced, however, about 5% otocephalics regularly for nearly twenty years (in 1934); one sub-branch of this strain produced an exceptional 27%. Wright wished to find out as accurately as possible the mode of inheritance of otocephaly and the role of environmental factors and to analyze the embryological sequence of events leading to the production of the monsters. In this last task he was aided by K. Wagner, a Norwegian anatomist who spent the academic year 1929–30 with Wright in Chicago.

The dominant theory at that time for the causation of otocephalic monsters in mammals attributed the effects to maternal toxemia of environmental origin or to maternal inheritance. Wright was able to disprove, in the case of guinea pigs, the theory of maternal toxemia. He found evidence for a gene complex (Ot_1) that increased, when homozygous, the frequency of otocephaly from the naturally occurring 0.05% to 5% and for a dominant gene (Ot_2) that caused when heterozygous the jump from 5% to 27%. Otocephaly also occurred twice as frequently in females (XX) as in males (XY), clearly implicating the X chromosome. But all these ascertainable genetic factors seemed to determine only a frequency of incidence in a strain. The other factors were sufficiently random that the chance differences (leading to otocephaly) were essentially the same between littermates as between nonlittermates. (Within a strain inbred for so many years, all individuals had almost the same genetic complements.)

Wright and Eaton (Wright 41; 1923) had earlier determined that damage to the zygotic embryo rather than to the egg or sperm caused the monsters, and that what occurred was a general inhibition to the most actively developing region of the embryo and not later than the early medullary plate stage. The different kinds of otocephalic monsters could then be attributed to the precise time of action of the inhibition of the anterior medullary plate, an exercise Wright and Wagner carried out in detail (Wright 77).

The most important conclusion to flow from this work on otocephaly was that small and simple quantitative changes in genetic structure could as a result of additional random correlative effects cause qualitative morphological changes. Wright's work on otocephaly thus contributed to the picture of genes in relation to development as it emerged in the mid-1930s (Wright 78).

Polydactylism is a very different characteristic of guinea pigs than otocephaly. The guinea pig normally has four digits on the front feet and three on the rear. By selection from an individual with a weakly formed fourth toe on the hind foot, Castle had formed a true-breeding polydactylous race in 1906. The difference between this stock and the normal did not appear to be a simple Mendelian determinant. Wright found polydactylism in one of his inbred families and chose to study this phenomenon in detail. In one branch of this family he found a simple Mendelian mutation that produced a pentadactyl foot, front and hind, in the heterozygous state but a monster in the homozygous state. In 1934 and early 1935 Wright published a series of four papers on the inheritance of polydactylism, the results of crossing inbred strains differing in the numbers of digits, and an analysis of the new mutation (Wright 79–81, 85).

Wright showed that nongenetic factors played a prominent role in the production of polydactylism in the inbred family he studied carefully. Age of mother was very important, determining between about 10% to 25% of the observed variance in the stock. This analysis effectively disproved contentions by Pictet and others that polydactylism was caused by a particular set of genetic factors. Individuals with the same genetic constitution might easily have differing numbers of digits. Crosses between Castle's pure-breeding four-toed stock and various stocks from the USDA showed that they differed in anywhere from one to three or more major Mendelian factors.

Perhaps the most interesting larger implications of this work on polydactylism relate to Wright's comparison with otocephaly. Although variations of a completely different morphological character, polydactylism and otocephaly shared certain attributes (a large role for nongenetic factors and multiple-factor determination), and in his comparison Wright tried to bring both phenomena under his general picture of physiological genetics. This he did by emphasizing the importance of threshold effects and inhibitory effects at crucial stages of embryonic development. Although this synthetic attempt was not easily accomplished, it indicated the drift of Wright's thinking and his belief that a comprehensive view of genetic control of development and physiological genetics should emerge.

Only a few geneticists were working on guinea pigs in the 1930s. One was the Swiss mammalian geneticist A. Pictet, who had a strong tendency to publish unproved assertions about the mode of inheritance of characters in guinea pigs. Wright, who was so careful in his own publications about mode of inheritance, was irritated that Pictet constantly published unproved or incorrect ideas about inheritance in guinea pigs. For years Pictet had disagreed with Wright's analysis of the inheritance of hair direction in guinea pigs as presented in Wright's doctoral thesis (Wright 5; 1916), but Pictet presented so little evidence that Wright could not even correlate their results. Then in 1934 Pictet and his colleague A. Ferrero published a paper with substantial data on inheritance of hair direction (Pictet and Ferrero 1934). With this more complete data, Wright was able to explain most of Pictet's results in terms of the factors Wright had isolated by 1916, admitted the probable presence of another factor postulated by Pictet, and said there was no real evidence for yet another. Wright's 1935 report was not in itself a major paper, but it marked the beginning of his new attempt to clarify experimentally the inheritance of hair direction (Wright 89). The results of this analysis were not published until 1949.

The last major project on guinea pigs undertaken by Wright in the mid-1930s concerned the inheritance of white spotting. In mice and rats, white spotting had early proved amenable to simple Mendelian analysis. Thus in rats Castle had demonstrated much heritable variability in the white spotting of the hooded rat, but the hooded pattern itself was clearly determined by a Mendelian recessive factor (though background modifiers could certainly alter the phenotype). The case was far different in guinea pigs. In his thesis Wright had simply attributed white spotting to ΣW, an assemblage of unanalyzed factors. Only a year later, however, he suggested the existence of a pair of alleles S,s that differentiated some of the inbred USDA stocks, but he published no detailed data in support of his theory. Then in his 1920 paper, "The Relative Importance of Heredity and Environment in Determining the Piebald Pattern of Guinea Pigs," in which he applied his method of path coefficients, Wright demonstrated that much nongenetic variability was present in the white spotting pattern; in 1926 he showed that some of the nongenetic variability came from maternal age.

Not all geneticists accepted Wright's analysis. These included Pictet, who hypothesized several major pairs of alleles to account for the variations in white spotting (and no emphasis upon nongenetic variation), and Heman Ibsen, a geneticist at the University of Kansas, who agreed with Wright about the existence of one major pair of alleles S,s (although Wright postulated no dominance relationship and Ibsen did) but also argued for a series of specific genetic modifiers to account for the variations in the spotting pattern. In the mid-1930s then, three different theories of white spotting in guinea pigs were well known in the literature. With the help of his graduate student Herman B. Chase, Wright attempted to clarify the problem by analysis of data from the USDA stocks and by additional critical experiments. A summary of the detailed biometric and genetic analysis is inappropriate here, but Wright and Chase were able to present a convincing case for four classes of factors affecting white spotting, summarized by them as follows: (1) a major pair of alleles S,s in which S (tending toward self) is usually incompletely dominant (statistically) over s (tending toward white); (2) a multiplicity of genes with individually small effects, additive on a suitably transformed scale (no dominance or epistasis); (3) an enormous amount of nonhereditary variability, not common for the most part even to littermates but including minor effects of common factors (for example increasing white in young with increasing age of mother); (4) a sex difference, females having slightly more white than males on the average in all strains. The evidence presented by Wright and Chase effectively disproved the theories of Pictet and Ibsen.

All of this work in physiological genetics in the period 1926–36 brought home to Wright how little was known about the fundamental mechanisms of gene control, but the progress of physiological genetics indicated at the same time that great steps forward should be possible. On October 25, 1937, Wright delivered a lecture to the Institute of Medicine of Chicago on the topic "The Hereditary Factor in Abnormal Development," where he concluded:

> The task of determining the exact nature of the primary action of genes and that of tracing the course of events between such primary

actions and the final complex of the developed monster are ones in which little more than a start has been made. There are some definite results, however, and geneticists and embryologists are actively engaged in a joint attack on these problems. I think that we may look confidently to great advances not only in the special problems of the origins of monstrosities but, along with this, progress in understanding the general nature of the developmental process. (Wright 93)

Here again Wright exhibits the optimism that the continued efforts of geneticists and embryologists, working together, would be able to elucidate in detail the developmental process.

Physiological Genetics, 1935–1961

Between 1937 and his retirement from the University of Chicago the proportion of Wright's published papers on physiological genetics decreased in relation to those on evolutionary biology. One might be tempted to conclude that Wright was moving his interests correspondingly from physiological genetics to theoretical population genetics. This conclusion would be wrong. The day-to-day work at the office and laboratory was devoted primarily to the guinea pig colony. By the mid-1930s Wright's experiments on hair direction and coat colors had become extremely complex. Not only did the experiments take much time but the results required extensive analysis. Although Wright had difficulty finding the time to fully analyse and write up his experiments for publication, his academic life was nevertheless tied closely to the lab. All his graduate students during the 1940s and 1950s, with one exception only, worked on physiological genetics. Wright certainly viewed himself as primarily a physiological geneticist during his entire University of Chicago tenure; I think this view is accurate, although there can be no doubt that well before his retirement from the University of Chicago his international reputation was based primarily upon his work in evolutionary theory.

In the two decades 1941–61 Wright published more than twenty papers on physiological genetics. This was indeed a time of great strides in the field, marked in the beginning by the Beadle and Tatum "one gene, one enzyme" hypothesis drawn from *Neurospora* and at the end by the decipherment of the genetic code. It had been Wright's intuition that genetic analysis of such features as pigmentation and hair direction in guinea pigs could be followed in greater and greater detail, leading eventually to a fundamental physiological genetics. In the period 1941–61 Wright sharpened his genetic analysis of color inheritance in guinea pigs to such a degree of complexity that it staggers the imagination of anyone who ventures to follow this work in detail.

The problem was that genetic analysis of the sort applied by Wright to the guinea pig would not be a major factor in the revolution that would occur in physiological genetics. It was, of course, the revolution in molecular biology and biochemistry that would provide physiological geneticists with the basis for understanding better the interaction of genetics and development.

Wright began the period 1941–61 as a well-known leader in physiological genetics, a spokesman for the entire field, and looked up to by persons such as George Wells Beadle. As his last papers analyzing the genetics of coat color in guinea pigs appeared in 1961, Wright's work was clearly anachronistic as a way of approaching the problems of genetics in relation to development. After 1961 Wright published almost nothing more on physiological genetics except in relation to his evolutionary theory, certainly a very important role (but see Wright's excellent summary of his experimental work on guinea pigs published in 1984 when he was almost ninety-five; Wright 210).

When Wright attended the International Congress of Genetics at Edinburgh in 1939 he could have given a paper in sessions on microevolution, or genetics in relation to evolution and systematics, or comparative genetics and evolution; he chose instead to give a summary paper in a session on physiological genetics along with Beadle, Fritz Baltzer, and Fritz von Wettstein, three international leaders in the field.

In 1941 and 1945 Wright published two major reviews of physiological genetics, "The Physiology of the Gene" (Wright 107) and "Physiological Aspects of Genetics" (Wright 117), and he also provided introductory general comments at a session on genes as physiological agents, which he organized for a meeting in 1944 (published as "Genes as Physiological Agents"; Wright 118). The fundamental message in each of these papers was the same: genes are material structures that replicate and code for enzymes that determine physiological processes in the organism. Wright had maintained this same theme for almost thirty years, but by 1945 the evidence was far greater than only a decade earlier.

Wright's summary papers on physiological genetics were widely read and influential at the time, but soon they were forgotten in the rush to molecular biology. The details about mechanisms of heredity changed dramatically between 1941 and 1945. In 1941 Wright stated:

> Nucleic acid is itself too simple a material and too uniform in nature to be responsible for the specificity of the genes. Similar statements have been made with respect to the protamines and histamines extracted from sperms. The possibilities of diversity among protein molecules through different possible arrangements of the amino acids and through the attachment of prosthetic groups is, however, so nearly infinite that there seems to be no theoretical difficulty in connection with gene specificity, even if only a minute portion of the visible chromosome is genic. (Wright 107, pp. 492–93)

In his outstanding essay "DNA" (Stent 1970), Gunther S. Stent argued that the view expressed here by Wright was so widespread and deep among geneticists that they were unable to appreciate the experimental work of Oswald T. Avery and his colleagues showing that DNA was the material that caused bacterial genetic transformation (Avery, MacLeod, and McCarty 1944).

Mayr (1982, 818–21) has argued cogently that Stent's thesis is untrue generally, but it certainly does not apply to Wright. In 1944 Wright organized a major symposium on genes as physiological agents for the annual meeting of the American Society of Zoologists. His speakers were Norman Horowitz (from Beadle's lab), Tracy Sonneborn, and Donald Poulson. At the meeting, he heard a paper given by Avery, the effects of which he described in a letter to Dobzhansky:

> The most interesting paper that I heard was one by Avery on the isolation of apparently pure desoxyribose nucleic acid as the active substance in transforming unencapsulated pneumococcus into the type from which the extract came. It looks almost like isolation of a gene from an organism and transferring it to another. I suppose that the most interesting point, however, is that a nucleic acid, with no protein whatever, can transmit the potentiality for synthesis of a specific carbohydrate as well as of itself. (Wright to Dobzhansky, November 3, 1944)

Dobzhansky was much less impressed by Avery's work than Wright, and as late as the third edition of *Genetics and the Origin of Species* (1951), Dobzhansky argued that just because DNA was a "transforming principle" did not mean it was the hereditary material. Wright was sufficiently impressed that he rewrote his introductory remarks for the symposium to include comment upon Avery's work and its implications:

> Most investigators find the condensed chromosomes of sperm cells to consist almost wholly of basic proteins (protamine or histone) combined with highly polymerized desoxyribose nucleic acid.
>
> It has been maintained that both of these are too simple in chemical composition to provide an adequate basis for the practically infinite diversity of genes. The possibilities of stereoisomerism are, however, almost unlimited. Moreover, there is now evidence of a more direct sort that desoxyribose nucleic acid actually can transmit from one organism to another the potentiality for indefinitely continued synthesis of itself and also of a chemically unrelated specific substance. Avery, MacLeod and McCarty have recently obtained an extract from one specific type of pneumococcus, apparently consisting of pure polymerized desoxyribose nucleic acid (molecular weight about 500,000), which can transform unencapsulated cells from another type into the type from which the extract was obtained. The desoxyribose nucleic acid not only conveys the capacity to produce the specific polysaccharide on which type specificity depends but also the capacity to produce more of itself. As to the protein constituent, the spacing of amino acids in a polypeptide chain is practically the same as that of nucleotides in nucleic acid and there is the possibility of corresponding specificities based on isomerism. (Wright 118, p. 290)

Moreover, Wright suggested several ways to overcome the obvious difficulty that if particular genes are coded for each kind of protein, then a single gene might have to code for millions of protein molecules during growth and cell division. There might, he said, be many copies of such a gene, and furthermore,

> The difficulty would be further obviated, if we suppose that there are proteins in the cytoplasm that can function as models on which more proteins of the same sort can be formed. It is known that large amounts of ribonucleic acid, closely related to the desoxyribose nucleic acid found only in the chromosomes, are present in rapidly dividing cells. This nucleic acid is located in the microsomes and mitochondria. Claude has suggested that these are self-duplicating bodies. (P. 296)

With his knowledge of physiological genetics and understanding of Avery's work, Wright was well prepared in 1944 to appreciate the coming revolution in molecular biology. He was not, however, with his experimental work on guinea pigs, in any position to play a significant role in this revolution. Indeed, with its extremely complex chromosome complement and largely unknown heredity, the guinea pig was a terrible organism to use for sorting out the detailed relations of genes and physiology.

Undaunted by the great successes of the molecular approach, of which he was keenly aware, Wright continued his experiments on the inheritance of hair direction, coat colors, and vital characters in the guinea pig. He found several new mutations in both coat colors and hair direction, which yielded ever more complex interactions and combinations with genetic characters he had already studied. When Wright retired from the University of Chicago at the end of December 1954, he took no guinea pigs with him to Madison. It took him, however, five years of concentrated work to analyse and publish the data he had accumulated before leaving the University of Chicago. The results of these labors appeared in seven substantial papers between 1959 and 1961 (154–156, 163, 165–167). With these papers Wright felt that he had finished his contribution to the physiological genetics of the guinea pig, and he turned his full attention to evolutionary theory.

The remainder of this book will be devoted primarily to Wright's theory of evolution in nature in relation to the whole field of evolutionary biology. Yet as I have argued earlier and will prove in what follows, the only way to fully understand his theory of evolution and his interactions with other evolutionists is to view it fully in the context of his work in physiological genetics. At all times Wright saw his work in physiological genetics as fundamentally related to and fully integrated with his ideas about the evolutionary process. Most of Wright's major addresses to general audiences after the early 1950s dealt specifically with the relation of gene to organism, going from physiological genetics to evolutionary biology. Notable among these were his presidential addresses for the American Society of Naturalists in 1952 ("Gene and Organism," Wright 144) and for the Tenth International Congress of Genetics in 1958 ("Genetics, the Gene, and the Hierarchy of the Biological Sciences," Wright 153), and his address to the Darwin Centennial Conference at the University of Chicago in 1959 ("Physiological Genetics, Ecology of Populations, and Natural Selection," Wright 158).

Thus although Wright finally ended his technical work on physiological genetics just before he threw himself into the work on what would emerge as a four-volume magnum opus on evolutionary biology, he was always and consciously using his background and understanding of physiological genetics. I now turn to Wright's evolutionary theory.

7
The Problem of Adaptation and Mechanisms of Evolution after Darwin

Introduction

My primary aim in the remaining chapters of this book is to present and analyze Sewall Wright's contributions to evolutionary biology. The best way to
accomplish this aim is to view Wright's work in the full context of evolutionary thought in his time. I will therefore examine closely Wright's interactions
with R. A. Fisher, E. B. Ford, Theodosius Dobzhansky, A. H. Sturtevant,
Ernst Mayr, Motoo Kimura, and many other evolutionists. But to understand,
for example, the conflict that developed between Wright and Fisher and the
immense ramifications this conflict had upon both evolutionary theory and
field research, further background in some aspects of evolutionary biology
before 1925 is essential. Providing this background is the purpose of this
chapter.

Evolutionary biology in the period 1859–1925 is extraordinarily complex
so I will focus only upon those few strands that are essential for understanding Wright's work and its impact. In particular, I will examine the basic tension that evolutionary biologists experienced over the problem of adaptation
in relation to mechanisms of evolution.

At the outset, a careful distinction must be made concerning the relation
of adaptation to mechanisms of evolution. We can speak about assessing
whether or not a particular character of an organism, such as a color pattern
or leaf shape or jaw structure, is adapted to the environment. Does the character enable the organism to leave more or fewer offspring than would otherwise be the case? Or we can ask whether the characters that distinguish two
species or other taxonomic categories are adaptive or nonadaptive. This is not
simply the earlier question applied twice but can mean that the assessment of
adaptation is applied to differences, for example, in homologous characters in
two different species. In this case, both characters might be highly adapted to
the two different environments, but neither contribute more than the other to
reproductive fitness. The adaptive difference between them would therefore
be none. In this sense, one could be an intense selectionist who believes that
all characters of a species are highly adapted to the environment and still believe that many constant differences in characters between closely related
species are nonadaptive. This distinction (one that I thank Malcolm Kottler

for bringing home to me) is important for the discussion of adaptation in relation to mechanisms of evolution in the rest of the book, not merely in this chapter.

To what extent are the characters of animals and plants adaptive to their environments? Are the measurable and constant differences in characters between closely allied varieties, species, or other taxonomic categories adaptive? The answers to these questions are directly related to the issue of mechanisms of evolution in nature. The two primary competing hypotheses for explaining the evolution of adaptive characters from the time of Darwin's *Origin* in 1859 to 1925 were natural selection and the inheritance of acquired characters.

Are there any nonadaptive characters? If so, what mechanisms of evolution can explain their origin in natural populations? This question was more difficult to answer than that of the origin of adaptive characters. A large number of competing hypotheses were vigorously debated in the period 1859–1925. No convincing primary hypothesis or hypotheses emerged, except that geographic isolation, clearly insufficient as a complete explanation, was believed to play an important role.

The basic tension between adaptive and nonadaptive views of the evolutionary process can be seen in Darwin's own work. A generation later the tension was well exemplified by the intense controversy that raged between Darwin's protégé, G. J. Romanes, and A. R. Wallace. John T. Gulick also played a major role in this controversy. At the turn of the century, "neo-Darwinians" such as E. Ray Lankester, E. B. Poulton, Karl Pearson, and W. F. R. Weldon argued strongly for adaptive evolution, whereas William Bateson, T. H. Morgan, and Hugo de Vries argued equally strongly for a major nonadaptive component to evolution in nature. In the first quarter of the twentieth century, many systematists and geneticists were drawn into the debate.

In light of this background, Wright's work in evolutionary biology will be far more understandable. This chapter is only the briefest of introductions. I recommend that readers wishing a broader background consult Peter Bowler's *The Eclipse of Darwinism* (1983), Vernon L. Kellogg's *Darwinism To-Day* (1907), and my "Adaptation and Mechanisms of Evolution after Darwin: A Study in Persistent Controversies" (Provine 1985), from which this chapter is primarily drawn.

Darwin on Adaptation, Mechanisms of Evolution, and Classification

In this section I have used the sixth edition of *The Origin of Species* with additions and corrections to 1872 (Darwin 1872). This is the edition most persons read. John Murray printed 9,750 copies of the first through fifth editions between 1859 and 1869; over 100,000 copies of the sixth edition came from Murray between 1872 and 1929. Appleton printed tens of thousands of copies of the sixth edition in New York City (Freeman 1977). The great emphasis now upon the first edition of 1859 reflects the mood of neo-Darwinians a century later who found that that edition fit their ideas of Darwinism better than did the sixth edition. One should observe that Ernst Mayr is the editor of the facsimile reprint of the first edition (Mayr 1964). The sixth edition of *Origin* is the one of greatest interest to those studying Darwin's influence after 1872, by which time the first edition was scarce.

Darwin's conception of natural selection, his basic mechanism of evolution, is too well known to require exposition here. For Darwin, natural selection of individual differences (those small ubiquitous variations found in every population) was the primary mechanism of evolution at every level of the evolutionary process. Certainly this was the dominant view neo-Darwinians attributed to Darwin.

Darwin constantly reminded the reader that natural selection did not operate to make all parts of an organism exquisitely adapted to its surroundings, as the natural theologians would have one believe: "Natural selection tends only to make each organic being as perfect as, or slightly more perfect than, the other inhabitants of the same country with which it comes into competition. And we see that this is the standard of perfection attained under nature" (Darwin 1872, 163). Thus Darwin explained that we should expect to see the contraptions natural selection had fitted to organisms and to observe that an introduced species might easily outcompete a native species, even though the native species appeared well adapted to its surroundings. Natural selection was not a mechanism of absolute perfection but of sufficient adaptation of organisms to their immediate environments.

All this would have sounded familiar to neo-Darwinians of 1959. A note of disquiet, however, is found in all editions of *Origin*. Darwin plainly states that use and disuse of parts is a substantial mechanism of evolution, right along with natural selection. As so many opponents of neo-Darwinism in the late nineteenth and early twentieth century pointed out, Darwin's introduction to *Origin* ends with the statement, "I am convinced that Natural Selection has been the most important, but not the exclusive, means of modification" (p. 4).

What indeed were Darwin's mechanisms of evolution in addition to natural selection, and what were their relationships to adaptive evolution? The list is longer, is more substantial, and has greater implications for nonadaptive evolution and classification than most neo-Darwinians after Wallace have realized.

1. *Use and disuse of parts*. Next to natural selection, Darwin considered this to be the most important mechanism of adaptive evolution. *Origin* is filled with references to the evolutionary effects of use and disuse, and Darwin provided a biological justification with his provisional hypothesis of pangenesis in *Variation of Animals and Plants under Domestication* in 1868.

2. *Sexual selection*. This mechanism was less rigorous than natural selection but was basically nonadaptive or maladaptive. Darwin in *Descent of Man* (1871) provided several examples of characters disadvantageous or nonadaptive in the general struggle for existence and concluded that "not one of the

external differences between the races of man are of any direct or special service to him" (1: 248–49). Darwin believed that sexual selection was the only mechanism that could adequately explain the existence of such nonadaptive characters. Natural selection prevented sexual selection from leading to highly disadvantageous characters.

3. *Directed variation*:

> Certain rather strongly marked variations, which no one would rank as mere individual differences, frequently recur owing to a similar organisation being similarly acted on,—of which fact numerous instances could be given with our domestic productions. In such cases, if the varying individual did not actually transmit to its offspring its newlyacquired character, it would undoubtedly transmit to them, as long as the existing conditions remained the same, a still stronger tendency to vary in the same manner. There can also be little doubt that the tendency to vary in the same manner has often been so strong that all the individuals of the same species have been similarly modified without the aid of any form of selection. (Darwin 1872, 72)

Directed variation was predominantly non or maladaptive.

4. *Correlated variation*. This was one of Darwin's favorite ways of explaining away maladaptive features. The argument was that the maladaptive character was correlated with another of adaptive value sufficiently high that their combination had positive adaptive value. Adaptationists since Darwin have loved this argument and used it frequently. The hypothesized linkage is in most cases very difficult to prove or disprove.

5. *Spontaneous variations*. These variations simply appeared spontaneously and then were passed on by heredity. Spontaneous variations were not induced by changed conditions, Darwin's favorite cause for the appearance of new increased variability. Darwin offered the "appearance of a mossrose on a common rose, or of a nectarine on a peach tree" as "good instances of spontaneous variations. . . . In the earlier editions of this work I underrated, as it now seems probable, the frequency and importance of modifications due to spontaneous variability. But it is impossible to attribute to this cause the innumerable structures which are so well adapted to the habits of life of each species" (p. 171). Spontaneous variations passed on by heredity accounted for many of the nonadaptive characters of organisms.

6. *Family selection*. Darwin invented this mechanism to explain the evolution of altruistic social behavior and neuter castes, apparently antithetical to individual natural selection. The difficulties of these examples, "though appearing insuperable, is lessened, or as I believe, disappears, when it is remembered that selection may be applied to the family, as well as to the individual, and may thus gain the desired end" (p. 230). To Darwin, familial selection might be nonadaptive or maladaptive as well as adaptive at the individual level but was always adaptive at the family level. Some biologists

have suggested that Darwin would have been pleased by W. D. Hamilton's calculus of inclusive fitness to explain such anomalous cases by individual rather than by familial selection, but others see Hamilton's inclusive fitness as a type of interfamily selection and not as individual selection (Hamilton 1964; D. S. Wilson 1980; Michod 1982; Ruse 1980; and Sober 1984, 216–19).

Darwin's mechanisms 2 through 5 above could all lead to nonadaptive differentiation in local populations, and by heredity after that to the levels of subspecies and species, and to even higher taxa. But did Darwin think that animals and plants in nature really exhibited very extensive nonadaptive characters?

The answer is unquestionably affirmative. Darwin's chapter 6 contains an important section entitled "Organs of little apparent Importance, as affected by Natural Selection." Such organs, Darwin said, presented for the theory of evolution by natural selection great difficulties, "almost as great, though of a very different kind, as in the case of the most perfect and complex organs" (Darwin 1872, 157). Darwin suggested that, in the first place, these apparently unimportant characters might have hidden adaptive value: "we are much too ignorant in regard to the whole economy of any one organic being, to say what slight modifications would be of importance or not" (p. 157). Then he further suggested that "Organs now of trifling importance have probably in some cases been of high importance to an early progenitor, and, after having been slowly perfected at a former period, have been transmitted to existing species in nearly the same state, although now of very slight use; but any actually injurious deviations in their structure would of course have been checked by natural selection" (p. 157). But having stated the caveats for natural selection Darwin added:

> In the second place, we may easily err in attributing importance to characters, and in believing that they have been developed through natural selection. We must by no means overlook the effects of the definite action of changed conditions of life,—of socalled spontaneous variations, which seem to depend in a quite subordinate degree on the nature of the conditions,—of the tendency to reversion to longlost characters,—of the complex laws of growth, such as of correlation, compensation, of the pressure of one part on another, etc., and finally of sexual selection, by which characters of use to one sex are often gained and then transmitted more or less perfectly to the other sex, though of no use to this sex. But structures thus indirectly gained, although at first of no advantage to a species, may subsequently have been taken advantage of by its modified descendants, under new conditions of life and newly acquired habits. (Pp. 157–58)

A primary example of nonadaptive variation, Darwin thought, could be seen in polymorphic genera and species:

There is one point connected with individual differences, which is extremely perplexing: I refer to those genera which have been called "protean" or "polymorphic," in which the species present an inordinate amount of variation. With respect to many of these forms, hardly two naturalists agree whether to rank them as species or as varieties. We may instance Rubus, Rosa, and Hieracium amongst plants, several genera of insects and of Brachiopod shells. In most polymorphic genera some of the species have fixed and definite characters. Genera which are polymorphic in one country seem to be, with a few exceptions, polymorphic in other countries, and likewise, judging from Brachiopod shells, at former periods of time. These facts are very perplexing, for they seem to show that this kind of variability is independent of the conditions of life. I am inclined to suspect that we see, at least in some of these polymorphic genera, variations which are of no service or disservice to the species, and which consequently have not been seized on and rendered definite by natural selection. (P. 35)

Special attention should be paid to Darwin's comments about polymorphism. At the centennial of Darwin's *Origin* in 1959, almost every neo-Darwinian believed that polymorphism was highly adaptive, rather than nonadaptive as Darwin himself clearly believed. As I will show in later chapters (particularly chapter 12), the shift among evolutionists toward an adaptive interpretation of polymorphism actually came very late, after 1940. Before that, with the notable exceptions of Ford and Fisher, most twentieth-century evolutionists indicated close agreement with Darwin as quoted above.

In chapter 7 Darwin gives a substantial list of nonadaptive characters of animals and plants. Why, one might ask, would someone clearly promoting the idea of natural selection as the primary mechanism of evolution spend considerable energy documenting the existence of nonadaptive variation that did not evolve by natural selection, at least in Darwin's opinion? The answer is that Darwin held allegiance not only to natural selection but to the idea of evolution by descent and to the implications of evolution by descent for the problem of classification:

From the fact of the above characters being unimportant for the welfare of the species, any slight variations which occurred in them would not have been accumulated and augmented through natural selection. A structure which has been developed through long-continued selection, when it ceases to be of service to a species, generally becomes variable, as we see with rudimentary organs; for it will no longer be regulated by this same power of selection. But when, from the nature of organism and of the conditions, modifications have been induced which are unimportant for the welfare of the species, they may be, and apparently often have been, transmitted in nearly the same state to numerous, otherwise modified, descendants. It cannot have been of much importance to the greater number of mammals, birds, or reptiles, whether they were clothed with hair, feathers, or scales; yet hair has been transmitted to almost all mammals, feathers to all birds, and scales to all true reptiles. A structure, whatever it may be, which is common to many allied forms, is ranked by us as of high systematic importance, and consequently is often assumed to be of high vital importance to the species. Thus, as I am inclined to believe, morphological differences, which we consider as important—such as the arrangement of the leaves, the divisions of the flower or of the ovarium, the position of the ovules, etc.—first appeared in many cases as fluctuating variations, which sooner or later became constant through the nature of the organism and of the surrounding conditions, as well as through the intercrossing of distinct individuals, but not through natural selection; for as these morphological characters do not affect the welfare of the species, any slight deviations in them could not have been governed or accumulated through this latter agency. It is a strange result which we thus arrive at, namely that characters of slight vital importance to the species, are the most important to the systematists; but, as we shall hereafter see when we treat of the genetic principle of classification, this is by no means so paradoxical as it may first appear. (Pp. 175–76)

Turning, as Darwin suggests, to his chapter on classification, the importance of the nonadaptive characters documented above becomes obvious. The purpose of classification to an evolutionist is to reveal community of descent; the true system of classification is based upon the evolutionary tree. Should not the classification be based upon those distinctive characters most clearly adapting the organism to its environment?

It might have been thought . . . that those parts of the structure which determined the habits of life, and the general place of each being in the economy of nature, would be of very high importance in classification. Nothing can be more false. No one regards the external similarity of a mouse to a shrew, of a dugong to a whale, or a whale to a fish, as of any importance. (P. 365)

And here, of course, is where the importance of nonadaptive characters enters, as revealed by the following quotes:

In formerly discussing certain morphological characters which are not functionally important, we have seen that they are often of the highest service in classification. This depends on their constancy throughout many allied groups; and their constancy chiefly depends on any slight deviations not having been preserved and accumulated by natural selection, which acts only on serviceable characters.

That the mere physiological importance of an organ does not determine its classificatory value, is almost proved by the fact, that in allied groups, in which the same organ, as we have every reason to suppose, has nearly the same physiological value, its classificatory value is widely different. (Pp. 366–67)

No one will say that rudimentary or atrophied organs are of high physiological or vital importance; yet, undoubtedly, organs in this condition are often of much value in classification. . . .

Numerous instances could be given of characters derived from parts which must be considered of very trifling physiological importance, but which are universally admitted as highly serviceable in the definition of whole groups.

On the view of characters being of real importance for classification, only in so far as they reveal descent, we can clearly understand why analogical or adaptive characters, although of the utmost importance of the welfare of the being, are almost valueless to the systematist. For animals, belonging to two most distinct lines of descent, may have become adapted to similar conditions, and thus have assumed a close external resemblance; but such resemblances will not reveal—will rather tend to conceal their blood-relationship. (P. 374)

So Darwin was actually glad that natural selection did not determine all species characters; in that case, the natural system of classification by descent would be impossible to establish. Nonadaptive characters were, to Darwin, the essential keys to accurate systematics (on this point compare with the discussion in Mayr 1969).

Moritz Wagner (1813–1887)

The reaction of the German naturalist Moritz Wagner to Darwin's views helps clarify the situation. Wagner had traveled extensively in North Africa, North, Central and South America, and in western Asia. He studied, among many other things, geographic distribution of plants and animals with an eye toward mechanisms of speciation. His first essay on Darwinism, *Die Darwin'sche Theorie und das Migrationsgesetz der Organismen* (1868), argued that geographic separation was essential for the evolutionary process. The proof that this process did not occur only by natural selection was his observation that closely allied species often differed by nonadaptive characters. He did not at this time rule out natural selection as a significant auxiliary mechanism in evolution. Darwin thought Wagner emphasized too much the necessity of geographic isolation for evolution, but he lauded Wagner's view where geographic speciation obviously had occurred.

Two years later, however, Wagner adopted his lifelong view that natural selection was totally unnecessary for speciation (Wagner 1870, 434). Darwin, pointing to clear adaptations in closely allied species, disagreed intensely with Wagner from 1870 on. This disagreement has obscured the substance of Darwin's agreement with Wagner—that geographic speciation could occur, leaving closely related species that differed by nonadaptive characters. (For substantive accounts of Wagner and Darwin on the question of isolation in relation to evolution, see Kottler 1976 and Sulloway 1979.)

Darwin frequently expressed in correspondence in his later years his belief in the importance of nonadaptive variation in natural populations. Thus in a letter to Moritz Wagner dated October 13, 1876, in which Darwin agreed with Wagner on the necessity of isolation for the splitting (not transformation) of species, he confessed:

> In my opinion the greatest error which I have committed, has not been allowing sufficient weight to the direct action of the environment, *i.e.* food, climate, etc., independently of natural selection. Modifications thus caused, which are neither of advantage nor disadvantage to the modified organism, would be especially favored, as I can now see chiefly through your observations, by isolation in a small area, where only a few individuals lived under nearly uniform conditions. (Darwin 1887, 3:159)

But Darwin was a complex man facing overwhelmingly complex data. Alfred Russel Wallace, for reasons I will clarify below, much preferred to quote Darwin's letter of November 30, 1878, to Karl Semper:

> As our knowledge advances, very slight differences, considered by systematists as of no importance in structure, are continually found to be functionally important; and I have been especially struck with this fact in the case of plants to which my observations have of late years been confined. Therefore it seems to me rather rash to consider the slight differences between representative species, for instance those inhabiting the different islands of the same archipelago, as of no functional importance, and as not in any way due to natural selection. With respect to all adapted structures, and these are innumerable, I cannot see how M. Wagner's view throws any light, nor indeed do I see at all more clearly than I did before, from the numerous cases that he has brought forward, how and why it is that a long isolated form should almost always become slightly modified.

In light of subsequent history, this quotation, rather than the previous one, appears to have been the clarion call to those who shouldered the mantle of "neo-Darwinians."

In the final sentence of this quotation Darwin raises a pertinent issue that would stimulate much thought and later play a prominent role in Wright's first formulation of his shifting balance theory of evolution. How and why, Darwin asked, did an isolated population continue to differentiate beyond the limits of variability in the mother populations? This was an excellent question, and a very difficult one, especially if the differentiation was nonadaptive.

Debate over Adaptation: Wallace, Romanes, and Gulick

Wallace, co-inventor of the concept of natural selection, became one of the two preeminent Darwinians after Darwin's death in 1882 (the other being Weismann). Wallace had far more direct acquaintance than Darwin with field research on natural populations (though Darwin certainly held the edge on knowledge of domestic populations). The overwhelming impression that Wallace gained from his work on natural populations was that Darwin had underestimated the effectiveness, rapidity, and importance of natural selection. In the 1889 summary of his evolutionary views, *Darwinism,* Wallace argued that since the sixth edition of *Origin* in 1872, much evidence had accumulated for greater heritable variability in natural populations than Darwin had imagined. The availability of this variability insured the effectiveness of natural selection. Except in the case of man (see Kottler 1974), Wallace was a far more thoroughgoing selectionist than Darwin. The differences in view between Wallace and Darwin emerged clearly on the issue of whether the characters used as taxonomic markers were adaptive or not.

Wallace's answer to Romane's theory of physiological selection is helpful in this analysis. (For an excellent introduction to this debate, see Lesch 1975.) Romanes, who had worked closely with Darwin on the evolution of mentality, argued in "Physiological Selection" (Romanes 1886) that natural selection was "not, strictly speaking, a theory of the origin of *species:* it is a theory of the origin—or rather of the cumulative development—of *adaptations*" (p. 345). Natural selection did not, he claimed, directly account for the mutual sterility of closely related species, nor of nonadaptive taxonomic markers:

> The features, even other than sterility *inter se,* which serve to distinguish allied species, are frequently, if not usually, of a kind with which natural selection can have had nothing whatever to do; for distinctions of specific value frequently have reference to structures which are without any utilitarian significance. It is not until we advance to the more important distinctions between genera, families, and orders that we begin to find, on any large or general scale, unmistakable evidence of utilitarian meaning. (P. 338)

> The only answer which Mr. Darwin makes to this difficulty is, that structures and instincts which appear to us useless may nevertheless be useful. But this seems to me a wholly inadequate answer. Although in many cases it may be true, as indeed it is shown to be by a number of selected illustrations furnished by Mr. Darwin, still it is impossible to believe that it is always, or even generally so. In other words, it is impossible to believe that in all, or even in most, cases where minute specific differences of structure or of instinct are to all appearance useless, they are nevertheless useful. . . . it surely becomes the reverse of reasonable so to pin our faith to natural selection as to conclude that all these peculiarities must be useful, whether or not we can perceive their utility. For by doing this we are but rea-

soning in a circle. . . . But I need not argue this point, because in the later editions of his works Mr. Darwin freely acknowledges that a large proportion of specific distinctions must be conceded to be useless to the species presenting them; and, therefore, that they resemble the great and general distinction of mutual sterility in not admitting of any explanation by the theory of natural selection. (Pp. 344–45)

Wallace vigorously denied that Darwin ever stated that the particular characters used by systematists to distinguish one species from another "are ever useless, much less that a 'large proportion of them' are so, as Mr. Romanes makes him 'freely acknowledge'" (Wallace 1889, 132). Wallace then gave several pages of recent evidence that taxonomic characters, formerly supposed to be nonadaptive, were strictly adaptive. He concluded:

On the whole, then, I submit, not only has it not been proved that an "enormous number of specific peculiarities" are useless, and that, as a logical result, natural selection is "not a theory of the origin of species, or, more commonly, of genera and families; but, I urge further, it has not even been proved that any truly "specific" characters—those which either singly or in combination distinguish each species from its nearest allies—are entirely unadaptive, useless, and meaningless; while a great body of facts on the one hand, and some weighty arguments on the other, alike prove that specific characters have been and could only have been, developed and fixed by natural selection because of their utility. We may admit, that among the great number of variations and sports which continually arise many are altogether useless without being hurtful; but no cause or influence has been adduced adequate to render such characters fixed and constant throughout the vast number of individuals which constitute any of the more dominant species. (Pp. 141–42)

Wallace and Romanes were basically arguing about the mechanism of speciation. Wallace said it was natural selection; Romanes said it had to be a mechanism other than natural selection, proposing his theory of "physiological selection" (selection for sterility factors without geographic isolation) as an alternative. If the differences between species were nonadaptive, then natural selection could not be the primary mechanism. Darwin, of course, had in part already finessed this argument with something like his "correlated" nonadaptive characters. In this instance natural selection could be the primary determinant of speciation; it was just that the adaptive characters by which the species had evolved were not as useful as the nonadaptive correlated characters for purposes of classification by descent.

John Thomas Gulick (1832–1923)

The third major participant in the ongoing debate between Romanes and Wallace was the missionary and naturalist John Thomas Gulick (1832–1923). He was one of the most influential evolutionists in the world at the turn of the century. Almost every evolutionist concerned with speciation knew and cited Gulick's work. Yet he has been curiously neglected by modern historians of science. His name does not appear in *Dictionary of Scientific Biography* (neither does Moritz Wagner's!), and he rates only one sentence in Peter Bowler's recent book, *The Eclipse of Darwinism*. John E. Lesch has written a useful article on the role of isolation and includes much material on Gulick, but by far the best accounts of his work are found in the little-read biography (consisting primarily of original documents) of Gulick by his son Addison Gulick (1932) and in Malcolm Kottler's doctoral thesis (1976).

Son of an American missionary who worked in Hawaii, Gulick spent most of his active life as a missionary in China and Japan. He was always interested in natural history, but in 1853, at age twenty-one, he decided that a serious study of the Achatinellid snails of the Hawaiian Islands (then called the Sandwich Islands) would give him greater insight into God's creative powers. In a mere eleven months he amassed a huge collection of snails, mostly by buying them from natives but also by personal collection and trading with a few other collectors. Nearly a century later this haphazard method of collection would encounter serious criticism, but in the meantime the conclusions Gulick drew from his collections carried great weight. All of Gulick's later publications on Achatinellid snails were based on this one huge collection made in 1853.

Between his education and missionary work in northern China, Gulick had no time to devote to an analysis of his snail collection. Beginning in 1871, however, Gulick took a break from his missionary work and in 1872 took his collection to England where he began a serious analysis of it. One problem, again of interest in contrast to modern methods of data collection on geographic and ecological distribution, was that Gulick discovered his place-names were on no map of the Hawaiian Islands available in England or in Honolulu. In 1873 he recreated his map of distribution by visiting Hawaii and "making a combination of oral historic lore with a personal check-up of some of the localities," this now being twenty years since the collections took place (Gulick 1932, 235).

By 1872 Gulick had reached the firm conclusion that natural selection could not account for the local differentiation of Achatinellid snails in Hawaii. The environments, he thought, were frequently indistinguishable, yet the snail populations were often distinctively and consistently different. Furthermore, the characters distinguishing the various species rarely appeared to have any adaptive value at all. Gulick presented this thesis at a meeting of the British Association for the Advancement of Science in 1872, and it was published, as Gulick later said, in *Journal of the Linnean Society* (Zoology) "though the kindness of Mr. Alfred Wallace" (Gulick 1872; 1888, 190). Not only that, Wallace was the one in 1887 who rather reluctantly communicated to the Linnean Society Gulick's very long and jargon-filled paper, "Divergent Evolution through Cumulative Segregation." The paper was published in the

Linnean Society *Journal* only because of a very enthusiastic review by Romanes, who recognized in Gulick a staunch ally in the fight against Wallace and the selectionist interpretation of speciation.

Wallace immediately recognized the challenge from Gulick and viewed the alliance of Romanes and Gulick with some alarm. He answered Gulick in *Darwinism,* which he was just writing at the time he read Gulick's paper:

> It need hardly be said that the views of Mr. Darwin and myself are inconsistent with the notion that, if the environment were absolutely similar for the two isolated portions of the species, any such necessary and constant divergence would take place. It is an error to assume that what seem to us identical conditions are really identical to such small and delicate organisms as these land molluscs, of whose needs and difficulties at each successive stage of their existence, from the freshly laid egg up to the adult animal, we are so profoundly ignorant. The exact proportions of the various species of plants, the numbers of each kind of insect or of bird, the peculiarities of more or less exposure to sunshine or to wind at certain critical epochs, and other slight differences which to us are absolutely immaterial and unrecognisable, may be of the highest significance to these humble creatures, and be quite sufficient to require some slight adjustments of size, form, or colour, which natural selection will bring about. All we know of the facts of variation leads us to believe that, without this action of natural selection, there would be produced over the whole area a series of inconstant varieties mingled together, not a distinct segregation of forms each confined to its own limited area. . . .
>
> . . . While isolation is an important factor in effecting some modification of species, it is so, not on account of any effect produced, or influence exerted by isolation *per se*, but because it is always and necessarily accompanied by a change of environment, both physical and biological. Natural selection will then begin to act in adapting the isolated portion to its new conditions, and will do this the more quickly and the more effectually because of the isolation. (Wallace 1889, 148, 150)

Although Romanes and Gulick would dismiss Wallace's attitude as mere knee-jerk selectionist rhetoric, it is sobering to consider that his statement here might easily have been said by Cain and Sheppard as they began their research on the snail *Cepaea nemoralis* in the late 1940s, and even their earliest results could not be dismissed as mere kneejerk selectionism (on this see chapter 12, "The Great Snail Debate").

Romanes and Gulick (then back in Japan) began a voluminous correspondence (see Gulick 1932, chapter 15), disagreeing on a number of points, especially the necessity of geographical isolation for speciation (Gulick) versus segregation within a population (Romanes's physiological selection), but agreeing on what they believed to be the observable fact of nonadaptive speciation. (On the differences between Romanes and Gulick see Kottler 1976.)

In answer to Wallace's arguments against Gulick's conclusions, Romanes in the third volume of *Darwin and After Darwin* praised Gulick's collection of shells and his analysis of it: "no one who examines this collection can wonder that Mr. Gulick attributes the results which he has observed to the influence of apogamy [prevention of breeding between populations irrespective of selective characters] alone, without any reference to utility or natural selection." Romanes added:

> To this solid array of remarkable facts Mr. Wallace has nothing further to oppose than his customary appeal to the argument from ignorance, grounded on the usual assumption that no principle other than natural selection *can* be responsible for even the minutest changes of form or colour. For my own part, I must confess that I have never been so deeply impressed by the dominating influence of the *a priori* method as I was on reading Mr. Wallace's criticism of Mr. Gulick's paper, after having seen the material on which this paper is founded. To argue that every one of some twenty contiguous valleys in the area of the same small island must necessarily present such differences of environment that all the shells in each are differently modified thereby, while in no one out of the hundreds of cases of modification in minute respects of form and colour can any human being suggest an adaptive reason thereof—to argue thus is merely to affirm an intrinsically improbable dogma in the presence of a great and consistent array of opposing facts. (Romanes 1897, 17)

Romanes and Gulick, as Lesch has accurately concluded, brought the issue of nonadaptive speciation to the wide attention of biologists, and by the turn of the century a growing number were supporting the idea that natural selection could not be the primary mechanism of speciation because too many differences between closely related species were apparently nonadaptive.

The Problem of Adaptation at the Turn of the Century

Many neo-Darwinians in England, however, followed Wallace's strong selectionist, adaptationist view. Prominent among them were E. Ray Lankester, Raphael Meldola, and E. B. Poulton. Wallace himself lived until 1913, and Poulton until 1943, by which time Julian Huxley, Fisher, Ford, Mayr, and others had firmly established the modern neo-Darwinism. The selectionist, adaptationist view of Wallace has had a continuous existence since 1889. Just how strongly the view was held by some at the turn of the century is well exemplified in Poulton's 1894 essay "Theories of Evolution," which he delivered in Boston in an attempt to bring Darwin's idea of natural selection to the American neo-Lamarckians.

The more we study the characters of animals in general, even though we at first can see no utility, the more we come to admit this principle, and to believe that either now or in some past time, the characters have been useful. I can certainly say of many characters which I have studied in some of my investigations, that at first they seemed to be meaningless, but afterwards appeared to be of much importance in the struggle for existence. I think we may safely assume with regards to many characters of which we can now see no explanation that ultimately the explanation will be forthcoming.

Being unable to prove utility does not invalidate Natural Selection. If inutility could be proved for any large class of characters, the theory would certainly be destroyed as a widereaching and significant process. I do not think, however, that any such evidence has been forthcoming. (Poulton 1908, 106–7)

If one grants Poulton's presumption of utility and adaptation, then his position becomes almost unassailable. He grants that if inutility could be proved for a character, then natural selection as an explanation for the character would be destroyed. But how would the proof proceed? How many hypotheses can an inventive person dream up for the possible adaptive value of an apparently useless character; or its net adaptive value when correlated with other characters; or its adaptive value at an earlier time in evolutionary history? Very many, indeed an almost unlimited number. And each one would have to be refuted. Granting the presumption of utility gives the adaptationist an insurmountable advantage; indeed, as Poulton says, inability to prove utility (so much easier than proving inutility) does not invalidate natural selection. William Bateson nicely summarized Poulton's kind of argument as being equivalent to the following: "'If,' say we with much circumlocution, 'the course of Nature followed the line we have suggested, then, in short, it did.' That is the sum of the argument" (Bateson 1894, v; for a modern review of this issue compare Gould and Lewontin 1979 with the answer of Mayr 1983).

If Wallace's selectionist view enjoyed a degree of continuity from 1889 on, it certainly did not gain continuous approval from other biologists. William Bateson had come to dislike the Darwinian view that evolution proceeded very gradually by natural selection working upon small individual differences; he advocated instead the view of Galton and Huxley that evolution proceeded by the selection of large discontinuous variations, of which Bateson's *Materials for the Study of Variation* (1894) was a catalog. In the introduction Bateson vigorously attacked the adaptationist view that natural selection was the mechanism of speciation:

The Study of Adaptation ceases to help us at the exact point at which help is most needed. We are seeking for the cause of the differences between species and species, and it is precisely on the utility of Specific Differences that the students of Adaptation are silent. For,

as Darwin and many others have often pointed out, the characters
which visibly differentiate species are not as a rule capital facts in the
constitution of vital organs, but more often they are just those features which seem to us useless and trivial, such as the patterns of
scales, the details of sculpture on chitin or shells, differences in number of hairs or spines, differences between the sexual prehensile organs, and so forth. These differences are often complex and are strikingly constant, but their utility is in almost every case problematical.
(P. 11)

An even deeper objection, Bateson said, was that even if a character could be
shown to be useful in some way, this fact was useless "unless we know also
the degree to which its presence is harmful; unless, in fact, we know how its
presence affects the profit and loss account of the organism" (p. 12).

Bateson therefore absolutely refused in his huge catalogue of 886 discontinuous variations to speculate on the usefulness or harmfulness of the variations.

Such speculation, whether applied to normal structures or to Variation, is barren and profitless. If any one is curious on these questions
of adaptation, he may easily thus exercise his imagination. In any
case of Variation there are a hundred ways in which it may be
beneficial, or detrimental. For instance, if the "hairy" variety of the
moor-hen became established on an island, as many strange varieties
have been, I do not doubt that ingenious persons would invite us to
see how the hairiness fitted the bird in some special way for life on
that island in particular. Their contention would be hard to deny, for
on this class of speculation the only limitations are those of the inge-
nuity of the author. While the only test of utility is the success of the
organism, even this does not indicate the utility of one part of the
economy, but rather the net fitness of the whole. (Pp. 79–80)

In 1894 it was Bateson rather than Poulton who sounded the dominant note of
the succeeding two decades. Bateson's fellow experimental biologists would
have little respect for the "just so" stories of Poulton or Wallace, and the neo-
Darwinian selectionist-adaptationist view would suffer its deepest decline in
the entire time between the first publication of *Origin of Species* and the
present.

The most prominent and influential critic of the neoDarwinian emphasis
upon the natural selection of small differences as the mechanism of speciation
was Hugo de Vries. He had clear experimental evidence for the sudden appearance of new true breeding varieties of *Oenothera* at a time when such experimental evidence was greatly admired; he published voluminously and
traveled widely; he was one of the rediscoverers of Mendelian heredity; and
his arguments just made good sense to experimental breeders and laboratory
scientists.

Perhaps the best place to turn in de Vries's published work for his view of the mechanism of evolution is *Species and Varieties: Their Origin by Mutation* (1905), composed of lectures originally delivered at the University of California (Berkeley) in the summer of 1904. Here de Vries argued that Darwin was right in making the analogy between artificial and natural selection; the problem was that Darwin had misunderstood what the most scientific breeders were really doing and therefore made incorrect inferences about the action of natural selection in nature.

Darwin thought breeders did their work primarily through individual selection. Giving evidence from many scientific breeders, de Vries challenged this view. The most successful breeders, he said, selected not individuals but "elementary species." Choosing the right variety as the foundation stock was the key to rapid success. True, breeders used individual selection once the elementary species was established, and it was important for fine-tuning of the population according to desire; but individual selection alone could not create the varieties produced by breeders.

The situation in nature was exactly analogous. Natural selection of individuals merely adapted local populations to the local conditions of their environment. Thus natural selection of individuals "produces the local races, the marks of which disappear as soon as the special external conditions cease to act. It is responsible only for the smallest lateral branches of the pedigree, but has nothing in common with the evolution of the main stems. It is of very subordinate importance" (de Vries 1905, 802).

Corresponding to the "variety testing" or selection of elementary species practiced by breeders was what de Vries called "survival of species" or "selection between species" in nature:

> The fact that recent types show large numbers, and in some instances even hundreds of minor constant forms, while the older genera are considerably reduced in this respect, is commonly explained by the assumption of extinction of species on a correspondingly large scale. This extinction is considered to affect the unfit in a higher measure than the fit. Consequently the former vanish, often without leaving any trace of their existence, and only those that prove to be adapted to the surrounding external conditions, resist and survive. (P. 799)

Both microevolution and speciation depended upon natural selection. Selection acted upon individual differences to produce geographical races, but to produce new species, natural selection acted upon "elementary species" originating by large mutations. Using Darwin's analogy for the similarity of artificial and natural selection, de Vries deduced a mechanism of speciation that Darwin pointedly denied (see Bowler 1978).

Here again is the view that the production of new species can be basically a nonadaptive or maladaptive process, yet by the genus level, adaptation was the rule. Thus a selective process had to occur at the species level. This process could not be the usual individual selection of small heritable differences, which was incapable of producing new species. In contrast, the basic position of the neo-Darwinians such as Wallace and Poulton was that individual selection produced new varieties, species, genera, and higher taxa. But was there really a "species selection" in nature corresponding to breeders consciously testing varieties to see which ones would be foundation stocks?

In the early twentieth century, when de Vries was so popular and influential, the concept of a "species selection" tied to the mutation theory was attractive to many biologists, especially the geneticists and experimental biologists. On the surface, however, this seems inconsistent. Experimentalists emphasized tangible, decisive experiments; yet who of them had ever observed species selection in action? I think de Vries's concept of species selection was not so much attractive per se as it was a conclusion to which experimentalists felt driven by the state of available knowledge in heredity and evolution.

One way to view the issue is to examine carefully Morgan's *Evolution and Adaptation* (1903). This book contains the most systematic and careful critique that I know of the Darwinian adaptationist view from the early twentieth century (and has been carefully analyzed by Allen 1968; 1978, 108–25). As a militant experimentalist, Morgan was antagonistic to adaptationist rhetoric, just as Bateson was. Morgan flatly rejected the inheritance of acquired characters and raised what he considered insuperable objections to gradual natural selection of small differences as the prevailing mechanism of evolution. Indeed, it cannot be emphasized enough that natural selection of individual differences did not appear intuitively to have sufficient power to create new species; this was a major objection to Darwinism raised by critics everywhere. Where then could Morgan turn for a mechanism of evolution? The only available alternative was to de Vries, whose theory was not only "experimental" but also fit the paleontological record much better than gradual evolution.

Morgan observed that taxonomists used nonadaptive characters for classifying species, so clearly these species did not arise by natural selection: "It is well known that the differences between related species consist largely in differences of unimportant organs, and this is in harmony with the mutation theory, but one of the real difficulties of the selection theory" (Morgan 1903, 299). But despite this statement and his extensive critique of adaptationist rhetoric, Morgan believed that above the species level most taxonomic features were adaptive, and indeed that adaptation was a fundamental aspect of animals and plants. If selection were not the key to the production of species ("Nature does not remodel old forms through a process of individual selection"), it still had to play a significant role at the next level. Here Morgan's reasoning and conclusions are worth quoting at some length:

> We find that the great majority of animals and plants show distinct
> evidence of being suited or adapted to live in a special environment,

i.e. their structure and their responses are such that they can live and leave descendants behind them. I can see but two ways in which to account for this condition, either (1) teleologically, by assuming that only adaptive variations arise, or (2) by the survival of only those mutations that are sufficiently adapted to get a foothold. Against the former view is to be urged that the evidence shows quite clearly that variations (mutations) arise that are not adaptive. On the latter view the dual nature of the problem that we have to deal with becomes evident, for we assume that, while the origin of the adaptive structures must be due to purely physical principles in the widest sense, yet whether an organism that arises in this way shall persist depends on whether it can find a suitable environment. This latter is in one sense selection, although the word has come to have a different significance, and, therefore, I prefer to use the term *survival of species*.

The origin of a new form and its survival after it has appeared have been often confused by the Darwinian school and have given the critics of this school a fair chance for ridiculing the selection theory. The Darwinian school has supposed that it could explain the origin of adaptations on the basis of their usefulness. In this it seems to me they are wrong. Their opponents, on the other hand, have, I believe gone too far when they state that the present condition of animals and plants can be explained without applying the test of survival, or in a broad sense the principle of selection amongst species.

It will be clear, therefore, in spite of the criticism that I have not hesitated to apply to many of the phases of the selection theory, especially in relation to the selection of the individuals of a species, that I am not unappreciative of the great value of the part of Darwin's idea which claims that the *condition* of the organic world, as we find it, cannot be accounted for entirely without applying the principle of selection in one form or another. This idea will remain, I think, a most important contribution to the theory of evolution. (Pp. 462–464)

Morgan had never seen his "principle of selection amongst species" in action in nature any more than the neo-Darwinians had ever seen natural selection in action in nature. Thus it is curious to see this avowed experimentalist turn, on the last two pages of his book, to a mechanism of evolution he not only had never seen, but for whose existence he proposed no possible experiments. I detect no enthusiasm from Morgan for the idea of species selection. The evidence before him drove him reluctantly to this idea. De Vries was delighted (de Vries 1905, 9).

The Rise of Genetics and the Problem of the Efficacy of Selection

Between them, Bateson and de Vries exerted an enormous influence on the rise of genetics in the early twentieth century. As shown earlier in chapter 2, most of the early geneticists believed that gradual natural selection operating upon small individual differences was an inadequate primary mechanism of evolution in nature. In particular, Wright's major advisor at Harvard, William Castle, had advocated the mutation theory of evolution before 1908. In a curious twist often exemplified in science, however, geneticists motivated by the mutation theory to conduct selection experiments soon produced firm evidence showing that, far from being contradictory, Mendelism and Darwinian selection theory were actually complementary. Elsewhere I have discussed this shift among geneticists to a belief in the efficacy of selection (Provine 1971, chapter 4).

What I wish to emphasize here are the implications of this shift for theories of evolution in nature. The most important thing to understand is that geneticists had shown the possibility that selection of small heritable variations was an efficacious force in nature, not that such selection was the only major mechanism of evolution in nature. Thus every student of Castle, East, Morgan, Baur, or Philipchenko knew that selection of small variations controlled by Mendelian heredity could be effective, but this was a far cry from the panselectionism of Wallace or Poulton.

The demonstration by geneticists of the effectiveness of selection set the stage for the quantitative synthesis of Mendelism and selection theory soon to be accomplished by Fisher, Wright, and Haldane. But all three knew that this model-building exercise was a theoretical venture not to be confused with the actual course of evolution in nature. The constants in their equations had to be determined from observations of natural populations themselves. Nothing that Wright learned from Castle about selection necessitated the conclusion that observed differences between species in nature had to be adaptive. This issue could be settled in one way only—direct observation. Thus what naturalists and systematists had to say about natural populations was crucial.

Systematists and the Problem of Adaptation, 1900–1925

As Mayr has well argued on many occasions (see especially Mayr and Provine 1980), naturalists/systematists knew far more than geneticists about natural populations in the early twentieth century. Most geneticists came from an experimentalist rather than a naturalist background. Whether or not differences between species were adaptive was a question systematists were best qualified to answer during the first three decades of the twentieth century. A geneticist wishing information about adaptation in natural populations necessarily had to turn to systematists.

In the 1880s Romanes, Gulick, and Wagner thought they were the only ones arguing for nonadaptive differentiation as a primary mode of speciation. In the early twentieth century, however, there was far more support for this view. Indeed, most of the prominent systematists of this period believed in nonadaptive speciation. A brief examination of the views of some systematists, all of whom were later cited by Wright for their expertise, follows.

David Starr Jordan (1851–1931) and Vernon L. Kellogg (1867–1937)

Jordan, one of the most prominent systematists and evolutionists in the United States, was known throughout the world for his taxonomic work on fish. In his writings on evolution Jordan constantly emphasized what he called "the survival of the existing" as a mechanism alternative to survival of the fittest in the origin of species, and he pointed out the implications of this mechanism for systematics:

> The process of natural selection has been summed up in the phrase "survival of the fittest." This, however, tells only part of the story. "Survival of the existing" in many cases covers more of the truth. For in hosts of cases the survival of characters rests not on any special usefulness or fitness, but on the fact that individuals possessing these characters have inhabited or invaded a certain area. The principle of utility explains survivals among competing structures. It rarely accounts for qualities associated with geographic distribution.
>
> The nature of the animals that first colonize a district must determine what the future fauna shall be. From their specific characters, which are neither useful nor harmful, will be derived, for the most part, the specific characters of their successors.
>
> It is not essential to the meadow lark that he should have a black blotch on the breast or the outer tailfeather white. Yet all meadow larks have these characters just as all shore larks have a tiny plume behind the ear. Those characters of the present stock, which may be harmful in the new relations, will be eliminated by natural selection. Those especially helpful will be intensified and modified, but the great body of characters, the marks by which we know the species, will be neither helpful nor hurtful. These will be meaningless streaks and spots, variation in size of parts, peculiar relations of scales or hair or feathers, little matters which can neither help nor hurt, but which have all the persistence heredity can give. (Jordan 1898, 218)

Kellogg was an entomologist whose specialties were the behavior and systematics of bird lice and silkworms. He taught for many years at Stanford with Jordan, and together they wrote three books, the most famous being *Evolution and Animal Life* (1907). Discussing the influence of geographical isolation, they examined the differences between the various species of orioles in North America and concluded:

> Not one of these varied traits is clearly related to any principle of utility. Adaptation is evident enough, but each species is as well fitted for its life as any other, and no transposition or change of the distinctive specific characters or any set of them would in any conceivable degree reduce this adaptation. No one can say that any one of the actual distinctive characters or any combination of them en-

ables their predecessors to survive in larger numbers than would otherwise be the case. (Jordan and Kellogg 1907, 129)

After discussing similar cases of nonadaptive specific differentiation on honeybees and ladybird beetles, they summarized the implications of isolation:

> In these characters, there is, therefore, no rigorous choice due to natural selection. Such specific characters, without individual utility, may be classed as indifferent, so far as natural selection is concerned, and the great mass of specific characters actually used in systematic classification are thus indifferent. . . . Adaptation is presumably the work of natural selection; the division of forms into species is the result of existence under new and diverse conditions. (P. 130)

Kellogg's most influential book was his review of the status of evolutionary theory in 1907, *Darwinism To-Day*. He drew upon a vast array of literature on evolution from all over the world, including Russia, France, England, the United States, and especially Germany. Of all the many books about Darwinism and evolution published in the early twentieth century, this one was deservedly the most widely read and sold. This book was assigned reading for almost every college-level course dealing with evolutionary biology in the decade after its publication. After Darwin's *Origin*, which Wright had read in high school, this was the book that introduced him to evolutionary theory.

Wright's well-worn copy of the book had first been given to his mother in 1909, and she gave it to him in his senior year at Lombard, where he read it in Wilhelmine Key's biology class. In a section with the heading "many species characters of no utility," carefully marked by Wright as important, Kellogg revealed his sentiments on nonadaptive differentiation:

> Every student of systematic zoology or botany has a keen realisation . . . of the fact that a majority of the distinguishing characters which he recognises in the various species and genera that come under his eye are of a sort that reveal to him no trace of particular utility or advantage. Indeed he can go farther and express, to himself at least, his conviction that many of these slight but constant specific differences can actually have no special advantageousness about them. One's experience as an observer of nature and one's common sense combine to protest against that easy and sweeping answer of the Darwinians: "shall 'poor blind man' say what characteristic, however slight and insignificant, is or is not of advantage in the great complex of nature?" . . . As a matter of fact the indifference of many specific characteristics of organisms is not denied by selectionists. Romanes was perhaps the first representative Darwinian, after Darwin himself, to admit this. . . . He [Darwin] admitted that these trivial, apparently non-useful, but constant specific characters could not be explained by natural selection, and must be due to a fixation

in the species of these characters at one time or another through the nature of the organism and the influence of extrinsic influences. (Kellogg 1907, 38–39)

Kellogg included a detailed chapter on isolation theories of species forming. He included substantive presentations of the views of Wagner, Romanes, and Jordan and a particularly precise and sympathetic analysis of Gulick's work on Achatinellid snails. Seventy-five years after he read this account of Gulick's data and conclusions, Wright had a keen recollection of how deeply impressed he had been with it.

Kellogg also placed his finger squarely upon the problem of how small populations, once geographically isolated, could continue to differentiate from the parent populations. The toughest question of all was what the mechanisms could be in the case of continued nonadaptive differentiation. This was a question that Wright would address in his first formulations of his theory of evolution in nature.

Henry Edward Crampton (1875–1956)

From the beginning of his research career, Crampton was extremely interested in the issues of selection and adaptation. In 1899 he initiated a series of selection experiments on larvae of silkworm moths. He used biometrical methods modeled after those used by H. C. Bumpus (1899) to measure selective death rate in the English sparrow and by Weldon (1895) in the shore crab. Reporting on his initial results in *Biometrika* (Crampton 1904), Crampton found selective elimination based upon certain characters in the larval stages. A serious difficulty was that the characters used were of no apparent use in the larval stages, but he was undeterred in his selectionist interpretation. A year later he clearly expressed a selectionist view in "On a General Theory of Adaptation and Selection" (Crampton 1905).

This was the very year in which Gulick finally published the detailed analysis of the data he had gathered back in 1853. Along with many other biologists, Crampton interpreted Gulick's work as the strongest attack upon the selectionist viewpoint that had yet appeared. He decided to go into the field to see for himself. In 1906 he began a major series of studies on the variation, geographical distribution, and evolution of the Polynesian land snails of the genus *Partula*. Working primarily on the islands of Tahiti and Moorea, Crampton found that the differences between the geographical races or closely related species of *Partula* could be correlated with no detectable environmental differences. Differentiation indeed followed upon isolation but not, he deduced, through the action of natural selection. Crampton's results, therefore, agreed perfectly with those reached by Gulick. The first of the three monographs that Crampton published on *Partula* appeared in 1916 in the same Carnegie Institution of Washington series in which Gulick's 1905 book had been published. This monograph, with its beautiful color plates of the snails, was widely known and cited by systematists (Crampton 1916). He also published a summary article in *American Naturalist* (Crampton 1925) in which he strongly advocated a nonadaptive interpretation of speciation. Many geneticists, including Wright, read this article and found it convincing. The hypothesis of nonadaptive speciation had a growing body of direct experimental evidence to support it.

Wilfred Osgood (1875–1947) and F. B. Sumner (1874-1945)

Osgood made his reputation as a systematist by studying the profusion of species and subspecies of the deer mouse, genus *Peromyscus*. His 1909 monograph on *Peromyscus* was much cited in the systematics literature through the 1950s and even later (Osgood 1909). In general, Osgood held an adaptationist interpretation of the differences between geographical races and between species, but he presented no detailed evidence for this interpretation nor did he seem terribly interested in the question of adaptation.

Sumner, however, initiated in 1913 a far more careful experimental and quantitative analysis of a few of the species and geographical races of *Peromyscus*. Until 1925 a neo-Lamarckian who (along with such budding young systematists as Ernst Mayr and Bernhard Rensch) believed in the direct action of the environment in producing the differences between geographical races, Sumner challenged Osgood's adaptationist outlook. Differences in intensity of pigment, for example, appeared to be correlated with humidity rather than the color of the background. Thus Sumner argued that the differences between geographical races did not result from natural selection and were nonadaptive. Because he enjoyed a well-deserved reputation as an exceedingly careful experimentalist, his reinterpretation of the adaptive value of the differences between geographical races carried much more weight with both systematists and geneticists than did Osgood's earlier adaptationist view (Provine 1979). Sewall Wright read all of Sumner's early papers on *Peromyscus* as they appeared, and although, like most geneticists, he rejected Sumner's neo-Lamarckism, he accepted Sumner's conclusions about nonadaptive evolution in *Peromyscus*.

Johannes Schmidt (1877–1933)

Schmidt was a well-known Danish marine biologist who, for a middle period in his career, concentrated upon morphological differences between geographical populations of fish. His best-known work was upon the viviparous blenny (*Zoarces viviparus*). Examining some 25,000 specimens from the coast of northern Europe, Schmidt discovered that local populations from fjord to fjord differed in the number of vertebrae and other characters. Heincke, having earlier found such geographical variation in the herring (Heincke 1898), had suggested that the differences were correlated with differences in salinity. Schmidt, although having no mechanism to account for observed racial differences, argued that neither selective forces nor direct action of the environment could account for the differences he found:

> Our investigations thus by no means support the hypothesis that the racial characters are determined exclusively by environment. On the contrary, they seem rather to indicate that differences of environment are not sufficient to explain the structural differences between the races, and that the importance of the salinity, especially, has doubtless been greatly over-estimated. (Schmidt 1918, 116)

A measure of how much stronger the nonadaptive interpretation of speciation had grown in the first quarter of the twentieth century can easily be seen in two publications by David Starr Jordan. Gulick died in 1923, and Jordan wrote a laudatory eulogy in *Science*. No longer was Jordan in any way defensive about nonadaptive evolution:

> No one considering the wealth of illustration given by Dr. Gulick can fail to recognize that isolation has been the immediate occasion of the moulding of each of the various forms; and while the evidence in most other groups of plants and animals is not so clearly visible, every competent field-worker finds the same factor in the origin of practically every species whatever. Adaptation is produced by Natural Selection: the final differential moulding by isolation and segregation. (Jordan 1923, 509)

And in the annual report of the Smithsonian Institution for 1925, Jordan stated flatly: "It is a matter of common knowledge among field naturalists that the minor differences which separate species and subspecies are due to some form of isolation with segregation. Selection produces adaptation, but the *distinctive characters of species are,* in general, *nonadaptive* (Jordan 1925, 323).

Conclusions

Neo-Darwinian selectionists in the tradition of Wallace did not disappear in the period 1900–1925. Lankester was still alive and influential. Among systematists, Karl Jordan and Poulton were strong selectionists. Indeed, this tradition gave rise to the highly influential work of Fisher and Ford. But the position of those advocating nonadaptive mechanisms of evolution in conjunction with isolation had greatly strengthened. The tension between the two ways of thinking had consequently increased to a high level.

Only with the full scope of this tension firmly in mind is it possible to fully appreciate the development of some crucial aspects of Wright's theory of evolution in nature and his interactions with other evolutionists.

8
Wright, Fisher, and the Theory of Evolution in Nature

Sewall Wright and R. A. Fisher, together with J. B. S. Haldane, Lancelot Hogben, Sergei Chetverikov, and other quantitative evolutionists, have had an important impact upon modern evolutionary biology (Provine 1978). They introduced the quantitative analysis and modeling of the evolutionary process. All the mathematical population geneticists agreed upon a number of specific points, such as the immense power of selection to change gene frequencies in a surprising small number of generations, the relative insignificance of mutation pressure in relation to selection pressure under most conditions, or the theory of balanced polymorphisms that flowed so obviously from the quantitative analysis of heterozygote advantage. The mathematical population geneticists also agreed for the most part upon which variables were the really important ones, such as selection rates, effective population size, or population structure. Finally, their work was a crucial element in the vast narrowing of the controversies over the mechanisms of evolution in nature.

The evolutionary synthesis of the 1930s and 1940s (Mayr and Provine 1980) certainly did not remove all controversy about mechanisms of microevolution or speciation, but what it did do, and resoundingly, was to greatly narrow the range of controversies that had existed before 1930. An evolutionist like Henry Fairfield Osborn was a very prestigious evolutionist before the synthesis, but who now talks about "Aristogenesis" or about any of the host of other theories that were so common and taken so seriously by one or another major school of thought before 1930 (Bowler 1983, Kellogg 1907, and Mayr 1982)? The mathematical population geneticists, especially Fisher, Haldane, and Wright, played a central role in this narrowing of the possible mechanisms of evolution, primarily by demonstrating quantitatively that some mechanisms were not as powerful as they seemed intuitively and that others were totally superfluous. Within this narrowed scope of the mechanisms of evolution in nature, the controversy between Wright and Fisher had an important and specifiable impact.

The intense controversy between Wright and Fisher, which lasted from 1929 until Fisher's death in 1962, was both highly visible and very influential in modern evolutionary biology. This controversy had a more fundamental, lasting, and stimulating effect upon evolutionary biology than any other controversy or rivalry in this century. Indeed, the controversy did not end with Fisher's death. Since 1962 Wright has produced a steady stream of papers and the four-volume *Evolution and the Genetics of Populations* (1968–78, 188, 193, 198, 200) in which he constantly contrasted his interpretations with those of Fisher. E. B. Ford has strongly carried the Fisherian banner in the successive editions of his important *Ecological Genetics* (1964 and later). Many others have joined the battle on both sides. Indeed, I would argue that many of the most fundamental issues now energizing modern evolutionary biologists are extensions of issues that were bones of contention between Wright and Fisher. There is no possible way to adequately understand Wright's influence in modern evolutionary biology in isolation from his interactions with Fisher.

The aim of this chapter is to elucidate the crucial issues in the Wright-Fisher interaction, including their agreements as well as their disagreements. I have reproduced here the complete correspondence between Wright and Fisher (all of it from the years 1929–32). (An excellent selection of this correspondence along with a critical commentary that differs in some respects from my own can be found in Bennett 1983.)

Background to Wright's Theory of Evolution in Nature

Sewall Wright's shifting balance theory of evolution in nature—seen in relation to the controversy with Fisher and including the developments that Wright himself saw as intimately tied to his theory of evolution, among them his theories of physiological genetics, inbreeding (including F-statistics), gene frequency distributions, isolation by distance, and fitness surfaces—is his most important contribution to the biological sciences. The direct and indirect influence of his evolutionary theory and associated ideas permeate modern evolutionary biology to perhaps a greater extent than the ideas of any other single evolutionary theorist of the twentieth century, the full impact not being felt until the last two decades. In this assertion I am not evaluating the relative accuracy of the models of evolution in nature flowing from Wright's evolutionary theory, only saying that he appears from the historical evidence to have been the most influential evolutionary theorist of the twentieth century. (I include as part of Wright's historical influence the evolutionary ideas that others attributed to him.) The origins of Wright's theory of evolution are therefore of considerable interest.

Wright never wrote about evolution in nature until his long typewritten manuscript of over one hundred pages written in 1925, in the months just before his move to the University of Chicago. His thirty-sixth birthday in December 1925 came shortly after he completed the manuscript.

In 1925 very little indeed was known about the genetics of natural populations, and even less about short- or long-term genetic change in natural populations. Only the year before Haldane had published the first case in which geneticists had actually calculated the rate of change in frequency of a Mendelian factor in a natural population; this was the case of the increase in the melanic form of the peppered moth *Biston betularia* in the area of

Manchester, England (Haldane 1924). Sumner had shown clearly that local races of the deer mouse *Peromyscus* differed not by the distinct Mendelian mutants studied by laboratory geneticists but by an accumulation of much smaller quantitative differences (Provine 1979). Wright read all of Sumner's papers on the *Peromyscus* studies as they appeared but from the beginning (in 1915) he took the view adopted reluctantly by Sumner after 1925—that the races of *Peromyscus* differed by quantitative characters determined by many Mendelian factors. Having spoken frequently with Richard Goldschmidt during his last year at the Bussey, Wright kept close track of the development of Goldschmidt's work on natural populations of the Gypsy moth *Lymantria dispar* (for a summary see Goldschmidt 1934). Very early, certainly by the time Wright and Goldschmidt talked in 1915, Goldschmidt had adopted a basically Darwinian approach to microevolution (although he later became skeptical that evolution above the species level could be explained by the same mechanism; see Goldschmidt 1940). The important point here is that the data produced by Sumner or Goldschmidt led by itself to no one consistent view of evolution in nature; instead, their data were consistent with many different views.

All Wright (or anyone else) knew in 1925 (or that any others knew at the time) about the genetics and evolution of natural populations was insufficient for the construction of a quantitative genetical theory of evolution in nature. Yet construction of such a theory was precisely Wright's aim—as was that of Fisher and Haldane at about the same time. The problem Wright faced was hardly new. Charles Darwin had faced a very similar situation when writing *On the Origin of Species*—he had no direct information about natural selection in natural populations, yet this was the pivotal theory in his book. Darwin's solution, as noted in the previous chapter, was to draw a fundamental analogy between artificial and natural selection, and to draw his concrete evidence from artificial selection. In 1925 it was still true that the best evidence for natural selection in nature came from its similarity to the efficacy of artificial selection, the evidence for which (as in the experiments of Hopkins, Castle, Sturtevant, Payne, and others) was overwhelming.

Wright was in a better position than Darwin for constructing a theory of evolution in nature. He understood well the new science of Mendelian genetics, including carefully controlled selection experiments and observation of interaction effects of Mendelian factors, and had more detailed information about inbreeding and outbreeding, even though this was one of Darwin's special interests. Wright also had the advantage of some sixty-five years of additional taxonomic work on species in nature. Finally, and crucially, he had an excellent quantitative mind and orientation that Darwin clearly lacked.

Faced with the difficulties of constructing a theory of evolution in nature with the evidence from natural populations alone, Wright made the same decision as had Darwin—he based his theory upon his firm knowledge and understanding of evolution and heredity in domestic and laboratory populations. Wright constructed his 1925 theory of evolution upon all of his knowledge of evolution and heredity in domestic, laboratory, and natural populations, but most specifically upon the following four major research projects, each of which I have addressed earlier in this book and which Wright has detailed in his 1978 paper "The Relation of Livestock Breeding to Theories of Evolution" (Wright 201). They are: (1) Castle's selection experiment with hooded rats, (2) Wright's thesis research on interaction effects of Mendelian color factors, (3) inbreeding, outbreeding, and selection in guinea pigs, and (4) analysis of the transformation of the Shorthorn breed of cattle over time.

From Castle's selection experiment on hooded rats Wright learned two crucial points—that mass selection of a merely quantitatively varying character could substantially and permanently change the expression of the character, and that this selection process had built-in limitations. The limitations stemmed primarily from a truth long observed by professional animal and plant breeders. Severe mass selection might indeed rapidly change a population, but at the cost of deleterious side effects, mostly expressed as loss of fecundity. In the breeding of large animals such as cattle, mass selection was a slow and tedious process, particularly when the character or characters being selected were not highly heritable. Many geneticists who in the 1930s and 1940s were influenced by the growing wave of neo-Darwinism tended to think of natural selection as only mass selection (simple direct selection upon all of the individuals in the population), as Darwin himself apparently did. Wright, while understanding the power of mass selection, was from the beginning of his work on evolution in nature also keenly aware of the limitations of mass selection in animal breeding, and he thought the same limitations would hold in nature.

His thesis research at Harvard in 1912–15 upon interaction effects in color characters in guinea pigs taught Wright clearly that organisms were built up of complex interaction systems rather than being, as Wright frequently said, a mere mosaic of unit characters each determined by a single gene. (I have already documented this assertion in chapter 3.) The same color gene might be expressed very differently in different genetic combinations; it followed that each gene had many multiple if indirect effects. To the animal breeder this meant that selection would be most effective by operating upon whole interaction systems rather than upon single genes. But in a large random breeding population, distinctive interaction systems of genes are rarely clearly expressed and therefore cannot be seized upon by the selection process. Thus in a large random breeding population the basic process of selection is limited to mass selection.

The start of a solution to this dilemma came from Wright's work in 1915–25 at the USDA with the highly inbred strains of guinea pigs. Because of the random fixation of genes caused by the many generations of intense inbreeding, each strain became fixed with a mostly homozygous genetic complement, so particular interaction systems were clearly expressed. Each inbred strain was easily distinguished from the others, and the wide range of variation in all characters between strains was striking. The inbreeding process had revealed the interaction systems so well hidden in the original random breeding population, making them available for the selection process. In actual animal breeding operations, intermediate rather than such intense inbreeding should be practiced to avoid the general decline in vigor and fecundity. Wright was aware also of the use of inbreeding in hybrid corn production.

Finally, from his analysis of the breeding history of Shorthorn cattle, Wright found that a major breed had indeed experienced intense inbreeding during its foundation period. Selection had accompanied the differentiation from the inbreeding, and diffusion from the selected few herds had then made over the entire breed. Mass selection had played a relatively minor role. By 1923 Wright had a comprehensive view of what he believed to be the optimum method of animal breeding.

Reasoning from his theory of animal breeding to his theory of evolution in nature, Wright proceeded upon the plausible but wholly unproved assumption that evolution in nature proceeded primarily by the three-level process utilized by the best animal breeders: (1) local mass selection and inbreeding, (2) dispersion from the more successful local populations, and (3) transformation of the whole species or breed. He had ample evidence from animal breeders who found that mass selection was frequently a slow, unsure, or even ineffective mechanism for changing gene frequencies; thus, Wright decided that evolution in nature must proceed from a more efficient and effective mechanism than mere mass selection. Judging from animal breeding, he thought that natural populations must be subdivided into small-enough partially isolated subgroups to cause random drifting of genes but large-enough subgroups to keep random drifting from leading directly to fixation of genes, for this was the road to degeneration and extinction. Mass selection within subgroups was followed by selective diffusion from subgroups with successful genetic combinations. The final step was the transformation of all subgroups by the immigration of organisms with a superior genotype and subsequent crossbreeding.

Population structure was the essential key. A breeder who practiced mass selection upon a very large randomly breeding herd of cattle made slow progress toward the desired type; but by inbreeding his herd, the breeder could soon reveal hidden variability and use this as a basis for selection. Then bulls from the most desirable herds could be used for breeding the cows in other herds. In nature, there was no breeder to artificially alter population structure or to import or export bulls. Wright's belief, based upon very little evidence, was that the assumption of large random breeding populations in nature was unwarranted; instead, populations were probably more or less subdivided into partially isolated subpopulations in which some random drifting of gene frequencies occurred in addition to mass selection. Migration from the more successful subpopulations then transformed by crossbreeding other subpopulations and in turn the whole species. Indeed, Wright's theory of evolution in nature was impossible if natural populations were large and random breeding.

By the same reasoning, the view championed by Fisher during the late 1920s and early 1930s, that evolution proceeded by mass selection of single genes in large random breeding populations, was impossible if natural populations were subdivided in the way Wright thought. Not surprisingly, so far as experimental evidence for their differing views is concerned, the disagreements between Fisher and Wright hinged upon analyses of population structure in nature and upon the importance of gene interactions.

The 1925 Version of "Evolution in Mendelian Populations"

No copy of the original version of Wright's typescript on evolution in Mendelian populations appears to have survived. I have, however, spoken at length with Wright about this version. In Wright's recollection, the 1925 version had the following structure. There was first a brief introduction that discussed theories of evolution, in particular the impact of the rediscovery of Mendelian heredity upon evolutionary theories. Wright concluded that any theory of evolution must be based on the quantitative consequences of Mendelian heredity. Second came a quantitative section on factors affecting the variation of gene frequency; the third section examined various combinations of these factors in relation to the statistical distribution of gene frequencies in the Mendelian population. Finally came a qualitative section on the evolution of Mendelian systems and a summary.

An attempt to recreate precisely the contents of the 1925 paper is tempting but too speculative. The published 1931 version, with its many references to papers published after 1925, obviously contains a number of additions or changes from the earlier version. The final manuscript was submitted only after Wright had made changes that resulted in part from a substantial interchange in letters and publications with Fisher on dominance and other issues. Some major differences and similarities between the two versions may, however, be stated with confidence. The basic theory of evolution was almost certainly the same. The 1925 version was too long for publication, so Wright deleted a quantitative section on the influence of dominance upon the distribution of gene frequencies. (He was unsatisfied with it and later published it in 1937 in Wright 92.) He also reduced considerably the quantitative analysis of the influence of selection because of the detailed series of papers on that topic published by Haldane in *Transactions* and *Proceedings of the Cambridge Philosophical Society*. Otherwise Wright recalls that these two versions were substantially the same and that any later additions or emendations were minor.

I see no reason to doubt Wright's firm recollection that his basic theory of evolution expressed in the 1931 paper was the same as in the earlier 1925 version. In both versions he argued that in populations small enough to cause pure random drift, random fixation would lead to deleterious effects (as in the inbred guinea pig families) and extinction would result. In very large random breeding populations, evolution was extremely slow and adaptive, giving little possibility of evolutionary advance. In intermediate-sized populations, where both selection and random drift were substantive factors, evolutionary advance was speeded up but was still rather slow. The most favorable conditions for evolutionary advance came with the population subdivided into partially isolated groups, with selection acting primarily at the level of these groups, as had occurred under domestication in the case of the selective diffusion of the favored subpopulation of Shorthorns.

Wright had almost no opportunity to work on his evolution paper during the first hectic years at the University of Chicago. With his limited time for research and writing, he concentrated upon publishing the results of guinea pig experiments carried on at the USDA. By 1928, however, Wright had begun again to think seriously about revising the 1925 typescript for publication. One stimulus in particular was the appearance of Fisher's first paper on the evolution of dominance in *American Naturalist* in April 1928.

Wright, Fisher, and the Evolution of Dominance

Fisher's life (1890–1962) and work is so well documented that repetition of this information is unnecessary here (see especially the biography by his daughter, Box 1978, and Bennett 1983). From early on, Fisher was a staunch Darwinian, believing that gradual natural selection was the primary determining factor of evolution in nature. Soon after 1910 he had focused upon statistics, genetics, eugenics, evolution, and design of experiments as his primary intellectual interests. His education was mainly in the physical sciences and statistics rather than in biology. He, like Wright, and with much better training in mathematics, was keenly interested in quantifying analysis of biological problems. And like Wright, he was easily attracted to Mendelian heredity as the dominant system of inheritance. Fisher and Wright both appreciated that the statistical regularities of Mendelian heredity could, with appropriate assumptions, be quantitatively extrapolated to bear upon questions of animal and plant breeding, eugenics, and evolution in nature. Fisher made many outstanding contributions to statistical theory; Wright contributed only his 1917 paper on the correlation between subgroups of a population and his method of path coefficients, which was really a contribution to the analysis of causation rather than a novel contribution to statistical theory per se. On the other hand, Wright had far more education and expertise in biology than did Fisher.

Fisher's first major contribution to the statistical analysis of Mendelian heredity came in his famous 1918 paper, "The Correlation between Relatives on the Supposition of Mendelian Inheritance," published in *Transactions of the Royal Society of Edinburgh*. By utilizing the analysis of variance (squared standard deviation) and taking account of dominance as Pearson had not, Fisher showed that Mendelian heredity was consistent with observed correlations between relatives. Important as this paper was, it did not gain a wide readership, as Fisher himself and many others (including I. Michael Lerner, personal communication) have noted.

Fisher had seen Wright's "Systems of Mating" series that appeared in *Genetics* in 1921. In that series, using his method of path coefficients, Wright had derived some of the same results Fisher had derived in 1918 by very different mathematical methods. Thus when Fisher attended the International Mathematical Congress in Toronto in the summer of 1924, and before returning to England visited various centers of learning on the eastern coast of the United States, he naturally asked to see Wright when in Washington, D.C. Fisher was at that time employed by the Rothamsted Experimental Station and was serving as a statistician for plant breeders just as Wright was performing a similar role for animal breeders at the USDA. Fisher and Wright had a long conversation. They talked about animal and plant breeding, the quantitative consequences of Mendelian heredity, and statistical techniques. Wright gave Fisher a copy of his 1917 paper and some other papers, and Fisher promised to send upon his return to England a copy of his 1922 paper, "On the Dominance Ratio."

Fisher's paper arrived as promised, and it had a strong effect upon Wright, who studied it carefully. In his 1918 paper Fisher had examined the statistical consequences of genic interaction (epistasis), dominance, assortative mating, multiple alleles, and linkage upon the correlations between relatives. In the 1922 paper he focused the analysis not on the correlation between relatives but on the statistical distribution of genes in the population. The changes in this statistical distribution over time constituted the evolutionary history of the population. Until reading Fisher's 1922 paper, it had not occurred to Wright to extend his own quantitative analysis to the statistical distribution of genes in natural populations.

The germinal ideas of many of Fisher's later works on evolution were contained in his 1922 paper. He discussed the influence of selection, dominance, mutation rate, random extinction of genes, and assortative mating upon the statistical distribution of genes in the population. First he treated the problem of equilibrium under selection. For a single locus with two alleles he showed that if selection favored one homozygote, the other allele would be eliminated. He then stated the possibility of a balanced polymorphism and its consequences:

> If, on the other hand, the selection favors the heterozygote, there is a condition of stable equilibrium, and the factor will continue in the stock. Such factors should therefore be commonly found, and may explain instances of heterozygote vigor, and to some extent the deleterious effects sometimes brought about by inbreeding. (Fisher 1922, 324)

Next he considered the problem of the survival of a new mutation. He found that individually a mutation had an extremely small chance of surviving. The survival of a rare mutation depended upon chance rather than selection. A mutation would be more likely to become fixed at low frequencies in a large instead of a small population simply because the mutation would more often occur in a large population. "Thus a numerous species, with the same frequency of mutation, will maintain a higher variability than will a less numerous species: in connection with this fact we cannot fail to remember the dictum of Charles Darwin that 'wide ranging, much diffused and common species vary most'" (p. 324). A consequence of this point of view was that a smaller mutation rate could bring new genes to equilibrium frequency (balancing the effects of adverse selection) in a large population more easily than in a small population. This idea was a fundamental tenet of Fisher's view of evolution.

In 1921 A. L. and A. C. Hagedoorn (p. 294) had published the theory that the random survival of genes in populations was more important than preferential survival as a result of selection (they were consciously following the lead of Gulick). Attacking this idea with vigor, Fisher demonstrated that even with the absence of new mutations in a population of what he considered minimal size, about 10,000 random breeding individuals, the rate of gene extinction was exceedingly small. He therefore rejected the importance of the chance elimination of genes as compared with the elimination by selection.

If the heterozygote were intermediate in selective value between the homozygotes at a locus, Fisher showed that selection could quickly eliminate one allele. But in the case of complete dominance, selection was ineffective in removing deleterious recessives present at low frequencies. Thus under the protection of dominance there was an accumulation of rare recessives in the population. This effect was heightened in large populations because a low frequency of mutation could inject new genes that could rise to equilibrium frequency by sheer stochastic processes.

In the 1918 paper Fisher had defined the quantity α^2 as the contribution which a single locus makes to the total variance. He now concluded on the basis of his calculations that one

> effect of selection is to remove preferentially those factors for which α is high, and to leave a predominating number in which α is low. In any factor α may be low for one of two reasons: (1) the effect of the factor on development may be very slight, or (2) the factor may effect changes of little adaptive importance. It is therefore to be expected that the large and easily recognised factors in natural organisms will be of little adaptive importance, and that the factors affecting important adaptations will be individually of very slight effect. We should thus expect that variation in organs of adaptive importance should be due to numerous factors, which individually are difficult to detect. (Fisher 1922, 334)

Fisher's basic ideas concerning the process of evolution were expressed in this paper. He believed, in accordance with his biometrical training, that evolution was primarily concerned with large populations where variability, because of storage of genes, was high. In such populations the deterministic results of selection acting upon single gene effects reigned supreme. Natural selection was slow but sure. Fisher even went so far as to compare the rules governing evolutionary change to the general laws of the behavior of gases.

> The investigation of natural selection may be compared to the analytic treatment of the Theory of Gases, in which it is possible to make the most varied assumptions as to the accidental circumstances, and even the essential nature of the individual molecules, and yet to develop the general laws as to the behavior of gases, leaving but a few fundamental constants to be determined by experiment. (Pp. 321–22)

Among the negligible assumptions as to the accidental circumstances in evolutionary theory were the effects of genic interaction and random genetic drift. Sewall Wright was to vigorously disagree with Fisher's judgment in these cases.

Wright was extremely interested in all that Fisher did in the 1922 paper. He agreed fully with Fisher's premise that the evolutionary process consisted primarily of the simultaneous change in gene frequency of a large number of alleles of individually small phenotypic effect. In this crucial sense, Wright and Fisher were both neo-Darwinians who emphasized small variations that Darwin had found so persuasive (his "individual differences"). Surely the shared interest and knowledge that Darwin, Fisher, and Wright had in domestic breeding is related to their common agreement on the heritable variability in natural populations.

Wright was particularly impressed by Fisher's attempt to incorporate the appropriate variables into a statistical distribution of genes in the population, although he had a difficult time at first understanding Fisher's mathematics— Fisher frequently used differential equations that Wright had never studied. Reading the paper was the primary stimulus leading to Wright's attempt to develop, using his own quantitative techniques, his version of the statistical distribution of genes in a population.

Although before the fall of 1924 Wright had not systematically developed his thoughts about evolution in nature, he was certain that Fisher had not identified the prevailing mechanism of evolution in nature. By no stretch of the imagination could the evolution of Shorthorns be viewed as the inevitable statistical outcome of a process governed by laws resembling the theory of gases. Wright strongly suspected that Fisher's hypothetical gigantic random breeding populations were rare if at all existent in nature, and that mass selection operating upon individual genes was not the mechanism to account for the evolution of complex interacting gene systems. He did not doubt Fisher's quantitative analysis (except on a few points) but questioned his fundamental premises. Fisher's "On the Dominance Ratio" (1922) was unquestionably a primary stimulus in inducing Wright to include the derivation of a statistical distribution of genes when he turned his attention to the problem of evolution in nature.

As Wright wrote the paper, one discrepancy was obvious. When Wright used his method of path coefficients to calculate the rate of decrease of heterozygosis in a population under no selection or mutation, he obtained the figure $1/2N$ where N was the effective population size. Using differential equations, Fisher arrived at the corresponding figure for the rate of decay of factor frequency as $1/4N$ (Fisher 1922, 330). Collins, who read Wright's evolution manuscript before his departure for Chicago, suggested that Wright send the manuscript on to Fisher to get the problem ironed out. Wright did not send the paper to Fisher at that time but tried occasionally and unsuccessfully to understand why Fisher had obtained a different result. This discrepancy was one reason Wright wanted to work the paper over again before publishing it.

Another reason was that Wright was dissatisfied with his section on the influence of dominance upon the distribution of gene frequencies. Anomalous results kept creeping in. This section was the one Wright removed from the manuscript before publication in 1931. He also had some doubts about the selection term in his statistical distribution. In the years 1925–29 he had hoped to resolve the remaining problems in the section before submission of the whole paper. Given Wright's busy teaching schedule, the status of the guinea pig research, and these problems with the manuscript, it is easily understandable why Wright hesitated to submit the evolution manuscript for publication.

During the years 1925–29 when Wright had little opportunity to develop further his ideas on quantitative evolution, both Haldane and Fisher forged ahead with theirs. Haldane continued to publish more in his series "On a Theory of Natural and Artificial Selection," rendering redundant some of the sections on selection in Wright's unpublished manuscript (some of the deleted pages on selection Wright reworked forty-four years later and published at that time [Wright 192]). Fisher, in addition to his many publications in statistics, including the first edition of *Statistical Methods for Research Workers* (1925), published papers amplifying or experimentally verifying aspects of the evolutionary view presented in his 1922 paper. In 1926 he and Ford published a study of thirty-five species of British moths (Fisher and Ford 1926, 1928). They found that, in accordance with Fisher's 1922 prediction, in one locality the abundant species exhibited much more variability than the rare species with respect to a continuously variable character. Fisher considered these data excellent verification of his theory.

A year later Fisher turned his attention to mimicry, a stronghold of the adherents to the discontinuous view of evolution so much opposed to his own Darwinian, gradualist view. Punnett had forcefully argued in *Mimicry in Butterflies* (1915) that a stable polymorphism in the Ceylonese butterfly *Papilio polytes* must have arisen discontinuously because the three forms of the polymorphism were distinctly discontinuous. Fisher (1927) challenged Punnett's assumption that the discontinuities in the polymorphism had to arise discontinuously. Citing Castle's experiments in which the expression of the Mendelian factor for the hooded pattern in rats had been significantly changed by the accumulation of modifiers, he argued that a similar accumulation of modifiers with small effect could have changed the expression of the Mendelian factors involved in the polymorphism in *Papilio polytes*. In accordance with the general view he had presented in 1922, Fisher provided a theory of mimicry based upon the deterministic effect of gradual selection acting upon small modifiers.

The Evolution of Dominance

As a logical extension of this viewpoint, Fisher next turned his attention to the problem of the evolution of dominance. Like mimicry, dominance was generally (but not always) complete or nearly complete. In the cases utilized by Mendel in peas, dominance was nearly complete. Moreover, mutations from the generally observed wild type in natural and laboratory populations were recessive in most cases. Fisher wanted a Darwinian interpretation of the recessivity of most mutations.

He began "The Possible Modification of the Response of the Wild Type to Recurrent Mutations" (1928a) with a table showing that the great majority of both autosomal and sex-linked mutations in *Drosophila melanogaster* (genetically by far the most studied organism at that time) were either completely or almost completely recessive. He then offered the following explanation: Most mutations, he argued, were recurrent, deleterious, and occurred at a finite rate. The mutations geneticists had observed in the laboratory in *Drosophila* must also have occurred in nature. Natural populations were very large and generally random breeding; a recurring deleterious mutant allele in such a population was likely to be held at low frequencies by reaching a state of equilibrium between adverse selection and recurrent mutation. If the mutant allele were not completely recessive, then natural selection would tend to make the heterozygote and mutant homozygote phenotypically resemble the homozygous wild type. Castle's experiment on hooded rats had shown how selection could accumulate modifiers of a mutant to change its appearance very significantly; dominance was no immutable property of the gene. Fisher concluded that since the heterozygotes were vastly more numerous in the population than the mutant homozygotes, that natural selection would tend over much time to make the heterozygotes resemble the homozygous wild type, thus accomplishing dominance.

Fisher suggested that the phenotype of the original heterozygote, when the mutation first occurred, was intermediate rather than exhibiting any dominance relation. If a wild-type allele had many recessive alleles, Fisher's view predicted that the heterozygotes of only the mutant alleles would present no dominance relation because they would occur together so rarely that selection could have little effect. For corroboration, he cited the albino series of alleles studied by Wright in guinea pigs (but see Wright's interpretation below).

Fisher was leaning toward the view that a recessive allele was a potential dominant wild-type allele and that the primary force required for the change was a minute selective advantage.

The whole picture of the evolution of dominance painted by Fisher harmonized perfectly with his mathematical conception of the evolutionary process expressed in the 1922 paper. Yet his tone throughout was suggestive and tentative rather than dogmatic. He specifically pointed out that his theory threw no light upon the usually dominant mutations that distinguished domesticated breeds of poultry and that the specific modifiers of eosin eye in *Drosophila melanogaster* appeared to have been fully recessive at first occurrence. And he recognized that the selective advantages causing the evolution of dominance were very small: "It may be calculated that with mutation rates of the order of one in a million the corresponding selection in the state of nature, though extremely slow, can not safely be neglected in the case of the heterozygotes" (Fisher 1928a, 126). This was an ingenious, highly suggestive paper; appearing in *American Naturalist,* it was read by most geneticists.

Fisher soon received from geneticists more corroborating evidence, and he himself invented a way to explain the unusual situation in fowl. He quickly wrote "Two Further Notes on the Origin of Dominance," beginning with the statement that his previous paper "had brought many further facts into the discussion, in the light of which the somewhat tentative tone of that paper seems to have been unnecessary" (Fisher 1928b, 571). The first note concerned the Crinkled Dwarf mutant confined to the Sea Island cotton plants, discovered by S. C. Harland to be fully recessive there but of intermediate dominance when bred to the New World cottons. Because New World and Sea Island cottons had long been reproductively isolated, Fisher deduced that dominance had evolved by selection of a number of modifying factors in the Sea Island cottons, and that similar modifiers had not evolved in the New World cottons in which the Crinkled Dwarf mutant was absent. This was a reasonable deduction that later proved dubious because the Crinkled Dwarf mutants were found to be genetically far more complex than Harland knew in 1928. (See Fisher's *Genetical Theory of Natural Selection* 1958, p. 63, where he retracts his earlier reliance upon the interpretation of Harland in light of newer evidence from Hutchinson's work on the genetics of cotton.)

The second note concerned dominance in poultry. Here Fisher argued that the selection for dominance of the mutants came from those who had domesticated the poultry and who tried over time to keep their prized mutant stocks dominant over genes from related natural populations, infiltrated into the domestic population by wild cocks. Fisher was obviously very pleased with his two notes and the added support for his theory of the evolution of dominance, and he concluded the paper:

The work of adaptation has hitherto seemed to be the only one upon which natural selection is engaged, and nothing could be more difficult to measure than achievement in this respect. It has now been shown that the same agency, as a minute by-product of its activity, must also tend to modify dominance, and, if the recessiveness of each several mutation be referred to this cause, the vast number of reactions which must have been so modified gives a measure of its efficacy, which might have startled even a Weismann. (Fisher 1928b, 574)

The tentative tone of the original presentation of the theory was gone, replaced by enthusiastic optimism.

Wright read both of Fisher's papers on dominance as they appeared, but he failed to share Fisher's optimism about the power of his theory. Wright thought that Fisher's general conception of evolution in nature was unconvincing, particularly its application to the evolution of dominance. Wright was more than a little surprised to see his albino series of alleles in guinea pigs, for which he had a detailed interpretation in terms of physiological genetics, used to support a theory of evolution to which he could not subscribe. So Wright wrote a reply to Fisher's two papers and sent it to *American Naturalist*.

In the reply Wright first calculated the size of the selection rate on the heterozygotes, the effects of which Fisher had described as "extremely slow." On the assumption favorable to Fisher's theory that the modifier immediately rendered the heterozygote as reproductively fit as the wild homozygote, Wright concluded that the selection rate was never more than half the rate of recurrent mutation. Wright considered this tiny mass selection rate too small for efficacy in natural populations for many reasons. The one he emphasized in this reply was that genes had multiple effects. Citing Theodosius Dobzhansky's 1927 paper on the manifold effects of some of the marker mutations in *Drosophila melanogaster,* Wright argued that any gene with a tiny net effect upon dominance would also have other effects that subjected the gene to larger selection rates.

The hypothesis that a selection pressure, of the order calculated here, can be the *general* factor making for dominance of wild type, depends upon the assumption that modifiers of dominance (assumed to be sufficiently abundant) are in general so nearly indifferent to selection that a force of the order of mutation pressure is the *major* factor controlling their fate. With the prevalence of multiple effects in mind it seems doubtful to the present writer whether there are many such genes. (Wright 56, p. 277)

Wright was careful to add that his analysis held only for the case used by Fisher in which the heterozygotes were rare: "If for any reason the proportion of heterozygous mutants reaches the same order as that of the type, selection of modifiers of dominance approaches the order of direct selection in its effects and might well become of evolutionary importance" (p. 277). This point would later figure strongly in the debate over the evolution of dominance, and Fisher moved perceptibly toward emphasizing the evolution of dominance in balanced polymorphisms in which heterozygotes were far more numerous than in the case to which Fisher first applied his theory. While rejecting the application of Fisher's theory to natural populations, Wright accepted in principle its validity applied to domestic populations, poultry especially.

Wright used his intimate familiarity with physiological genetics to suggest an alternative hypothesis to explain the prevalence of recessivity in deleterious mutations. In line with the views of Goldschmidt, Bateson (who had popularized the work of Garrod in *Mendel's Principles of Heredity* in 1913), and many other geneticists, Wright suggested that genes act indirectly as catalysts and explained that lower levels of activity of the gene would show more or less imperfect dominance, corresponding to the usual case among intermediate allelomorphs in multiple series. He concluded "that in the hypotheses that mutations are most frequently in the direction of inactivation and that for physiological reasons inactivation should generally behave as recessive, at least among factors with major effects, may be found the explanation of the prevalence of recessiveness among observed mutations" (p. 278).

What Wright did not say in this first reply to Fisher's theory of the evolution of dominance is as much or more interesting than what he did say. Wright did not mention that not only was he well aware that dominance relations could be changed by selection but that he had himself produced a reversal of dominance in his guinea pigs and published the results only two years earlier. That dominance could be altered by selection seemed so obvious and well known to Wright that he saw no need to even mention the fact; he would later regret not doing so after others frequently accused him of believing dominance fixed by physiology and not alterable by selection. A second omission was any hint that Wright held a well-articulated theory of evolution already in manuscript, a theory which differed fundamentally from Fisher's theory. Under the theory Wright had already developed, random breeding natural populations were nowhere near the giant size hypothesized by Fisher; indeed they were sufficiently small that random drift due to inbreeding alone would have swamped the minute selection rates of Fisher's theory of dominance. Yet there was no mention at all of effective population size in nature, or random drift, or Wright's theory of evolution.

Wright's reply appeared in the May-June 1929 issue of *American Naturalist*. Fisher was eager to defend his theory, but he was not really sure what Wright was arguing. His best guess, apparently, was that Wright had failed to grasp the quantitative argument that the effect of a small selection rate was as great as that of a large selection rate if it acted for a proportionally longer time. He decided to write Wright and see whether his surmise was correct. How Fisher could have thought that the same person who wrote the "Systems of Mating" papers of 1921 would miss this simple quantitative point is unclear. On June 6 Fisher wrote his first letter to Wright.

> I was much interested in your note in the *American Naturalist* on the evolution of dominance, though of course, sorry that you should consider the numerical values too small to be effective.
>
> I do not think there is any use in controversy except when the point at issue is perfectly clear to both parties, and I should therefore like to have your opinion of the enclosed, which is the kind of thing I should now be inclined to write, before publishing anything on the matter.
>
> Perhaps you would find it worth while to work out the case you cite making allowance for the effect of the more favourable factors on the frequency of the heterozygotes, and dropping the assumption that the modifier is dominant.
>
> What I mainly want to know however, is whether you agree with me that a very slight selective effect acting for a correspondingly long time will be equivalent to a much greater effect acting for a proportionately shorter time. Or, whether, on the other hand you think I have underestimated the ratio of the selective intensities, or overestimated the ratio of the times. I cannot see how a conclusion can be reached without considering the latter.

Wright replied on June 28.

> I am in receipt of your letter of June 6th, and was naturally very much interested in its contents. I hope that I made it clear in my discussion that I was not criticising your suggestion in a spirit of controversy but because I am very anxious to reach an answer to the question as to how far dominance (and combination effects of factors) ordinarily express simple physiological relations and how far complex, and from the physiological viewpoint, rather artificial situations built up by natural selection. This is in connection with attempts to measure more nearly quantitatively than before the amounts of pigment in various color varieties of the guinea-pig.
>
> I am not sure that I entirely understand the first suggestion in your letter. I assumed that my modifier M was dominant partly for simplicity (if M is relatively infrequent, MM will be infrequent to the second order and correspondingly negligible) but largely because it seemed to me that the greater the effect of Mm (over mm) the more effective the selection for fixation of M would be. The assumption of dominance could then not give too small an estimate of this selection. The assumption that Mm makes Aa appear fully like wild type seems to be the most favorable case for selection and MM could hardly do more. Even in this extreme case it appears that the pressure toward fixation of M is only of the order of mutation pressure which I take to be ordinarily a second order process in relation to a direct

selection pressure. If *M* has only a slight effect on dominance of *A*, the selection pressure toward fixation of *M* becomes of the third order.

The main question at issue seems to be clearly whether a very slight selective effect acting for a correspondingly long time will be equivalent to a much greater effect acting for proportionately shorter time. I assumed that the relative frequencies of any factor pair would be affected by many evolutionary pressures of varying orders of magnitude. As the resultant of all such pressures there would be a certain equilibrium point but this point would practically be dependent (in this sense of approximate fixation of *MM* or of *mm*) on only the one or two most important pressures. Thus it seemed to me that selection and mutation pressures of a lower order would have virtually no effect, however long the period of time.

Factors *Mm* would simply reach their equilibrium frequency as determined by the major selection pressure, and stay there, unless as might well happen new major selection pressures appeared. I considered the matter merely for the case of a pair of allelomorphs but, if as seems likely, mutations are occurring in many directions in each gene, a complete analysis would require consideration of an indefinitely extended multiple allelomorphic series and would require various assumptions for any definite solution, for which there seems to be as yet no adequate factual basis. This complicates the matter, but it seems to me that it would still be the major evolutionary pressures which would control the moving equilibrium of the frequencies in such a system and that minor pressures would be negligible.

My argument rests of course on the assumption that there would nearly always be selection pressures acting on important modifiers of dominance of more importance than of the order of mutation pressure. How far this is true may be a matter of opinion.

In this first reply to Fisher, Wright still did not mention his well-developed theory of evolution that really constituted the foundation stone of his rejection of Fisher's theory of evolution of dominance. With Wright's clarification of his position in hand, Fisher proceeded with his rebuttal and sent a copy to Wright with this letter.

I was very glad to get your letter, and see what your point really is. As others besides myself may have missed it, and fancied that you desired to establish insufficiency of selective intensity in relation to time available, I think it will be worth while to reply, little though either of us can know on the real point at issue.

I enclose what I am sending to the *American Naturalist* so that, if you think it desirable, you can have another go, in the same issue as mine.

I do not think dominance modification need be complex, though perhaps artificial; if the response curve to some substance is like that suggested by Stern for *bobbed*

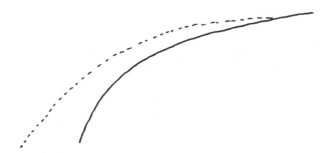

Figure 8.1. Response curves. Redrawn from Fisher's letter.

you need only shift the whole curve so as to get the heterozygote on to the flat part, in order to deal with the fact, on this view, that nearly every wild gene does twice as much as it need do. P.S. I am glad to know about rough in guinea-pigs. Are you certain that it has not been intensified by fanciers?

In his note (I have only the published version, but internal evidence from the Fisher/Wright correspondence indicates the draft note was never altered) Fisher began by stating that Wright's "primary formulas differ in no essential respects from my own and that the selective intensity which inclines Professor Wright to reject the theory is in fact the same that originally led me to adopt it" (Fisher 1929, 553).

No statement could be more true or accurate, or revealing of the qualitative, as opposed to quantitative, differences in the fundamental views of evolution of Wright and Fisher. They would in the future frequently accomplish the same end point—first, using their different quantitative methods, arrive at the same quantitative figures, only to then disagree fundamentally about the significance or interpretation of the agreed-upon quantitative figures. In the note Fisher pointed out that Wright had miscalculated when he stated that the selection pressure upon the heterozygote would never exceed half the mutation rate. He showed that the correct conclusion was never more than twice the mutation rate. But he added: "I do not in the least wish to dispute that the selective intensity will be proportional to, and generally of the order of, the mutation rate, . . . though the fact that the evolution of dominance by selection proceeds with increasing speed as dominance becomes more complete is an essential point stressed in my original note" (p. 555). Fisher clearly understood Wright's objection—"Where he really differs from me is in my assumption that a small selective intensity of say 1/50,000 the magnitude of a larger one will produce the same effect in 50,000 times the time"—and appeared to understand that the reasoning behind Wright's objection had nothing to do with Fisher's arithmetic but with biological interaction effects. Fisher evaluated Wright's biological objection as a weak one because it contained no "obstate fact of arithmetic" opposing Fisher's theory.

Although random genetic drift was an essential aspect of Wright's theory of evolution in domestic and natural populations, he had not yet mentioned it in connection with his criticism of Fisher's theory of dominance. Thus imagine his surprise to find Fisher using selectively neutral genes to buttress his selectionist argument:

> As to ratios having neutral stability, there is one reason for thinking that the factors suffering the feeblest selective action will at any one time be the most numerous. The fate of those powerfully selected is quickly settled; they do not long contribute to the variance. It is the idlers that make the crowd, and very slight attractions may determine their drift. On the whole, it seems that the most reasonable assumption which we can make, on an obscure subject, is that the effect is approximately equal to the cause (P. 556).

In his letter accompanying the draft note, Fisher had invited Wright to have another go in the same issue of *American Naturalist*. How could Wright refuse, now that Fisher had dismissed Wright's major objection to Fisher's theory of the evolution of dominance as being weak, and perhaps more important, now that Fisher had used genes so neutral as to be influenced by random drift to argue for the mass selectionist viewpoint?

Wright, however, faced a major problem. He had written his long paper on evolution but had not yet submitted it for publication. What he really needed to present a strong rebuttal to Fisher was the evolutionary theory in that paper. Fisher's letter had been dated July 10, 1929. On July 23, 1929, Wright sent his comment on Fisher's reply to J. McKeen Cattell, editor of *American Naturalist*. He chose the following strategy in his comment. First, he admitted his error in relation to selection rate and the rate of recurrent mutation, then argued that his basic objection to Fisher's theory was in no way compromised by the error. After a restatement of his previous objection, using the gene interaction argument, Wright turned to the issue of random drift.

> I do not hold, however, that even the most important selective action on a gene is necessarily the controlling factor . . . factors which are almost neutral to all other evolutionary forces should be highly unstable with respect to gene frequency. I can not accept the view that this instability is favorable to the success of feeble selection. In fact, I would say that it sets a lower limit below which selection is not effective. (Wright 57, p. 559)

For the first time in the interaction with Fisher, Wright now emphasized the great importance of effective population size (in this published note designated as small n; later Wright used the notation N, which I will use throughout to avoid confusion) and indicated that N in natural populations was smaller than Fisher assumed, with obvious consequences for random drift. Wright then presented his own general formula for the statistical distribution of genes in the population developed in his unpublished paper, saying that "the probability of different values of q," later termed $\phi(q)$, equaled approximately

$$e^{2Nsq}q^{4Nv-1}(1 - q)^{4Nu-1},$$

where s was the selection rate favoring the gene with frequency q, u the mutation rate of the gene, and v the reverse mutation rate. "The form of the curve under different conditions," Wright said, "indicates whether selection (s), isolation effect ($1/2N$), or mutation (u,v) is the factor which controls the fate of the gene and thus in a sense adjudicates between the principles of Darwin, Wagner, and de Vries, respectively" (p. 560). Wright showed, according to this distribution, that with a relatively small N, random drift became more important than selection upon the gene in question.

The crucial question was How small did N have to be in order for random drift to swamp the effects of selection rates of the size hypothesized by Fisher? Wright stated:

> Unfortunately it is difficult to estimate N in animal and plant populations. In the calculations, it refers to a population breeding at random, a condition not realized in natural species as wholes. In most cases random interbreeding is more or less restricted to small localities. These and other conditions such as violent seasonal oscillation in numbers may well reduce N to moderate size, which for the present purpose may be taken as anything less than a million. If mutation rate is of the order of one in a million per locus, an interbreeding group of less than a million can show little effect of selection of the type which Dr. Fisher postulates even though there be no more important selection process and time be unlimited. (P. 560)

Wright believed that one million was far higher than the effective population sizes of most populations in nature. Thus selection rates of the order required by Fisher's theory of the evolution of dominance could not determine the fates of genes in the population.

Another phrase in Wright's quote above deserves careful attention because it helps pinpoint a major difference between his theory of evolution in nature and Fisher's theory. Wright says clearly that he does not believe that whole natural species are random breeding populations. Instead, he considered random breeding to be "more or less restricted to small localities" over the entire range of the species. Thus the effective population size of a whole species was vastly less than the number of breeding individuals in the species. Indeed, Wright thought the effective sizes of the random breeding populations in a species in nature were small enough that random drift could be a stronger force determining gene frequencies than the prevailing magnitude of natural selection affecting the same gene frequencies. Fisher would disagree strongly, as the following pages demonstrate.

After arguing that random drift would overpower the selection rates hypothesized by Fisher in the evolution of dominance, Wright suggested that the explanation must be sought elsewhere. He offered again: "The suggestion that mutations most frequently represent inactivations of genes, and that, for simple physiological reasons, inactivation should generally behave as a recessive, still seems adequate as a positive alternative hypothesis" (p. 561).

On the day after Wright sent his comment to *American Naturalist,* he drafted a letter to Fisher:

> You are clearly right about the incorrectness of my statement that selection pressure is never more than half that of mutation pressure and I have written a note to send to *The American Naturalist* acknowledging this, but otherwise maintaining my position. There is a minor correction of your statement in this connection. I compared $\delta p = 2up^2(1 - p)$ intended to measure selection pressure with $\delta p = up$ for mutation pressure. Thus I had in mind the comparison (at $p = 2/3$ of your $1 - p' = 2/3$, not $p' = 2/3$) of $8/27\ k$ with $2/3\ k$ (not k as you state). My statement was, however, rather loose at best apart from the primary error.
>
> I am still unable to see why my comparison of the two values of δp is not correct in principle and hence do not see the significance of your use of the selection intensity i. My new formula checks your value of i except for some confusion in your use of p'. I used p for the frequency of the allelomorph of the modifier and not of the modifier itself. As to the main point at issue, I think that the difference of opinion is at least brought clearly to the surface.
>
> My reference to rough fur in guinea-pigs was perhaps not an altogether happy one, since there is no question but that the fanciers have enormously strengthened it. They do not, however, seem to have strengthened its dominance. The two wild species—*Cavia rufescens* of Brazil and *Cavia cutleri* of Peru, the latter perfectly fertile with the guinea pig and quite certainly its wild ancestor, both are smooth furred *rrMM*. The factor *R* of rough guinea-pigs seems to be completely dominant in all combinations, but by itself only reverses the hair direction on the hind toes. The imperfectly recessive mutation *m*, adds a pair of dorsal rosettes, or at least a dorsal crest, when heterozygous and several others on the back and head when homozygous. The two factor ratios came out in a perfectly orthodox fashion in crosses between full rough guinea-pigs and the wild species, indicating that the wild species though lacking *M* must nevertheless have already had whatever modifiers of dominance there may be. There is probably somewhat greater dominance of *M* in the wild species than in the smooth strains possessing it, and it is possible that intensive selection of heterozygotes might make *m* completely dominant over *M* in time.

Although originally dated July 24, 1929, in Wright's carbon copy the date had been replaced by hand as July 30. A handwritten note, of which Wright kept a rough copy, appears to have been included with this letter when it was actually sent off to Fisher. Apparently Wright was rather sensitive that he had used small effective population size and random drift in his comment on Fisher's reply, without earlier having used these arguments against Fisher's position. He thus added to the letter above:

> The question of the effect of random variation in factor frequency is one of the reasons that has long made me doubtful of the effect of feeble selection. I did not go into the matter in my first discussion of your hypothesis as being rather complicated, involving an unpublished formula and especially because of the uncertainty in the application of the N in my formula. I am sorry that I forgot this point at the moment in answering your previous letter as you may wish to add to your note on this subject now that I have discussed it.
>
> I have a manuscript on the subject which was stimulated by your extremely interesting 1922 paper which I studied carefully at the time when you sent me a reprint shortly after your return from this country [summer 1924]. A discrepancy between your result for the rate of decay $(1/4N)$ of factor frequency (no mutation) and the figure for the decrease of heterozygosis by my path coefficient method $(1/2N)$ led me to look for the cause. I reached the somewhat different formula for factor frequency which I mentioned. It is of course a matter of two approximate formulae. I am not wholly clear yet as to the cause of the discrepancy. I would like very much to get your comment on it. I will try to send a copy in a few days.

This letter indicates that Wright thought his argument about random drift to be sufficiently powerful that Fisher would want to revise his note and answer it. The letter also indicates that Wright wanted Fisher to read the draft of his evolution manuscript for two purposes—so Fisher could see how Wright had derived the formula for gene distribution, used in Wright's draft comment but not derived there, and so Fisher could look for the source of the discrepancy in their different values for the rate of decay of heterozygosis in finite populations. It is clear to me that Wright was completely confident of his own figure but did not understand Fisher's derivation well enough to spot the error in it.

On August 13, 1929, Wright sent the copy of his evolution manuscript to Fisher with this covering letter.

> I am enclosing a copy of the manuscript to which I referred in my previous letter. It was written some time ago and I shall probably revise somewhat the concluding sections before publication. As it may be some time before I have leisure to reconsider it carefully, I am sending it in its present form.
>
> I would be very glad to get any criticism which you would care to make, especially of course, with regard to my statements on your results. I am not yet entirely clear as to the cause of the differences

though I have supposed it to be merely a case of different approximation resulting from rather different modes of attack. As far as application to evolutionary theory is concerned, it makes little difference whether $1/2N$ or $1/4N$ of the factors tend to be fixed per generation in a recently isolated group, or whether the equilibrium distribution (no selection, infrequent mutation) is of the type

$$\phi(q) = \frac{C}{q(1-q)}$$

I was not clear about your derivation of $1/4N$ as the rate of decay (p. 330 in your 1922 paper). In my method, I started from $1/2N$, deduced from considerations not connected with the form of the distribution of factor frequencies and which in fact I had obtained by the path coefficients method some time before I had read your 1922 paper. The discrepancy led me to try to work out the distribution in a different way, using the value $1/2N$ in doing this.

In looking over the note on your reply which I sent, I see that the sentence on p. 4 "mutation pressure is relatively unimportant if $2u$ and $2v$ are less than $1/2N$" while true of form of distribution of frequencies which I had in mind at the moment, is not true of the mean and should be erased.

On the same day, Fisher wrote to Wright about his letter and draft comment sent some two weeks earlier.

Many thanks for your interesting letter and the copy of your comment on my reply. I am inclined to think your comment carries the discussion of your main point as far as it can be usefully carried in the present state of our knowledge, and I do not see that I can usefully add anything.

The point about using selective intensity

$$i = \frac{\delta p}{p(1-p)}$$

was of course aimed at comparisons with the selective value of "multiple effects," in which also δp will contain the factor $p(1-p)$ depending on the gene ratio. From this point of view counter-mutation is infinitely powerful against the prevalent type of gene, as is illustrated by the power of mutation to keep a gene in existence against powerful selections.

You see, of course, that the principle of multiple effects, if carried far enough, greatly increases the number of factors available for modifying dominance, though possibly it does not increase the number whose fate will be settled by the effect in modifying dominance.

I am not sure that I agree with you as to the magnitude of the population number N. To reduce it to the number in a district requires that there shall be *no* diffusions even over the number of generations considered. For the relevant purpose I believe N must usually be the total population on the planet, enumerated at sexual maturity, and at the minimum of the annual or other periodic fluctuation. For birds twice the number of nests would be good. I am glad, however, that you stress the importance of this number.

Have you ever bred back "reversed toed" guinea pigs, to the wild stock for a few generations so as to be sure that the heterozygote and two homozygotes were in other respects genetically alike? A quantitative comparison with such material would seem to be of the greatest interest.

Fisher's letter contains a number of revealing points. He saw no need whatever to revise his own note in the light of Wright's new argument about random drift and effective population size—Fisher apparently did not see it as a threat. Wright's insistence upon wide-ranging interaction effects would merely increase the number of factors for modifying dominance and thus support Fisher's view rather than Wright's. Actually, Fisher's argument here makes little sense because the selection rates upon multiple modifiers of dominance would necessarily be less than those upon a single modifier yielding complete dominance, as Wright had assumed in his calculations. Thus the multiple modifiers would be even less likely to have their fates determined by selection toward dominance.

Most interesting is Fisher's comment upon effective population size in nature. He clearly believed at this time that the effective population size was the "total population on the planet," properly enumerated at sexual maturity and at population minimums. In other words, an entire species for the purpose of the statistical analysis of evolution was a random breeding Mendelian population. This view fit perfectly with Fisher's belief that the process of evolution by natural selection was as sure and deterministic as the gas laws or the second law of thermodynamics. One can legitimately wonder how Fisher thought speciation occurred in such a world of species, each of which were random breeding populations. The difference between Fisher's and Wright's conception of effective population size in nature is extremely important for understanding their differences on other issues, as we shall see.

Fisher and Wright, whatever their differences at this time, agreed perfectly that the effective population size, or population number as Fisher called it, was fundamentally important in the statistical analysis of evolution in nature. Much of the history of population genetics in the ensuing decades would indeed hinge upon theoretical and practical field investigations of effective population size.

Fisher's letter set Wright to thinking about effective population size in relation to migration. If there were sufficient migration, then Fisher would be right that the effective population size was that found on the entire planet. Wright's first quantitative attempt to take diffusion wholly into account was somewhat disappointing and pushed Wright in Fisher's direction of greater effective population size, as Wright's reply of August 28, 1929, to Fisher indicates.

> My dear Doctor Fisher:
> I was much interested in your comment on the population number in your letter of August 13th. Since I wrote, I have been trying to get a clearer idea of the effect of diffusion and I see, at least, that isolation in districts must be much more nearly complete than I realized at first, to permit random fixation of strains. Assuming that the total population of the species is divided into districts each with population N and that there is interchange to the proportion m between each district and a random sample of the whole (mean gene frequency $\bar{q}$) the effect on the gene frequency of districts would be $\delta q = -m(q - \bar{q})$. This can be written $-m(1 - \bar{q})q + m\bar{q}(1 - q)$ which throws it into the same form as mutation pressure, $\delta q = -uq + v(1 - q)$. If my formula for the effect of mutation on the equilibrium distribution of factor frequencies is correct, the formula for partial isolation (small m) can be written at once by analogy:
>
> $$(1) \quad \phi(q) = Cq^{4Nmq-1}(1 - q)^{4Nm(1-\bar{q})-1}$$
>
> If genic selection and mutation be taken into account this becomes:
>
> $$(2) \quad Ce^{2Nsq}q^{4N(v+m\bar{q})-1}(1 - q)^{4N[u+m(1 - q)]-1}$$
>
> From (1) (negligible mutation and selection) and taking $\bar{q} = \frac{1}{2}$ there is an equal frequency of all values of q if $m = \frac{1}{2}N$, interchange of one individual every other generation. The standard deviation of q among the districts seems to be
>
> $$\sqrt{\frac{\bar{q}(1 - \bar{q})}{4Nm + 1}}$$
>
> for small values of m.
>   For completely random replacement ($m = 1$) I suppose it should be
>
> $$\sqrt{\frac{\bar{q}(1 - \bar{q})}{2N}}$$
>
> by the ordinary formula for standard deviation of a percentage. I am not clear about this discrepancy except that small values of m are assumed in the derivation.

[The above paragraph was lightly scratched out in Wright's carbon with the added note: "variation of q not allowed for—σq should be zero." I do not know if the paragraph was in the original sent to Fisher or removed from it.]

I must admit that rather strict isolation is necessary on this basis to maintain appreciable variation of q and I recognize of course that variation in q does not interfere much with selection unless it is so great as to bring about a U-shaped piling up close to $q = 0$ and $q = 1$. I am not entirely clear as to the effects of interchange between adjacent districts with similar q's. Presumably there could be considerably more such interchange without preventing a drifting apart of the q's for remote districts. However, it seems clear that N must be based on the entire species, unless isolation in districts is substantially complete, in considering the interference with selection.

Have you written or do you know of any discussion of the effects of diffusion? I haven't run across such a discussion but I may readily have overlooked it.

In regard to the rough furred guinea pigs, I have not made the crosses which you suggest and unfortunately do not have any of the wild cavies at present. I think it is likely that the modifiers would be such that the factor R (for rough fur) would produce little if any visible effect on introduction into practically pure cavies—i.e., $RRMM$, $RrMM$ and $rrMM$ would probably all be smooth. In tame guinea pigs and wild hybrids, whatever the status with respect to M I have never been able to see any difference between RR and Rr.

Fisher's theory of the evolution of dominance and Wright's initial reply attracted much attention from geneticists. One of the first attracted to the issue was Haldane, whose abilities in quantitative population genetics were widely recognized by 1929. Haldane's first paper on the evolution of dominance did not appear in *American Naturalist* until the January-February 1930 issue, but at least several months earlier Fisher knew of his position. Haldane criticized Fisher's theory both theoretically and using apparent counterexamples, one of them being the generally dominant mutants found in the grasshoppers *Apotettix* and *Paratettix*, at that time being carefully studied by the American geneticist R. K. Nabours of Kansas State Agricultural College in Manhattan, Kansas. Fisher seized upon this example because Nabours's grasshoppers apparently exhibited stable polymorphisms, thus greatly increasing the number of heterozygotes upon which selection might act in modifying dominance. Similar polymorphisms were found in the fish *Lebistes* and the snail *Helix*. Fisher indicates in his reply to Wright's letter of August 28 how eager he was to look into stable polymorphisms as a way to account for a certain class of dominance effects and like Wright, Fisher offered a conciliatory gesture by suggesting a case in which small selective advantages were ineffective.

Many thanks for your letter of August 28th., which is not only exceedingly interesting in itself, but helps me to understand the larger paper, which I have been puzzling over occasionally for some time.

I have so far published nothing on the diffusion problem, but have in the Press a book on 'The Genetic Theory of Natural Selection,' which has part of a chapter on the cohesion of species in relation to the problem of their fission. I think it must be generally true that the ancestry of all individuals of a species is practically the same except for the last 100 or perhaps 10,000 generations, and that a gene frequency gradient is maintained by selection between different parts of a species' range. So that well marked local variations may or may not be incipient species, according as real fission, cessation of diffusion, ultimately supervenes. My discussion of this point is necessarily superficial and qualitative, but may have some points to interest you.

Haldane has brought up *Apotettix* and its allies *in re* the dominance problem, and I fancy that discussion of the group may in the end be fruitful. Do you know Nabours? and if he has, or would be willing to obtain the relative frequencies in nature of his dominant types? I am inclined to suggest that they must be great enough to imply a selective stability of gene ratio due to heterozygous advantage.

By the way, here is a case in which a small selective advantage seems to be totally ineffective. Suppose there is only one (haploid) chromosome with no crossing over. Occasionally, advantageous mutations arise giving advantage a; it will take something of the order of $1/a$ generations for one of these to establish itself effectively, and it will fail to do so if in the mean-time a better mutation occurs elsewhere, i.e. in the unimproved part of the population, for the improvements have to "queue up" to pass through a single door, and the more valuable press the less valuable aside. If $v\ da$ is the frequency of mutation as good as a, there will be a value of a (say b) given by something like

$$b = \int_0^\infty v\ da$$

such that mutations giving advantage $>b$ have a reasonable chance, while those giving advantage $<b$ will certainly fail to establish themselves.

With several chromosomes and crossing over there are many and wider doors, and I do not know how low the qualification may become, but the point is relevant owing to the small amount of crossing over in *Apotettix, Lebistes, Helix* etc., all I think, polymorphic forms in nature. (Fisher to Wright, September 9, 1929)

By the middle of October Fisher had been able to locate the source of his error in the discrepancy about which Wright had written him, and sent Wright a very nice letter.

I have reason to be immensely grateful to you for sending me your paper, which, I fear, I have kept all too long, as I have now fully convinced myself that your solution is the right one. It may be of some interest that my original error lay in the differential equation

$$\frac{\delta y}{\delta t} = \frac{1}{4N} \frac{\delta^2 y}{\delta \theta^2}$$

which ought to have been

$$\frac{\delta y}{\delta t} = \frac{1}{4N} \frac{\delta}{\delta \theta} (y \cot \theta) + \frac{1}{4N} \frac{\delta^2 y}{\delta \theta^2}$$

(You might care to give this correction from me when you publish) the new term coming in from the fact that the mean value of δp in any generation from a group of factors with gene fraction p, is exactly zero, and consequently the mean value of $\delta \theta$ is not exactly zero but involves a minute term $-1/4N \cot \theta$.

With this correction I find myself in entire agreement with your value $2N$ for the time of relaxation and with your corrected distribution for factors in the absence of selection. Re-examining the whole work has been a great gain to me in clarifying my ideas, and I appreciate what I had not realised before, that selection, except when directed to an optimum value, is not important in keeping down the variance.

I have done a good deal of work on the terminal conditions, which when it is fit to be seen, will, I hope, be of interest to you. A very striking result is that a mutation can only be regarded as effectively neutral if the selective intensity multiplied by the population number is small, so that the zone of effective neutrality is exceedingly narrow, and must be passed over, one way or the other, quite quickly in the course of evolutionary change. (Fisher to Wright, October 15, 1929)

In later years, after he and Fisher no longer communicated with each other, Wright was suspicious of one phrase in this letter. When Fisher said he agreed with Wright's figure for the rate of loss of heterozygosis, he added "and with your corrected distribution for factors in the absence of selection." But Wright also had of course developed a formula that included the effects of selection, as had Fisher. Had Fisher disagreed with Wright when selection was included but hesitated to say so in the letter? Fisher knew that Wright intended to publish the paper; Wright thinks now that Fisher knew Wright's formulation with selection was wrong, but did not say so because Fisher intended to rederive his own formulation and publish it. It is possible that Fisher had not gotten this far at the time of writing and had not spotted the discrepancy; or Wright's surmise may be correct.

Fisher did publish a paper that started with Wright's correction and proceeded to derive a new formulation of the distribution of gene frequencies under selection. Wright eventually caught his own mistake and corrected it when his big evolution paper was in proof. He was unhappy at the time that Fisher had not told him of the problem after he had sent the paper to Fisher for comments. The whole experience of sending the paper to Fisher for comment was rather soured by this perception, and Wright never again sent his own papers out for comment before submission for publication. His reply at the time revealed no hint of discontent.

> I was glad to see that it is possible to reconcile exactly the two modes of approach to the distribution of factor frequencies. I had puzzled over it a good deal but had concluded that the correction would be something more complicated than the term which you found. I was much interested in the point which you made (in your letter of September 9th) of the ineffectiveness of selection where there is complete linkage. It seems to me that this would hold to a large extent even with a considerable number of chromosomes since the fate of each of these would be controlled by only the few most important selection pressures on its genes. On the other hand, even slight crossing-over would not interfere greatly with the formation of every possible combination of genes in the course of geologic time (cf. Robbins) and would thus permit effective selection of all of them. In fact there would seem to be a distinct advantage in the resistance offered by incomplete linkage to the immediate breaking up of the gene combination since it is the organism as a whole that is really the object of selection and not the separate gene effects. Perhaps a theory for the prevailing degree of linkage relative to chromosome length could be built up along this line.
>
> I expect to see Nabours at the scientific meetings this Christmas and will try to find out about the situation in nature among his grasshoppers, and urge the importance of statistical studies of the occurrence of the difference type in different parts of their range. (Wright to Fisher, October 31, 1929)

Fisher's Genetical Theory of Natural Selection

Between October 1929 and May 1930 Wright and Fisher did not correspond. Wright returned to his evolution paper, revised it, and submitted it for publication in *Genetics* on January 20, 1930. He delivered a much shortened version at the annual AAAS meeting in December 1929 and published a very brief (one paragraph) summary in *Anatomical Record* (Wright 59). During the same months Fisher saw *The Genetical Theory of Natural Selection* through Oxford University Press. It was the first book to explore in sophisticated quantitative detail the synthesis of Mendelian genetics with evolutionary theory and is a landmark in the history of twentieth-century evolutionary theory. Wright received two copies of the book—one from *Journal of Hered-*

ity with a request for Wright to review it, and one from Fisher, sent along with a letter via the American office of Oxford University Press. Fisher's letter was dated March 19, 1930, but it and the book were not forwarded to Wright until May 26.

> I am sending herewith a complimentary copy of my new book "The Genetical Theory of Natural Selection." It was written too soon to include the later developments of dominance theory, which threaten to be extensive. This is really an advantage for it would be a pity if the interest of this special development were to draw attention away from the more general questions.
>
> In some ways the first chapter is the most important, and in some the second. The sixth chapter and the group on Man will attract very different sorts of readers. However, I am sure you will think it an attempt worth making, and should you happen to review it anywhere, remember that I shall be most interested to see your opinion.

Wright replied to Fisher on June 10:

> I wish to thank you very much for sending me a copy of your recent book. I have found it extremely interesting and stimulating. I presented my paper on the subject before the American Association for the Advancement of Science last December. It should appear soon in Genetics. In reading your book I have naturally attempted comparison at every point with the views which I had reached. Our basic assumptions are, of course, very similar.
>
> Certain differences in detail are of a rather superficial nature and can doubtless easily be ironed out. There appear to be some rather important differences in emphasis, however. You would probably not approve at all of the conclusions which I gave in the abstract of my paper which was published (*Anatomical Record* 44: 267, 1929). This somewhat exaggerates the difference, since I was forced by limitation of space to express my views in a bolder and more unqualified form than I would care to maintain fully. The main differences all seem to trace to the greater role which I have attributed to random differences among local strains of a species brought about by local inbreeding.
>
> I have not yet been able to follow the mathematics in Chapter 4 to my satisfaction but hope to be able to do so. There appears to be substantially complete agreement with the results of my method in the case of no mutation and slight mutation. Your determination of the exact character of the terminal frequencies seems to agree well with the conclusions which I had drawn from consideration of very small populations. There may be a trifling discrepancy at the bottom of page 86. I obtained $1/2N$ as the exact rate of decay in the case of a population of monoecious organisms with completely random combinations of gametes and a formula for the case of separate sexes which does not seem to be exactly the same as yours, but which ap-

plies exactly to the cases of brother-sister mating. In the case of low mutation rates, my formula for the number of genes maintained by a given mutation rate $[2(.577 + \log (2N - 1)]$ in the case of one mutation per generation differs only slightly from yours.

I was a good deal troubled by the difference between your formula for the selection effect (page 92) and that which I had reached—$e^{2CNp}(C_1/p + C_2/q)$ in your symbols.

I had not considered the exact case which you give, flux equilibrium (because of the general difference in viewpoint) but on solving for it, I find a ratio of C_1 to C_2 in the above formula which gives results in close agreement up to a certain point ($q < 1/2N$) but widely divergent beyond this. Your approximation is clearly a better one in this region, indeed, mine rapidly becomes wholly valueless in cases in which the terminal frequencies are large. Have you a general demonstration that the chance of fixation is $2a$? The example given on page 76 for $a = .01$ seems to depend on repetition of a formula for the case in question. I have not, however, as yet gone carefully through the reasoning.

I liked very much your opening chapter with its comparison of the consequences of blending and particulate heredity, also the chapters on sexual selection, mimicry and human evolution.

I have been asked to review the book for the *Journal of Heredity*.

It is important to note in this letter Wright's keen interest in comparing Fisher's views with his own in quantitative detail, assumptions, and emphasis. He was clearly concerned at the time of this writing with his formula for the effect of selection upon the statistical distribution of genes and its differences from Fisher's corresponding formula. He also wanted from Fisher an exact derivation of the figure for the chance of survival of a mutation in a large population.

Fisher sent this last derivation to Wright on June 23, 1930.

Many thanks for your letter. I have not the summary from the *Anatomical Record*, so will await the appearance in *Genetics* before going into some of the small discrepancies you mention.

The method by which I should relate selective advantage when not necessarily small to chance of survival in a large population would be to say that the substitution of $f(x) = e^{C(1-x)}$ for x is without effect only if $x = e^{-C(1-x)}$; writing the solution of this equation in the form $1 - P$, P will be the limiting probability of survival, and $-\log(1 - P) = P + 1/2P^2 + 1/3P^3 + \ldots = CP$, whence $P = 2(C - 1)$ approximately, or if a is the selective advantage

$$C = e^a$$

$$P = 2a - \frac{5}{3}a^2 + \frac{4}{9}a^3 - \frac{131}{540}a^4$$

as far as I have worked it (See sheet enclosed).

I do not think the equation has any biological interest except when a is small.

Did I tell you that the cases of polymorphism mentioned by Haldane in connection with dominance theory, really fit in exceedingly well? I am publishing a note on them primarily to encourage workers on these species to pay attention to the further predictions of the theory.

I shall be very much interested in your review, and hope you will give yourself space enough to deal with the many different aspects of the book on which I want to know your opinion. I am particularly glad you like Chapter I, as I suspect many biologists will be tempted to leave it out (i) because they will naturally expect a first chapter to be trite as well as elementary (ii) because they are tired of introductory expositions of Mendelism, and (iii) because they have believed almost since boyhood that they know all about what Darwin thought.

Wright did not reply to this letter until October 15. In the meantime Fisher had published his ideas about the evolution of dominance in polymorphic species in which many of the Mendelian mutants were partially or wholly cominant. In the *American Naturalist* of January-February 1930, Haldane had published his comments about such species as counterexamples to Fisher's theory of the evolution of dominance. Fisher very neatly showed how Haldane's counterexamples could be incorporated into his theory. Fisher's summary describes the argument succinctly:

Polymorphism in wild populations must usually imply a balance of selective agencies, of which the simplest type is a selective advantage of the heterozygote over both homozygotes. Such a condition should not be confused with the maintenance of a rare mutant type against counter-selection by means of repeated mutations. While such mutations should on the theory of selective modification of dominance tend to become recessive, heterozygotes in polymorphic species will tend to resemble in external appearance whichever homozygote it is most advantageous to resemble. The selective balance must then be maintained by some constitutional disadvantage of the homozygous dominant.

The modification of dominance should in such cases be especially rapid; partly by reason of the far greater frequencies of the heterozygotes exposed to selection, and partly, if any tract of chromatin is permanently associated with the dominant gene, from the fact that the evolutionary modification of such a tract will be reserved for the improvement of the heterozygote, and in less degree of the corresponding homozygote. (Fisher 1930b, 405)

Wright read this article in the September-October issue of *American Naturalist,* just before answering Fisher's letter of June 23. Between the time of his previous letter to Fisher (June 10) and this one on October 15, Wright had worked out the difficulties with his formula for the distribution of genes under selection and had been able to make the necessary corrections in the proofs of "Evolution in Mendelian Populations" in press at *Genetics.*

I should have thanked you long ago for your letter of June 23rd, which entirely cleared up for me the derivation of your value $2a$ for the chance of survival of a mutation in a large population. I think that I have cleared up the apparent discrepancy between the result which I gave for the distribution of genes under selection ($s = -a$) and irreversible mutation (at a rate u such that $4Nu$ is negligibly small), viz.,

$$y = C \frac{e^{2Nsq}}{1 - q}$$

and your value

$$\frac{2dp}{pq} \frac{1 - e^{-4aNq}}{1 - e^{-4aN}}$$

which seems clearly to be correct. The two formulae agree (with proper choice of coefficient) when ns is less than 1 but diverge rapidly above this. I had been aware of the limited range of applicability of my formula, which in fact I first reached in the form

$$y = C \frac{1 + 2Nsq}{1 - q},$$

but had not seen how to deal with second order terms involving Ns^2, N^2s^3, etc. in the derivation. I find now that these condense into a simple expression the inclusion of which gives identically your formula in this case. In the case of reversible mutation, however, the corrected formula appears to be

$$y = C \frac{e^{4Nsq}}{q(1 - q)}$$

for all values of Ns (up to the point at which Ns^2 approaches 1) in place of my previous formula

$$y = C\frac{e^{2Nsq}}{q(1-q)}$$

and for mutation rates (u,v) which are not negligible in comparison with $1/4N$, the formula seems to become

$$y = Ce^{4Nsq}q^{4Nv-1}(1-q)^{4Nu-1}$$

to at least a much better approximation than the result which I gave in one of my papers in the *American Naturalist* last fall, viz.,

$$y = Ce^{2Nsq}q^{4Nv-1}(1-q)^{4Nu-1}$$

Fortunately (assuming my present formula to be sufficiently accurate) I have merely had to make all my statements on interpretation in my forthcoming paper in *Genetics* apply to intensities of selection just half as great as before and my graphs merely needed relabelling. I have included these corrections to my formula in the review of your book for the *Journal of Heredity* (which should appear next month) to show that there is now no mathematical difference between our results in the cases which can be compared. I have discussed at some length the rather different interpretations of the role of selection which we have reached and will be much interested in getting your criticism of my view. I was much interested in your discussion of dominance in *Paratettix,* etc. The situation certainly seems to conform well to the expectation from your theory, and the objections which I made in the case of ordinary recessive mutations do not seem to hold here.

This letter indicates that Wright had now been able to reconcile all of his formulas with those in Fisher's *Genetical Theory of Natural Selection,* and he again warned Fisher that despite this quantitative agreement he disagreed substantively with Fisher's view of evolution in nature. Fisher must have been gratified that Wright had no objection to his theory of the evolution of dominance in polymorphic species. Fisher replied:

Thanks for your letter. I am glad to hear the little discrepancies are clearing themselves up. With respect to the polymorphism work the important thing from the mathematical standpoint is to ascertain in what manner the chance of success depends on selective advantage in the case of restricted recombination discussed in the last section. As far as I can see this might be a matter of great difficulty, but this may be merely because I have not spotted some simple way of looking at it. It would evidently include the problem, the quantitative treatment of which I shirked at the beginning of Chapter VI, and would certainly throw light on the equally elusive problem of the effect of a stream of gene substitution in loosening the linkage to which I refer in Chapter V.

Mathematicians always tend to assume that the hardest mathematics will be the most important, and this is perhaps true enough in the well worn topics. It is certainly not true of my book, where the apparently non-mathematical parts, where I have left the mathematics *undone,* are often of the greatest ultimate interest.

I shall be much interested to see your review for the *Journal of Heredity.* (Fisher to Wright, October 25, 1930)

Since Wright lent me the letter in 1978, I have always considered this letter important, at first for the wrong reason—as I will explain below. As I view the letter now, Fisher appeared pleased that his quantitative results and Wright's were in agreement, but he attributed no great significance to this fact. Indeed, in the second paragraph he says that it is the "apparently non-mathematical parts" of his book that have the "greatest ultimate interest."

Fisher's insight here is accurate for Wright's work as well as for his own work. Without a clear understanding of Fisher's point here, one cannot appreciate the tension between the evolutionary views of Wright and Fisher. With this letter, both Wright and Fisher knew that their most quantitative results agreed in all significant particulars. What divided them was decidedly not the differences in quantitative analysis but their qualitative views about the process of evolution. For both of them, the level of quantitative analysis available was insufficient to encompass their qualitative ideas of evolution in nature, on which they differed greatly.

I first considered this letter important because it was the last letter in the Fisher/Wright series given me by Wright in 1978. The last sentence in which Fisher said again that he was looking forward to Wright's review, and then the cessation of correspondence, indicated to me that Fisher was not pleased with the review. I knew that in the early 1930s Fisher and Wright had broken off communication except through professional journals, and the occasion of Wright's review was the logical beginning.

Wright's eight-page review appeared in the August 1930 issue of *Journal of Heredity* (which actually appeared several months after the nominal publication date). He began with the historically accurate observation that after its rediscovery in 1900, Mendelism was most closely associated with the discontinuous mutation theory of evolution. Fisher, he said, had taken a leading part in demonstrating that the statistical consequences of Mendelian heredity pointed to a more continuous or Darwinian view of the process of evolution. Fisher's book "is certain to take rank as one of the major contributions to the theory of evolution" (Wright 60, p. 349). Regarding Fisher's view of the role of mutation in evolution in nature, Wright stated his agreement, but sounded the theme he would expand in the review:

I am in accord with Dr. Fisher on the role of mutation, except that I would perhaps allow occasional significance to chromosome aberration, and to hybridization, as direct species forming agencies. It appears to me, however, that . . . throughout the book, he overlooks the role of inbreeding as a factor leading to nonadaptive differentiation of local strains, through selection of which adaptive evolution of the species as a whole may be brought about more effectively than through mass selection of individuals. (P. 350)

In a long section on the distribution of gene frequencies, Wright presented a detailed account of his interaction with Fisher and concluded that "our mathematical results on the distribution of gene frequencies are now in complete agreement as far as comparable, although based on very different methods of attack" (p. 352). This statement of agreement in quantitative analysis was, however, immediately followed by the section "Differences in Interpretation."

Here Wright criticized Fisher's assumption of very large random breeding populations and his statistical approach based upon that assumption. In the second chapter of the book, Fisher developed his "fundamental theorem of natural selection"—that "the rate of increase in fitness of any organism at any time is equal to its genetic variance in fitness at that time" (Fisher 1930a, 35). The theorem is actually poorly stated by Fisher because he did not at all mean to refer to a single organism but more precisely to an idealized infinitely large random breeding population of organisms in which random drift had no role whatsoever. Of course Fisher knew that natural populations were finite, but he thought they were big enough that the random fluctuations suggested by Wright were negligibly important. Fisher analogized his fundamental theorem of natural selection to the second law of thermodynamics, the regular and inevitable increase in entropy in a closed physical system. To Eddington's statement that the second law of thermodynamics held "the supreme position among the laws of nature," Fisher added the revealing comment: "It is not a little instructive that so similar a law should hold the supreme position among the biological sciences" (p. 37).

Agreeing immediately that Fisher's fundamental theorem did apply to large panmictic ideal populations, Wright challenged the theorem on two primary fronts. He first questioned the validity of Fisher's "genetic variance."

He uses "genetic variance" in a special sense. It does not include all variability due to differences in genetic constitution of individuals. He assumes that each gene is assigned a constant value, measuring its contribution to the character of the individual (here fitness) in such a way that the sums of the contributions of all genes will equal as closely as possible the actual measures of the character in the individuals of the population. Obviously there could be exact agreement in all cases only if dominance and epistatic relationships were completely lacking. Actually, dominance is very common and with respect to such a character as fitness, it may safely be assumed that there are always important epistatic effects. Genes favorable in one combination, are, for example, extremely likely to be unfavorable in another. Thus allelomorphs which are held in equilibrium by a balance of opposing selection tendencies may contribute a great deal to the total genetically determined variance but not at all to the genetic variance in Fisher's special sense, since at equilibrium there is no difference in their contributions. (Wright 60, p. 353)

Second, Wright challenged Fisher's assumption of large populations in the theorem and his noninclusion of terms for migration or mutation in it. On Fisher's scheme, Wright said, evolution was exceedingly slow, depending upon a constant fresh supply of new favorable mutations in the absence of environmental change. He questioned whether such a supply of mutations was available. Other assumptions about population structure than the one used by Fisher, Wright argued, could lead to very different conceptions of evolution in nature. Thus,

If the population is not too large, the effects of random sampling of gametes in each generation brings about a random drifting of the gene frequencies about their mean positions of equilibrium. In such a population we can not speak of single equilibrium values but of probability arrays for each gene, even under constant external conditions. If the population is too small, this random drifting about leads inevitably to fixation of one or the other allelomorph, loss of variance, and degeneration. At a certain intermediate size of population, however (relative to prevailing mutation and selection rates), there will be a continuous kaleidescopic shifting of the prevailing gene combinations, not adaptive itself, but providing an opportunity for the occasional appearance of new adaptive combinations of types which would never be reached by a direct selection process. There would follow thorough-going changes in the system of selection coefficients, changes in the probability arrays themselves of the various genes and in the long run an essentially irreversible adaptive advance of the species. It has seemed to me that the conditions for evolution would be more favorable here than in the indefinitely large population of Dr. Fisher's scheme. It would, however, be very slow, even in terms of geologic time, since it can be shown to be limited by mutation rate.

Wright then presented the view of evolution in nature that he found most appealing:

A much more favorable condition would be that of a large population, broken up into imperfectly isolated local strains. . . . The rate of evolutionary change depends primarily on the balance between the effective size of population in the local strain (N) and the amount of interchange of individuals with the species as a whole (m) and is therefore not limited by mutation rates. The consequence would seem to be a rapid differentiation of local strains, in itself non-adaptive, but permitting selective increase or decrease of the numbers in different strains and thus leading to relatively rapid adaptive advance of the species as a whole. Thus I would hold that a condition of subdivision of the species is important in evolution not merely as an occasional precursor of fission, but also as an essential factor in its evolution as a single group. (Wright 60, pp. 354–55)

This statement adequately summarizes the view that Wright later called the "three phase shifting balance theory," involving random drift, intrademe, and interdeme selection, as he saw it in the summer of 1930. Here was a clear alternative to Fisher's whole view of evolution in nature.

If, as I thought, Fisher was irritated by this review, it was also true that Wright had reason to be irritated at Fisher at about the same time. When Fisher discovered that Wright's figure of $1/2N$ was correct for the rate of decrease in heterozygosis in a population of effective size N instead of Fisher's figure $1/4N$, he not only found the error (as indicated above in the correspondence) but also rederived his formulas for the distribution of genes under several special cases, as he had done earlier in his 1922 paper. These results Fisher presented both in *The Genetical Theory of Natural Selection* and in a technical paper, "The Distribution of Gene Ratios for Rare Mutations," (Fisher 1930c) published in August 1930.

What disturbed Wright is that of the three special cases treated by Fisher (steady decay of variability in the absence of mutations, variability maintained constant by mutations in the absence of selection, and effects of a small selection pressure), Wright had already achieved Fisher's exact results for the first two in the manuscript read by Fisher; yet Fisher did not mention Wright's earlier derivation of the same results. Using the method of functional equations, Fisher made a more sophisticated analysis of the effects of selection than had Wright; but the distribution given by Wright in his manuscript was also more sophisticated by far than the one in Fisher's 1922 paper. By the time Wright's manuscript appeared in *Genetics* in 1931, as Wright explained in his letter to Fisher of October 15, 1930 (quoted above), he had worked out a distribution including selection of the same degree of approximation as in Fisher's distribution. Knowing this much about the relationship of Fisher and Wright, I hypothesized that Wright's review of Fisher's book, generally positive but very critical in several ways, was the last straw in their relationship. When I first tried the hypothesis on Wright in our interviews, he indicated agreement but with none of the certainty he typically exhibited when he clearly recalled an event. I was convinced, and shared the hypothesis with audiences at several universities. When about a year later Wright uncovered more Fisher/Wright correspondence, Fisher's actual reaction to Wright's review was revealed:

I was delighted to see your review of my book in *The Journal of Heredity,* for August last, which for some reason has only just appeared in this country. Your opening paragraphs especially will be most valuable in getting the less genetical sorts of biologists to see that the evolutionary bearings of genetical discussions are not at all what they were supposed to be; but indeed I ought not to praise one part rather than another for I liked it all heartily. It is in fact the most understanding review of my book which has yet appeared anywhere, and apart from personal vanity, which will of course absorb any amount of mere praise, that is really what an author craves for.

I was extremely interested in your more critical discussion, but what a shame that they should have printed your formulae so illegibly. You must really take some later opportunity to set out your views more fully, for I am willing to be convinced, not of the importance of sub-division into relatively isolated local colonies, which I should agree to at once, but that I have overlooked here a major factor in adaptive modification, which is what at present I am not convinced of. The point is very well worth going into in detail. I fear though that an adequate discussion will be above the heads of many biologists.

I hear that I have recently been attacked in the Zoological Society for daring to *intrude* in biological discussions; perhaps you have had occasionally a similar experience. I do not think it is this kind of thing which does any real harm; it makes a few old pundits feel more comfortable on their perches, but it carries mighty little weight with the younger men.

I had not intended to take up any special point in this letter, but I am tempted to mention this one (p. 353) "The formula itself seems to need revision in the case of another important class of genes, ones slightly deleterious in effect but maintained at a certain equilibrium in frequency by recurrent mutation" (I can leave migration aside here). The point here is that the average fitness *is* continually being increased by selection, at exactly the same rate as it is being decreased by mutation. This cause of deterioration of adaptedness, due to mutations of the organism, is, in my treatment, classed with the parallel deterioration due to changes in the environments. This supplies an amendment to the corresponding statement on p. 352, "The only effective offset to undeviating increase in fitness, which he recognizes, is change of environment." I think, if you happen to re-read p. 4l, you will see that I class deleterious mutations equally as an offset.

I wonder if you would agree that in attributing somewhat less weight than I to what selection always is doing, you are *ipso facto* attributing more to what it has already done. I mean that that the situation sketched at the end of p. 353 would be undoubtedly right if selection had in the recent past been infinitely effective, or infinitely rapid, as a means of modification, and is only therefore ineffective now. This is what I was driving at in saying that the difficulties encountered by natural selection were chiefly of its own making, i.e.

the high perfection of existing adaptation. When the spirit moves you, I should be exceedingly interested to hear if you think this is rightly put. (Fisher to Wright, January 19, 1931)

I could scarcely have been more wrong in my hypothesis about Fisher's response to Wright's review; yet at the same time I think the hypothesis was reasonable and even compelling with the available evidence. I have had at my disposal an extraordinary array of sources of information in writing this book and have emphasized the value of these sources for writing a more accurate account of Wright's life. I tell this particular story here to illustrate graphically that these excellent source materials by no means guarantee historical accuracy.

One passage in Fisher's letter might be confusing. In the second paragraph he says that he agrees at once to Wright's emphasis upon the subdivision of a population into partially isolated local colonies; this may sound strange after telling Wright earlier that the effective population size of a species was the total population of the species on earth (reduced appropriately for sexual maturity, population minimums, etc.). I think Fisher had in fact changed his mind as a result of thinking about "fission" while writing *The Genetical Theory of Natural Selection*. A huge panmictic population does not split, but species obviously had. Thus Fisher had to move in the direction of some kind of local isolation. But he was careful to add in his letter that this subdivision was not a major factor in adaptive modification, as Wright insisted. I think Fisher meant that the differences in the subpopulations resulted from differential selection rather than from random drift.

Wright's answer to this letter is interesting particularly, for its restatement of Fisher's fundamental theorem of natural selection and for Wright's first presentation with a figure of his multidimensional fitness space, here in one dimension (compared to two in the published version of 1932.

I was very glad to hear that you were pleased with my review. I wished to define as well as I could the differences in our conclusions, as well as to express my fundamental agreement with your mode of approach to the problem, and I was afraid that I might perhaps have emphasized the former too much. My final conclusions on the relative importance of evolutionary factors can be stated so as to appear closer to those reached by Karl Pearson recently, than to yours, but in the former case, the reasoning involved such antagonistic theories of heredity and particularly of inbreeding that any similarity seems of no more value than a coincidence. In the latter case, what divergence there is, comes near the end of the train of reasoning instead of at the beginning, and correspondingly offers much more hope of being cleared up.

I am very sorry that I overlooked your treatment of mutations on page 41. The essential difficulty which I felt with your conclusions on pages 34–37 still remains, but I should have worded it differently.

I recognized at once that the discussion on these pages was meant to apply only under constancy of external conditions, somewhat as the law of increase of entropy is meant only to apply to a system insulated from outside disturbances, but I assumed that all internal factors (including recurrent mutation under what would be considered constant environmental conditions) were intended to be taken into account. I do not know how I managed to overlook your clear statement to the contrary on page 41. I think that if the theorem on page 35 had been stated with its qualifications: "The rate of increase in fitness of any population at any time is equal to its genetic variance in fitness at that time, except as affected by mutation, migration, change of environment and the effects of random sampling," I would not have been confused by the discussion in the following pages. Put in this form, the theorem does not lend such force to the idea of minute control of the course of evolution by natural selection as it seems to without the qualifications. Some aspects of the idea which I tried to express in pages 353 to 355 of my review might be visualized as follows: Think of the field of visible joint frequencies of all genes as spread out in a multidimensional space. Add another dimension measuring degree of fitness. The field would be very humpy in relation to the latter because of epistatic relations, groups of mutations which were deleterious individually producing a harmonious result in combination. In the figure below this field (very imperfectly represented by a single line) is plotted against fitness.

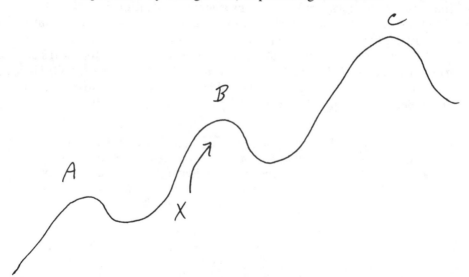

Figure 8.2. Two-dimensional fitness surface. Redrawn from Wright's letter.

A species, at point X with respect to its system of gene frequencies, will tend to move steadily toward some point near B (not B exactly if there is recurrent mutation) under the influence of selection, but will tend to stay there even though C would be a position of greater fitness if it could be reached. Something other than the steady pressure of selection is needed. I recognized four factors in my review. First, an irregularly changing environment. This by continually changing the system of humps themselves would keep the species moving, the immediate effect being downward as a rule but in the long run permitting some species to reach positions of adaptation which could never have been reached by a direct process. Life as a whole would advance at least in complexity of adaptation. Second, novel mutations in indefinitely extended multiple series, creating new dimensions and occasionally new paths of advance upwards. I interpreted your idea of the mechanism for indefinite continuance of the evolutionary process as being of this nature. Third, limitation in the size of the population to such a figure that random variation in gene frequency becomes important (but not overwhelmingly important as under close inbreeding). The position of the species would move at random about B and occasionally might go so far that the upward pressure of selection would tend to carry it toward A or C instead of to B. Since it could more easily escape from A than C, the general tendency would be toward increasing adaptation even though the process would be very slow. Fourth, subdivision of an indefinitely large species into many small, not quite completely isolated groups. The positions of these groups would move at random about B. Some would reach such positions as A, others C. The latter being better adapted would tend to increase in population and become the major source of migrants to other groups. The position of the mean of the species should thus approach C.

I certainly do not profess to have any complete mathematical analysis of this situation, but as far as my figures go they have seemed to indicate that the fourth scheme is the most effective. In a sense selection plays as great a role as in your theory, the selection of individuals to which I have attributed less complete control being supplemented by selection between local races.

My main paper should come out next month in *Genetics*. I will be very much interested in seeing your criticism. Perhaps I am all wrong with regard to the humpiness of the field of gene frequencies in relation to adaptation. (Wright to Fisher, February 3, 1931)

Fisher was unquestionably interested in Wright's idea of the multidimensional fitness space with its humps. He had been invited to spend the summer of 1931 in the United States, mostly at Iowa State in Ames, Iowa, with George W. Snedecor and Jay L. Lush. Fisher preferred to talk with another person about complex issues rather than to correspond and wanted to see Wright sometime during the visit for a long talk. Wright in general preferred to correspond because he liked to think issues through carefully before taking a position on them. Fisher wrote on February 17:

I very much hope I shall have a chance of seeing you again during the summer. . . .

If you can forecast your own movements during the summer, I think I ought to be able to catch you for a good long talk some time. I do think that differential selective action in different stations or regions may be exceedingly important, even if there is a steady diffusion of germ plasm between them. Let me know what the prospects are of meeting you.

Wright replied on March 6:

I was very glad to hear that your plans for a trip to this country have taken definite form. I expect to spend the summer in Chicago since I am scheduled to give two courses during the summer quarter. I shall look forward to seeing you at whatever time is most convenient to you.

My paper in *Genetics* has come out at last and I shall naturally be much interested in your criticism. I regret very much that two typographical errors got into a formula which I quoted from you. They were correct in page proof, as were a few other errors which I have found.

Fisher and Wright did indeed work out a time for a visit. On the day after his arrival in the United States, May 31, Fisher wrote to Wright to propose June 27 as the day for him to come from Ames to Chicago, adding this substantive point:

Your letter of February 3rd contains a point about non-optimal points of genetic stability which I should like to take up with you. In one dimension, a curve gives a series of alternate maxima and minima, but in two dimensions two inequalities must be satisfied for a true maximum, and I suppose that only about one fourth of the stationary points will satisfy both. Roughly I would guess that with n factors only 2^{-n} of the stationary points would be stable for all types of displacement, and any new mutation will have a half chance of destroying the stability. This suggests that true stability in the case of many interacting genes may be of rare occurrence, though its consequence when it does occur is especially interesting and important.

This was Fisher's first handwritten letter to Wright, who literally had to write out a translation of Fisher's minuscule and difficult writing. Wright's translation, I might add, was very useful to me in transcribing Fisher's letter.

Wright wrote back on June 5 to confirm the date and to invite Fisher to stay at the Wright home. As for Fisher's substantive comment, Wright replied:

In regard to the theoretical point which you raised, I appreciate that with increase in the number of dimensions the chance that one may pass by a continuously upward path from one point to another increases. I have recently received some reprints from J. B. S. Haldane in which he seems to have reached views somewhat like mine and in which he represents populations in multidimensional space in ways which I suggested in my letter. He points out that with m genes it would be possible for alternate apices in his hypercube to be stable. I do not suppose, of course, that there would actually be anything approaching this number of maxima. I suppose that most species would carry many thousands of genes in an unfixed state. Presumably all of the four hundred more or less conspicuous mutations of Drosophila picked up in laboratory cultures since 1910 would be found present in nature if every wild fly could be examined and tested genetically. There are doubtless many more inconspicuous mutations. 2^{1000} is an infinity of such a high order that the maxima may be very widely scattered and still permit a practically infinite number of them. Moreover, even where it is not a case of evolution from one maximum to another, but merely movement along a very slight gradient, it has seemed to me that a trial and error mechanism relating to the organism as a whole and not merely to genes (as in the case of mutation) would enormously speed up the process of evolution.

Wright was certainly referring here to part 7 of Haldane's "Mathematical Theory of Natural and Artificial Selection" on metastable populations (Haldane 1931). Not only had Haldane anticipated Wright's idea of representing a population in a multidimensional space with positions of stable equilibria, but in one sentence he also had suggested that rupture of the equilibrium "will be specially likely where small communities are isolated" (p. 142). Haldane of course had no idea how fully Wright had developed this view by 1925.

Personal meetings are disasters for historians. Tracing accurately the exchange of ideas between Fisher and Ford is an example in point. Although Ford was in Oxford and Fisher in Cambridge, they wrote only to arrange meetings for substantive discussion, reserved for visits. I would love to have direct contemporary evidence for their influence upon each other, but none exists to my knowledge (although Ford has been most kind in sharing his recollections with me). Thus I cannot help wishing for the impossible: to have sat in upon Fisher's visit with the Wrights. Wright recalls little about the substantive discussions during the visit. From Fisher's side there is only the undated handwritten letter clearly penned soon after the visit:

> This is just a note to thank you and Mrs. Wright for your kindness and hospitality to me in Chicago. I wish I could better understand your views on those points on which I differ from you, but on the points I have discussed with Lush, I see little chance that I shall ever do so. However, there is a substantial body of theory on which I think we do agree and that after all is of infinitely more interest to the world at large than the very obscure points still in dispute.

Wright's letter of June 5 and this one from Fisher were their last such communications, so far as I can determine. Their series of letters, reproduced here very nearly in full, speaks for itself. Fisher and Wright each learned a great deal from the other, and each modified his own view to some extent toward that of the other. Although Fisher's last letter has a tone of finality and some frustration, by no means does it give any hint of the later rancor that he freely exhibited toward Wright, nor does Wright's last letter to Fisher indicate any of Wright's later strongly defensive attitude toward Fisher. The differences in viewpoint between Wright and Fisher are central in later theoretical and field population genetics and in evolutionary theory in general—a theme I will develop in detail in the succeeding chapters.

9
Wright's Shifting Balance Theory of Evolution

Wright's "Evolution in Mendelian Populations"

Wright's paper on evolution in nature was finally published in *Genetics* in March 1931, over five years after he wrote most of it and fourteen months after submission for publication. It occupied sixty-two printed pages. No other geneticist in the United States had the quantitative and biological background to even attempt such an analysis; thus Wright's paper was immediately viewed by geneticists as a major achievement. Indeed most geneticists at that time had so little preparation in quantitative analysis that Wright's paper took on a mystical tinge. Theodosius Dobzhansky expressed this point well in his oral memoir recorded in 1962; he can speak with some authority, having been much influenced by Wright in the 1930s and 1940s.

> Genetics is the first biological science which got in the position in which physics has been in for many years. One can justifiably speak about such a thing as theoretical mathematical genetics, and experimental genetics, just as in physics. There are some mathematical geniuses who work out what to an ordinary person seems a fantastic kind of theory. This fantastic kind of theory nevertheless leads to experimentally verifiable prediction, which an experimental physicist then has to test the validity of. Since the times of Wright, Haldane, and Fisher, evolutionary genetics has been in a similar position. (Dobzhansky 1962, 500–501)

Wright's "Evolution in Mendelian Populations" soon led to his admission to the National Academy of Sciences and to a much wider international reputation than he had previously enjoyed. When the paper was published, Wright still thought of himself primarily as a physiological geneticist. His reputation as a quantitative evolutionary theorist, however, rapidly outgrew his reputation as a physiological geneticist.

The 1931 paper is divided clearly (a table of contents is printed on the first page) into five sections: (1) theories of evolution in historical perspective; (2) variation of gene frequency; (3) distribution of gene frequencies and its immediate consequences; (4) evolution of Mendelian systems; and (5) summary. While this all appears easily understood in outline, the paper is difficult reading.

The introductory section is particularly interesting because Wright views the relationship of Mendelism to Darwinism in the same basic way as Fisher. One can easily see why Wright and Fisher could say to each other in correspondence that each shared fundamental assumptions of the other, while at the same time disagreeing intensely about basic processes of evolution. Both viewed (accurately, in my opinion) the history of genetics in relation to evolution in the early twentieth century as strongly influenced or even dominated by the de Vriesian view of discontinuous evolution. Each saw his own work as an attempt to demonstrate that the quantitative consequences of Mendelian heredity in conjunction with the conditions of populations in nature led to a gradual process of evolutionary change as Charles Darwin had envisioned. Fisher and Wright shared this vision despite their differences on the specific mechanisms of gradual evolutionary change. They also, of course, shared the view that the quantitative analysis of the consequences of Mendelian heredity was essential for an understanding of evolution.

After this brief introductory section came forty-two pages of highly quantitative analysis. I suspect few geneticists worked their way through all of the mathematics, most of which is not so difficult per se, but the pages look imposing and uninviting. The second section, on variation of gene frequency, presented a brief analysis of factors affecting gene frequency. These included the effects of mutation, migration, and selection; equilibria with and without selection; multiple alleles; random effects; the rate of decrease of heterozygosis; and population number or effective population size.

Two points in this otherwise straightforward section should be noticed. In the discussion of multiple alleles, Wright points out a fundamental limitation of most of the quantitative analysis he and Fisher had done on the statistical distribution of gene frequencies as well as on their discussions of factors affecting gene frequency. The problem was that the available quantitative methods could easily be applied only to pairs of alleles at each locus, and frequently, as in the entire section on the distribution of gene frequencies, only one locus could be treated formally. Yet Wright's whole approach to evolution in domestic and natural populations was deeply tied to multiple alleles at interacting loci. Wright specifically addressed the problem in discussing multiple alleles: "The foregoing discussion has dealt formally only with pairs of allelomorphs, a wholly inadequate basis for consideration of the evolutionary process unless extension can be made to multiple allelomorphs" (Wright 64, p. 104). He added that most or all loci may have multiple alleles, making this a very important factor. One consequence of interacting loci with multiple alleles was that constant selection pressures of the sort emphasized by Fisher in evolution in nature were unlikely:

> The selection coefficient, s, relating to a gene A cannot be expected to be constant if the alternative term a includes more than one gene. The coefficient should rise to a maximum positive value as A replaces less useful genes but should fall off and ultimately become negative as the group of allelomorphs comes to include still more

useful genes. But as already discussed, even if A has only one allelomorph, the dependence of the selection coefficient on the frequencies and selection coefficients of non-allelomorphs keeps it from being constant. (P. 106)

The tension between the quantitative formal theory and the qualitative theory in this section is real and deep. I think the tension is even greater between the quantitative section on distribution of gene frequencies and its immediate consequences and the qualitative section on the evolution of Mendelian systems; but here Wright does not explicitly raise the issue, though I am certain he felt the tension strongly.

The second point of particular interest in the section on variation of gene frequency concerns population number. Wright emphasized that in natural populations the effective size of the population for the purposes of evolutionary change might be far less than the number of individuals that meet the eye of the observing naturalist.

Obviously N applies only to the breeding population and not to the total number of individuals of all ages. If the population fluctuates greatly, the effective N is much closer to the minimum number than to the maximum number. If there is a great difference between the number of mature males and females, it is closer to the smaller number than to the larger. . . .

The conditions of random sampling of gametes will seldom be closely approached. The number of surviving offspring left by different parents may vary tremendously either through selection or merely accidental causes, a condition which tends to reduce the effective N far below the actual number of parents or even of grandparents. (Pp. 110–11).

Wright would certainly have liked to support this argument by examples taken from natural populations, but he could cite only the cases of the Clydesdale breed of horses and the Shorthorn breed of cattle.

By far the longest (thirty-one pages) and quantitatively most difficult section of the paper was that on the distribution of gene frequencies and its immediate consequences. Throughout the section Wright compared his derivations and results with those of Fisher and found basic agreement of results, as their correspondence has already revealed. Wright's aim was to develop a single formula for the statistical distribution of gene frequencies in a population, including as many factors affecting gene frequency as possible. He finally combined all the pertinent factors into the single distribution

$$\phi(q) = C e^{4Nsq} q^{4N(mq_m + v)-1} (1 - q)^{4N[m(1-q_m) + u]-1} \qquad \text{(p.134)}$$

This formula explicitly takes into account mutation in both directions and migration, neither explicitly incorporated by Fisher into his corresponding formula. Wright then constructed curves depicting distribution of gene frequencies under a series of different assumptions. Dominance was not included in the formal distribution because, as explained earlier, Wright found that both his and Fisher's earlier (1922) attempts to include the effects of dominance on the statistical distribution of genes led to some impossible results. Wright did include, however, a substantial qualitative section on the influence of dominance upon the distribution. By the time the paper was published he had found a way to incorporate dominance, but he did not publish the result until 1937 (Wright 91).

The obvious and substantial limitation of the entire analysis in this section and in Fisher's corresponding analysis in 1922 and in *The Genetical Theory of Natural Selection* was that only a single locus with two alleles was under consideration. Since interaction between loci was integral to Wright's whole view of the process of evolution, one might guess that in turning from this section to the qualitative presentation of his actual theory of evolution in nature, he would mention the limitation directly and show how the qualitative view related to a much narrower quantitative base. Neither Wright nor Fisher did a careful job of explaining this limitation at the crucial stage of advancing beyond the strictly formal models to discussions of evolution in nature, although Wright certainly did more in this direction than did Fisher. Fisher, however, in his letter to Wright of October 25, 1930 (quoted above), was quite explicit that the nonmathematical parts of *The Genetical Theory of Natural Selection* (or rather the parts where he said the mathematics was "undone") were the most important. Surely Wright would have agreed with regard to his own 1931 paper. I have frequently thought that the later antagonism of such major figures as Mayr (see Mayr 1959) or C. H. Waddington (see Waddington 1953) would have been much lessened if Fisher, Wright, and Haldane had taken more seriously the probable doubts of their much less quantitative colleagues and had discussed more frankly and fully the limitations of their quantitative analyses and the precise relations of these analyses to their qualitative theories of evolution.

Wright's section on his qualitative theory of evolution begins with a useful clarification of the factors of evolution into those that promote genetic homogeneity or that promote genetic heterogeneity (Wright 64, p. 143).

Factors of Genetic Homogeneity	Factors of Genetic Heterogeneity
Gene duplication	Gene mutation (u, v)
Gene aggregation	Random division of aggregate
Mitosis	Chromosome aberrabtion
Conjugation	Reduction (meiosis)
Linkage	Crossing over
Restriction of population size $(1/2N)$	Hybridization (m)
Environmental pressure (s)	Individual adaptability
Crossbreeding among subgroups (m_1)	Subdivision of group $(1/2N_1)$
Individual adaptability	Local environments of subgroups (s_1)

After careful discussion of each pair of factors, Wright turned to an analysis of the effects of the most important of these interacting factors by examining the consequences of his basic formula, somewhat simplified.

In a very small population, $1/4N$ much greater than u and than s, random drift determined the direction of the process and led to "nearly complete fixation, little variation, little effect of selection and thus a static condition modified occasionally by chance fixation of new mutations leading inevitably to degeneration and extinction" (p. 157). Wright had said much the same thing in his 1929 abstract.

Next Wright considered the opposite extreme of a very large random breeding population, s much greater than u, and u much greater than $1/4N$. Here Wright argued that favored genes were quickly fixed, yielding a "complete equilibrium under uniform conditions if the number of allelomorphs at each locus were limited" (p. 149), relieved only by the occurrence of new favorable mutations that took the place of the previous wild-type genes. Wright ascribed this view of evolution to Fisher in a footnote, but this is not quite fair because Fisher clearly emphasized the constant degeneration of the environment in his *Genetical Theory of Natural Selection*. Wright did address this possibility in the next paragraph but did not change his conclusion: "At best an extremely slow, adaptive, and hence probably orthogenetic advance is to be expected from new mutations and from the effects of shifting conditions." Even when gene frequencies did change rapidly after a change of selection pressures, he stated that the changes were essentially reversible and the situation was "distinctly unfavorable for a continuing evolutionary process" (p. 149). On this particular point, Wright would dramatically change his conclusion before publishing his 1932 paper (Wright 70), after realizing that the great multiplicity of adaptive possibilities made genetic reversibility highly unlikely.

In an intermediate-sized population, random drift in conjunction with selection led to nondegenerative continuing and irreversible changes of gene frequencies, but rate of change was limited by mutation pressure and thus very slow (but not so slow as in very large populations).

The optimum conditions for evolutionary advance, the process Wright believed would be found to predominate in nature, he summarized in the last paragraph of the paper:

> Finally in a large population, divided and subdivided into partially isolated local races of small size, there is a continually shifting differentiation among the latter (intensified by local differences in selection but occurring under uniform and static conditions) which inevitably brings about an indefinitely continuing, irreversible, adaptive, and much more rapid evolution of the species. Complete isolation in this case, and more slowly in the preceding, originates new species differing for the most part in nonadaptive respects but is capable of initiating an adaptive radiation as well as of parallel orthogenetic lines, in accordance with the conditions. It is suggested, in

conclusion, that the differing statistical situations to be expected among natural species are adequate to account for the different sorts of evolutionary processes which have been described, and that, in particular, conditions in nature are often such as to bring about the state of poise among opposing tendencies on which an indefinitely continuing evolutionary process depends. (Wright 64, p. 158)

Each of the three phases—random drift, and intrademic and interdemic selection—caused gene frequencies to shift, and the three phases were poised in balance. This process, later called "three phase shifting balance" by Wright, he saw in 1931 as belonging in the neo-Darwinian tradition of gradual evolution. He described the results of the process, "apparent continuity as the rule, discontinuity the rare exception" (p. 153), and he concluded that "the enormous recent additions to knowledge of heredity have merely strengthened the general conception of the evolutionary process reached by Darwin in his exhaustive analysis of the data available 70 years ago" (p. 154).

Wright's "Evolution in Mendelian Populations" immediately attracted the attention of geneticists. Muller wrote, "Please let me take this occasion to send you my warmest compliments on your monumental work on evolution which has just appeared in *Genetics*" (Muller to Wright, March 16, 1931); several months later he wrote again to request two more copies because of the intense use his own copy had endured at the University of Texas. Alexander Weinstein, then teaching at Johns Hopkins University, wrote that he had tried Wright's paper on his students and even the less intelligent ones learned much from it, even though they could not follow the mathematics. Weinstein added: "Thanks to the mathematical work on selection and the x-ray work on mutation we have at last emerged from what I call the KatydidKatydidn't period when evolutionary discussion consisted largely of statements that selection could or couldn't accomplish results" (Weinstein to Wright, November 10, 1931).

Muller and Weinstein (who had studied physics) both had substantial quantitative sophistication, but even geneticists who had little or none learned from Wright's paper. Thus A. Franklin Shull at the University of Michigan, who in 1936 wrote a popular book on evolution (Shull 1936), wrote to Wright twice in the spring of 1931 to ask Wright for simpler, less quantitative explanations for nonmathematical persons like himself. This was important, he said, because

I am trying to recast my ideas of evolution in accordance with your conclusions from the mathematical considerations in your article, and without the mathematics I am having to dig pretty hard to see what it all means. Some of my attitudes on evolutionary questions are having to be pretty radically changed because of it, and I am anxious to be certain that I understand what you have concluded. (Shull to Wright, May 13, 1931)

Wright's paper certainly was widely read and aroused much interest among geneticists, but I think few geneticists were able to follow the details of his quantitative arguments and wished, like Shull, that Wright would present his ideas in a briefer and more accessible form.

Wright's 1932 Paper on Evolution

Edward Murray East, organizer of the 1932 International Congress of Genetics in Ithaca, New York, was well acquainted with the work of Wright, Fisher, and Haldane. He also knew that their published papers on evolution were difficult to follow. So he asked all three to share a session at the congress (chaired by Goldschmidt) and to each deliver a short and accessible paper to the assembled geneticists. In particular, Wright was requested to present his evolutionary theory concisely and nontechnically; the published version for the *Proceedings* could occupy no more than ten pages of text. Because of teaching responsibilities at the University of Chicago, Wright came late for the congress (he was not present for the group picture), but he did have a pleasant luncheon with Fisher and Wright's former teacher, Wilhelmine Entemann Key who arranged it. The session itself attracted an overflow audience and was, according to persons present, a highly stimulating occasion. Fisher and Haldane used no mathematics in their presentations. Fisher's address, at least in its published form, was basically a defense of his theory of the evolution of dominance.

Wright found preparation of his address difficult. He was not in the habit of reducing his complex quantitative ideas to the level of brief qualitative presentations and did not feel comfortable with his assignment. He decided to develop the idea of a fitness surface that he had first expressed in his letter to Fisher (February 3, 1931, quoted above). Wright's idea was to take the entire field of possible gene combinations of a population and grade it (one gene combination at a time) with respect to adaptive value (reproductive fitness) under a specified set of conditions. Each point on the surface was thus the fitness of a particular genotype. According to Wright's estimate, a population might have a thousand or more dimensions in its field of gene combinations, and of course an extra dimension would have to be added to represent the level of adaptive values. Here Wright provided a diagrammatic representation of his adaptive "landscape" in two dimensions rather than in one, as presented in his letter to Fisher, and in place of the actual surface of 1,000 dimensions or more (fig. 9.1).

The basic idea was that a huge number (perhaps 10^{800}) of peaks existed on the surface, each separated by valleys. The peaks varied in height, meaning that some were more adaptive than others. Natural selection tended to drive the population up the nearest adaptive peak. Thus for Wright,

the central problem of evolution as I see it is that of a mechanism by which the species may continually find its way from lower to higher

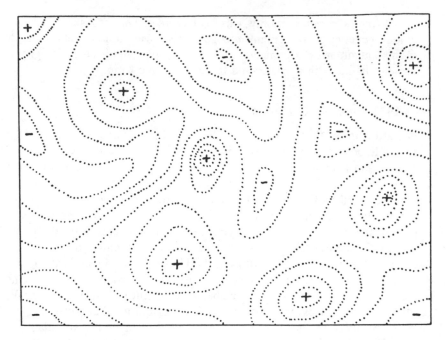

Figure 9.1. Adaptive landscape.

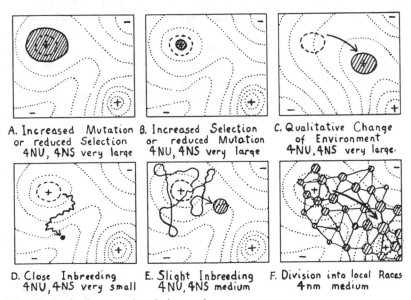

A. Increased Mutation or reduced Selection 4NU, 4NS very large

B. Increased Selection or reduced Mutation 4NU, 4NS very large

C. Qualitative Change of Environment 4NU, 4NS very large.

D. Close Inbreeding 4NU, 4NS very small

E. Slight Inbreeding 4NU, 4NS medium

F. Division into local Races 4nm medium

Figure 9.2. Adaptive landscape and evolutionary change

peaks in such a field. In order that this may occur, there must be some trial and error mechanism on a grand scale by which the species may explore the region surrounding the small portion of the field which it occupies. To evolve, the species must not be under strict control of natural selection. (Wright 70, pp. 358–59)

Next Wright exhibited (without derivation) the distribution curves of his basic formula of the statistical distribution of gene frequencies under various assumptions. These came direct from the 1931 paper. Then he presented a series of diagrams depicting these various distribution curves of gene frequencies in relation to the fitness surface discussed above. The results were diagrams of evolutionary change familiar to almost every student of evolutionary theory from the early 1930s to the present.

The six diagrams in figure 9.2 represent the same four situations already presented in the 1931 paper and discussed above: very large panmictic populations (diagrams A, B, and C), very small inbred populations (D), intermediate-sized populations (E), and division of a large population into semi-isolated subgroups (F).

Wright's interpretation remained the same as it had in 1931 with one significant exception. In 1931 Wright had argued that evolution in very large populations was extremely slow under all circumstances, including change of the environment. By 1932 Wright had changed his mind as evidenced by his discussion of diagram C.

The environment, living and non-living, of any species is actually in continual change. In terms of our diagram this means that certain of the high places are gradually being depressed and certain of the low places are becoming higher. A species occupying a small field under influence of severe selection is likely to be left in a pit and become extinct, the victim of extreme specialization to conditions which have ceased, but if under sufficiently moderate selection to occupy a wide field, it will merely be kept continually on the move. Here we undoubtedly have an important evolutionary process and one which has been generally recognized. It consists largely of change without advance in adaptation. The mechanism is, however, one which shuffles the species about in the general field as a whole. Since the species will be shuffled out of low peaks more easily than high ones, it should gradually find its way to the higher general regions of the field as a whole. (P. 362)

What Wright did not say here, although he did point out the change in 1967 (Wright 184, p. 252), was that he had, in a paper published only the year before, given no importance at all to this process. Substantive changes between the 1931 and 1932 papers should come as no surprise because Wright had for the most part written the 1931 paper seven years before he wrote the 1932 paper.

A fair question to ask, however, is what led Wright to change his mind and who had "generally recognized" the process depicted in diagram C? This process was in fact the one emphasized strongly by Fisher (1930, 41–42). Fisher's letter to Wright of January 19, 1931 raised Wright's neglect of the degrading effects of mutation and inevitable environmental change upon fitness in his review of Fisher's book, and Wright acknowledged these points in his reply of February 3. I suspect that the interaction with Fisher played a substantial part in Wright's change of view on this issue, though certainly Wright's conception of multiple selective peaks also led to the conclusion that evolution by this process was essentially nonreversible.

It is important to understand that from 1932 on, Wright did not see Fisher's theory of evolution as fundamentally incorrect. He gave much weight to the Fisherian scheme of evolution in natural populations. He believed that Fisher's scheme dominated in the long periods of evolutionary stasis, whereas he thought the shifting balance theory became the dominant scheme in periods of rapid evolutionary change. This general view adhered closely to the view offered by Darwin in *Origin*, where he argued that evolution was characterized by long periods of stasis interspersed by periods of evolutionary change dominated by the force of natural selection (see Rhodes 1983). Over fifty years later Wright would apply his distinction between periods of evolution, dominated by Fisherian modes of change, and shorter periods of intense evolutionary change, dominated by his shifting balance process, to the theory of punctuated equilibrium put forward by Eldredge and Gould (Eldredge and Gould 1972; Wright 207).

The process depicted by diagram F was of course Wright's three-phase shifting balance theory, upon which he placed greatest emphasis in the paper. It was the most efficient method by which a population could explore the adaptive surface and move relatively quickly toward and up higher adaptive peaks.

Wright's presentation at the congress and his later published version, "The Roles of Mutation, Inbreeding, Crossbreeding, and Selection in Evolution," were highly successful and influential. With its absence of difficult-looking mathematics, brevity, intriguing diagrams, and simple explanation of Wright's views of the evolutionary process, the paper could be easily read by anyone interested in evolution. Dobzhansky was in the audience for Wright's presentation at the congress and was immediately persuaded of the great importance of Wright's views. He incorporated these views and diagrams into the genetics course he taught at the California Institute of Technology and into all three editions of his very famous and influential *Genetics and the Origin of Species* (1937, 1941, 1951). Looking back in 1962, Dobzhansky said,

Wright gave a splendid paper at the Genetics Congress in 1932. In a sense, that is still his best paper. He is a remarkably difficult writer. In most cases, he writes with so much profound and esoteric mathematics that common mortals cannot read him anyhow. Even when he attempts to write without esoteric mathematics, he is often rather hard to follow. This 1932 paper is an exception. (Dobzhansky 1962, 355)

Many others besides Dobzhansky were much impressed by Wright's 1932 paper. His adaptive surface diagrams appeared and reappeared frequently in papers and books and were constantly referenced. Wright ordered an ample supply of reprints, but he soon ran out of them. Since he liked to hand this paper out to his students in his evolution course, the Department of Zoology had the paper professionally reprinted with a very large number of copies, and Wright again filled reprint requests, continuing even after his departure from Chicago.

Ambiguities of Wright's 1931 and 1932 Papers on Evolution

If Wright's 1931 and 1932 papers on evolution were highly visible and influential during the 1930s and 1940s, they were not thereby free of ambiguity or inconsistency on some crucial issues. Such problems are, indeed, usual in the initial expression of new scientific theories.

An important ambiguity relates to the problem of the levels or units of selection in the shifting balance theory of evolution. (For a modern analysis of this complex problem, see Lewontin 1970 and Brandon and Burian 1984.) Consider the following two quotations from the 1932 paper, taken from two successive paragraphs (Wright 70, p. 363):

> With many local races, each spreading over a considerable field and moving relatively rapidly in the more general field about the controlling peak, the chances are good that one at least will come under the influence of another peak. If a higher peak, this race will expand in numbers and by crossbreeding with the others will pull the whole species toward the new position. The average adaptiveness of the species thus advances under intergroup selection, an enormously more effective process than intragroup selection.

> The effective intergroup competition leading to adaptive advance may be between species rather than races.

I suggest that these two uses of "intergroup" selection, while appearing the same, actually require wholly different evolutionary mechanisms and for this reason lead to ambiguity and confusion.

The first usage of intergroup selection corresponds to what Wright later called "interdemic" selection. The very term is ambiguous because it conveys, especially when contrasted with "intrademic" selection, the sense that selection is operating *between* populations or demes, in a manner perhaps similar to selection operating between individuals. But the process that Wright actually defines in the first quote as intergroup is nothing of this sort.

The process is an interaction of individual selection (meaning intragroup or intrademic selection) with population structure and migration. What happens in the process, according to Wright, is that a subpopulation of a species divided into many subpopulations develops by random genetic drift an especially favorable interaction system of genes. The population expands because of the favorable genotype and sends out migrants that interbreed with members of the other subpopulations, thus genetically transforming them (by individual selection) into populations genetically similar to the favored subpopulation that sent out the migrants. Throughout the entire process the real mechanism is individual selection in relation to population subdivision and migration. To me, it is confusing to call this process intergroup or interdemic selection.

The second quote, however, speaks of intergroup competition between species. Here there can be no process as Wright describes above because one species cannot be transformed by migrants from another species that is reproductively isolated from it. This kind of intergroup selection has to be based upon species level (intergroup) differences, or in other words is like the "species selection" that Stanley and Gould have recently advocated. Whether any such process actually exists in nature is a hotly debated question at the present time, but at least it makes sense to call it a form of intergroup selection.

Wright's shifting balance theory of evolution requires no true intergroup selection of any kind at all below the level of reproductively isolated populations, where he generally applied the theory. Throughout the corpus of Wright's work on evolution after 1932 I recommend reading his term interdemic selection as individual selection in relation to population subdivision into demes and migration.

Another area of much confusion concerns the problem of adaptation and the shifting balance theory. When I first interviewed him, Wright clearly expressed the view that his ideas on evolution had been much misinterpreted over the years and continued to be at that time. In particular, he felt that many biologists interpreted his shifting balance theory of evolution to mean that evolution occurred primarily by random drift rather than by natural selection. Wright emphasized to me that he had always believed that in his shifting balance theory the role played by random drift was only to shift gene frequencies to the extent of creating novel interaction systems of genes upon which selection would then act. The process as a whole was not only adaptive but far more rapidly adaptive than evolution in large panmictic populations. I understood him to mean that his shifting balance theory insured adaptive advance by the subspecies level, and certainly always at the species level, with random drift as the major factor generally only at the level of very local populations.

Wright has expressed these views in print and in conversation on many occasions. Thus in his 1967 paper on the foundations of population genetics,

Wright stated:

> Many critics have seized on the concept of random drift that was pro-
> posed and have asserted that I have advocated this as a significant *al-
> ternative* to natural selection. Actually, I have never attributed any
> evolutionary significance to random drift except as a trigger that may
> release selection toward a higher selective peak through accidental
> crossing of a threshold. (Wright 184, pp. 254–55)

Wright was referring here specifically to the criticisms of random drift by
Fisher and Ford in their 1947 paper and in their reply (1950) to Wright's rebuttal. Certainly this is the view of the shifting balance theory that Wright
presents throughout his four-volume *Evolution and the Genetics of Populations* (1968–78).

One is left wondering why Wright has been so misunderstood. His argument about the role of random drift is not so difficult to understand. Wright's
view as he expressed it to me was that Fisher, Ford, Julian Huxley, and a few
other influential evolutionists had initially misinterpreted him, with the historical inevitability that others followed their lead, creating a snowball effect.

This much is certain. In 1929, 1931, and 1932 Wright argued without
ambiguity that random drift in very small isolated populations led inevitably
to adaptive decline and extinction, not to progressive evolution of any kind.
Wright has referred frequently to these passages in trying since 1948 to clarify perceptions of his views on random drift, saying specifically that he had
always rejected the evolutionary significance of random drift in very small
natural populations.

The basic question then reduces to this: At what level precisely did
Wright believe that random drift in the shifting balance process caused nonadaptive shifts in gene frequencies, and at what level did selection of the gene
combinations produced by random drift cause adaptive evolutionary change?
In domestic populations, the corresponding process that Wright had investigated in detail was highly "adaptive" (i.e., conformed to the breeders ideals)
at the level of the breed, certainly a far more restricted taxonomic unit than
the species. A subordinate but obviously related question is what role random
drift and nonadaptive differentiation play in subspeciation and speciation.
The most instructive approach to these questions is simply to document what
Wright actually said in the years 1929–32.

(1) The nonadaptive nature of the differences which usually seem to
    characterize local races, subspecies, and even species of the same
    genus indicates that this factor of isolation is in fact of first impor-
    tance in the evolutionary origin of such groups, a point on which
    field naturalists (e.g., Wagner, Gulick, Jordan, Osborn, and Cramp-
    ton) have long insisted. (57, pp. 560–61)

(2) The actual differences among natural geographical races and subspecies are to a large extent of the nonadaptive sort expected from random drifting apart. (64, p. 127)

(3) [Fisher's] theory is one of complete and direct control by natural selection while I attribute greatest immediate importance to the effects of incomplete isolation. (64, p. 149n)

(4) The direction of evolution of the species as a whole will be closely responsive to the prevailing conditions, orthogenetic as long as these are constant, but changing with sufficiently long continued environmental change. (64, p. 151)

(5) Adaptive orthogenetic advances for moderate periods of geologic time, a winding course in the long run, nonadaptive branching following isolation as the usual mode of origin of subspecies, species, perhaps even genera, adaptive branching giving rise occasionally to species which may originate new families, orders, etc. . . . are all in harmony with this interpretation. (64, p. 153)

(6) [Under the shifting balance process] complete isolation originates new species differing for the most part in nonadaptive respects but is capable of initiating an adaptive radiation as well as of parallel orthogenetic lines, in accordance with the conditions. (64, p. 158)

(7) Complete isolation of a portion of a species should result relatively rapidly in specific differentiation, and one that is not necessarily adaptive. The effect intergroup competition leading to adaptive advance may be between species rather than races. Such isolation is doubtless usually geographic in character at the outset but may be clinched by the development of hybrid sterility. (70, p. 363)

(8) That evolution involves nonadaptive differentiation to a large extent at the subspecies and even the species level is indicated by the kinds of differences by which such groups are actually distinguished by systematists. It is only at the subfamily and family levels that clearcut adaptive differences become the rule (Robson 1928; Jacot 1932). The principal evolutionary mechanism in the origin of species must then be an essentially nonadaptive one. (70, pp. 363–64)

(9) Subdivision into numerous local races whose differences are largely nonadaptive has been recorded in other organisms wherever a

sufficiently detailed study has been made. [There follows citations of the work of Gulick, Crampton, David Starr Jordan, Ruthven, Kellogg, Osgood, Kinsey, Osborn, Rensch, Schmidt, David Thompson, and Sumner.] (70, pp. 364–65)

And finally, Alfred C. Kinsey, who was a taxonomist specializing in the gall wasp *Cynips* before turning his attention to human sexuality, wrote to Wright for his 1931 paper and enclosed his own monograph on *Cynips*. Wright, looking for corroboration of his theory, hoped that Kinsey's wasps might show random differentiation:

(10) I am especially interested in the question as to how far there is subdivision of species into small local strains differentiated in the *random* fashion expected of inbreeding (instead of in adaptive ways by natural selection). My results seem to indicate that such a condition is the most favorable for progressive evolution of the species as a single group. (Wright to Kinsey, April 14, 1931)

Kinsey replied, incidentally, that his team routinely searched over a wide area "in order to avoid such local strains and to obtain a more complete idea of the species as a whole" (Kinsey to Wright, April 22, 1931). Thus Kinsey's fieldwork could not furnish the precise corroboration Wright desired. (For a critique of Kinsey's taxonomic work on *Cynips,* see Goldschmidt 1937.)

Viewed together, these ten citations shed some light upon the question of why Wright was so much misunderstood during the 1930s and later. "The direction of evolution of the species as a whole will be closely responsive to the prevailing conditions" (4) appears inconsistent with the statement that the evolutionary process consists of "nonadaptive branching following isolation as the usual mode of origin of subspecies, species, and even genera" (5), or with "the principal evolutionary mechanism in the origin of species must thus be an essentially nonadaptive one" (8). The careful reader in 1932 would almost certainly conclude that nonadaptive random drift following isolation was a primary mechanism in the origin of races, subspecies, species, and perhaps genera. Wright's more recent view that the shifting balance theory should lead to adaptive responses at least by the subspecies level is found nowhere in the 1931 and 1932 papers or in the letter to Kinsey. In any case, with these citations in mind one can easily understand why some biologists understood Wright to be saying that random drift played the dominant role in the origin of subspecies and species.

Influence of Systematics upon the Shifting Balance Theory

The only way to understand Wright's presentation in the 1930s of the role of random genetic drift in the balance of evolutionary factors that affected gene frequencies is to adequately understand what Wright believed systematists were telling him at the time. I have already provided in chapter 7 sufficient historical background to show that many leading systematists in the mid-1920s leaned strongly toward the view that in nature closely related species differed more in nonadaptive characters than in adaptive ones. Wright was familiar with this view, and he knew it had a distinguished history going back to Gulick, Romanes, and Darwin. Furthermore, during the late 1920s and early 1930s the belief that closely related species differed by nonadaptive rather than adaptive traits gained strength among systematists whose work Wright admired. (Although, as also emphasized in chapter 7, a tradition in taxonomy that viewed most differences between species as adaptive also existed at the same time, affecting such taxonomists as Poulton, K. Jordan, Rensch, Stresemann, and Mayr.)

In his 1932 paper, Wright cited a number of systematists whose recent work he found persuasive. Among them were G. C. Robson (1888–1945), O. W. Richards (b. 1901), A. C. Kinsey (1894–1956), D. H. Thompson (b. 1897), and A. P. Jacot (b. 1890). All argued strongly for the nonadaptive interpretation of species differences.

More than any other taxonomists in the late 1920s and 1930s, Robson and Richards promoted the concept of nonadaptive differentiation at the level of geographical races and species. In his 1932 paper, Wright cited Robson's *The Species Problem* (1928), which appeared in the prestigious Oliver and Boyd Biological Monographs series. I suspect that Wright had not read the earlier Richards and Robson paper, "The Species Problem and Evolution," which had appeared in *Nature* two years earlier (Richards and Robson 1926). This paper made an eloquent plea for evolutionists to give up the charade of inventing adaptationist stories to account for differences between geographical races and allied species. These apparently nonadaptive differences were, they argued, just what they appeared. After reviewing the subject, they concluded:

> It thus seems that the direct utility of specific characters has rarely been proved and is at any rate unlikely to be common. Furthermore, since the correlation of structure, etc., with other characters shown to be useful does not at present rest on many well-proved examples, it cannot yet be assumed that most specific characters are indirectly useful. Thus the rôle of Natural Selection in the production of closely allied species, so far as it is known at present, seems to be limited. This statement is not to be taken as a wholesale denial of the power of Natural Selection. The latter is not in question when structural differences of a size likely to affect survival are involved. It is only the capacity of selection to use on a large scale the small differences between closely allied species that is unproved. (P. 384)

In his 1928 book, Robson not only provided evidence of nonadaptive differentiation but also a very careful critique of selection experiments purporting to exhibit natural selection in action. Natural selection, Robson argued, could not be the mechanism leading to the ubiquitous nonadaptive differentiation found at the species level in nature.

In 1928 Robson and Richards began work on a major book on animal taxonomy in relation to observable variation in natural populations and to mechanisms of evolution in nature. Although not published until 1936 (three years after it was almost complete in manuscript), its general contents were well known several years before actual publication. This book has been in disrepute since the late 1940s because of its antagonism to natural selection, but I would strongly emphasize the lack of historicity in this modern judgment. When the book appeared, it was the best-known general work on animal taxonomy, at least until Julian Huxley's edited volume, *The New Systematics*, which appeared in 1940 and was not really replaced until the publication of Mayr's *Systematics and the Origin of Species* in 1942. *The Variation of Animals in Nature* by Robson and Richards was a very influential book.

After reviewing a wide range of evidence from systematists, they concluded:

> A survey of the characters which differentiate species (and to a less extent genera) reveals that in the vast majority of cases the specific characters have no known adaptive significance. . . . It may be conceded that in a number of instances structures apparently useless may in the future be found to play an important part in the life of the species; further, many "useless" characters may be correlated with less obvious features which are of real use, but, even allowing for this, the number of apparently useless specific characters is so large that any theory which merely *assumes* that they are indirectly adaptive is bound to be more a matter of predilection than of scientific reasoning. (Robson and Richards 1936, 314–15)

I disagree strongly with Mayr's assessment that the view represented in this quote was "very much in the minority" (Mayr and Provine 1980, 132), as compared with the adaptive view of other systematists at the same time. I think instead that Wright had excellent reason to believe what Robson and Richards were saying, both because they were following a strong tradition and because contemporary evidence supported their view (Provine 1983).

Kinsey, Thompson, and Jacot all published works, between 1930 and 1932, indicating nonadaptive differentiation at lower taxonomic levels. Kinsey (1930) had published the first part of his work on the taxonomy of the gall wasp, genus *Cynips*. One of the basic taxonomic characters used by Kinsey was wing length:

> There seems no basis for believing the shortened wings or any of the concomitant variations of any adaptive value to any of these insects. The short wings are not confined to warmer of colder climates, and

long- and short-winged forms of various species are active at the same season in the same localities. The field data suggest nothing as to the survival value of these outstandingly basic modifications of structure. (Kinsey 1930, 34)

Thompson had studied variation in fishes in Illinois as a function of distance both within a single stream and between streams. He found that even great environmental differences along the length of a single stream led to far less racial differentiation than found in the same species in different streams but in apparently similar environments. In general terms, Thompson explained his view:

Every species that has been studied genetically has shown heritable variations of greater or less degree. Furthermore, these heritable variations are continually arising *de novo* by changes in the germ plasm due to causes entirely unrelated to ordinary variations in the physical environment, or at least not as adaptive responses to them. While most of these heritable variations are disadvantageous to the animal and are rapidly eliminated, many of them are of indifferent selective value and may persist. (Thompson 1931, 278)

Jacot, a taxonomist who worked primarily upon small worms in the soil, published in the July-August 1932 issue of *American Naturalist* an attack upon the adaptationist view. He cited Robson's 1928 book as evidence that no proof existed for the argument that natural selection caused interspecific differences. If evidence for the adaptationist argument did exist, Jacot asked, "after so much observation and experimentation, should it not be plainly evident?" (Jacot 1932, 351).

To be sure, three of the sixteen systematists cited by Wright in the years 1929–30—Henry Fairfield Osborn, Bernhard Rensch, and Alexander Ruthven—argued for a primarily adaptive view of evolutionary change. One can easily see, however, why Wright evaluated their views as being less compelling than the authorities previously discussed. Ruthven's study of speciation in garter snakes used dwarfing and scutellation for taxonomic markers. Scutellation was highly correlated with degree of dwarfing. Ruthven admitted that the amount of dwarfing "does not seem to be directly associated with the nature of the environment," but then concluded that the dwarfing was "associated in some way with the environment" in an adaptive manner (Ruthven 1908, 193). But on what authority did Ruthven make this surmise? The answer was Tower's 1906 monograph on the inheritance of acquired characters in Colorado potato beetles, work revealed by 1915 to have been faked (Tower 1906; Weinstein 1980).

Although Osborn and Rensch both emphasized adaptive evolution, each left considerable room for nonadaptive differentiation. In his 1927 summary, Osborn had divided speciation into gradual adaptive and mutational nonadaptive. His concluding statement was, that "Speciation is a normal and continuous process; it governs the greater part of the origin of species; it is apparently always adaptive. Mutation is an abnormal and irregular mode of origin, which while not infrequently occurring in nature is not essentially an adaptive process; it is, rather, a disturbance of the regular course of speciation" (Osborn 1927, 42). Rensch's 1929 book, which Wright read carefully, had a substantial section devoted to nonadaptive differentiation, which Rensch considered to be of two types: (1) characters that arose apparently randomly; and (2) characters exhibiting geographic change by a gradual succession of steps, but apparently in the absence of selection.

Wright drew the obvious conclusion from the writings of his experts in systematics. Having no independent first-hand knowledge of the careful and long-term study of natural populations required for research in systematics, he accepted from those who had studied natural populations that geographical races and closely allied species most frequently differed by nonadaptive characters.

I am not arguing that the systematists upon whom Wright relied were justified in concluding that evolution at the lower taxonomic levels was primarily nonadaptive. Indeed, these systematists suggested a confusing plethora of mechanisms to account for nonadaptive evolution, from geographical isolation to direct action of the environment to random mutation to sampling effects in small populations derived from larger ones. But I can say that Wright accepted their conclusion that there was much nonadaptive differentiation between geographical races and closely allied species.

Furthermore, many evolutionists were influenced by these same systematists to believe that closely related species frequently differed by nonadaptive characters. Haldane, for example, argued in *The Causes of Evolution* (1932) that natural selection was the primary mechanism of evolution in nature, yet added this caveat: But when we have pushed our analysis as far as possible, there is no doubt that innumerable characters show no sign of possessing selective value, and, moreover, these are exactly the characters which enable a taxonomist to distinguish one species from another. This has led many able zoologists and botanists to give up Darwinism" (Haldane 1932, pp. 113–14).

When the ecologist Charles Elton published his first book, *Animal Ecology* (1927), it appeared in a series edited by Julian Huxley and had a glowing introduction by him. In the book, Elton, who had worked with Richards as well as Huxley, summarized with approval the Richards and Robson article that had appeared in *Nature* (1926):

> The gist of their conclusions is that very closely allied species practically never differ in characters which can by any stretch of the imagination be called adaptive. If natural selection exercises any important influence upon the divergence of species, we should expect to find that the characters separating species would in many cases be of obvious survival value. But the odd thing is that although the charac-

ters which distinguish genera or distantly allied species from one another are often obviously adaptive, those separating closely allied
species are nearly always quite trivial and apparently meaningless.
(Elton 1927, 184)

Huxley himself, certainly a champion of Darwin's idea of evolution by natural selection, accepted the argument of Richards and Robson that many of the
differences between closely allied species were nonadaptive. Indeed, Huxley's work on allometry in the 1920s and early 1930s (summarized in Huxley
1932) provided explicit proof of correlation that could lead to the evolution of
nonadaptive characters.

The most telling proof of the seriousness with which the arguments of
Richards and Robson were taken comes from Ford. Fisher and Ford were perhaps the two most influential and vocal adherents of a purely selectionist view
of the evolutionary process during the 1930s. Fisher's theoretical view rejected the significance of all mechanisms of evolution with the sole exception
of natural selection. Ford's background and research on natural populations
led him also to become a staunch selectionist (Ford 1980). In 1931 he published the first edition of his widely read book, *Mendelism and Evolution,*
which by 1960 had gone through a total of seven editions. After arguing that
the effects of genes were probably multiple, Ford stated:

This consideration may throw some light on the nature of the characters which separate local races and closely allied species. That these
are sometimes entirely non-adaptive has been demonstrated, we believe successfully, by Richards and Robson (1926). It is evident that
certain genes which either initially or ultimately have beneficial effects may at the same time produce characters of a non-adaptive
type, which will therefore be established with them. Such characters
may sometimes serve most easily to distinguish different races or
species; indeed, they may be the only ones ordinarily available,
when the advantages with which they are associated are of a physiological nature. Further, it may happen that the chain of reactions
which a gene sets going is of advantage, while the end-product to
which this gives rise, say a character in a juvenile or the adult stage,
is of no adaptive significance. . . .
    J. S. Huxley has pointed out another way in which nonadaptive
specific differences may arise. For he has shown that changes in absolute body-size, in themselves probably adaptive, may automatically lead to disproportionate growth in a variety of structures, such
as horns and antlers in Mammalia and the appendages in Arthropoda.
The effects so produced may be very striking, but, as they are the inevitable result of alteration in size, they can rarely have an adaptive
significance.
    It is not perhaps always recognized how complete has been the
demonstration provided by the above authors that the characters

available to systematists for the separation of allied species may be of a wholly non-adaptive kind. (Ford 1931, 78–79)

Of course, Ford went on to argue that one was not justified in assuming that the nonadaptive characters were produced by a process of nonadaptive evolutionary change—quite the contrary. The nonadaptive characters used by systematists were simply the necessary, correlated by-products of adaptive evolution, not the result of a nonadaptive mechanism such as random drift. Ford's statements about nonadaptive differences between local races and allied species quoted above remained unchanged through the fifth edition of *Mendelism and Evolution* (1949).

The evidence is overwhelming that Sewall Wright and many other evolutionists accepted in the early 1930s the view of prominent systematists that geographical races and closely allied species differed primarily by nonadaptive characters by which such races and species were most easily distinguished.

Wright's Random Genetic Drift and Systematics

While it is true that Wright was much influenced by systematists, he in turn influenced their thinking on the mechanism of nonadaptive change. As mentioned earlier, the systematists who agreed that geographical races and closely allied species differed by primarily nonadaptive characters proposed a confusing and unconvincing array of mechanisms to account for such apparently nonadaptive evolution. The two favored explanations were correlation of parts (including Huxley's allometry) and sampling from larger populations (Gulick's favorite explanation). Elton, who had studied systematically the subject of fluctuating population sizes, argued strongly for sampling effect as the cause of nonadaptive differentiation:

> Many animals periodically undergo rapid increase with practically no checks at all. In fact, the struggle for existence sometimes tends to disappear almost entirely. During the expansion in numbers from a minimum, almost every animal survives, or at any rate a very high proportion of them do so, and an immeasurably larger number survives than when the population remains constant. If therefore a heritable variation were to occur in the small nucleus of animals left at a minimum of numbers, it would spread very quickly and automatically, so that a very large proportion of numbers of individuals would possess it when the species had regained its normal numbers. In this way it would be possible for non-adaptive (indifferent) characters to spread in the population, and we should have a partial explanation of the puzzling facts about closely allied species, and of the existence of so many apparently non-adaptive characters in animals. (Elton 1927, 187)

The major problem with Elton's scenario was that even very small samples of most natural populations were not very different from the parent populations, nor was it clear why a small isolated population would diverge even further in nonadaptive characters from its parent population. The problem with the allometry argument was that it could be proved or disproved only with great difficulty, if at all. In light of this dilemma in explaining the origin of nonadaptive differentiation, the impact of Wright's theory of random genetic drift upon systematists becomes more understandable.

The ten quotations given earlier from Wright's published papers for the years 1929–32 show clearly that Wright was suggesting isolation plus random genetic drift as the most plausible mechanism for nonadaptive differences at the lower taxonomic levels. Wright's concept of random drift was unquestionably a conceptual advance over the mechanisms offered previously by systematists. Wagner, Gulick, D. S. Jordan, Kellogg, Elton, and many others had emphasized the importance of geographical isolation or fluctuation of population size in producing nonadaptive differentiation by isolation of a nonrepresentative sample of the parent population. Yet most of these same systematists were uneasy about the power of such sampling effects to produce the observed results. It was also true that the Hagedoorns (1921) and Fisher (1922) had pointed out the possibility of random genetic drift. But it was Wright who convincingly connected random drift with observable nonadaptive differentiation in nature. His concept of random genetic drift, a statistical consequence of enforced inbreeding in a small population, was precisely the mechanism required to produce the observed differentiation. Even in the absence of initial nonrepresentative sampling in the isolated population, random drift could cause increased differentiation generation after generation, as long as the population remained small enough—up to the limit of total homozygosity, when, of course, differentiation must cease.

Wright's contention in 1929–32 that random genetic drift could explain nonadaptive differentiation as observed by systematists did not, in fact, fit well with his general view of evolution. He originally saw random drift as valuable in creating novel genic interaction systems in highly inbred domestic populations, thus enabling the breeders to select the most desirable combinations. Random genetic drift was, therefore, crucial in producing variation, not in providing an evolutionary mechanism as an alternative to selection. Indeed, as Wright had observed in his experimental populations of highly inbred strains of guinea pigs, the effect of random drift in a population of very small effective size was nearly always deleterious to the fitness of the population. Very small isolated populations were, according to Wright in 1931 and 1932, doomed to extinction.

Wright's shifting balance theory of evolution emphasized large but subdivided populations as the key to rapid adaptive evolution in nature. There the balance between random drift, selection, mutation, migration, and other factors produced the optimum conditions for adaptive advance. From the earliest attempt Wright made to devise a theory of evolution in nature, he stressed the proper balance of all the variables:

> I have attempted to form a judgment as to the conditions for evolution based on the statistical consequences of Mendelian heredity. The most general conclusion is that evolution depends on a certain balance among its factors. There must be gene mutation, but an excessive rate gives an array of freaks, not evolution; there must be selection, but too severe a process destroys the field of variability, and thus the basis for further advance; prevalence of local inbreeding within a species has extremely important evolutionary consequences, but too close inbreeding leads merely to extinction. A certain amount of crossbreeding is favorable but not too much. In this dependence on balance the species is like a living organism. At all levels of organization life depends on the maintenance of a certain balance among its factors. (Wright 70, p. 365)

Why then did Wright in the years 1929–32 emphasize so strongly the role of random drift in the evolution of geographical races and species, when at the same time, his general shifting balance theory of evolution in nature did not require an imbalance favoring random drift as a primary variable at the racial and species level?

The answer has already been provided. Wright thought from what he read in the systematics literature that the differences between geographical races and allied species were primarily nonadaptive, so he interpreted the balance in his balance theory to favor random drift at the lower taxonomic levels. Wright's theory of evolution would actually have fit better with a more adaptive view of evolution than systematists provided in the early 1930s.

Even though Wright's connection of systematists' observations and his concept of random genetic drift may not have harmonized well with his own shifting balance theory of evolution in nature, the connection certainly solved for systematists and evolutionists, in general, the problem of the unknown primary mechanism of the evolution of nonadaptive differences between related races and species. After 1932 Wright's name and the concept of random drift were invoked constantly whenever the evolution of nonadaptive characters came under discussion. Proof for this assertion will be provided in the beginning of chapter 12.

Wright, Fisher, and the Evolution of Dominance Revisited

Fisher's original paper on the evolution of dominance and Wright's alternative explanation for the origin of dominance, combined with their published interchanges, stimulated a lively debate that has not been wholly resolved even at the present time. (For a fascinating recent review of the evolution of dominance controversy read chapter 15, "The Evolution of Dominance," in

Wright 198, pp. 498–526.) Haldane's paper (Haldane 1930) has already been mentioned as raising the existence of polymorphic populations as a problem for Fisher's theory of the evolution of dominance. Fisher responded by arguing that stable polymorphisms were ideal for his idea of the evolution of dominance because the heterozygotes, upon which his proposed selective pressure acted, were far more plentiful than heterozygotes in the case of rare deleterious mutations.

At this point Ford entered into the debate. Julian Huxley had brought Ford and Fisher together in 1923, and they became very close friends; the scientific work of each was in some ways dependent upon the work of the other. Ford had strong backgrounds in field natural history and the physiology of gene action, both areas of weakness in Fisher's education; Ford, on the other hand, had no significant abilities in quantitative analysis. Both were staunch Darwinians from very early on. Together they constituted a powerful team in the analysis of evolutionary problems. (For a more detailed account of Ford's development, see his comments in Mayr and Provine 1980, 334–42.)

Ford was greatly impressed by Fisher's papers on the evolution of mimicry (1927) and the evolution of dominance (1928a, 1928b) and often defended Fisher's extreme selectionist view of evolution in nature with whatever evidence he could find in natural populations. Ford entered into the evolution-of-dominance debate with a brief paper in *American Naturalist* (by now obviously the locus of choice for views on the evolution of dominance), published in November–December 1930. He argued that Wright had merely resuscitated Bateson's old presence and absence hypothesis, to which Ford raised a series of objections—reverse mutation, dominance of white in some cases, deletions that lead to a dominant mutation, and others. Most significant, in light of his later research Ford stated that "my friend Dr. Fisher" had enlightened him with regard to the problem Haldane had raised about polymorphisms. Ford repeated, with some examples from nature, Fisher's thesis about polymorphism and dominance already discussed earlier. Ford would soon focus a major part of his energies on the investigation of stable polymorphisms in natural populations. (For the results of a decade of this work see Ford 1940.)

Fisher and Ford were clearly closely allied in their opposition to Wright's ideas about the origin of dominance. This alliance of opposition to Wright's views not only on dominance but on the whole process of evolution in nature was to continue until Fisher's death in 1962, when Ford actively continued the tradition on his own. The picture, of course, is symmetrical. Wright was every bit as opposed to their view of the mechanism of evolution as they were to his, and he actively participated in the debate, defending his own views and attacking theirs. Fisher's position was in my opinion much strengthened by Ford's collaboration and friendship. One reason why Wright readily agreed to actively collaborate with Dobzhansky's research on natural populations of *Drosophila pseudoobscura* was to have a more symmetrical theoretical quantitative evolutionist/field naturalist combination on both sides of the Atlantic.

Returning to the debate over the evolution of dominance, C. R. Plunkett and H. J. Muller independently expressed similar views in presentations at the International Congress of Genetics at Ithaca in 1932 (Plunkett 1932a, 1932b; Muller 1932, 237–42). Plunkett had observed that wild type homozygotes in *Drosophila* were far more stable in response to both environmental or genetic perturbation than mutants. This, he said, resulted in "a generally smaller variability of wild-type as compared with mutant characters and, incidentally, to a usual, though not necessarily universal, dominance of wild-type to mutant allelomorphs" (Plunkett 1932b, 84). Plunkett argued that dominance thus resulted from selection, "but, in general, merely an incidental result of selection in the direction of enhancement and stabilization of favorable characters" (p. 85), thus diminishing the significance of Fisher's idea of selection of the genetic modifiers of heterozygotes of rare deleterious mutations in the evolution of dominance. Muller agreed with the essentials of Plunkett's argument.

Ever since working on his thesis research, Wright had been keenly interested in the problem of dominance. In the short time since Wright's interchange with Fisher in 1929, others had moved in with relatively well-developed theories of the evolution of dominance. Thus when Richard Goldschmidt invited Wright to participate in a session on physiological genetics at the Symposium of the American Society of Naturalists at the Century of Progress meeting of the AAAS in Chicago in June 1933, Wright elected to present his views in a substantive paper entitled "Physiological and Evolutionary Theories of Dominance." Wright extended the paper for publication but had to cut it down somewhat when the editor Jacques Cattell objected to its length. The paper appeared in *American Naturalist* in January–February 1934.

The body of the paper was a detailed synthesis of Wright's well-developed views on the physiology of the gene, with the problem of dominance and its origin. He concluded that this deeply biological conception of dominance led him to agree basically with the Plunkett/Muller model. In both the introductory and concluding sections Wright defended his views against the criticisms of Fisher and Ford. He reviewed again in part the development of the argument with Fisher and again concluded that Fisher's theory of the evolution of dominance by selection of the heterozygotes of rare deleterious mutations was untenable. He did not, however, even mention Fisher's thesis, supported by Ford, that the evolution of dominance would proceed rapidly under Fisher's scheme with stable polymorphisms. Wright, it should be recalled, had agreed with Fisher on this thesis in his letter to Fisher of October 30, 1930, and in his first reply to Fisher's theory of dominance (Wright 56, p. 277).

In a particularly revealing section, Wright told why the argument with

Fisher about the evolution of dominance had a greater significance than at first meets the eye:

> From the standpoint of the theory of dominance it may seem of little importance which mechanism is accepted if it be granted that selection has been an important factor. This is not at all the case, however, with the implication of Fisher's and Plunkett's selection theories, for the theory of evolution. Fisher used the observed frequency of dominance as evidence for his conception of evolution as a process under complete control of selection pressure, however small the magnitude of the latter.
>
> My interest in his theory of dominance was based in part on the fact that I had reached a very different conception of evolution (1931) and one to which his theory of dominance seemed fatal if correct. As I saw it, selection could exercise only a loose control over the momentary evolutionary trend of populations. A large part of the differentiation of local races and even of species was held to be due to the cumulative effects of accidents of sampling in populations of limited size. Adaptive advance was attributed more to intergroup than intragroup selection. (Pp. 50–51)

I think Wright is correct in saying that what really was at stake in the argument with Fisher over the evolution of dominance was not the particular problem of dominance but their differing conceptions of evolution. If either was correct on the evolution of dominance, it was perceived by the other as fatal to his entire conception of evolution. No wonder each defended his theory of the evolution of dominance with such vigor. There are many reasons why Fisher might have been angered by Wright's paper. He saw that his whole conception of evolution was under attack. Perhaps he was offended that Wright had raised again in print an argument Fisher thought they should settle privately by correspondence. From Fisher's reply it appears that he felt slighted by Wright's account of the history of their interchange and that he disliked Wright agreeing with all his calculations but still disagreeing with the theory on other grounds. And he pointed out that Wright had not raised the issue of stable polymorphisms, the most favorable case for Fisher's theory.

There is also the factor of Fisher's personality. He was always a contentious and difficult person who was constantly embroiled in one scientific fight or another, pouring out acrid criticism at scientific meetings and in print (Savage 1976; Box 1978; Kruskal 1980). He could also be an endearing and loyal friend, an extraordinary scientific helper to persons such as Ford. But Fisher did not take criticism lightly, particularly if he thought the criticism ill founded or repeated again after proper refutation.

For whatever reasons, Fisher unleashed a harsh polemic against Wright's paper. The reply was sarcastic throughout and ended with this paragraph:

> Professor Wright's recent allusion to the subject evolution of dominance was but a preface to his own interesting speculations on the

physiological causation of dominance. It had, perhaps, achieved its purpose when he could write, "If this hypothesis is untenable what alternative is there?" If, however, Professor Wright's views can only be made plausible, by the exclusion of all alternatives, he must find other objections to the selection theory more weighty than those he has revived. (Fisher 1934, 374)

Wright was genuinely surprised and dismayed by Fisher's performance.

In 1959, when Ernst Mayr questioned (in a spirit of provoking substantive discussion) the significance of the contributions of Fisher, Haldane, and Wright to evolutionary theory, Haldane later said in "Defense of Beanbag Genetics" (1964) that "Fisher is dead, but when alive preferred attack to defense. Wright is one of the gentlest men I have ever met, and if he defends himself, will not counterattack. This leaves me to hold the fort" (p. 344). Haldane is right about Fisher but wrong about Wright. Wright has always been in person mild mannered and quiet, but from the beginning, he has defended his views vigorously enough that his defenses could easily be taken for counterattacks. For example, Wright answered Mayr's 1959 foray immediately (Wright 167) and continued to answer it over the years, up to and including 1984.

Wright did not allow Fisher's note to go unanswered. His reply in *American Naturalist* was a brief but vigorous defense that raised the question of why Fisher had not given proper credit to Wright for the statistical distributions of gene frequencies developed in the manuscript of the 1931 paper, a neglect that obviously rankled Wright. The closing paragraph of Wright's reply was conciliatory: "Finally in view of the tone of Dr. Fisher's note let me say here that I have the greatest admiration for his contributions to statistical genetics and biometry. If I have devoted considerable space to their criticism it is only because his views have seemed to me worthy of exhaustive study" (Wright 83, p. 564).

The mutually beneficial interchanges between Wright and Fisher had ended. They would never again be on friendly terms, never again corresponded, and never again have a pleasant conversation despite being thrown together on several occasions. They did, however, exchange reprints.

In 1928 and 1929, when Fisher and Wright were publishing their interchanges on the evolution of dominance, Fisher was only beginning to establish his reputation as a theoretical quantitative evolutionist and Wright had no such reputation. By the time of their 1934 interchange on the evolution of dominance, Fisher and Wright were unquestionably known as the world's best theoretical population geneticists.

I am sure Wright personally felt dismayed by the breakdown of all communication between them after 1934. But I am equally certain that the tension between them and their views of the process of evolution served as an enormous stimulus to both theoretical and field research in evolutionary biology in the ensuing decades.

Evolutionary Theory, 1934–1939

As I have shown earlier in this chapter, Wright was faced with a fundamental problem regarding the relationship of his quantitative and qualitative views of the evolutionary process. The statistical distribution of genes that he had reached in the 1931 and 1932 papers included a selection term, terms for mutation and migration in both directions, and ways to account for effective population size. But the distribution was confined to one locus with two alleles, assumed one level of intermediate dominance, and had only one kind of selection included in the selection term; it did not, therefore, take account of the effects of interaction between loci. Wright's qualitative theory, on the other hand, assumed universal gene interaction, grades of dominance ranging from none to complete, frequency dependent selection, and a variety of population structures. In short, Wright's quantitative population genetics fell far short of his qualitative shifting balance theory of evolution in nature.

No one felt this problem more keenly than Wright himself. It was Haldane, however, who called part of the problem to the attention of evolutionists. In *The Causes of Evolution* he stated:

> Wright arrives at formulae analogous and often equivalent to those of Fisher for the distribution of gene ratios in populations under the simultaneous influences of selection, mutation, random survival, and migration. Unfortunately the type of selection considered is almost always one involving no dominance, *i.e.* in which (under the influence of selection alone) the values of [gene frequency q] in successive generations form a geometrical progression. I suspect that some of his most important theoretical conclusions would no longer hold if dominance were allowed for. This would greatly complicate the mathematical treatment, but I believe that it must be done before full weight can be given to Wright's results. (Haldane 1932, 212–13)

Wright certainly agreed with Haldane's point that the quantitative work had definite limitations. He consciously set for himself the agenda of bringing his statistical distribution closer to the level of the shifting balance theory. Indeed, one may view Wright's work in quantitative population genetics after 1932 into the 1980s as a sustained attempt to bring the mathematical analysis more closely in line with a qualitative theory that changed very little after 1932, at least as far as the mathematical analysis itself was concerned.

Wright made major progress on this agenda in the years 1935–39; he published a total of nine papers on theoretical population genetics during these years (86,87,91,92,94,95,96,97,99). I will attempt in this section to indicate just how he extended the quantitative analysis as it stood in 1931–32.

Gene interaction involving many loci was a fiendishly difficult problem that simply could not be attacked head on. Even the largest computers today cannot handle the interactions of thousands of genes, or at least such calculations would require astronomical periods of time. Wright chose to develop further an idea already suggested and worked out in some detail by Fisher (1930a, 104–11) and Haldane (1932, 196–98). The basic idea is apparent in the titles of two of Wright's 1935 papers (86, 87), "The Analysis of Variance and the Correlations between Relatives with Respect to Deviations from an Optimum" and "Evolution in Populations in Approximate Equilibrium" (Wright originally wrote these papers as one). The trick was to examine measurable characters for which the selective optimum was at or near the mean and to quantitatively assess the drop in selective values according to the square of the deviations above or below the mean. This conception guaranteed the maximum amount of interaction with respect to selection of the character. Yet so far as the genetic determination of the character itself was concerned, Wright could conveniently treat that quantitatively as wholly additive and linear, enabling him to bring in multiple factors without making the analysis too complex for mathematical treatment.

In the first paper Wright applied this approach to the correlations between relatives, and in the second to the problem of evolution in populations. Only in the second paper do the really important implications of the approach become apparent. Wright shows that the concept of average adaptive value of a character determined by a set (of any size) of gene frequencies can be used to incorporate the idea of intermediate selective optima and also to incorporate different assumptions about the degree of dominance. But average adaptive value was precisely the quantity (when extended to the whole population instead of to a single character) that Wright had decided to use in constructing his surface of selective values. The surface of selective values, in turn, was intimately tied to Wright's qualitative conceptions of the evolutionary process, as his 1932 paper demonstrated so dramatically. The tantalizing prospect held out by this paper was that Wright could possibly put the "average adaptive value" term into the statistical distribution of genes, thus in one major fell swoop tying much closer together his quantitative and qualitative views of the evolutionary process. Given Haldane's criticism in 1932, it should come as no surprise that Wright sent these two papers to him for publication in *Journal of Genetics*. (Haldane and Punnett were at that time joint editors.)

In 1937 Wright took the step of incorporating in two papers average adaptive value of a population ($\overline{W}$) into his statistical distribution of genes (91 was an abstract of 92, but 91 actually contained the most general version of the statistical distribution, including the migration factors). Mutation and migration had not been major problems in the earlier version of the distribution, so Wright replaced the limited selection term (e^{4Nsq}) of the earlier version with the vastly richer term $\overline{W}^{2N}$ in the new distribution. Since $\overline{W}$ was the surface of selective values, it could incorporate any degree of dominance, any amount of interaction, and any number of loci or alleles. The cost of all this added flexibility was of course the greater abstractness of the term $\overline{W}$. Actually calculating it for a natural population for more than one locus with two alleles was practically impossible. But this was theoretical population genetics living up to its name. Theoretical physicists did this sort of maneuver constantly.

The version of the new distribution for a single locus with two alleles thus became

$$\phi(qi) = C\overline{W}^{2N} q_i{}^{4Nv_i-1}(1 - q_i)^{4Nu_i-1}$$

and the version for multiple genes was

$$\phi(q_1, q_2, \ldots, q_n) = C\overline{W}^{2N} \prod_{i=1}^{n} q^{4Nv_i-1}(1 - qi)^{4nu_i-1}$$

Factors for migration could be of course added to the mutation terms.

This version of the statistical distribution does not look on inspection much different from the 1931 version, but I want to strongly emphasize that the advance is actually dramatic. Now Wright had a version that related directly to his shifting balance theory, something the 1931 version lacked completely. Wright was justly very proud of this 1937 paper, which he published under the title "The Distribution of Gene Frequencies in Populations" in *Proceedings of the National Academy of Sciences*.

It is true, however, that Wright did not help the reader to understand the importance of what he had accomplished in the paper. He sent a copy to Dobzhansky, who tried to read it but could not. I suspect that to his dying day Dobzhansky never understood the essential modification of the statistical distribution that Wright made in 1937.

The three papers, "The Distribution of Gene Frequencies under Irreversible Mutation" (95), "The Distribution of Gene Frequencies in Populations of Polyploids" (96), and "The Distribution of Self-Sterility Alleles in Populations" (97), were all attempts to assess the consequences of various assumptions upon the statistical distribution of the 1937 paper. The last of these stemmed from data collected by Sterling Emerson about the distributions of self-sterility alleles in *Oenothera organensis* (Emerson, 1938). Fisher and Wright later clashed over the interpretation of Emerson's data, which I will discuss in some detail in chapter 13.

A central feature of Wright's shifting balance theory of evolution in nature was its emphasis upon the breeding structure of the population. The evolutionary history of a population might vary dramatically, depending upon its breeding structure. Thus in order to quantify his shifting balance theory, Wright had to reveal the effects of various breeding structures upon the statistical distribution of genes in the population. His first paper on this topic came in a very brief note in *Science* in 1938. Entitled "Size of Population and Breeding Structure in Relation to Evolution" (Wright 94), this paper briefly pointed out that population size and breeding structure were crucially important variables for any clear understanding of evolution in natural populations.

Here for the first time Wright introduced the idea of isolation by distance, which he would develop into a major theoretical model in the 1940s. This little paper was certainly a sign of the direction that Wright's theoretical and collaborative work would take in the 1940s.

The final paper in this series of nine, "Statistical Genetics in Relation to Evolution" (99) was an extensive summary of Wright's evolutionary theory, both qualitative and quantitative. Georges Teissier had encouraged Wright in the mid-1930s to write a general essay on his quantitative theory of evolution for a series on biometry and statistical biology that Teissier edited for Hermann & Cie, publishers in Paris. This paper was an appropriate summary for a decade of publications on evolutionary theory. The paper contained substantial sections on the distribution of gene frequencies (complete with a clear derivation of the 1937 version of the statistical distribution of gene frequencies), effects of changes in conditions upon the distribution, breeding structure of populations (including effective size of populations, partial isolation, and local differences in selection), frequencies of interacting factors, and the biometric properties of populations. The longest section of all was a twenty-four-page qualitative discussion of Wright's shifting balance theory of evolution. In this section Wright still strongly emphasized the role of random genetic drift in accounting for observed nonadaptive differentiation at low taxonomic levels. He also stressed that under other conditions (large population size, etc.) the shifting balance process could lead to adaptive change at these same taxonomic levels.

I assess this paper as being the clearest, most accessible, and most comprehensive statement of Wright's shifting balance theory that he published until the publication of his magnum opus beginning in 1968. With the exception of the greater emphasis upon the nonadaptive differentiation that Wright would later downplay, this paper points directly toward the basic contents of the magnum opus. But as it turned out, this was also one of the least read of Wright's papers. World War II broke out before Wright himself got the handful of reprints that finally arrived several years late. I am aware of no evolutionist other than James F. Crow in the United States or England who has even read the paper. In France (the paper was published in English) the only person I know was influenced by it was Maxime Lamotte (see chapter 12).

If Wright had published this paper as a small and easily accessible monograph in the United States, it would have been the most widely read and perhaps most influential of all his published work on evolutionary biology before the appearance of his magnum opus. The Hermann & Cie monograph is now so rare I have been unable to locate a copy for my own collection.

Wrinkles in the Surface of Selective Value

I have already stated that Wright's concept of a "fitness" surface, or a surface of selective values, was one of his single most influential contributions to modern evolutionary biology. Few evolutionary biologists would question this assertion because most textbooks on evolution since the first edition (1937) of Dobzhansky's *Genetics and the Origin of Species* (with the notable exceptions of those of Mayr, Fisher, Ford, Cain, and Sheppard) have included representations of the famous fitness surfaces. Hundreds of published papers refer to the surfaces. I travel to many universities and centers where population genetics is taught, and at most places graduate students talk as if natural populations lived on fitness surfaces rather than on the earth's surface in ecological settings. These surfaces are generally referred to as "Wright's fitness surface," "Wright's adaptive landscape," "Wright's surface of selective values," or some similar term invoking Wright's name. Allegiance to the surfaces is intense. Whenever I venture to question the value of the surfaces in evolutionary biology, as I have on a few occasions, the reaction is vigorously negative. Wright also has been keenly attached to his fitness surfaces since 1932.

Wright's concept of the surface of selective value has become famous because of its apparently great heuristic value in conveying graphically the relationship between organisms, mechanisms of evolution, and adaptation. Despite its great attractiveness and apparent ease of interpretation, the surface of selective value is one of Wright's most confusing and misunderstood contributions to evolutionary biology. Wright himself has contributed to the confusion.

In the first place, Wright contributed not just one conception of his surface but two very different versions that mathematically were wholly incompatible. Worse, he frequently treated these two incommensurable conceptions as if they were together a unitary conception, and he switched back and forth between them often. Finally, when Simpson proposed and attributed to Wright a third version of the adaptive surface that mathematically was wholly incompatible with his two versions, Wright objected not at all, leading others to believe that this third version was just another way of expressing Wright's concept of the fitness surface.

A brief word of introduction about functions and surfaces may help here. With some restrictions, most functions of independent variables determine a surface. Consider a simple function such as Boyle's law of gases, $T = kPV$, where T is the absolute temperature, P the pressure, V the volume, and k a constant determined by the gas being used. In this nice example, where all relationships are linear (we will ignore such inconveniences as the gas becoming a liquid), T can be visualized as a three-dimensional surface, the axes of which are gradations of pressure, volume, and finally temperature. One can visualize this scenario easily by thinking of the plane determined by P and V as being on the floor; temperature then becomes the height off the floor. For each pair of values of P and V, there is a corresponding point on the temperature surface, which is three dimensional. There are some intuitive advantages to visualizing temperature as a function of pressure and volume in this way. For example, at a given temperature the surface reduces to a line figure on the temperature plane, an isothermal, meaning all those combinations of pressure and volume possible at one temperature. The temperature surface is an intuitively satisfying way of thinking about the relationship of temperature with pressure and volume. I emphasize, however, that the temperature surface is simply a mathematical consequence of the equation relating temperature to pressure and volume, and it adds nothing to what is already in the equation. The value of the temperature surface is purely heuristic. In this case, its value is apparent.

Any number of independent variables can be included in the function, and the surface has correspondingly more dimensions. There is basically no problem manipulating finite dimensional vector or manifold spaces. Some mathematicians love to do this. When the number of dimensions rises above three, the surface becomes more difficult to visualize intuitively. This can be a problem because the only motivation for constructing the surface in the first place is to aid the intuitive understanding of the relationships embodied in the original equation. A surface of several hundred thousand dimensions, or even ten, does not have the easily visualized contours of a three-dimensional surface. There are two ways of escaping this difficulty. One is to project the n-dimensional surface on a mere three dimensions, in mathematically the same way that we collapsed the three-dimensional temperature surface above onto a two-dimensional plane, yielding an isothermal. The projection then becomes as easily visualized as any three-dimensional surface, but inferring the shape of a surface of thousands of dimensions from its reflection in a three-dimensional surface presents rather formidable obstacles. The other way to escape the difficulty of imagining n-dimensional surfaces is to focus upon very small parts of them, determining curvature and slope, and thus whether this part of the surface is at a local maximum, minimum, saddle point, or whatever. The surfaces generated by very nice linear equations can be shown to be in turn nicely continuous, so even if the n-dimensional surfaces cannot be visualized directly, at least they can be grasped to some extent locally if only by analogy with similar properties in three dimensions or less.

Yet even with nice linear equations, the heuristic value of generating the surface may be limited. It may be easier for someone to understand that some quantity is a specified function of ten variables with certain maxima or minima than it is to help the understanding along by picturing the contours of an eleven-dimensional surface. With equations that involve interaction between what were formerly independent variables so that the equation no longer is linear, the surface becomes far more difficult to imagine. It may or may not be differentiable (may have discontinuities) and may well be simply unintelligible. When this happens, the surface becomes heuristically completely useless. The recourse in such a case may be to approximate the nonlinear equation with one that is linear and which therefore has a more interpretable surface. In other words, we can simplify the equation in order to generate a more understandable surface that can in turn help one to visualize better the relationships in the equation. This can be a vicious circle. It may be better to concentrate on understanding the relationships of the variables in the equation

than to spend time simplifying the equation and generating the surface associated with the simplified equation in hopes that it will intuitively clarify the relationships in the original equation. With this discussion in mind, let us turn
back to Wright's two conceptions of the surfaces of selective value.

In the original published version in 1932, as I have shown earlier in this
chapter, Wright was absolutely clear that the surface was generated by the
field of gene combinations, with one fitness value assigned to each genotype.
Each gene combination represented the genotype of a single organism. Thus
each point on the surface was the adaptive value of the genotype of a single
organism, making it the genotypic individual fitness surface. There were as
many dimensions to the surface as there were gene combinations, plus one
for adaptive value. Since the number of gene combinations in a simple
genome is probably larger than the number of electrons in the universe, these
surfaces had very many dimensions, as Wright was well aware. A whole interbreeding population occupied an area of the adaptive landscape determined
by all the actual genotypes represented in the population, augmented by the
vastly greater number of possible genotypes in the population. This scenario
has a direct intuitive appeal, resting upon the notion that each genotype
should have a reasonably well-defined adaptive value or fitness, even if it
could not be determined experimentally. (Actually, in a sexually breeding
population it is not so clear that each genotype should have a well-defined
adaptive value since each genotype occurs only once and parents cannot pass
their same genotype to their offspring.) Wright's famous diagram of six
specified conditions in the 1932 paper was properly labeled "field of gene
combinations occupied by a population within the general field of possible
combinations" (Wright 70, p. 361). Exactly the same diagrams and label appeared in the 1939 Hermann & Cie monograph (although I will soon show
that the situation was more complicated in that publication).

The first and most important thing to notice about Wright's first published version of his fitness "surface" is that his construction does not in fact
produce a continuous surface at all. Each axis is simply a gene combination;
there are no gradations along the axis. There is no indication of what the units
along the axis might be or where along the axis the gene combination should
be placed. No intelligible surface can be generated by this procedure. By no
stretch of the imagination can Wright's famous diagrams of the 1932 paper be
constructed by his method of utilizing gene combinations. The diagrams represent a nicely continuous surface of selective value of individual genotypic
combinations; the method Wright used to generate this surface actually yields
an unintelligible result. Thus the famous diagrams of Wright's 1932 paper,
certainly the most popular of all graphic representations of evolutionary biology in the twentieth century, are meaningless in any precise sense.

So far as I can determine, the only difficulty that Wright himself had with
this version of the fitness surface was that it required such a great number of
dimensions, one for each possible genetic combination. For illustration of the
shifting balance theory of evolution, Wright always in his publications relied

upon this original version of his diagrams. Not until the final months (spring 1985) of revising this book did I come to the realization that this version of the fitness surface did not make sense. When I spoke with Wright about the problem, he thought it over for several days, concluded that my objection was justified, and suggested that the only way he could see to save something of his original version of the individual fitness surface was to use continously varying phenotypic characters as the axes of the surface. Wright reproduced his 1932 diagrams throughout his career but never realized until 1985 that the method he used for generating the surfaces in 1932 led to unintelligible results.

Wright was, however, uneasy about his original diagrams at some level. Compare, for example, his 1932 conception of the surface of adaptive value with the one that he presented in 1978 when he was describing the origin of his shifting balance theory of evolution (Wright 201, p. 1198). Here again were the same familiar diagrams. But now the label reads "hypothetical multidimensional field of gene frequencies (represented in two dimensions) with fitness contours," and the description in the text reads "if a selective value is attributed to each of the millions upon millions of possible sets of gene frequencies, the 'surface' of selective values will have many peaks, each separated from the neighboring ones by saddles." In this conception the surface is generated by axes, each of which represents the frequency of a single gene. Each set of gene frequencies represents one genetic constitution of an entire population, and this is graded for mean fitness. Thus each point on the surface represents an entire population, and the surface is one of populational fitness rather than of individual fitness as before. There are as many dimensions as there are possible sets of gene frequencies, plus one of course for selective value.

The immediate advantage of this way of constructing the fitness surface is that each axis, graded between 0 and 1 for gene frequency, is for practical purposes (in a large population) relatively continuous. Thus now the surface becomes, at least initially, intelligible and relatively continuous. But this conception of the fitness surface has difficulties also. Unlike a single genotype, a set of gene frequencies has no unique adaptive value assignable to it. A single set of gene frequencies might have a very high adaptive value in one array of gene combinations and a very low one in another array of gene combinations. Thus each set of gene frequencies is represented by an average adaptive value, assessed over all arrays of gene combinations possible with that set of gene frequencies. This is generally an astronomical number of combinations.

A further complication is that a natural population has only one set of gene frequencies at a time. Thus experimental determination of the genotypic fitness of a population can determine only one point, not a whole fitness surface or even a small area of it. Different sets of gene frequencies are required to generate more points on the fitness surface and thus give it contours.

Given that Wright did not see the difficulties with his 1932 version of the fitness surfaces, why, in the face of the difficulties just pointed out, did he change the parameters of his surface of selective values to gene frequencies from gene combinations? The enormous appeal of the "sets of gene frequencies" version becomes clear in light of the developments in Wright's quantitative theory of evolution described in the previous section. Wright's strong desire was to relate elements of his qualitative theory of evolution, such as interaction effects, grades of dominance, and multiple factors, to the statistical distribution of genes. The advantages of a fitness surface generated from sets of gene frequencies instead of gene combinations is obvious from the requirements of a statistical distribution of genes. The fitness surface was now simply the surface of $\overline{W}$, which was the key variable in Wright's statistical distribution of gene frequencies in a population.

The questions now become, how and when did Wright get the idea for this new conception of the surface of selective values, and what was their relationship to each other? One plausible answer to the first question is that Wright got the idea from Haldane's 1931 paper on "metastable populations," where Haldane envisioned a population with m genes as being situated on a hypercube of m dimensions. Support for this suggestion comes from Wright's 1935 paper "Evolution in Populations in Approximate Equilibrium," in which Wright first mentions in a publication the idea of a fitness surface based upon sets of gene frequencies. When Wright outlined the idea, he referenced Haldane's 1931 paper (Wright 87, p. 257). But I have already presented incontrovertible evidence that Wright had the idea of basing the surface of selective values upon sets of gene frequencies well before he had read Haldane's paper. In Wright's letter to Fisher of February 3, 1931, is found Wright's first known attempt to express the idea of a surface of selective values: "Think of the field of visible joint frequencies of all genes as spread out in a multidimensional space. Add another dimension measuring degree of fitness." This is a fitness surface with axes of gene frequencies, not individual genotypes.

Then why in his first published account of the fitness surface idea in 1932 did Wright switch to the individual fitness surface? He wanted the fitness surface to express as clearly as possible the qualitative aspect of the shifting balance theory of evolution, his assignment for the paper at the international congress. Only an individual fitness surface could show a population that occupied an area around a selective peak contracting under selection or expanding under mutation. Besides, the direct intuitive appeal of each individual in the population being a point on the surface enabled one to think of the population of organisms as being a corresponding population on the fitness surface.

On the other hand, the advantages of relating directly the statistical distribution of genes to the populational fitness surface determined by sets of gene frequencies were great. Faced with the problem of which version of the fitness surface to use, Wright's solution was to use both of them, depending upon the needs at the time. When he was trying to illustrate his qualitative shifting balance theory of evolution, he tended to use the individual fitness surface, switching to the populational surface when he wanted to become quantitative and to relate his equations to the surface.

A revealing example of the two competing conceptions of fitness surfaces comes in the 1939 Hermann & Cie monograph. Here Wright presented both conceptions of the surface in detail and switched back and forth between them, at one point suggesting their equivalence: "The multidimensional surface of selective values for all possible genotypes and also the surface of mean selective values for all possible populations should be of a very rugged character, with innumerable peaks at different heights, connected by saddles of varying depth" (Wright 99, p. 42). Although the famous diagrams were labeled "field of gene combinations," and each of the six diagrams were explained in detail in the paper, Wright clearly wished to relate the mean selective value of an entire population $\overline{W}$ in his statistical distribution to the fitness surface, so he also presented the populational surface. If there were any difficulty in going from one of these surfaces to the other, or in thinking that they were basically equivalent, Wright gave no indication.

Are Wright's two conceptions of the fitness surface somehow equivalent or interchangeable? One reaction I have frequently heard when I have raised this question with evolutionists is some version of the glib response, "Oh, all you have to do is sum up or integrate over the surface of individual selective values and that will give you the populational surface. No problem." But there really is a problem. One of Wright's two versions of the fitness surface is unintelligible, and even if one were able to escape this problem and put the gene combinations on continuous axes, the two versions would be mathematically wholly incompatible and incommensurable, and there would be no way to transform one into the other.

Take Wright's surface of individual selective values with its axes of gene combinations and integrate over the surface, or average out the fitness values, or take the average genotype, whichever you prefer. The result will be that the surface collapses into one point on the plane determined by the fitness axis and the last remaining genotypic axis, that generated by the average individual genotype. Trying to map individual genotypes on to the surface of populational fitness does not even begin to work. An individual genotype is unintelligible on a surface whose axes are sets of gene frequencies, characteristic of populations, not individuals.

In later years Wright spread confusion about his two versions of the fitness surfaces. In the texts of his papers and books, he almost consistently described the gene frequencies version of his fitness surface. But he did not give up the famous diagrams of the 1932 paper. In addition to the reproduction of the 1932 diagrams in Wright 201 cited above (where Wright labeled the surfaces incorrectly as being determined by sets of gene frequencies), he reproduced them in Wright 184, p. 257 (where the diagrams were similarly mislabeled) and in Wright 198, p. 452 (where the diagrams were correctly labeled as a "multidimensional field of gene combinations," but in the text referred to only as the $\overline{W}$ surface, of "mean selective values relative to the frequencies of genes at all loci"). In other words, Wright treated his two different conceptions of the fitness surfaces as equivalent and interchangeable, to the extent of using the same diagrams for one version as perfectly appropriate for the other.

Many evolutionists were confused by the switch Wright made from a "gene combinations" based surface to a "sets of gene frequencies" surface. Dobzhansky, through all three editions of *Genetics and the Origin of Species* (1937, 1941, 1951) and in *Genetics of the Evolutionary Process* (1970, 26–27), continued to describe the surface as based upon gene combinations. I think this was because, for him, the intuitive appeal of the 1932 diagrams was great; I doubt he ever even understood the quantitative advantages of the other version. In his *Introduction to Modern Genetics* (1939), Waddington, who had calculated and drawn the figure of a metastable population for Haldane's 1931 paper, described in the text Wright's surface as based upon sets of gene frequencies and in the caption to his reproduction of the famous diagrams as based upon the field of gene combinations (pp. 293–94).

In his influential 1944 book, *Tempo and Mode in Evolution,* Simpson introduced an apparently new wrinkle on the adaptive surface. In his own words:

> Wright (1931) [*sic*] has suggested a figure of speech and a pictorial representation that graphically portray the relationship between selection, structure, and adaptation. The field of possible structural variation is pictured as a landscape with hills and valleys, and the extent and directions of variation in a population can be represented by outlining an area and a shape on the field. Each elevation represents some particular adaptive optimum for the characters and groups under consideration, sharper and higher or broader and lower, according as the adaptation is more or less specific. The direction of positive selection is uphill, of negative selection downhill, and its intensity is proportional to the gradient. The surface may be represented in two dimensions by using contour lines as in topographic maps. (P. 89)

Here the axes are no longer gene combinations or gene frequencies but measurements of phenotypic characters, such as elements of body structure. It is unclear from the definition whether Simpson intended the surface to be one of individual fitness or of populational fitness, although the diagrams he used for illustration are clearly ones like those of Wright in 1932 that required interpretation in terms of individual fitness. This surface, with its axes of phenotypic measurement, cannot be transformed mathematically into either of Wright's two versions with their axes of genotypic combinations or sets of gene frequencies.

From his review of Simpson's book it is clear that Wright read the section dealing with fitness surfaces, but he did not say that his own versions differed substantially except that "there is a difference in point of view" and that Simpson had not carried the humpiness of the landscape far enough down the taxonomic level (Wright 119, p. 419). He did call Simpson's version "obviously related" to his own.

The subsequent history of the fitness surface idea in evolutionary biology is impossible to analyze here except for a few brief comments. All three versions have enjoyed a lively existence, and all have gone under the name of "Wright's fitness surfaces" or "Wright's adaptive landscape."

The first published version, that with parameters of gene combinations, has appeared over and over in the literature of evolutionary biology, frequently labeled with no parameters at all, or with gene frequencies as axes. Following Wright, this version is used primarily to illustrate qualitative descriptions of different mechanisms of evolution. So far as I am aware, no one has tried to push the quantitative analysis of this version any further than Wright took it in 1932, and it has remained in its original unintelligible state. But almost all evolutionary biologists are familiar with Wright's 1932 diagrams or redrawn versions of them. I have discovered that, when pressed, most evolutionary biologists who have used and taught the idea of fitness surfaces mistakenly think that Wright's 1932 diagrams have gene frequencies, rather than gene combinations, as parameters, although as I have already pointed out, these diagrams make sense only if the axes are gene combinations and the surface that of individual fitness.

Wright's version of the surface with gene frequencies as axes has been the one most used by population geneticists. The nice connection between $\overline{W}$ as a fitness surface, as connected with the statistical distribution of gene frequencies in a population, has been the great attraction. Also, constructing a fitness surface using data from field research or from simple hypothetical situations is relatively easy. A lot of points can be generated in theory, even on a fitness surface of only three dimensions, because each axis, being a gene frequency, can take on many values. Generally, the data from nature and from the hypothetical examples are chosen so that two loci each with two alleles are under consideration. Thus only two gene frequencies are required, and one can construct a fitness surface of three dimensions. The best-known example of this from a natural population comes from Lewontin and White (1960) in their study of the grasshopper *Moraba scurra*. They concluded that all ten populations tested were on saddle points, a surprising result. Of course the fitness surface depends upon the way that the mean fitness is calculated from the set of gene frequencies, and others using different assumptions (such as small amounts of inbreeding) found that with their altered surfaces the populations examined by Lewontin and White would be found all on selective peaks (Allard and Wehrhahn 1964; Wright 200, pp. 127–45), or about equally on peaks or saddle points (Turner 1972).

The quantitative theory of the gene frequencies version of the fitness surface has been considerably developed since Wright's presentation in 1935. One problem with the surface is that it becomes unintelligible if the fitnesses of the genes (the set of whose frequencies are graded to obtain the fitness on the surface) are frequency dependent. Wright conceived a partial way around this difficulty by proposing that instead of constructing a fitness surface, he would instead construct a fitness function that could take frequency dependent selection or other complications into account. (For summary see Wright 158, 193, pp. 131–62; but the surface of fitness function also becomes unintelligible under many circumstances.) Recently the gene frequencies version has been nicely developed by Curtsinger (1984a, 1984b) to include most of one-locus population genetics theory. Application of Curtsinger's approach to multiple loci, however, appears to face formidable obstacles.

Simpson's idea (which was earlier and, I am sure, independently conceived by Pearson in 1903) of a fitness surface based upon quantitative variations in phenotypic characters has been much developed in recent years by Lande, Arnold, and Wade (Lande 1976, 1979; Lande and Arnold 1983; Arnold, Wade, and Lande 1985). The great advantage of this surface is that phenotypic characters can be measured with some certainty, but the disadvantages are that the surface is systematically unrelated to population genetics theory based upon distributions of gene frequencies, and the genetic variances and covariances necessary to connect the trajectory of the population on the surface to genetic evolutionary change are not easily calculated in natural populations. Because this $\overline{W}$ surface is unrelated—except by unrealistic assumptions such as phenotypic characters being monotonically related to gene frequencies, with one gene per phenotypic character—to the $\overline{W}$ in population genetics theory, it should not be called a Wrightian fitness surface. Instead it should be known as a Pearsonian or Simpsonian surface, even though Simpson got the stimulation from Wright.

My evaluation of Wright's concept of surfaces of selective values, in addition to pointing out its great popularity in modern evolutionary biology, also sounds two cautionary notes. Much confusion has been associated with the parameters of the surfaces, both in terms of whether the axes are gene combinations, gene frequencies, phenotypic characters, or something else, and in terms of how the fitnesses are calculated. The evolutionary literature is literally jammed with representations of selective surfaces whose construction is obscure. Wright himself has contributed substantially to this confusion.

The other cautionary note concerns the heuristic value of the surfaces. It should give pause to consider that for over fifty years the majority of evolutionary biologists have believed Wright's 1932 diagrams of the adaptive landscape to be the most heuristically valuable diagrams in all of evolutionary biology, yet to discover that the surface as he conceived it is unintelligible. Although the surfaces have in very simple cases conveyed the ideas of multiple adaptive peaks and of the irreversibility of evolution, and occasionally have given a pictorial representation of the direction and intensity of selection on gene frequencies or phenotypic characters, evolutionary biologists have generally overestimated their heuristic value. The trend in modern evolutionary biology has been to move beyond extremely simplistic assumptions to more realistic ones. The assumptions that yield easily interpretable fitness surfaces (as in the examples usually presented) are generally far below the level of sophistication that evolutionists now achieve in other ways.

I would emphasize in conclusion that Wright's shifting balance theory of evolution in no material way depends upon the usefulness of his fitness surfaces as heuristic devices. The shifting balance theory gains its place among the few really robust theories of evolutionary change as a result of its close relation to known effective methods of animal breeding and increasing understanding of population structure in natural populations, which in many organisms more closely resembles Wright's concept of subdivision rather than Fisher's conception of panmictic population structure.

Wright's Influence upon Animal Breeding Theory

I once gave a talk to the animal breeding department at Cornell University on the subject of Sewall Wright and animal breeding theory. I planned to talk about the influence that animal breeding theory had had upon the origin of Wright's shifting balance theory of evolution in nature. The animal breeders assumed, however, that what I would talk about was the influence that Wright's shifting balance theory had exerted on animal breeding theory. The influences indeed worked in both directions. I have already shown the first half of the circle earlier in the biography and will complete it here in this section.

At the time Wright moved to Chicago, he had been closely associated with animal breeding theory and animal breeders for more than ten years. His teacher Castle was keenly interested in animal breeding, and a temporary fellow student at the Bussey with Wright, E. N. Wentworth, went directly into the field. Wright's major responsibilities and research at the USDA were tied to animal breeding. The analysis of inbreeding and outbreeding in guinea pigs, the application of path coefficients to systems of mating, the development of a useful quantitative measure of inbreeding in animals, the analysis of breeding history in the shorthorn cattle—all of these major projects were contributions to animal breeding theory. During the USDA days, Wright attended regularly the annual meetings of the American Society of Animal Production, the major society of practical animal breeding in the Western world. He kept in close touch with influential animal breeding theorists such as Wentworth, H. D. Goodale, and L. J. Cole.

Despite the importance of animal breeding to Wright and to his reputation as it existed in 1925, he made the conscious decision to turn away from the field of animal breeding and back to the fields of physiological genetics and evolutionary theory. He thought such fields more appropriate to a faculty position at a research institution like the University of Chicago; these fields were, after all, ones in which he had long been involved. He certainly did not lose all interest in animal breeding, however. He continued to attend meetings of the American Society of Animal Production and was frequently asked to deliver papers at the meetings. Except for the papers on animal breeding begun before Wright left for Chicago, he published very little on animal breeding theory after 1925. One notable exception was a paper Wright delivered at the 1931 meeting of the American Society of Animal Production, and published in their proceedings in early 1932, entitled "On the Evaluation of Dairy Sires."

Wright first suggested that many questions were really conflated in the attempt to determine the breeding potentials of bulls, and distinguished and analyzed the three most important of these. He settled upon one as the most important for the evaluation of a sire: "What record would be expected of daughters of a bull, if the latter were mated with a random sample of cows of the breed?" He then evaluated critically the formulas proposed by Yapp or Goodale and suggested his own, which reduced to that of Yapp or Goodale under certain assumptions but which also took account of factors not included in theirs. This little paper with its proposed formula had considerable impact upon animal breeding theory through its extension by Jay L. Lush, whose work is discussed below.

Although Wright turned away from animal breeding as a major research interest, his reputation in the field increased rather than decreased. One reason is that the few existing modern textbooks on animal breeding incorporated Wright's thinking and work. The most popular of these books, *Animal Breeding* by L. M. Winters of the University of Minnesota, in its second edition of 1930 contained long sections on inbreeding, outbreeding, and selection, based explicitly and in detail upon Wright's work. Winters refers to Wright more frequently and at far greater length than to any other thinker in the entire book. Any student or animal breeder reading Winters's book could not help but come away with the thought that Wright was the most prominent animal breeding theorist in the world.

Wright's international reputation in animal breeding theory is attested to by the breadth of his correspondence files. His work was apparently well known in Russia, particularly by A. S. Serebrovsky of the Zootechnical Institute of Moscow. Serebrovsky was the most prominent animal breeder in Russia at the time and had much responsibility for government-sponsored research in improvement of domestic farm animals. He wrote to Wright in 1926 to ask for help in the quantitative analysis of the best method for improving herds using the technology of artificial insemination (in 1926 Serebrovsky said it was possible for one bull to sire 3,000 offspring in a single season). The letter ended: "Being a great mathematic we are sure that the solution of this problem which we think is a very important will also interest you very much." Serebrovsky also wrote to Wright in 1927 about Wright's paper on maternal age and white spotting in guinea pigs.In 1930 Wright received a letter from I. I. Nasarenko, specialist on genetics of the Academy of Agricultural Sciences, saying that Nasarenko had just translated Wright's pamphlet "Principles of Livestock Breeding" into Ukrainian for publication in that language.

More revealing of the situation in animal breeding in Russia, and of the esteem in which Wright was held there and in the United States in the field, can be seen in a long handwritten letter from H. J. Muller in 1929. It is a rambling letter but worth quoting in full for what it reveals of Muller's attitude toward science and culture in Russia as well as the opportunities he saw there for Wright.

Dr. S. G. Levit, a Russian geneticist who has been with us for nearly a year as fellow of the Rockefeller Foundation, has just received a letter from Dr. Serebrovsky, asking whether the former could find some American geneticists who would furnish advice and aid in the building up of the animal industry of Russia. Serebrovsky, who is primarily a "zootechnical" specialist, was recently given an important post in such work. He proposes to produce several hundred thousand cattle, sheep, etc., by artificial insemination during the coming year (one or more thousand from each father, by Ivanov's method), and will also make species crosses—as cow-bison, cow-zebu, cow-yak, etc.—on a very large scale, conducting the work as a series of genetic experiments in addition to utilizing any advantageous ones as possible. It will be the first really large scale attempt to apply modern genetics to animal improvement, and at the same time to gather data that might be of use to genetics. In addition, pigs are to be X-rayed, and, there will no doubt be much other work.

Serebrovsky would like to find good geneticists who would be of aid in several different directions, if necessary getting a man for each line, but alternatively dividing the lines among the men in a different way. The lines, as he mentions them, are (1) Interspecies crosses (a man with a "zootechnical" background, i.e. knowledge of animal husbandry, breeds of domestic animals × their good × bad points × uses; a man willing to take a chance in view of large possibilities); (2) theory and practice of selection; (3) influence of mother on young, as in temporary transference of immunity (this I believe should not be stressed as the effects are too transient to be of much account); (4) genetics and selection of disease resistance; (5) the finding of the best genotypes already existing in America, so that, where possible, they—or samples of them—may be purchased and sent to Russia.

Levit referred the letter to me for advice concerning possible men and I of course gave your name as the best fitted to cope with the whole combination of problems, and especially number 2. I then found that Serebrovsky had himself mentioned your name and asked whether it might be possible to get your help. He would really like to get men to come to Russia (except in the case of number 5, I suppose) for one or more years, for a definite or indefinite term, or permanently (according to the circumstances of the man, etc.). I do not know whether you could or would accept an appointment for a year or more, but, as I said to Levit, you might be able to go for at least a quarter or a half, and the half of a man who really knew the subject, for a few months would be far more valuable than that of another for some years. Or you might go for one quarter, and then one or two years later for another, etc. to catch the successive generations. Possibly you could be of some help even without going, as a consultant, although S. did not mention that possibility. And your knowledge of which other men might be suitable would be of a very great value. As you know, I am not acquainted with the animal industry field, and hence am not fitted to appraise men, for most of the lines he has mentioned, particularly #5 and #4. The man for #5 would I suppose stay at least most of the time in the U.S. as the head of an American bureau for the obtaining of data concerning the most suitable breeds for given Russian conditions (preferably visiting Russia too to study the conditions and for follow-up work) and the best animals within these breeds, and for the actual obtaining of such animals. Would you care to mention any names as suitable candidates? What would you think of Lush, for example, for this work? Or Landauer? Who else could give competent advice on the selection of a man here, if not you? And who might be suitable for #4? It seems to me that the work of #1 and #4, inclusive could be largely under the man who directs #2, or at least subject to his advice and correction, and that #2 should be the most important, and should therefore be a #1 in genetics as well as have the proclivity of tackling the animal breeding problems.

Things are now in the formative stage and any advice or suggestions you might give as to personnel or plans would be greatly appreciated. Levit is writing S. that you are being consulted and he will probably have an answer to that letter by the time of the New Orleans meeting. He and I are both going there and perhaps we can have a conference then, but would also like to get some word from you sooner as things seem to be moving fast now in Russia.

I do not, perhaps, have to mention the fact that foreign specialists are well paid in Russia and an effort is made to provide them with suitable living conditions, etc. And nowhere else is there such an opportunity for the scientific man really to put his knowledge into effect. At the same time, he has the chance to carry on purely theoretical work. I do not have any personal interest in the matter of which I am writing, but I should like to see a good demonstration made of the way in which genetics can be applied.

With best regards, Yours sincerely, H. J. Muller

P.S. Levit is returning to Russia at the end of January. He is not an animal husbandry specialist but a physician-head of the Moscow Medico-biological Institute, which is engaged principally in studying human genetics. They have a staff of over 100 physicians, who are being trained in genetics, have a school for twins, a tissue-culture department, cytological dept. etc.

Wright clearly was less enamored of this opportunity than was Muller and replied in part:

> I was much interested in your account of the Russian animal breeding program as well as flattered by being considered in connection with it. It seems to me, however, that the position requires more executive ability and detailed knowledge of Animal Husbandry than knowledge of theoretical genetics, which would be my only claim to consideration, if any. I should think that men like Serebrovsky could supply as much theory as could be used for a long time. Live stock improvement is a long project and failure is more likely due to lack of consistency of administration policy over a period of years than to unsound initial plan.

Complying with Muller's other requests, Wright suggested a number of animal breeders who might possibly be of aid. I have been unable to find evidence that any of them went to Russia to help Serebrovsky. Given Muller's wide acquaintance with geneticists all over the world, it is significant that he thought of Wright as perhaps the foremost quantitative theorist of animal breeding; he certainly thought of Wright also as one of the foremost evolutionary theorists, as I showed earlier.

S. S. Munro (Canadian Department of Agriculture) and I. Michael Lerner (Department of Poultry Husbandry at the University of British Columbia) each wrote several times to request Wright's aid. A. D. Buchanon Smith wrote many times from F. A. E. Crew's Institute of Animal Genetics in Edinburgh to request help in his work on construction of the modern Shorthorn (Wright would spend a year there in 1949). And in the United States, Winters, Goodale, W. L. Gaines (Illinois Agricultural Experiment Station), Arthur Chapman (then a graduate student at Madison), Cole, Elmer Roberts (University of Illinois), and Wright's old friend G. N. Collins (asking about a plant breeding problem), and many others wrote to ask Wright's help with their work.

It should be emphasized that Wright's work in animal breeding was influential primarily among animal breeding theorists rather than among practical animal breeders, who generally could not profitably read Wright's papers. Wright was well aware that most animal breeders could not put into effect his theory of the "shifting balance" evolution of domestic breeds; but his great interest was the theory, not practical breeding.

Great though Wright's direct influence upon animal breeding theory might have been in the period from about 1915 to the early 1930s, his most pervasive and lasting influence in the field of animal breeding theory and practice came indirectly through the work and teaching of Jay L. Lush, who is widely acknowledged to be the single most influential animal breeding theorist since the early 1930s. Lush came from a farm background and majored in animal husbandry as an undergraduate at Kansas State Agricultural College in Manhattan, Kansas. There he came under the influence of Wright's friend Wentworth, who was a professor of animal breeding there from 1914 to 1917. Lush graduated in 1916 and went on to take an M. S. degree in 1918, doing research on heredity in swine with Wentworth. Through Wentworth, Lush met Wright, who early in 1918 had published his paper on color inheritance in swine. Lush and Wright spoke frequently at meetings of the American Society of Animal Production. They had begun to correspond not later than early 1918.

The earliest letter from Lush in Wright's extensive correspondence file with him concerned a problem of inheritance in swine.

> Professor Wentworth and I are now getting out a preliminary report on the work done here in cross breeding swine. One of the points under investigation is, "The Dish of Face of the Berkshire Swine as Compared with the Much More Moderate Dish of the Duroc-Jersey." Professor Wentworth tells me that Mr. C. C. Little has done a good deal of work on this subject, but he does not know whether any of it was ever published. Mr. Little is in the army now and we cannot reach him, but think that perhaps you would know either where we might get hold of that work, or the substance of the conclusions he reached. I am writing you for what help you can give us. (Lush to Wright, April 9, 1918)

Wright of course knew Little well, but knew only that Little had done much research on the genetics of mice and, his favorite subject, dogs. So Wright wrote to Little, who was a captain in the office of the chief signal officer in Washington, D. C. Little replied: "I suggest a possible confusion between hogs and dogs. I never studied the former, much less being guilty of writing about them. I hope that this letter does not reach the official files, as I would probably be made subject for a court of inquiry as being mentally unsound" (Little to Wright, April 22, 1918). I do not know how Wentworth got the idea that Little had worked on inheritance in swine unless Little's suggestion is correct.

Lush published his work on inheritance in swine beginning with a substantial article, "Inheritance in Swine," in *Journal of Heredity*. The section on color inheritance was based explicitly upon Wright's earlier paper on color inheritance in swine, but Lush added some new interesting data from the crosses made at Kansas State.

In 1918, of course, Wright had not yet begun to publish work on systems of mating, inbreeding and outbreeding, animal breeding, or the general theory of path coefficients with its many applications. The correspondence between Wright and Lush reveals that Lush kept up with this work as it appeared. In 1922 Lush took his doctorate working with Cole at the University of Wisconsin with a thesis entitled "The Possibility of Sex Control by Artificial Insemination with Centrifuged Spermatozoa" (published 1925), and from 1922 to 1930 was an animal husbandman at the Texas Agricultural Experiment Station (Texas A&M University) in College Station, Texas. Beginning in 1923, the correspondence shows that Lush was keenly interested in the method of path coefficients and its application to animal breeding theory. As early as July 1923, in answer to questions from Lush, Wright wrote him a five-page letter describing in detail the aim and uses of path coefficients. Sometimes Lush put questions to Wright at meetings of the American Society of Animal Production, and Wright would (characteristically) later answer them by letter after thinking a few weeks about the questions. Almost all of the correspondence consists of Lush asking quantitative questions of Wright and Wright's detailed answers. There can be no doubt that Lush relied heavily upon Wright, not only for help on quantitative questions, but even more for Wright's general formulation and approach to the basic problems of animal breeding. Wright's published papers most influential in Lush's thinking were those on systems of mating, path coefficients, effects of inbreeding and outbreeding in guinea pigs, analysis of the role of inbreeding in the development of the Shorthorns, coefficient of inbreeding, and evolution papers of 1931 and 1932.

When Lush took a position in the Department of Animal Husbandry at Iowa State University in the fall of 1930, he immediately arranged to spend the spring quarter 1931 in Chicago, taking Wright's evolution course and talking extensively to Wright about quantitative analysis. It is indeed interesting that Lush took the evolution course rather than the biometry course which Wright did not offer that spring. Lush always viewed the optimal method for improvement of a whole breed (but not the most practical method) as essentially the shifting balance theory he had learned from Wright primarily in that course.

Returning to Iowa State, Lush and his students used the method developed by Wright and McPhee to analyze the history of inbreeding in many different farm animals, particularly swine breeds. They found much less inbreeding than Wright had found in Shorthorns. Far more than this, Lush took what he had learned from Wright, the geneticists at Iowa State (E. W. Lindstrom was then leading the genetics group), the statisticians (led by G. W. Snedecor), and Fisher, whose *Statistical Methods for Research Workers* was used by many at Iowa State and who visited Iowa State in 1932 and 1936, and synthesized all of this into a highly organized and influential theory of animal breeding oriented to a much more practical level than Wright's publications. Lush attracted to Iowa State an outstanding group of graduate and postdoctoral students who took his general view of animal breeding theory literally all over the world. Lush's synthesis of animal breeding theory, combined with his making Iowa State a center of thought in animal breeding, made him the acknowledged leader in the field in the world.

I am not arguing that Lush simply took Wright's thoughts and used them for himself or that Lush's way of developing animal breeding theory is the only or best way to do so. I do think that Lush incorporated many of Wright's ideas and methods and frequently relied upon Wright's critical analysis in building his own synthesis of animal breeding theory, and that Lush was in fact the most influential animal breeding theorist. Lush has always clearly acknowledged his debt to Wright. When in 1937 Lush published *Animal Breeding Plans,* drawn primarily from his course in animal breeding at Iowa State, he said in the preface: "The ideas in this book have been drawn freely from the published works of many persons. I wish to acknowledge especially my indebtedness to Sewall Wright for many published and unpublished ideas upon which I have drawn, and for his friendly counsel about many detailed problems during more than a dozen years" (p. vi). The book indeed draws very heavily upon Wright's ideas and methods. The index shows Wright to have thirty-one references, Gowen next with eleven, followed by Wriedt with eight.

I will give here only one striking example from a section entitled "Ideal Breeding Systems for Rapid Improvement of the Whole Breed." In this section Lush was carefully distinguishing between the plan an individual breeder might follow and the plan for rapid improvement of the entire breed. Wright is not referenced in this section at all.

An ideal breeding system for the most rapid improvement of the breed as a whole would be about as follows: Each breed would be divided into many small groups, each such group rarely introducing any breeding animals from other groups and then only with caution. Each group would be large enough for the use of from three to five breeding males at all times and, of course, would include a much larger number of females. If the groups were much smaller than this, the rate of fixation of genes on account of the inevitable inbreeding might be too high to be kept under control by selection. If the groups were much larger than this, progress toward uniformity within each group and toward distinctness from group to group would be needlessly slow. Such a system is pictured diagrammatically in fig. [9.3], where the large area represents a whole breed and each small area within it means a partially isolated subgroup of the breed into which individuals from other subgroups are rarely introduced. Naturally the few introductions which are made would usually be from the neighboring subgroups and only rarely from a distant subgroup. Groups of subgroups or major geographical subdivisions might thus tend to form within the breed.

The consequence of such a separation into groups, each breeding very largely within itself, would be that each such group would quickly become more uniform than herds are today and that each group would become different from other groups. Selection between the groups would then be effective to an extent impossible today and probably never attainable in the selection of individual animals, no matter how much the animal is studied nor how skilled is the man who does the selection.

Many of these subgroups would begin to show undesired traits varying in severity. Side by side with these, they would show other highly desired traits more uniformly than present herds do. Groups showing many desired and few undesired traits would make mild

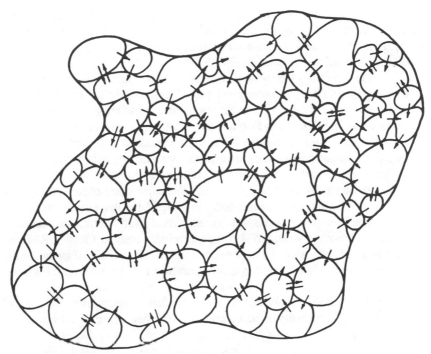

Figure 9.3. Subdivision of a breed into small local groups.

outcrosses to neighboring groups which were strong where they themselves were weak. Then by renewed linebreeding with rigid selection for the traits they wished to introduce by the outcross, the breeders would attempt to fix the introduced desired traits without losing the desired traits they already had. Groups showing few desired and many undesired traits would either be discarded altogether or would by graded up by the continued use of sires from the most successful groups until their individual merit was restored or even exceeded that of the most successful group. Then the breeders would start breeding within this group to find and fix some one of the almost infinite number of desirable new combinations of genes which would be possible. The general rule would be that the more successful each subgroup was, the less readily would it do any outcrossing and the milder such outcrossing would be.

The general picture thus presented is an alternation of mild linebreeding with tentative outcrosses, both accompanied at all times by intense selection. The most of the linebreeding would be done in the best of the herds with extreme outcrossing being confined to the poorer herds. (Pp. 340–42)

What Lush advocates in this section is the same breeding plan for improvement of entire breeds emphasized by Wright in the closing paragraphs of his 1923 paper analyzing the Duchess family of Shorthorns as bred by Thomas

Bates, which Wright emphasized in his 1931 and 1932 papers (particularly the 1932 paper) on evolution. This section remained unchanged in all three editions of Lush's book (3d ed., 1945), which went through at least fifteen different printings up into the 1970s.

Lush also in 1949 republished, under the title *Systems of Mating and Other Papers,* Wright's five 1921 papers on systems of mating, "Correlation and Causation" (the 1921 paper on path coefficients), the big 1931 paper "Evolution in Mendelian Populations," and the 1934 paper "The Method of Path Coefficients," primarily for his own students. The collection proved so popular that Iowa State kept it in print for over thirty-five years.

Perhaps because of Lush's admiration and debt to Wright, and Lush's great influence in the field, I can say from personal experience that animal breeding theorists as a group hold Wright in greater esteem than any other group of scientists, including geneticists and evolutionists. Even animal breeding theorists who, like Lush's prominent student Charles R. Henderson of Cornell University, have moved away from the use of path coefficients, still have the greatest respect and admiration for Wright as a thinker and as a person.

The influence that Wright's shifting balance theory of evolution had upon animal breeding theory is of course only the beginning of its impact in modern biology.

10
Wright, Dobzhansky, and the Genetics of Natural Populations

Theodosius Dobzhansky was a central figure in evolutionary thought from the mid-1930s until his death on December 18, 1975. His *Genetics and the Origin of Species* (1937, 1941, 1951) was a widely read and influential work, perhaps the most influential single book on evolutionary biology during the period from 1937 through the 1950s. I see no good reason to doubt the assessment of the editors of the 1981 reprint edition of Dobzhansky's Genetics of Natural Populations (GNP) series: "Taken as a whole, the Genetics of Natural Populations is the most important single corpus in modern evolutionary biology. It is the armature on which has been built a large body of evolutionary genetics . . . this series is and has been central to evolutionary genetics" (Lewontin, Moore, Provine, and Wallace 1981, xi). This series, begun in the spring of 1937, consisted of forty-two published papers when Dobzhansky died. The forty-third and final paper was published posthumously.

Sewall Wright had a great influence upon *Genetics and the Origin of Species* and upon the GNP series. Indeed, perhaps Wright's greatest influence upon evolutionary biology from 1937 through the 1950s came not from his own publications but from those of Dobzhansky or those on which he and Dobzhansky collaborated.

The collaboration between Wright and Dobzhansky reveals important facets of the interaction between theoretical population genetics and field research. Understanding this interaction is crucial for the historical analysis of the development of evolutionary biology after the mid-1930s. Theoretical population geneticists and field naturalists needed each other greatly. Fisher, Haldane, and Wright all were acutely aware of their lack of knowledge of the genetics of natural populations. They knew that discrimination between their theoretical schemes could come only from the genetic analysis of natural populations.

Those who studied natural populations, on the other hand, needed the models of the theoretical population geneticists as a guide to meaningful research programs. The quantitative models identified the pertinent parameters of microevolution, such as selection rates, mutation rates, dominance, gene interaction, linkage, balanced polymorphisms, and random processes. The field research of Dobzhansky, Ford, and their students after the early 1930s was inconceivable except in the context of the theoretical visions of Wright and Fisher. This chapter and the two that follow will provide the supporting evidence for this assertion.

In the two preceding chapters, I have asserted that the tension between Fisherian and Wrightian views of evolution in nature was a central creative factor in the development of evolutionary biology after the early 1930s. Analysis in this chapter of the collaboration between Wright and Dobzhansky gives this assertion strong support, which I will further strengthen in the next chapter. The tension between the evolutionary views of Wright and Fisher focused attention upon population size, breeding structure and genetic variation in natural populations, and the relative importance of natural selection and random genetic drift. These were precisely the primary questions that stimulated Dobzhansky, Ford, and a great many others to pursue their research on natural populations. The tension between the Fisherian and Wrightian views also greatly influenced the development of quantitative theories of evolution.

Collaborations are notoriously difficult for historians to analyze because documentation is generally lacking. Fisher and Ford, for example, communicated by talking rather than by substantive letters, even though Ford was in Oxford and Fisher in Cambridge. They found it easier to let difficult issues wait for the next visit. Although I find post hoc oral interviews extremely useful in many ways, they are in my experience unreliable for recreating the development of collaborations.

Thus from the historian's viewpoint the collaboration between Dobzhansky and Wright is particularly fortunate. They were routinely separated by long distances, with Wright in Chicago and Dobzhansky either in California or (after 1940) in New York City. They did have substantive talks at meetings or on occasional visits, but this loss to historians is more than compensated for by the existence of over 500 pages of detailed correspondence dating from 1937 until Dobzhansky's death. Concentrated in the decade 1937–47, this set of correspondence provides extraordinary insight into the Wright-Dobzhansky collaboration and more generally into the interaction of theoretical population genetics and field research. This is the single most valuable set of correspondence for understanding the development of the evolutionary synthesis after 1937. I have accordingly utilized this correspondence liberally in this chapter.

I have already published (Lewontin et al. 1981, 1–76) a long essay on the origins of the GNP series. Only a bare summary of that analysis is possible here. For a fuller account, especially of the collaboration of Dobzhansky and Sturtevant (and pertinent references), I refer the reader to that essay.

Theodosius Dobzhansky and the Study of Natural Populations

Wright and Dobzhansky first met in August 1932 at the Sixth International Congress of Genetics in Ithaca, New York. Here Wright delivered his famous 1932 paper with the first published diagrammatic account of his fitness surfaces, with their peaks and valleys. Dobzhansky was captivated. Shortly before he died, Dobzhansky wrote me that he "fell in love" with Wright at the 1932 meeting, finding Wright's shifting balance theory of evolution in nature irresistible.

In 1932 Dobzhansky was already an accomplished field naturalist, having published papers on natural populations for over a decade. As a boy, Dobzhansky had been a keen butterfly collector. He decided, at age sixteen (in 1916) to concentrate on the study of ladybird beetles, family Coccinellidae. At about the same time he read and discussed Darwin's *On the Origin of Species* (in Russian translation), a book he greatly enjoyed. He decided to become a biologist. After completing his undergraduate studies at the University of Kiev in the spring of 1921, he took a position as assistant (roughly equivalent to an instructor in the United States) on the faculty of agriculture, Polytechnic Institute of Kiev. There he taught general biology and its cultural implications until 1924.

Dobzhansky immediately began serious, systematic collecting of Coccinellidae in the area around Kiev. He now had use of well-equipped laboratories, which enabled him to examine microscopically the anatomy and development of the ladybird beetles. He also began to read more widely about other groups of insects. In 1922 and 1923 he published three papers on the distribution, anatomy, and new species of Coccinellidae. When Dobzhansky assumed his teaching duties at the Polytechnic Institute, he had never taken a course in genetics. None had been offered at the University of Kiev. Furthermore, he had read little literature about genetics, partly because the literature containing the new genetic discoveries being made in Germany, England, and the United States was not imported into Russia during the difficult years between 1914 and 1921.

Beginning in 1921, Dobzhansky began his acquaintance with modern genetics. Using the resources of Nikolai Vavilov's Institute of Applied Botany in Petrograd and Nikolai Kol'tsov's Institute of Experimental Biology in Moscow (Adams 1980), Dobzhansky quickly learned enough genetics to begin laboratory experiments on *Drosophila* in 1922.

To understand fully Dobzhansky's excitement when he discovered *Drosophila pseudoobscura* in 1932 and began to use it as an experimental animal, one must understand his overall research program in the years 1922–24. First, from the time of his reading of *Origin of Species,* what really interested Dobzhansky most about biology were the larger questions of evolution and their meaning for the evolution of humans, human society, and human ethics. Always, though he did not address these questions explicitly in his writings until the 1950s, these fundamental questions were foremost in his mind. This attitude of mind he emphasizes in his oral memoir (1962). But to understand these larger questions, which he debated endlessly with colleagues and friends in Russia, Dobzhansky believed he needed to understand evolution at the microlevel. Knowledge of evolution in Coccinellidae or *Drosophila* was not so much an end in itself for Dobzhansky; it was a means to know more about the most interesting questions, involving the implications for evolution in humans.

To understand microevolution, Dobzhansky already believed in 1922 that one must have accurate knowledge about geographical distribution and variability between populations, individual variability within populations, and inheritance of variability. A research program based upon these suppositions would involve both the study of variability in wild populations and laboratory genetics. In 1922–24, he actively pursued all parts of this research program in evolution; but a basic division remained. In the field he was studying the variability within and between species, races, and populations of Coccinellidae, and in the laboratory he was studying the genetics of *Drosophila melanogaster*. Not until he began work upon *Drosophila pseudoobscura* would he put together his whole research program for microevolution.

The research on Coccinellidae in the field and on *Drosophila melanogaster* in the laboratory were, however, related. In a paper published in German in 1924 on geographic and individual variability in a species of ladybird beetles, Dobzhansky reached two very fundamental conclusions: (1) there was no essential difference between the variation between geographical races and the variation within a single population; but (2) variation within a population was often caused by single gene differences, whereas geographical races generally differed by complexes of genes. In *Drosophila melanogaster* Dobzhansky discovered that, using his background in anatomy and microscopy, he could find measurable differences in the shape of the spermatheca (the sperm storage sac in the female *Drosophila*) in the different mutant stocks he had obtained at Kol'tsov's institute. His general conclusion from this specific observation was that every gene probably acts upon all parts of the body, or in the language of geneticists, has pleiotropic or manifold effects. Thus Dobzhansky brought his two researches together by conjecturing that the gene complexes by which geographical races differed were harmoniously interrelated, a result he attributed to natural selection.

Unfortunately for Dobzhansky, the genetics of Coccinellidae was extremely complicated. Not only did the simplest differences in color pattern in ladybird beetles often depend upon two, three, or more genes, but Coccinellidae had nearly four times as many chromosomes as *Drosophila melanogaster*. The family Coccinellidae was simply not the ideal organism for Dobzhansky's two-pronged attack upon evolution in natural populations.

Late in 1923, after hearing about Dobzhansky's work on the pleiotropic effects of genes in *Drosophila melanogaster*, Filipchenko invited him to join the Department of Genetics at the University of Petrograd (the city became Leningrad the day after Dobzhansky arrived there in January 1924). Dobzhansky learned more about genetics from Filipchenko and Vavilov, had his imagination fired by long talks with Leo Berg, creator of the theory of "nomogenesis" (Berg 1926), and continued his research. In the years 1924–27 he published fifteen papers on the morphology, variability, and systematics of Coccinellidae, and in 1927, a much expanded version of his 1924 paper on the manifold effects of genes in *Drosophila melanogaster* (Dobzhansky 1927). One other quite different study contributed to Dobzhansky's development as a field population geneticist. In 1925 he married Natalie (known usually as "Natasha") Sivertzev. To support the family, he took a second job as leader of a research team whose purpose was to examine the domesticated animals of peasants in central Asia in hope of discovering ways of genetically improving domesticated breeds. Dobzhansky spent the summers of 1926 and 1927 in the field in central Asia, measuring and attempting to specify hereditary traits in horses, cattle, yaks (a different genus than domesticated cattle, but capable of hybridizing with them), and marals (a near relative of the elk). The marals especially interested Dobzhansky because they had only recently been domesticated, and the effects of artificial selection were thus more clearly demarcated. These studies were unsatisfactory because so little was known about the genetics of the domesticated animals that few conclusions about genetic improvements were possible. But Dobzhansky loved the field work.

By 1927 Dobzhansky had reached a frustrating plateau in his efforts to understand evolution in nature. He knew a great deal about the systematics of Coccinellidae but almost nothing of its genetics. He knew some *Drosophila melanogaster* genetics but nothing about natural populations of *Drosophila*. What little he knew of the genetics of domesticated animals was of no obvious use in understanding evolution in wild populations. At this crucial time, Filipchenko arranged through the International Education Board (a part of the Rockefeller Foundation after 1928) for Dobzhansky visit Morgan's lab for one year or possibly two.

Sturtevant, Morgan's former student and an outstanding geneticist, took Dobzhansky under his protective wing. A very fruitful period of formal and informal collaboration followed. Sturtevant's wide and deep interest in all aspects of genetics appealed to Dobzhansky. Sturtevant was not only a pioneer in chromosome mapping and proving the existence of chromosomal inversions but also had performed a set of significant selection experiments in *Drosophila*, was the foremost *Drosophila* taxonomist in the United States, and was very interested in evolution in general. He was knowledgeable about all aspects of genetics. Wright told me on several occasions that he considered Sturtevant to be the one member of the Morgan group in the 1920s and early 1930s who was really sophisticated about evolutionary problems. Wright said that Sturtevant read and understood Wright's own papers on evolution. Sturtevant had always been far more at ease than Morgan with quantitative reasoning. Dobzhansky, as I will show later, admired quantitative reasoning but was a novice at anything much more complicated than calculating chi-square tests of significance. Dobzhansky naturally gravitated toward Sturtevant as his intellectual mentor.

Under Sturtevant's tutelage, Dobzhansky began a series of original contributions. He was sole author of thirteen papers on *Drosophila melanogaster* between 1928 and 1932 and coauthored two with Sturtevant. He cytologically examined translocations in *Drosophila melanogaster* and constructed "cytological" maps of the chromosomes. He used reciprocal translocations to suggest cytological reasons for the observed decrease of crossing over in translocations and, with Sturtevant (who was doing active research in *Oenothera* genetics at this time), pointed out how reciprocal translocations in

Drosophila corroborated John Belling's interpretation of *Oenothera* cytogenetics. In 1932 he published the first of three important genetic studies of the effects of translocations, inversions, and duplications on chromosome behavior during meiosis. With the fine-structure analysis made possible by the discovery of giant salivary gland chromosomes in 1933, most of these cytological studies were quickly outmoded. But certainly geneticists considered them important at the time of publication.

One striking thing about all the genetic studies Dobzhansky had done between the time of joining Morgan's group in 1927 and 1932 is the total absence of any explicit relationship between the genetic studies and a theory of evolution in nature. Yet evolutionary questions were certainly on Dobzhansky's mind in 1932. In that year he finally wrote, in English, a major article on geographical variation and evolution in ladybird beetles for publication in *American Naturalist* (Dobzhansky 1933). In this very modern-sounding paper, Dobzhansky argued that speciation generally resulted from geographic isolation and that a species was merely a population that had genetically differentiated to such an extent that it became reproductively isolated from other populations.

The 1933 paper on Coccinellidae still exhibited two major problems that would have to be solved before Dobzhansky would be able to put together *Genetics and the Origin of Species*. First, he says little about the dynamic mechanisms that cause the creation of races or species, except to present three very simplistic suggestions in the last paragraph. I am sure Dobzhansky had completed the draft of his paper before attending the Sixth International Congress of Genetics at Ithaca in August 1932, where he heard Wright's address on mechanisms of evolution. But Dobzhansky put his finger directly upon a crucial point.

> An analysis of the mechanism of the formation of the geographical races and species ought to begin with a study of the behavior of the single characters distinguishing the different forms from each other. Only subsequently can one study the interaction of the unit-characters in the complex systems representing the types with which taxonomy is primarily concerned. (Dobzhansky 1933, 124)

This statement, which is prophetic with regard to the next four years of Dobzhansky's research, raises the second problem: Dobzhansky was still a victim of scientific schizophrenia. It was fine to say that one must understand the genetics of race and species differences before studying the dynamic processes of race and species formation, but Dobzhansky made the point in a paper on coccinellids, whose genetics was virtually unknown. He was forced into using hypothetical examples of genetic determination of coat patterns in the paper! He could not possibly study in detail the genetics of race and species differences in coccinellids. Even in *Drosophila,* genetic analysis of species differences was impossible because no two species could be interbred to produce fertile hybrids, and thus no genetic experiments could be conducted. The best that had so far been accomplished was Sturtevant's hybridization of *D. simulans* with *D. melanogaster,* but the hybrids were completely sterile (Sturtevant 1920). Dobzhansky, obviously wanting to study evolution, found no beginning handle with his two experimental organisms, Coccinellidae and *Drosophila.*

The answer to Dobzhansky's dilemma came in the form of *Drosophila pseudoobscura.* Donald E. Lancefield, a student of Morgan and Sturtevant, had begun a serious study of this species in 1919. In 1922 he discovered a morphologically indistinguishable stock of *D. pseudoobscura* (until 1929 still mistakenly called *D. obscura*) that was nearly sterile with previous stocks. He soon discovered that F_1 hybrid males were sterile but that F_1 hybrid females were partially fertile and could be backcrossed with either parental stock. In his major 1929 paper on *Drosophila pseudoobscura,* Lancefield pointed out the significance of these two "races," which he named *A* and *B*: "The two races of *D. obscura* are less differentiated from each other than are *D. melanogaster* and *D. simulans* and may represent an earlier step in the evolutionary process. The fact that the hybrid F_1 females may produce offspring makes it possible to carry out a further analysis of the situation" (Lancefield 1929, 288). Lancefield indeed carried out a series of preliminary genetic and cytological analyses of the basis for the sterility.

It is difficult to imagine a more stimulating paper for a Drosophilist interested in evolution. Here at last was the opportunity to study the genetic basis of high but incomplete sterility in an organism whose chromosomes were few and whose genetics was accessible to study. Lancefield moved slowly in pursuing the study of *D. pseudoobscura.* In 1932 the young Hungarian cytologist, Pio Ch. Koller, published another study of the factors affecting fertility between races *A* and *B*. Like Lancefield, Koller was clearly sensitive to the evolutionary implications of the research: "The real interest of this material lies in the fact, that it furnishes a possible example of the method of origin of a new species. . . . Support is given to Lancefield's suggestion that the relation of the two races is a study in the origin of species" (Koller 1932, 150). How tantalizing this conclusion must have seemed to Dobzhansky.

Dobzhansky wanted to begin work immediately on *D. pseudoobscura* in the spring of 1932, but Sturtevant only had stocks of race *A* at California Institute of Technology. In the spring quarter of 1932, Dobzhansky taught a general course in genetics. Discovering that one of the students, Robert D. Boche, lived in the Seattle area where both races *A* and *B* existed in nature, Dobzhansky suggested that he collect *D. pseudoobscura* during the summer and bring them back to Cal Tech in the fall. Boche happily agreed, and by midsummer he had collected seven different strains.

Dobzhansky was sufficiently intrigued that when he, G. W. Beadle, and Jack Schultz found that they could travel to the Sixth International Congress of Genetics in Ithaca, New York, just as cheaply by way of Seattle, they decided to visit Boche and see his stocks of *D. pseudoobscura* on the way.

They spent an afternoon and evening with Boche, who describes Dobzhansky's reaction as enthusiastic. Dobzhansky clearly wanted to begin work on *D. pseudoobscura* when Boche returned with the stocks to Cal Tech for the fall quarter.

Despite having his imagination fired by Wright's talk on mechanisms of evolution in nature at the Ithaca congress, upon his return Dobzhansky initiated a series of laboratory studies analyzing the genetics of the hybrid sterility between races *A* and *B*. He was able to discover where the sterility factors were on the chromosomes, how they interacted with each other and with the cytoplasm, and that each chromosome carried several genetic factors for sterility.

This stunning experimental work on the genetics of hybrid sterility was closely tied to the problem of species differences, which Dobzhansky addressed in two papers (Dobzhansky 1934, 1935). Here he argued for the "biological species concept," namely, that a population properly assumes the status of a species when it becomes reproductively isolated for genetic reasons from other similar populations. This was not a new concept of species, having been earlier advocated by such taxonomists as Poulton, K. Jordan, and Stresemann (Mayr 1982, 273–75). The concept was significant because it tied together taxonomy with the genetical conception of a population as a gene pool. Dobzhansky's biological species concept, which would lead him to classify races *A* and *B* of *Drosophila pseudoobscura* as different species because of their reproductive isolation, was not accepted by Sturtevant, who was a more traditional taxonomist who relied upon morphological differences. When Dobzhansky and Epling eventually classified race *B* as the species *D. persimilis* (Dobzhansky and Epling 1944), Sturtevant criticized this classification in his review (Sturtevant 1944), arguing that if the two species were morphologically indistinguishable, they should be only one species.

Despite Dobzhansky's long-standing interest in evolution in nature, and although he had begun to avidly collect *D. pseudoobscura* every summer after 1933, he did not by himself envision or initiate the series of studies on the genetics of natural populations. Indeed, the start of the GNP series unquestionably came from a collaboration with Sturtevant.

Soon after Sturtevant's return from a sabbatical leave in England during the academic year 1932–33, when Dobzhansky had begun the genetic studies on sterility between races *A* and *B*, Sturtevant began to devote less attention to *D. melanogaster* and more to *D. pseudoobscura*. In 1929 Sturtevant had found in *D. pseudoobscura* a "sex ratio" gene that caused a male carrying it to produce nearly all X sperm, instead of the usual nearly equal X and Y sperm. Thus about 95% of the offspring of the male carrying the gene were female. Using the stocks that Dobzhansky had gathered in the wild, Sturtevant analyzed the geographical distribution of the "sex ratio" gene. Dobzhansky, meanwhile, analyzed spermatogenesis in the males to pinpoint the mechanical cause of the unusual predominance of X sperm. Dobzhansky and

Sturtevant had a delightful time working on this project and published a joint paper on the topic (Sturtevant and Dobzhansky 1936a).

The collaboration of Sturtevant and Dobzhansky entered a new and deeper phase with the arrival in 1934 of the Chinese biologist C. C. Tan. Shortly before Tan's arrival at Cal Tech, the giant salivary chromosomes in *Drosophila* had been discovered. Dobzhansky and Sturtevant gave Tan the job of making genetic and cytological maps of the chromosomes of *D. pseudoobscura* in races *A* and *B* to see if there were any systematic differences between them. Sturtevant had already discovered that inversions could be found in the various laboratory stocks of races *A* and *B;* the question for Tan was whether races *A* and *B* differed consistently by the same inversions.

Tan discovered that populations within either race often differed by one or more inversions but that none of these inversions found within a single race appeared to be identical with six inversions by which the two races differed consistently. From these results, Tan came to two very exciting conclusions. First, "Since a rather large number of strains coming from different parts of the distribution area were involved in the hybrids in which the inversions were discovered, it is reasonable to assume that these inversions are racial characters" (Tan 1935, 400). In other words, these six inversions could be used as taxonomic markers in distinguishing races of *A* and *B* of *D. pseudoobscura*. Before the use of the salivary giant chromosome technique, the suggestion would have been ridiculous; now, it was an exciting prospect. Second, Tan argued that his results showed that "rearrangements of the genic material within the chromosome took place during the processes that led to the separation of these two races." Clearly, as Tan pointed out, the evolutionary differentiation of races *A* and *B* involved variability at both the genic and chromosomal levels.

Tan's results galvanized Sturtevant and Dobzhansky into action. They immediately began mapping out the geographical distributions of the inversions in both races of *D. pseudoobscura,* finding new ones Tan had not seen. Dobzhansky excitedly wrote Demerec on January 26, 1936:

> Sturtevant and myself are gone crazy with the geography of inversions in *pseudoobscura,* and working on this whole days—he with crosses and myself with the microscope. No short inversions are found. . . . As to our inversions, Mexico seems to be an inexhaustible source of them, and I am beginning to regret that last year only relatively few Mexican strains were collected.

On February 17, Dobzhansky reported to Demerec:

> Sturtevant and myself are spending the whole time studying the inversions in the third chromosome in geographical strains of *pseudoobscura*. We are constructing *phylogenies* of these strains, believe

it or not. This is the first time in my life that I believe in constructing phylogenies, and I have to eat some of my previous words in this connection. But the thing is so interesting that both Sturtevant and myself are in a state of continuous excitement equal to which we did not experience for a very long time.

By the middle of March 1936, Dobzhansky and Sturtevant, working extremely well together, were envisioning an extraordinarily ambitious project on the genetics and evolution of *Drosophila pseudoobscura*. One problem was acute. Dobzhansky had little expertise in quantitative data analysis in Mendelian populations, and while Sturtevant was more at home in this area, he too was uneasy about the prospect of working without expert advice on quantitative matters. Knowing that Sewall Wright would be the perfect person and that Wright had little contact with evolution in nature despite being the major theoretician of quantitative evolution in the United States, Sturtevant sent him on March 18 an outline of the proposed work on *D. pseudoobscura* and asked him to come to Pasadena for a year. "We feel the need of your help, not only in analyzing the data but also in planning sensible experiments." Sturtevant appears to be the primary architect of the outline. This is the earliest document of which I am aware to outline the Genetics of Natural Populations series of studies. It deserves to be quoted in full.

Status and Prospects of the *Drosophila pseudoobscura* Analysis

General

Strains are available from many regions scattered from central British Columbia to southern Mexico, and from California to the Black Hills of South Dakota and to central Texas—probably nearly the entire range of species. The collection of mutant stocks is constantly improving, and is already adequate to make possible the efficient analysis of most problems that arise.

Drosophila pseudoobscura was first selected for study because, unlike most *Drosophila,* it has easily studied spermatogenesis. Despite the salivary gland technique, this advantage is still important, although less now than before. Mr. Tan has now gotten fairly extensive correlation between the genetic maps and the cytological salivary chromosome maps for the autosomes, and he is pretty certain to be in a position to extend this work during another year to the X.

The physiology of the animal has been studied somewhat. Such things as developmental rate, rate of oviposition, temperature toleration, and oxygen consumption have been determined for certain conditions. The ecology of wild strains is open to study, but has not been seriously attacked yet. It is known that wild females may mate more than once outdoors, and that they may carry enough sperm to produce 1000 offspring without further mating when brought to the laboratory. Further work along these lines is planned.

II. Races *A* and *B*

The "strength" has been determined for a large number of strains in both races, and a geographical story can be made of it. [In crosses of *B* females with *A* males, "strong" yielded hybrid males with smaller testes and "weak" yielded larger testes. ED.] The inheritance of strength can be studied; some data are already available. The analysis of the *A*-*B* difference by a study of the genetic behavior of the hybrids is progressing. The immediate project here is a systematic study of the results of replacing each chromosome or section of chromosome of each race by the homologous section of the other race.

It is uncertain whether crossing between the two races occurs outdoors and leads to the production of fertile hybrids through which material may be transferred from one race to the other. This problem may be approached through detailed analysis of strains of *A* and *B* living together, or through the setting up in the laboratory of artificially mixed populations. These are projects for the future.

III. Inversions in Chromosome III

Salivary chromosomes show that at least 15 different sequences occur in wild strains. These mostly have fairly wide geographical distributions, but definitely limited ones. Many localities contain 2 or 3 sequences. The salivary technique makes possible the exact analysis of these inversions. If the "standard" sequence is called (1) *ABCDEF*, it is possible to recognize (2) *ADCBEF* and to distinguish it from (3) *ABEDCF*. All 15 of the observed sequences may be so described (using more letters than here), though not all of them are yet fully analyzed. In the case given, it is clear that the historical relation is 3 1 2; i.e., 2 and 3 are related only through 1. This gives the possibility of constructing an honest-to-God air-tight phylogeny—though of course this method alone gives no clue as to where one is to start the pedigree. Starting at any given point will, however, determine exactly the necessary order in which the others must have arisen. We already think we know, on other grounds, what was the original sequence—one that occurs chiefly along the Pacific coast.

IV. Lethals

About 15% of the III chromosomes of freshly collected wild stocks carry lethals. The same frequency (approximately) is found in stocks that have been in the laboratory for a year or more. So far as tested, these lethals are all different—in no case has the same lethal been found in the descendants of two different wild females. They occur in various ones of the different sequences mentioned above. Cytological study of several of them has not shown any deficiencies. The frequency appears to be the same in southern Mexico and in the Rocky Mountain region. If it be concluded that 15% represents an equilibrium, then the rate of elimination of lethals must equal the rate of occurrence of new lethal mutations. The latter can be determined experimentally. There should result a method of measuring the degree of inbreeding in wild populations, since that must be the chief element fixing the rate of elimination.

The frequency of lethals in other chromosomes is unknown; but in chromosome II the mutant genes available make the determination possible, and experiments are under way. Preliminary results will be available in a few weeks. It is hoped that here one can get a check on the determinations planned for chromosome III. This chromosome (II) rarely shows inversions; it will be interesting to see whether the frequency of lethals and of inversions is correlated.

V. "Sex-Ratio" Gene

This is a gene, present in wild populations of race A, race B, and at least three of the related species. Its effect is to induce, in the meiosis in males, two equational divisions of the X. The result is that a "sex-ratio" male produces about 95% daughters, with no increase in mortality. The gene should automatically increase in frequency at a rapid rate—which would be fatal to the race. It evidently does not do so; the reason for this failure to increase is unknown, but should be possible to work out experimentally. The algebra of populations carrying "sex-ratio" is difficult, but needs doing in order to judge the magnitude of the necessary unknown counter agent.

In all three cases where the point has been checked (race A, race B, *affinis*) "sex-ratio" is associated with an inversion; the only other inversion in this arm of the X that has turned up is that differentiating race A and race B. The "normal" sequence of race A is identical with the "sex-ratio" sequence in race B. The relation of the "sex-ratio" of these inversions is unclear, but seems to be significant.

VI. Y-Chromosomes

Seven visibly different types of Y occur in wild strains, each having a definite geographic range. These ranges show some correlation with those of the inversions in III, though by no means a complete one. From this source we hope to get valuable checks on the historical problems raised by the inversions.

VII. Modifiers

The different wild stocks differ in the modifiers they carry for various measurable or countable mutant characters. The analysis of these has only begun, but should be of interest in connection with such population problems as those raised by Fisher's theory of dominance. It should be possible also to construct distribution maps that will help with the historical problems.

VIII. Mutant Genes

Not infrequently wild stocks are found to be heterozygous for visible mutant characters (recessives). Some of these recur often enough so that adequate measurements of their frequency may be possible. One of them, orange eye-color, appears to be more frequent along the Pacific coast than in Mexico or the Rocky Mountains. Another (cinnabar eye-color) has appeared in several localities in race *B*, but never in race *A*.

IX. Related Species

There are at least 12 fairly closely related species (7 American, 5 European, 5 in the eastern states). We have 5 in the laboratory now, and several of the others are known to be breedable. There are two known species hybrids *(pseudoobscura-miranda* and a new one from an Alaskan species and one of Mexico), and others are obtainable— especially if one uses the gonad-implantation technique. The northwestern species referred to above is interesting in that it appears to be ecologically equivalent to *pseudoobscura,* the two replacing each other as one moves away from the rather narrow zone (British Columbia to Colorado) where they occur together.

Here in this document are seeds of the GNP series: chromosomal and genic variability, and use of the variability to deduce population parameters or to discriminate between the hypotheses of theoretical population geneticists. Two differences with the series should also be noted. The stocks of *D. pseudoobscura* already collected are considered adequate, and no mention is made of either comparisons of variability in local populations or of such population dynamics as seasonal changes in frequencies of genes or inversions.

Wright replied to Sturtevant on April 10: "There are few things I would like to do more than spend a year on such a project. . . . I am tremendously interested in the work with *Drosophila obscura* as it seems to be much the most favorable animal material for genetic study of problems of speciation." But Wright could get only one academic quarter off and decided to spend the fall quarter, September to December, in Pasadena with Sturtevant and Dobzhansky. The plan to have Sturtevant, Dobzhansky, and Wright all collaborate on the ambitious GNP project never materialized for two primary reasons. First, Dobzhansky received an offer from Dunn at Columbia to give the prestigious Jesup Lectures during the fall quarter when Wright was planning to be in residence at Cal Tech. Dobzhansky pleaded with Dunn to change the timing of the Jesup Lectures ("Dr. S. Wright of Chicago is coming here for the fall term [September-December], and I would like very much to be present in order to learn from him all I can"), but Dunn could not do it. The other reason is that in May 1936 Dobzhansky and Sturtevant experienced a deep rift in their friendship and collaboration, and they initiated no new joint research projects after that time (for detailed explanations, see Provine 1981).

The two papers that resulted from their research already completed at that point were, however, the real beginnings of the whole GNP series. The first, sent to press on June 8, 1936, announced in a preliminary note the surprisingly high number of third-chromosome inversions they had found—fourteen. Even more exciting was their announcement that the overlapping inversions could be used to suggest phylogenies, and included was the first published diagram of *D. pseudoobscura* phylogeny. *Drosophila miranda* was not included in this first diagram, and only seven of the fourteen inversions were included. Finally, in this brief three-page paper they reported that their preliminary studies indicated that "from 15 to 29 percent of the third chromosomes tested from wild stocks have been found to carry lethal genes. This raises the question as to whether such stocks are now in a condition approaching that of balanced lethals" (Sturtevant and Dobzhansky 1936b, 50). Certainly this little paper was filled with suggestion.

The full report was not submitted for publication until August 23, 1937. Dobzhansky was now the senior author, although Sturtevant had been in the preliminary version. The research was essentially completed but not written up at an earlier time because Dobzhansky had reported most of the data in the paper in the manuscript of *Genetics and the Origin of Species,* submitted to Dunn on April 26, 1937. The famous diagram of *D. pseudoobscura* phylogeny, which appeared on page 93 of the first edition, is reproduced exactly in the 1938 paper (Dobzhansky and Sturtevant 1938, 51), where the diagram encompasses eighteen inversions, including the "hypothetical A" of *D. miranda* in the center. The paper is wonderfully detailed, with careful geographical distribution maps of the inversions and a foldout page with drawings of the configurations observed in ten inversion heterozygotes. This is the foundation paper referred to in the first seven papers (excepting VI, on *Linanthus parryae,* a plant) of the GNP series as the basic source of information about using inversions to study natural populations. This paper became the source to which geneticists turned when they wanted to study inversions in *D. pseudoobscura.*

This final joint paper of Dobzhansky and Sturtevant was not Sturtevant's last effort in the direction of the 1936 outline sent to Sewall Wright. It was Sturtevant who had been studying genic variation in the stocks brought by Dobzhansky into the laboratory, and his paper on the subject, "Autosomal Lethals in Wild Populations of *Drosophila pseudoobscura,*" was published in December 1937. The first paragraph pointed out why the study of lethals was useful and important in the study of natural populations.

It has been recognized for some years, following the work of Muller, that lethals are especially convenient material for the study of mutation rates. This is because they occur with a frequency that is great enough to be measured, and because their occurrence can be detected by a technique that is independent of the personal equation of the observed. These same two advantages apply to the use of lethals in the study of the constitution of wild populations. Two other advantages are also to be noted in this field. Owing to extensive studies on the mutation rates of lethals, there is available a large body of data on the frequency of the occurrence of new lethals under a variety of conditions. On the other hand, the rate of elimination of lethals from a population furnishes the minimum possible difficulty of quantitative estimations. (Sturtevant 1937, 542)

Sturtevant was saying that studies on the frequency of lethals could make possible the estimation or calculation of the parameters of populations, such as the effective population size. Intellectually speaking, the GNP series could just as easily have begun with Sturtevant's 1937 paper as with GNP I by Dobzhansky and Queal.

As Sturtevant turned away from the study of natural populations of *Drosophila pseudoobscura*, Dobzhansky increasingly directed his attention to this research. But if Sturtevant and Dobzhansky together thought Wright was essential for the success of the GNP project, one can easily imagine how much Dobzhansky, working without Sturtevant, wanted Wright to join the project.

Dobzhansky and Wright: Genetics and the Origin of Species

Even as Dobzhansky began the research on the first paper of the GNP series, Wright had already exerted a strong influence upon him. Going back to the spring of 1936, Dobzhansky had accepted Demerec's offer to come to Cold Spring Harbor for the summer and Dunn's offer to give the Jesup Lectures at Columbia University in the fall. He therefore had two major topics on which to consult with Sewall Wright: the proposed study of the genetics of natural populations, and the evolution theory, which would be an essential part of his book from the Jesup Lectures. Thus when Dobzhansky left for Cold Spring Harbor in late June, he spent a day with Sewall Wright on the way. This was the first of many visits he would pay to Wright during the following ten years.

After delivering the Jesup Lectures in October and early November 1936, Dobzhansky returned to Pasadena in the middle of the month, only to discover (as he lamented in a November 18 letter to Demerec) that Wright had gone to Berkeley for a week. Nevertheless, in a period of less than three weeks, Dobzhansky began to crystallize his general view on the relation of theoretical population genetics to experimental genetics, especially the genetic study of wild populations. He then wrote the book in four months, sending the complete manuscript to Dunn in late April, and had the first published copy in hand in September—less than one year from the time he delivered the first Jesup Lecture. This was a thoughtful, formative year for Dobzhansky.

For the first time, he had been compelled to take time away from the daily routine of experimental work and to reflect upon the large evolutionary questions that had so interested him from boyhood on.

Dobzhansky consciously modeled *Genetics and the Origin of Species* after Darwin's *Origin*. Like Darwin, Dobzhansky began his book with a study of variation. Most of Dobzhansky's analysis dealt with gene mutation or chromosomal changes (translocations, inversions, duplications, etc.) about which Darwin knew nothing. The argument accompanying the recital of details was that these same variations found within populations were the raw materials for microevolution (below the species level), speciation, and all higher evolution. Dobzhansky had been making this argument since 1924, and he presented it forcefully.

The heart of Dobzhansky's evolutionary argument can be found in his chapters 5 ("Variation in Natural Populations") and 6 ("Selection"). To the extent that Dobzhansky makes in these two chapters an argument about mechanisms of evolution in nature, he relies almost totally upon Sewall Wright for the substance and terms of the argument.

From its title, "Variation in Natural Populations," one might expect Dobzhansky's chapter 5 to examine genic and chromosomal variations in natural populations. Dobzhansky actually assumes the presence of this variability from the previous chapters and instead examines the fate of this preexistent variability in natural populations. Thus the chapter is really about differentiation in local populations leading to racial differentiation. A theory of microevolution is therefore required.

Beginning with a very elementary presentation of Wright's gene frequencies, Dobzhansky goes directly to the centrality of the effective population size, N: "The great significance of the population number, N, comes from the fact, emphasized by Wright, that in a population of N breeding individuals, $1/2N$ genes either reach fixation or are lost in every generation" (Dobzhansky 1937, 132). The argument continues:

> Wright considers the situation that may present itself in a species whose population is subdivided into numerous isolated colonies of different size, with the exchange of individuals between the colonies prevented by some natural barriers or other agents. As we shall attempt to show below, such a situation is by no means imaginary; on the contrary, it is very frequently encountered in nature. (P. 133)

The consequence, according to Dobzhansky, was differentiation of local populations by random drift, leading in time to similar differences between races. The amount of differentiation, of course, was dependent upon the severity of population constrictions:

> The conclusion arrived at is an important one: the differentiation of a species into local or other races may take place without the action of natural selection. A subdivision of the species into isolated popula-

tions, plus time to allow a sufficient number of generations to elapse (the number of generations being a function of the populations size), is all that is necessary for race formation. This statement is not to be construed to imply a denial of the importance of selection. It means only that racial differentiation need not necessarily or in every case be due to the effects of selection. (P. 134)

Every example of variation in natural populations presented in the chapter is designed to show the direct effect of random genetic drift in small populations. Wright's favorite examples from Gulick, Crampton, Kinsey, and Schmidt appear, along with Edgar Anderson's *Iris* conclusions (Anderson had worked closely with Wright on this research) and Dubinin's work on wild populations of *Drosophila melanogaster*.

In a section on population size, Dobzhansky emphasized that N is generally far smaller than the number of individuals observed in a population and concludes the chapter with the statement: "With the present status of our knowledge, the supposition that the restriction of population size through the formation of numerous semi-isolated colonies is an important evolutionary agent seems to be a fruitful working hypothesis" (P. 148). Dobzhansky was not arguing in this chapter that random drift was the single primary mechanism. Fisher, after all, had denied *any* significance to random drift as a mechanism of evolution in nature. Dobzhansky understood well the effect of N upon selection and drift:

> The foregoing discussion shows how important is the effective size of the breeding population of a species for its evolutionary perspectives. If population sizes in most species tend to be small on the average, the scattering of the variability and the random variations of the gene frequencies will loom large as evolutionary agents. If, on the other hand, the population sizes are usually so large that they may be regarded for all practical purposes as infinite, the evolutionary role of these agents is negligible. (P. 138)

Both random drift and selection could be expected to play major roles in the evolutionary process.

The next chapter, "Selection," must be viewed from the perspective of the chapter just discussed. "Selection" begins with a recital of documented cases of natural selection (not a very convincing recital, in my opinion, but not Dobzhansky's fault since the evidence did not exist in 1937). The intellectual meat of the chapter begins, however, with the section entitled "Selection in Populations of Different Size," in which Dobzhansky discusses the interaction of selection and random drift. He drew explicitly upon Wright's concept of a balance of mechanisms of evolutionary change:

> Although in the abstract the differentiation may be pictured as taking place under the influence of the restriction of the population size alone, or under the influence of selection alone, in nature the process

is going on because of both these factors, one or the other gaining the upper hand probably only temporarily. The genetic equilibrium in a living species population seems to be a delicately balanced system which can be modified by a number of agents. (P. 184)

The question then becomes What is the relative efficacy of the mechanisms involved in the "balance" theory? In the conclusion to the chapter, Dobzhansky presented and discussed Wright's diagrams from the 1932 paper, showing the results of evolution in populations under a variety of conditions. Then he clearly opted for Wright's shifting balance theory:

> Wright (1931, 1932) argues very convincingly that . . . a differentiation into numerous semi-isolated colonies is the most favorable one for a progressive evolution. Indeed, a number of considerations speak in its favor. The present writer is impressed by the fact that his scheme is best able to explain the old and familiar observation that races and species frequently differ in characteristics to which it is very hard to ascribe any adaptive value. Since in a semi-isolated colony of a species the fixation or loss of genes is to a certain degree independent of their adaptive values (owing to the restriction of the population size), a colony may become different from others simultaneously in several characters. One or a few of the latter may be adaptive, and may enable the population to conquer new territories or ecological situations. The rest of the characters may be neutral with respect to adaptation, and yet they may spread concomitantly with the adaptive ones. For example, the chromosome structures that are so variable in *Drosophila pseudoobscura* can hardly be regarded as anything other than neutral characters, although some of them have become racial characteristics in subdivisions of the species populations. (Pp. 191–92)

Dobzhansky then ends the chapter with a long quote from the conclusion of Wright's 1932 paper, "The most general conclusion is that evolution depends on a certain balance of its factors . . . " It should be obvious that Wright in 1931–32, and Dobzhansky in 1937, believed that random genetic drift was a prominent factor in the balance.

I will argue later that both Dobzhansky and Wright became more selectionist in their outlooks through the 1940s and 1950s, so the "balance" had more selection and less random drift at levels above that of local populations, where drift might still be of central importance (thus supporting Gould's thesis of the hardening of the synthesis: Gould 1980, 1982, 1983). To give just one example here, when Dobzhansky introduced the section on random drift in *The Genetics of the Evolutionary Process* (1970; originally intended to be the fourth edition of *Genetics and the Origin of Species*), he stated: "Sometimes referred to as the 'Sewall Wright principle,' it has been misused in a way Wright himself never intended, namely, as a spurious explanation of evolutionary changes that seem to be devoid of adaptive significance, and therefore hard to explain by natural selection" (Dobzhansky 1970, 231; compare with Wright 204, p. 839).

I emphasize that Wright as well as Dobzhansky shifted views on this issue. In March 1937 Dobzhansky sent the draft of his chapter on selection to Wright for comment before publication. Wright wrote back on April 9 to say: "I have gone over the draft of Chapter 6 of your book a couple of times. It seems to me a most interesting and in all ways satisfactory account. I am, of course, glad that you found so much value in my theoretical approach to the matter." Wright then went on to say that Dobzhansky's treatment of the evolution of dominance was too simplistic and that his analysis of the evolutionary effects of selection under fluctuating conditions should take into account the advances in Wright's view since the big 1931 paper. Dobzhansky revised the manuscript accordingly. But Wright objected not at all to Dobzhansky's characterization of the action of random drift, although he had every opportunity to do so.

Looking toward the GNP series, *Genetics and the Origin of Species* offers three primary insights: Wright's influence upon Dobzhansky's thinking was already enormous before the GNP series was officially begun. Dobzhansky's theory of evolution was Wright's theory of evolution. Dobzhansky indeed has never claimed any originality for the theory of evolution he presented in *Genetics and the Origin of Species,* as he explained in his oral memoir:

> The reason why the book had whatever success it did was that, strange as it may seem, it was the first general book presenting what is nowadays called . . . "the synthetic theory of evolution." I prefer to call it "biological theory of evolution." I certainly don't mean to make a preposterous claim that I invented the synthetic or biological theory of evolution. It was, so to speak, in the air. People who contributed most to it I believe were R. A. Fisher, Sewall Wright, and J. B. S. Haldane; their predecessor was Chetverikov. What that book of mine, however, did was, in a sense, to popularize this theory. Wright is very hard to read. (Dobzhansky 1962, 398–99)

Although Dobzhansky mentioned both Fisher and Haldane as well as Wright in this comment, I can find no evidence that Dobzhansky ever actually read or was influenced in any way by either Fisher or Haldane before writing *Genetics and the Origin of Species*. His conception of their work appears drawn entirely from Wright.

At Ernst Mayr's conference on the evolutionary synthesis in May 1974, Lewontin asked Dobzhansky a very pertinent question. He asked how Dobzhansky had been able to popularize and incorporate so deeply into his book an evolutionary theory the quantitative details of which he could not read or understand.

Dobzhansky answered this same question in the oral memoir in the section quoted above, immediately after saying, "Wright is hard to read":

> He has a lot of extremely abstruse, in fact almost esoteric mathematics. Mathematics, incidentally, of a kind which I certainly do not claim to understand. I am not a mathematician at all. My way of reading Sewall Wright's papers, which I still think is perfectly defensible, is to examine the biological assumptions the man is making, and to read the conclusions he arrives at, and hope to goodness that what comes in between is correct. "Papa knows best" is a reasonable assumption, because if the mathematics were incorrect, some mathematician would have found it out. (Dobzhansky 1962, 399)

The evidence from the Dobzhansky-Wright correspondence in the 1930s supports the accuracy of this statement by Dobzhansky. Thus when Wright published in 1937 his important extension of the stochastic distribution of genes to include gene interaction, grades of dominance, and different kinds of selective forces (Wright 92), Dobzhansky read it and replied:

> Just read (or tried to read?) your paper. . . . I am delighted to see it, although my mathematical understanding is far too insufficient to read and understand it completely. But, I have done the same thing that I have with other papers: read the part of the text preceding and following the mathematics, skipped the latter in assurance that to it the expression "papa knows how" is applicable. (Dobzhansky to Wright, July 25, 1937)

I see nothing in the letter to indicate that Dobzhansky actually understood what Wright accomplished in the paper.

In *Genetics and the Origin of Species* Dobzhansky took the final (qualitative) section, "The Evolution of Mendelian Systems," from Wright's 1931 paper and his 1932 paper, using them as the theoretical structure for his whole book. Fisher and Haldane are mentioned occasionally, but it is Wright's theory that is the backbone of Dobzhansky's presentation of evolutionary theory. The theoretical work of Fisher, Haldane, and Wright had set for Dobzhansky the terms of the debate about mechanisms of evolution in nature. Was Fisher's panselectionism correct, or Wright's shifting balance theory with its emphasis upon population subdivision and random genetic drift?

I point out this tension between Fisher and Wright here because it permeates the theoretical aspects of *Genetics and the Origin of Species* and in particular forms the basis for research problems on natural populations. The study of effective population size, genic and chromosomal variability, and selection were meaningful in the context of the Wright-Fisher debate. To choose one example, Dobzhansky says of selection: "The manner of action of selection has been dealt with only theoretically, by means of mathematical analysis. The results of this theoretical work (Haldane, Fisher, Wright) are however invaluable as a guide for any future experimental attack upon the problem" (Dobzhansky 1937, 176).

Writing *Genetics and the Origin of Species* crystallized in Dobzhansky's mind the relation between Wright's theory of evolution and the experimental study of natural populations. From the spring of 1937 on, Dobzhansky had little use for any experimental study on natural populations unless a substantive point of evolutionary theory were at stake. Data, no matter how interesting, were worthwhile only if they elucidated, refined, tuned, disproved, or discriminated between theoretical models of evolutionary change. The models of such persons as Fisher, Haldane, and Wright should be the guides to experimental study of natural populations. Dobzhansky virtually dismissed, late in 1937, two papers he had published on the geographical variation of the Y chromosome in *Drosophila pseudoobscura* only a short time before. The data were interesting, yes, but unimportant because no connection between the data and evolutionary theory could be made. Dobzhansky hated wasting his time gathering data that had no decisive meaning for theory. A view that Dobzhansky expressed many times after 1937 was that

> Genetics is the first biological science which got in the position in which physics has been for many years. One can justifiably speak about such a thing as theoretical mathematical genetics, and experimental genetics, just as in physics. There are some mathematical geniuses who work out what to an ordinary person seems a fantastic kind of theory. This fantastic kind of theory nevertheless leads to experimentally verifiable prediction, which an experimental physicist then has to test the validity of. Since the times of Wright, Haldane, and Fisher, evolutionary genetics has been in a similar position. (Dobzhansky 1962, 500–501; see also 354)

And finally, *Genetics and the Origin of Species* sounded again and again the clarion call for research on natural populations. Regarding N, Dobzhansky implored:

> It is no exaggeration to say that the conclusions which eventually may be reached on the dynamics of the evolutionary process will depend in no small degree on the information bearing on the problem of population numbers. The dearth of pertinent data in the existing biological literature is, therefore, most unfortunate. Anything like accurate statistics even on the total number of existing individuals of a species is a rarity, and such data are far from being what one would desire to know for the determination of population numbers. (Dobzhansky 1937, 138–39)

Genetics and the Origin of Species unmistakably carried the advice to get out and test or discriminate between the models of the theoreticians by studying natural populations. Even before the book was published, Dobzhansky was following his own advice.

When *Genetics and the Origin of Species* was published in September 1937, Dobzhansky immediately sent a copy to Wright, who responded in a letter of October 22, 1937:

> I want to thank you for sending me a copy of your book and to congratulate you most heartily on it. I have just finished reading it. It seems to me to be by far the best synthesis that has come out.
>
> I have not used any text hitherto in an advanced course on evolution that I give, but your book is just what I need.

Dobzhansky replied: "Needless to say, I was much elated over your letter and your compliments regarding my book. The plain fact is that there is nobody anywhere whose compliments in such matters I would value as highly as yours" (Dobzhansky to Wright, November 2, 1937).

Neither Wright nor Dobzhansky appears to be exaggerating here. Wright used *Genetics and the Origin of Species* as the major textbook of his evolution course every time he taught it from 1937 until his retirement in December 1954, using the second and third editions as they appeared. And Dobzhansky unquestionably valued Wright's compliments very highly, as the following analysis of the GNP series will demonstrate.

Dobzhansky and Wright: Origins of the Genetics of Natural Populations Series

The Jesup Lectures and *Genetics and the Origin of Species* brought Dobzhansky financial support. In the spring of 1937 the Carnegie Institution of Washington, spurred by the recommendation of Demerec, awarded Dobzhansky a grant of about $2,000 to continue work on natural populations of *Drosophila pseudoobscura*. Dobzhansky, who was incapacitated by a severely broken leg in the spring of 1937, hired a young assistant named Marion Queal to help with the collections and analysis. Prodigious labor was required for Dobzhansky's experiments. The Carnegie Institution of Washington continued to support Dobzhansky's work every year up to the early 1950s; without this help, there is some doubt whether Dobzhansky could have produced the twenty papers in the GNP series between 1938 and 1952.

In the spring and summer of 1937, when *Genetics and the Origin of Species* was in press, Queal, Dobzhansky, and others collected from isolated mountain ranges in the Death Valley region the material that would form the basis of GNP I and II. The plan was basically to use the same material to study the geographical distribution of inversions and the genic variability. The new data left Dobzhansky with a major problem. He wanted to gather from natural populations data that were important for evolutionary theory, particularly the breeding structure of populations. But Sturtevant was the only one at Cal Tech who could even begin to grasp the quantitative theory of breeding structure and who had substantive thoughts about the interpretation of data in terms of theory. Dobzhansky therefore turned to Wright, who was a better choice anyway, as Sturtevant himself recognized.

GNP I presented little problem in this regard. The data on the relative frequencies of three different third-chromosome inversions indicated that, in the eleven populations studied, wide differences between populations were uncorrelated with their geographical distribution. This was exactly the result Dobzhansky expected and hoped to see. He and Sturtevant both thought the inversions were selectively neutral. If local populations were small, or went through seasonal populational bottlenecks, Dobzhansky expected that, from what he knew of random drift, it should lead to different frequencies of the inversions even in closely adjacent populations.

Dobzhansky's expectations were amply fulfilled by his initial analysis of the new data. On July 25, 1937, he wrote to Wright:

> I am now working on the project which we have discussed with you last winter, i.e., populations from the isolated mountain ranges in the Death Valley region. Thus far it was nothing but a lot of hard work for four persons, but just a few days ago I have taken time to make a few summaries and calculations, and can now see that something of interest seems to be emerging. All ranges (eleven of them studied) proved to have the same four chromosome structures in the populations. But the frequencies of these four are different on almost every mountain, and, what seems to me especially interesting, *adjacent mountain ranges show no tendency to have more similar populations than relatively more remote ones*. It seems to me that this is exactly what one might expect on the basis of your views on the role of isolation in differentiation of local populations.

Wright did not reply to this letter until October 22, 1937, when he wrote to thank Dobzhansky for having sent a copy of *Genetics and the Origin of Species*. Wright's comments about the research were simply: "I was much interested to hear last summer of the results which you are getting from the Death Valley populations."

On September 3, Dobzhansky wrote to Dunn about the work:

> The summer is about over. Although I am very tired, I can enjoy the interesting things which have emerged from the summer work on the genetic structure of wild populations. The results are only now starting to come in, but unless the first data are deceiving, we have a proof of the genetic differentiation of the population due to isolation only without, and even despite, the influence of natural selection. But—it is still uncertain, and it might be wiser for me to refrain from talking too much about these things before the rest of the material comes in, which now, fortunately, is not too long to wait.

Dobzhansky was fully prepared to see Wright's concept of random drift at work in his populations.

It is interesting, however, that Dobzhansky and Queal stated in the published paper that the random drift explanation was not a "proof" (as in the letter to Dunn above) but a hypothesis requiring further verification:

> The factor that is decisive for an evaluation of the above hypothesis is how large is the effective size of the breeding population in the colonies of the flies inhabiting each mountain range? The greater the size, the shorter the time interval during which this differentiation may be supposed to have taken place, the less probable is the hypothesis, and vice versa. We hope to be able to present some information on this problem at a later date. (Dobzhansky and Queal 1938a, 250)

Dobzhansky dearly wished to have an estimate of effective population size in natural populations. Indeed, determination of N was crucial in theory, as Wright and Fisher had made abundantly clear. What Dobzhansky unquestionably intended to do was supplement the data on recessive lethals from GNP II and use them to calculate the effective population size. Dobzhansky had learned this possibility from Sturtevant, who had indicated in his 1937 paper on lethals in populations of *D. pseudoobscura* that one could easily calculate the rate of elimination of lethals from a population. By comparing the measured rate of occurrence of lethals with the measured rate of origin of lethals, Sturtevant thought an estimation of effective population size was possible. The suggestion was that if the rate of occurrence of lethals were less than expectation based upon the rate of origin of lethals, the cause was probably the more rapid elimination of lethals when the population was small in the beginning of the season due to the inbreeding effect of small effective population size. In other words, the lethals had been eliminated more rapidly than expectation because of increased homozygosity from the inbreeding inevitable in small populations.

Although Dobzhansky himself did not have any clear idea of exactly how to compute effective population sizes from his data, he was convinced, on the basis of Sturtevant's confidence, that Sewall Wright could do it. Accordingly, Dobzhansky (on his way to the East Coast) arranged to visit with Wright in late November 1937. During this visit, Wright apparently expressed an interest in helping Dobzhansky relate his data on natural populations to evolutionary aspects of breeding structure. For his part, Dobzhansky planned experiments to obtain data on the mutation rates of lethals in third chromosomes of *D. pseudoobscura* and also on the allelism of lethals. He thought these data were all that were additionally necessary for Wright to calculate effective population sizes.

In a letter of April 14, 1938, Dobzhansky sent Wright the just-completed manuscript of GNP II, and said he had almost completed the work on allelism of lethals and was beginning that on the mutation rate of lethals. "After that, if the data will be reasonable, we might perhaps put in effect the plan of writing a joint opus with you—the analysis of the data obviously better be done by yourself."

Dobzhansky submitted the draft of GNP II (Dobzhansky and Queal 1938b) for publication at the same time he sent it to Wright. By the time Wright replied on June 2, 1938, Dobzhansky had happily announced that the next step was to gather data on allelism of lethals and mutation rate of lethals and, when added to the data in GNP II on frequencies of lethals, to calculate effective population sizes. On the face of it, this was an exciting and very important prospect.

Wright, however, raised some important objections in his letter. Dobzhansky had assumed (as had Sturtevant in 1937) that the lethals (as opposed to semilethals) were wholly recessive in heterozygous combinations. Wright calculated from Dobzhansky's own data that heterozygous lethals were selected against. Since the number of heterozygous lethals had to be vastly more numerous than homozygous lethals, this adverse selection "is going to make it difficult to make any estimates of effective size of populations from comparisons of the rate of occurrence of lethals and their accumulation. Nevertheless I think that data on the mutation rate of lethals comparable to those accumulating in nature is of great importance for its own sake" (Wright to Dobzhansky, June 2, 1938). Wright was suggesting that the measured rate of occurrence of lethals was less than expected on the basis of the rate of origin of lethals because selection against lethals in heterozygous combinations had decreased the frequency of lethals, not that inbreeding was the cause of the disparity. A slight selection against the heterozygote was far more effective in eliminating lethals from the population than even complete selection against the homozygous lethals. Wright added that the compound coefficient Nm (effective population size × migration index = effective number of immigrants) was much more easily determined, and he made some sample calculations. But he emphasized the impossibility of obtaining a separate estimate of N by itself.

Dobzhansky was upset by Wright's letter. He had gone into the GNP series intending to produce data that would inform theory. In particular, he was sure the data on lethals would yield an estimate of effective population size. No, said Wright, but the data were interesting in their own right. Dobzhansky could not accept this. The result was an extraordinary series of letters involving Sturtevant and Waddington in addition to Wright and Dobzhansky. This episode would profoundly alter Dobzhansky's conception of acceptable experimental design. It would have a great effect upon the GNP series, upon the collaboration of Wright and Dobzhansky, and, through them and the series, upon all of evolutionary biology. I consider this episode to be one of the most crucial for understanding the origins of modern genetic analysis of natural populations.

For the time being, Dobzhansky simply excised from the manuscript of GNP II his statement about estimating effective population size. Replying to Wright's June 2 letter on June 6, Dobzhansky said that he did not think the lethals were incompletely recessive and hoped to have some further data to prove that soon. He provided his available data on the allelism of lethals and formulas from Muller, Morgan Ward, and K. Mather for calculating the number of mutable loci from the allelism of lethals. "I am more than a little lost in the woods," he said to Wright, "and beg your authoritative opinion." Wright replied on June 22 that he could not see where Muller, Ward, or Mather had gotten their formulas; but when Muller had published his in 1929, Wright had written him to propose a better formula, which Muller thought was correct. Wright rederived this formula for Dobzhansky in three tough-looking pages and ended by saying that Muller had used it to calculate the number of mutable loci for all chromosomes in *Drosophila melanogaster*. The result was a minimum of 200 and a maximum of infinity; Wright told Dobzhansky that that result "is not very illuminating." Undaunted, Dobzhansky still believed the data on lethals could be used to estimate something of importance about breeding structure.

In early August Dobzhansky spent a day with Wright in Chicago. Wright undoubtedly told Dobzhansky again that he could not use his data on lethals to estimate effective population size. And again Dobzhansky was unsatisfied. He really thought Wright should be able to turn the trick. Back in Pasadena, Dobzhansky consulted Sturtevant who, after all, had put the idea in his head in the first place. Dobzhansky was unable to convey Wright's objections to Sturtevant, who wrote to Wright on September 16: "I've been figuring on this business of elimination of lethals, and the calculation seems to me to be extremely simple and easy. Have I made some fool slip, or have I missed the point of what you are after?"

Sturtevant's argument is worth examination. The frequency of third-chromosome lethals could be determined as already described above. The frequency of homozygote lethals (that is, zygotes with two third chromosomes, each bearing at least one lethal) was then the frequency of lethals × frequency of lethals. Multiplying the frequency of homozygote lethals by the frequency of allelism of lethals then should yield the frequency of zygotes homozygous for the same lethal, this being the frequency of elimination of lethals from the population. This frequency could then be compared with the experimentally determined frequency of mutations causing new lethals. Citing the data of Dobzhansky and Dubinin as indicating a much higher frequency of mutations than frequency of elimination, Sturtevant concluded that close inbreeding must occur at times of population minima.

The argument was, as Sturtevant had suggested, "extremely simple and easy." The final paragraph in Sturtevant's letter to Wright indicates, however, that Dobzhansky had not hitherto grasped the argument: "Probably all this is perfectly familiar to you, and I've missed the point you're trying to get at—but when I tried it on Dobzhansky he seemed to think it was new stuff to him." Still, all along Dobzhansky had thought that an argument like Sturtevant's made possible the deduction of effective population sizes. So now Dobzhansky's former mentor, whom he respected greatly, had pointed out clearly how Wright could begin the calculation of effective population sizes. Before turning to Wright's reply, one element of the view shared by Sturtevant and Dobzhansky must be made absolutely clear. They both believed that a population bottleneck, with its concomitant increase in inbreeding, necessarily caused a decrease in effective population size. That was why they deduced a small effective population size when the frequency of lethals was less than expected from measured mutation rates. In taking this view, both believed they were following the spirit and the letter of Wright's evolutionary theory as presented in his 1931 and 1932 papers. Indeed, they thought the evidence from *D. pseudoobscura* was a direct confirmation of Wright's shifting balance theory of evolution in nature.

Wright's reply (dated October 6, 1938), addressed to both Sturtevant and Dobzhansky, was therefore a shock to both of them. He began his eight-page, single-spaced typewritten letter by stating that on the question of deducing population structure from the frequency of lethals, he would be "disappointingly negative." Sturtevant, he said, had assumed in his argument that the lethals were "1) complete recessives accumulating in 2) indefinitely large 3) random breeding populations in nature." But Dobzhansky's data were inconsistent with each of these three assumptions, and alteration of any of the three was sufficient to account for the discrepancy between the frequency of elimination and the frequency of occurrence of lethals.

First, if the lethals were not wholly recessive, then selection against the heterozygotes could easily explain the lower-than-expected frequency of lethals. Wright had already made this argument to Dobzhansky in an earlier letter as explained above, but he developed it more fully here. Second,

> Another possible reason for deficient accumulation of lethals is a departure from random mating *within* local populations. If matings between close relatives, especially brother with sister, occur more frequently than expected by chance, the proportion of homozygotes will obviously rise above that expected from the binomial square rule, giving more rapid elimination. *This effect of inbreeding has nothing to do with the effective size of population in relation to the differentiation of populations.* [emphasis added].

Here the surprise comes in the final sentence, since both Sturtevant and Dobzhansky had assumed that the hypothesized inbreeding resulted in smaller effective population sizes. I will explain this apparent paradox below, as Wright did later in his letter. Finally, Wright demonstrated that in order to significantly reduce the frequency of lethals in a population, the effective size would have to be surprisingly small. Using approximate figures from the data of Dobzhansky and Sturtevant, who had suggested a very low migration rate,

Wright calculated that the effective size of one of Dobzhansky's populations would, to reduce the frequency of lethals to the observed level, have to be about 170. Wright added: "This seems an absurdly small figure for the effective sizes of populations sufficiently isolated that immigration is much less important than mutation as a source of lethals." From these three arguments Wright concluded: "I think that inferences on effective population size must be based almost wholly on observed differentiation of populations, for which of course the data from the inversions is much better than those from lethals."

The question that Sturtevant and Dobzhansky undoubtedly raised, at this point in Wright's letter (after five pages) was why, if inbreeding to the extent of brother-sister mating were occurring at some stage of the population cycle, was the population thereby not reduced to a very small effective size? Wright had a ready answer. From Dobzhansky's data from the canyons of Death Valley he tentatively concluded that "the differentiation among the canyon populations is probably an annually shifting affair, not a semipermanent drift—as is probable between whole ranges and certainly between state-wide regions." In other words, random drift might occur especially at times of population constriction, but the differentiation was not permanent.

> With very small effective values of N (the canyon populations) should differentiate very much more than observed unless there is much cross breeding at some phase of the annual cycle (making separate estimates of N of little significance). The same process may occur on a grander scale. The populations of whole canyons (in certain years) may start late because of local unfavorable conditions and be swamped by migrants from regions which have got an early start. Under these conditions, differentiation of a reasonably permanent sort would be encountered only between more isolated regions (such as the whole ranges).

Even so, Wright argued, with a breeding structure like that expected in the canyons of Death Valley, "the effective N even of an entire range may be surprisingly small." The conclusion therefore retained the possibility that random genetic drift might play an important role in the permanent differentiation in *D. pseudoobscura* seen in different mountain ranges, if not in the different canyons.

He reemphasized, however, that

> the deficiency of accumulated lethals cannot be largely explained by the smallness of this N . . . since the scheme still requires a great deal of brother-sister mating which would necessarily be much more effective in cutting down the frequency of lethals than the small total N. It accordingly seems that the most plausible hypothesis at present is that the deficiency of accumulated lethals is not due to any important extent to a small N of the semi-permanently differentiated populations (N much larger than 170, probably many thousands) but

mainly to prevalent brother-sister mating at some time in the annual cycle, aided by some selection against heterozygotes.

With this argument, Wright undercut the thrust of the GNP series as it existed in October 1938.

He was telling Dobzhansky in no uncertain terms that the data on frequency of lethals, mutation rates of lethals, and allelism of lethals, all of which Dobzhansky had painstakingly gathered for the ultimate purpose of deducing effective population size and other features of population structure, were in fact nearly useless for that purpose. Moreover, he totally contradicted the conclusions reached in GNP III (Koller 1939). With Dobzhansky's urging and guidance, P. C. Koller (who had earlier worked on the cytology of *D. pseudoobscura*) had examined the frequencies of inversions in populations inhabiting a single mountain range, instead of comparing frequencies between mountain ranges as Dobzhansky and Queal had done in GNP I. Finding differences in frequencies of inversions between populations only four miles apart, Koller concluded that the third-chromosome inversions were most probably neutral in adaptive value, and that

> the whole complex of facts at hand can consistently be accounted for on the assumption of random variations in the frequencies of the gene arrangements and without the intervention of selection. . . . The genetic differences between the canyon colonies as well as the yearly variations in the frequencies of the gene arrangements, are probably due to the smallness of the effective breeding size of the colonies. (Koller 1939, 32–33)

Thus far Wright would probably have agreed. But in extending the mechanism of differentiation of canyon populations to previously observed differentiation between mountain ranges, as Koller did (certainly with Dobzhansky's agreement), he went beyond what Wright said in his long letter. The differentiation between canyons was probably not permanent, according to Wright, whereas that between mountain ranges was at least semipermanent. Wright indeed suggested in his letter that "it would be very desirable to repeat Koller's observations for the same 6 canyons for one or two years more, to get a better idea of the degree of permanence in their differentiation."

Wright was accurate in describing his long letter to Sturtevant and Dobzhansky as "disappointingly negative." A scientist with less determination than Dobzhansky would have been discouraged. Dobzhansky, however, was spurred to greater efforts and remained fully determined to have Wright estimate N from the *D. pseudoobscura* data. The data in GNP II had indicated a weak dominant effect of the lethals (they were not fully recessive), as Wright had pointed out in his letter. Dobzhansky repeated this experiment in greater detail and discovered no statistically significant indication that the lethals were anything but fully recessive. Thus, he thought he had some experimental evidence against one of Wright's major objections to using the data on lethals to deduce effective population size.

On December 12, 1938, Dobzhansky wrote to Wright for the first time since Wright's long letter of October 6. "Your letter of October 6, addressed to Sturtevant and myself," Dobzhansky wrote, "was *studied* very carefully, and is being preserved as guide for the future." After telling Wright that his new experiment to detect any deviation from complete recessivity was totally negative, Dobzhansky admitted that a selection rate too small for experimental detection was quite sufficient to cause the deficient frequency of lethals observed. He suggested, however, that perhaps an upper limit for this selection rate could be estimated. Then he optimistically asked Wright again to estimate the value of N from his data:

> Now having the figures for the total frequency of lethals, for mutation rate, can one calculate N? In your letter you give an estimate of N based on the assumption that the selection rate is zero. This is 170. Why is this figure absurdly small? And the main point: can this problem be worked out to take into account the selection rate? If so, what kind of an estimate of N is arrived at?
>
> I am still working on the assumption that by the spring time when the figure for the mutation frequency will be ready we shall be able to write a joint paper with you, giving, with as many reservations as possible, an estimate of the value of N. Clearly, nothing of this kind could I do alone, and although I fully realize the extreme uncertainty that will be attached to this figure, I hope that you will not refuse to do it.

Dobzhansky simply would not give up. He was determined to get an estimate of effective population size from Wright.

It was almost four months before Wright again responded. Meanwhile, Waddington visited Cal Tech, talked to Dobzhansky about the calculation of effective population size, and could not understand why Wright felt any hesitancy about making the calculation. He wrote on February 22, 1939, to Wright, suggesting that the estimation of N should not be difficult. Wright on March 8 wrote Dobzhansky a forceful letter:

> Your new figures certainly indicate that there is extremely little dominant effect of the lethals found in nature. I think that this is a very important thing to know about such populations but I am pretty thoroughly convinced that the effective size of populations (as a factor in the differentiation of populations) can only be estimated by more direct means than the lethal situation—viz. the actual amount of differentiation, supplemented where practicable by actual enumeration of individuals.
>
> As I stated in my last letter I am fully in agreement with Sturtevant's view that the apparent deficiency of lethals in nature is probably due primarily to local inbreeding of a sort which brings about no

permanent differentiation of populations. The great changes in the frequencies of inversions from one year to the next at least in certain years indicate derivation from very few individuals, while the small amount of differentiation of canyon populations compared with that between ranges indicates that the canyon differentiation is not permanent. Since the temporary non-differentiating inbreeding is more effective in reducing lethal frequencies than the restriction of numbers responsible for differentiation, it follows that the lethal situation can not be used to determine the latter. In this connection, is it not possible that the high frequency of lethals in Mexican strains is due to the relatively small amount of temporary inbreeding? It would be interesting to find the correlation between the frequency of lethals and the frequency of the sex ratio gene in a large number of localities since both should increase with reduction in the amount of inbreeding.

The probability of temporary inbreeding is enough to prevent the lethal frequencies from being used to calculate effective size of population. The possible effects of partial dominance of lethals is also enough it seems to me. . . .

It seems to me that the lethal situation like the frequency of sex ratio is an important part of the total picture but that it can't be used to estimate effective size of population.

With respect to the latter, it seems to me that the most important thing is to establish the smallest units of population that have sufficient permanence of differentiation to indicate a cumulative inbreeding effect. Present indications seem to be that it isn't much smaller than the whole mountain range, but perhaps canyon populations are in some cases units of this sort even though in other cases their differentiation is merely an annually renewed consequence of local inbreeding.

Having distinguished the limits of cumulative and non-cumulative inbreeding, the next thing is to estimate the amount of the latter in various localities (Death Valley, Mount Whitney, San Gabriel, Mexican localities, etc.). . . . Some idea should be obtainable from the differentiation between successive years at the same locality in comparison with the differentiation between adjacent localities in the same year. The lethal situation may help here in suggesting limiting values. Finally comes the estimation of at least Nm for the permanently differentiated population units. Whether any separate estimates of N and m will be possible seems doubtful.

Not only did Wright with a high degree of finality say that the lethal data could not be used to estimate N but, even with much additional data on inversions or other indications of genetic differentiation, an estimation of N was doubtful. He did, however, hold out some hope of estimates of Nm together. Finally, Wright made an impression upon Dobzhansky, who replied:

Frankly, I must say that your letter of March 8th was discouraging to me. I did hope that at least an idea about the order of magnitude of the population number may emerge from studies on lethals, etc. But anyway I am thankful to you for having taken the trouble to put me straight, because evidently it is better to know that one is on an entirely wrong track and wasting one's time than to have illusions on the subject.

Since however I am deeply involved in this business, and must try to salvage what there is to salvage from the wreckage, I would like to ask you to help me again to clarify my ideas about the situation. I realize that you are a busy man and have not much time to waste writing letters, but you are the only person whom I may ask these questions, since, as you know, I do not understand the mathematics involved to be able to work without guidance. (Dobzhansky to Wright, March 12, 1939)

This letter signals a very significant development in Dobzhansky's thinking and in the progression of the GNP series. He had begun the work on lethals in *D. pseudoobscura* with the good and reasonable idea that the experimental data would lead to an important contribution to evolutionary theory—namely, an estimation of effective population size. After untold hours of intense labor, he discovered that his data could not be used for their intended purpose. What Dobzhansky now clearly realized, and never forgot, was that he must be certain *before* gathering any data about the theoretical relationships of the data to theory. But since Dobzhansky had little facility in quantitative evolutionary theory, his new realization meant that he had to check with Wright, or someone like him, before going into the field to gather data. When Sturtevant had originally written to Wright to ask for help in the study of natural populations (in March 1936), he said, "We need your help, not only in analyzing the data but also in planning sensible experiments." Dobzhansky now took the second half of Sturtevant's request seriously. He did not want to go into the field again and waste time gathering data, no matter how interesting, unless Wright said the data generated were really worthwhile and could serve the purpose Dobzhansky envisioned. Just how important Dobzhansky thought it was to check with Wright before commencing with data gathering will be obvious in the coming pages. Dobzhansky went to amazing lengths to talk with Wright about possible experiments. Only if one appreciates Wright's role in experimental design in the GNP series can the series be fully understood.

Dobzhansky began his careful checking with Wright about experimental design in the March 12 letter quoted above. Wright had mentioned in an earlier letter (the long one of October 6) that isolation of breeding populations might occur even in a continuous habitat and population. Dobzhansky proposed an experiment to test whether samples taken from a single small continuous territory were differentiated and whether the differentiation was ephemeral or semipermanent. He had obviously designed the experiment to test Wright's suggestion that the canyon populations were differentiated probably only very early in each breeding season. Dobzhansky wrote:

My present idea, which may prove all wrong to be sure, is that the population is semi-permanently differentiated into a very fine mosaic of colonies, "semi"—rather than simply permanently—in view of the drastic changes from year to year. Now, here again is a point on which you may set me straight. Taking your scheme of a continuously inhabited territory with small migration radii for individuals, is it possible that the population will represent a sort of a mosaic in which each constituent stone is constantly changing its size and its properties? In other words, could in such a mosaic conditions be found that would permit each small locality to have a population evolving independently from others, not only in the static sense of becoming different and remaining so for ever, but also in the sense of independently undergoing violent fluctuations *not connected* however with *seasonal* expansions and contractions? I have in mind the picture of a territory inhabited by birds with a strong homing instinct or whatever homing is. Imagine a village of swallows with a population of, say, ten pairs; there may be other villages in the near neighborhood, but if every bird and its offspring nests for a long time in the same village, exchange of individuals being rare between villages, could each village become distinct in its population, distinct in the sense that the frequency of a certain gene would go up this year in village A and next year in village B?

You may say that comparison of Drosophilas with swallows is far fetched. But who knows that this comparison is absurd? At any rate the question seems open to an experimental approach. Suppose, I am to take samples of flies in several canyons in May or June and find them to be different; sampling is repeated in August or September. If the differences found in June are ephemeral, due to local inbreeding in spring, they must exist no longer in September. Please, let me know whether you think this approach is valid, since this is not an imaginary plan but a plan of the actual work to start as soon as the flies appear outdoors. (Dobzhansky to Wright, March 12, 1939)

Dobzhansky also raised the question of his mutation-rate experiments, describing two techniques that would serve to check each other on the issue of recessivity. He asked Wright: "Do you think data so collected will tell something about the recessivity of newly arising lethals? The experiment has been planned on the assumption that they will, but your opinion will be appreciated." Who could blame Dobzhansky for asking the questions about experimental design after his recent experience?

Now we shall see, however, the other side of the coin. It is true that Dobzhansky needed Wright, but it is also true that Wright needed Dobzhansky. (I emphasize that this is my assessment and *not* Wright's during our interviews; he consistently denied that he "needed" the collaboration with Dobzhansky in any way and asserted that he was merely reactive in all his collaboration with Dobzhansky. I think the contemporary evidence supports my interpretation.) In the laboratory at the University of Chicago, Wright had cages upon cages of guinea pigs. Each offspring had to be carefully recorded. The crossing experiments by the early 1940s had become exceedingly complex. Wright was, after all, at this time primarily a physiological geneticist with a full research program. He could not possibly have undertaken a program of field research even vaguely resembling that of Dobzhansky. Yet Wright was keenly aware that his theoretical work on population genetics had to be complemented with observations from natural populations before it could be meaningful to evolutionary biologists in general. In the letter of October 22, 1937, in which Wright thanked Dobzhansky for sending a copy of *Genetics and the Origin of Species,* he added: "I also started a book on the Genetics of Populations intended to be a fairly complete review of the mathematical theory but got rather disgusted with my ratio of mathematics to observations when stretched over a whole book."

Fisher, Haldane, and Wright all said that population genetics theory needed to be based more upon knowledge of the genetics of natural populations. What Dobzhansky offered Wright was the chance to develop new theory in relation to intimate knowledge of the population genetics of at least one organism in the wild. Thus Wright's reply to Dobzhansky's letter of discouragement on the lethals question was positive and in many ways encouraging:

> I am sorry that my letter seemed discouraging. I did not mean at all that I thought the data on lethals worthless. I think that the accumulation of lethals in nature and the rate of occurrence are extremely important things to know about a population. Any interpretation of breeding structure must be consistent with these data. The analysis of breeding structure is so complicated a matter at best that every fact that bears on it at all is desirable. I am convinced, however, that there are so many unknown quantities connected with the lethal situation that by itself it can give little information of those aspects of breeding structure that are responsible for permanent differentiation. These can only be deduced from facts about differentiation itself where actual censuses at times of least numbers and migration statistics are lacking. (Wright to Dobzhansky, March 31, 1939)

Although Wright at this time remained adamant that the data on lethals alone could give little insight into population structure, he clearly did not wish in any way to discourage Dobzhansky from field research. Indeed, the letter as a whole was quite encouraging about what might be discovered concerning structure in natural populations. In particular, Wright encouraged Dobzhansky to pursue his hunch (expressed in his most recent letter to Wright) that even small units of populations exhibited semipermanent differentiation, saying that "the establishment of the size of unit at which there is reasonably permanent differentiation seems to me the most essential point to determine." He also said, "it is very desirable to test for persistence from May to September," though here Wright's suspicion was that the really important point was persistence of gene or chromosomal frequencies from year to year rather than from month to month. Neither he nor Dobzhansky suspected the great changes in these frequencies soon to be discovered over even a single season.

The relationship of Wright and Dobzhansky soon settled into a fairly consistent pattern. Dobzhansky would suggest a field research project. He would describe the project to Wright and ask whether the data produced by the project would be pertinent to population genetics theory and whether the experimental design was adequate. In return, Wright often suggested modifications of experimental procedures. Whenever substantive quantitative analysis of the field data in relation to breeding structure or selection theory was required, Dobzhansky asked Wright to collaborate. Although Wright was extremely busy with his teaching responsibilities and guinea pig research, and later with war work, he found it difficult to refuse Dobzhansky's invitations to collaborate because Dobzhansky was forceful and the work was deeply related to population genetics theory.

One good way to see why Wright needed Dobzhansky is to look at Wright's 1940 paper, "Breeding Structure of Populations in Relation to Speciation," (Wright 100). Dobzhansky, who was the organizer of the speciation symposium sponsored by the American Society of Zoologists, the Genetics Society of America, and the American Association for the Advancement of Science held in Columbus, Ohio, December 28, 1939, invited Wright to give the paper: "your name is first on the list, since without you the symposium will be like a body without the head" (Dobzhansky to Wright, May 12, 1939). Wright proposed to discuss three aspects of his topic:

> First, there is the observational problem of determining what the breeding structures of representative species actually are. Naturalists are only beginning to collect the detailed information which turns out to be necessary, but that which we have indicates situations of great complexity. The second step is that of constructing a mathematical model which represents adequately the essential features of the actual situation while disregarding all unimportant complications. The third step is the determination of the evolutionary implications of a given breeding structure in relation to mutation and selection. (Wright 100, p. 233)

Wright's entire argument depended directly upon the quality and quantity of observational data related to breeding structures in natural populations. Dobzhansky was the preeminent person in the world in 1939 gathering such data, so Wright could not hope for a more ideal collaborator.

The collaboration with Dobzhansky on the GNP series would have a direct impact upon the "balance" of Wright's shifting balance theory of evolution in nature. In one sense, Wright's theory has never changed substantially since he first conceived it in 1925. He has always argued that a certain "balance" among the various factors affecting the evolutionary process exists and that generally all the factors are acting in the balance. Thus, to say that natural selection or random drift is the primary determinant of the evolutionary process, makes no sense in Wright's scheme. Both are working, and it is the balance of their interaction that (along with all the other factors, of course) determines the course of evolution. Wright has never veered from emphasizing the "balance" of his shifting balance theory.

Within the "balance," however, significant shifts of emphasis are possible. The novelty of Wright's shifting balance theory was in his emphasis upon random drift as a mechanism for generating new genetic interaction systems, upon which natural selection then acted through "intergroup" selection. The essential question is At what level do the significant nonadaptive effects of random drift cease, to be succeeded by intergroup adaptive selection? Wright of course did not deny the existence of cases in which natural selection was the primary determinant to the near total exclusion of effects of random drift.

I will argue in the two succeeding chapters that during the 1940s and 1950s Wright altered the emphasis in the "balance" of his shifting balance theory toward lesser emphasis upon random drift above the level of local populations. That is, probably by the end of the 1940s, and certainly by the time of the Darwin centennial in 1959, Wright was arguing that random drift was only important in very local populations; thus, intergroup selection led to adaptive differences at the very lowest taxonomic levels of interest to systematists. This represents a substantial change of emphasis from the views Wright expressed in the years 1929–40, when he said that random drift was the preferred explanation for the nonadaptive differences observed by systematists at the subspecies, species, and even genus levels.

This is not to say that random drift was in any way less important or central to Wright's shifting balance theory of evolution in nature. Random drift still played the crucial role of generating novel interaction systems in local populations, selective diffusion from the most successful of these being the next step of the process. The point is not that Wright lessened the role of random drift but that he saw it acting effectively only at the level of local populations, whereas earlier he thought it acted effectively to produce observable nonadaptive differences between subspecies, species, and even genera.

Before Wright collaborated with Dobzhansky on the GNP series, Wright's views had changed very little from those he had expressed in 1931 and 1932. One of his most substantive papers published in the early 1940s was that written for Julian Huxley's edited volume, *The New Systematics* (published 1940, but Wright's essay reached Huxley in July 1938). Entitled "The Statistical Consequences of Mendelian Heredity in Relation to Speciation," Wright presented briefly the quantitative models he had developed during the 1930s, showing how all the determinants of the evolutionary process could be fit into a stochastic distribution of gene frequencies. The model itself, of course, gave no indication of what the "balance" of the factors might be in natural populations. What is most interesting for the present analysis, therefore, is how Wright related his theoretical model to conditions in nature.

At the outset he dismissed the determination of selection rates in nature as hopeless: "It is probable that most of the mutations which are important in evolution have much smaller selection-coefficients than it is practicable to demonstrate in the laboratory." He strongly supported, however, the general direction of Dobzhansky's research in the GNP series:

> The phase of the theory that is most open to investigation in nature is that of breeding structure. It should be possible in many cases to estimate the effective size of the randomly breeding units, and the effective amount with sufficient accuracy to form some judgement of the role which can be played by partial isolation. The distribution of frequencies for a single approximately neutral gene in a species gives an index of the amount of random differentiation. (Wright 101, p. 178)

Wright stated flatly that "Sturtevant and Dobzhansky (1936) have shown that chromosome-inversions in *Drosophila pseudoobscura* behave as approximately neutral Mendelian units" (p. 178). Citing the data of Dobzhansky and Queal he concluded:

> There is the possibility of a great deal of non-adaptive differentiation in *Drosophila pseudoobscura*. In the human species, the blood-group alleles are neutral as far as known. The frequencies vary widely from region to region and in such a way as to indicate that the historical factor (i.e. partial isolation) is the determining factor. The frequency distribution indicates a considerable amount of random differentiation even among the largest populations. (P. 179)

Citing the theory of Goldschmidt on speciation in the gypsy moth *Lymantria dispar* ("hopeful monsters") and that of Cuénot in rodents ("preadaptations"), Wright countered:

> One may sympathize with these difficulties, but question whether it is necessary to bring in an unknown factor. Kinsey finds little indication of adaptiveness in the trivial taxonomic differences between adjacent populations of *Cynips*. But these apparently random differences accumulate along the species-chains and lead ultimately to differences as qualitative (in character of galls, for example) as those which distinguish species of *Lymantria*. (Pp. 180–81)

But the emphasis was not solely upon random drift, rather upon the "balance" of factors. Wright said evolution could occur primarily by selection, or by random drift (evolution "may be determined by random differentiation of small populations, with or without intergroup selection"), or even by mutation pressure in very special cases. In general, however, "the most favorable conditions for a continuing evolutionary process are those in which there is, to a first-order, balanced action of all of the statistical evolutionary factors. It is consequently to be expected that in most actual cases indications can be found of simultaneous action of all of them" (p. 181).

What I want to emphasize, of course, is that when Wright wrote this paper he was still ascribing to random drift a major role in the origin of observable taxonomic differences. He argued:

> Even from the first, certain authors (e.g. Gulick and Romanes) maintained that it was futile to look for a selective mechanism back of many of the differences between isolated populations living under substantially identical conditions. The majority of systematists have probably been skeptical of the adaptive significance of all taxonomic differences. (P. 179)

In support of this assertion, Wright cited Robson and Richards's *The Variation of Animals in Nature* (1936), which emphasized strongly the nonadaptive interpretation of taxonomic differences.

In relation to the impending collaboration with Dobzhansky on the GNP series, Wright's statement that the inversions in *Drosophila pseudoobscura* are "approximately neutral" is particularly noteworthy. In the first draft of the paper sent to Huxley, the phrase actually read "essentially neutral." Huxley sent a copy of the galley proof of Wright's paper to four or five other contributors to the volume, among them C. D. Darlington, who objected strongly to Wright's characterization of the inversions as "essentially neutral." Not only were the inversions probably subject to selection pressures, Darlington argued, but treating them as anything but selectively neutral would make Wright's quantitative model unworkable. Huxley was sufficiently disturbed by this criticism that, having carefully instructed contributors to make only absolutely necessary and minimal changes in the galley proofs, he wrote to Wright, enclosing Darlington's remarks, saying that they "seem to require" Wright's attention (Huxley to Wright, December 29, 1938).

Wright's answer to Huxley concerning Darlington's criticisms helps a great deal in understanding Wright's reasoning in making the statement about the neutrality of the inversions:

> In am not sure that I understand Darlington's criticism of my reference to the inversions of *Drosophila pseudoobscura* as essentially neutral. I would admit that these inversions are probably not absolutely neutral. The absence of crossing over in the male and the relegation of crossover strands to the polar bodies in oogenesis prevent selection due to the production of single crossovers. There should be some selection against the rarer inversions due to the occasional production of double crossovers in the inverted region. It is very likely that this is more than compensated for by selection favoring heterozygosis such as Darlington discusses in his chapter [in *The New Systematics*] and which Sturtevant and Mather have discussed re-

cently. This is very important in itself as tending to bring about an accumulation of inversions but in the context in which I was using it, it is not important provided that the selective advantage of the heterozygote is small. The variations in the frequency of a given inversion among more or less isolated regions, give an indication of the breeding structure (i.e. the value of *Nm*) largely irrespective of the factors (such as selection) that determine the average gene frequency. The data show that the trinomial square rule for the frequencies of the various homozygotes and heterozygotes is so nearly realized in nature that I feel that I can use the data safely for this purpose. It is possible that "essentially" is too strong a word and "approximately" [Wright had crossed out *substantially*] might be substituted. (Wright to Huxley, January 11, 1939)

Thus Wright's reasoning was not that the inversions were almost completely selectively neutral but that the effective population sizes in *D. pseudoobscura* were so small that, whatever the selective forces acting upon the inversions, they were *effectively* neutral. Collaboration with Dobzhansky on the GNP series would dramatically change Wright's assessment of effective population sizes in *D. pseudoobscura*. Indeed, the collaboration with Dobzhansky signaled the beginning of Wright's restriction of the role of random drift in the "balance" of his shifting balance theory of evolution in nature to the level of local populations and his abandonment of the argument that random genetic drift was the cause of nonadaptive character differences at the level of subspecies, species, and genera.

11
The Genetics of Natural Populations Series

The collaboration of Wright and Dobzhansky on the Genetics of Natural Populations series unfolds in their correspondence in a natural manner that could be best re-created by including the correspondence itself. Unfortunately the approximately 500 pages of correspondence cannot fit in this book. I will therefore touch systematically upon the highlights of their collaboration by dividing it into sections by subject matter, mostly but not wholly in chronological order. These sections are (1) GNP V and VII: lethals in relation to population structure in Death Valley and San Jacinto; (2) GNP VI: *Linanthus parryae* and its implications for isolation by distance; (3) GNP X: dispersion and population structure; (4) the Brazilian *Drosophila* venture; and (5) GNP XII and GNP XV: selection of chromosomal arrangements in population cages and dispersion revisited.

GNP V and VII: Lethals in Relation to Population Structure

When Dobzhansky initiated the research for GNP I and II in the spring of 1937, he clearly wanted his data on the frequency of lethals and allelism of lethals to be used for deducing parameters of population structure, effective population size in particular. But Wright consistently told him that the data on lethals were by themselves nearly useless for deducing parameters of population structure. Dobzhansky reacted by attempting to obtain experimental data to answer Wright's objections and persistently encouraged Wright to collaborate on what was, after all, the real reason why Dobzhansky bothered to gather the data in the first place. As the data accumulated (by the efforts of Dobzhansky and his associates, very rapidly), Wright's resistance to collaboration declined. The turning point came in the last week of August 1939 at the Seventh International Congress of Genetics held in Edinburgh. Dobzhansky and Wright both attended the congress, and here they agreed to collaborate. Dobzhansky was to write up the presentation of the data and send this to Wright, who then was to analyze the data and deduce the effective population size and other parameters of population structure.

On November 16, 1939, Dobzhansky sent to Wright his preliminary manuscript, with the suggestion that the resulting collaborative paper be published as part of the GNP series. With characteristic underestimation of the speed with which Wright could do all of his other work and the analysis of Dobzhansky's data, Dobzhansky encouraged Wright to do his part immediately so that the entire manuscript could be completed by New Year's 1940. Wright wrote back immediately to say that he couldn't even begin work on the manuscript until after January 1, and he actually sent the completed manuscript to Dobzhansky on July 3, 1940.

Wright of course knew that he faced a formidable and in some ways an impossible task of data analysis. By April 1940, however, Wright wrote Dobzhansky that despite having "a briefcase full of calculations, most of which will probably have to be scrapped because of some theoretical complication that I have not entirely cleared up," he still was now confident "that a lot more can be got out of the data than I thought at first" (Wright to Dobzhansky, April 18, 1940). Dobzhansky had gathered four kinds of data: frequency of third-chromosome lethals, the mutation rate at which these lethals occurred, and the rates of allelism of lethals within and between populations. Wright's job was to fit these data into the quantitative model he had presented in his big 1931 paper. The problem was twofold. Wright's shifting balance theory of evolution in nature was far more complex than was his quantitative model. Also, his model suggested no less than nine important population constants (not all independent), those most important for population structure being the effective breeding size (N), the inbreeding coefficient of the locality (F), and the immigration index (m). But only six equations were possible (four came from Dobzhansky's data and two from internal relations). Determinate solutions to the equations therefore required at least three additional data determinations (assuming no more internal relations were possible). These data were unavailable, so Wright was forced to estimate unknown factors in order to deduce the desired parameters of population structure. For example, in order to estimate N, he could solve the equations only for Nm together, and then only if he made preliminary estimates of N.

Wright and Dobzhansky were absolutely clear about the tentativeness of their estimates under this procedure:

> In conclusion it may be stated that we entertain no illusions regarding the lack of precision in the results obtained. At best it is hoped that the order of magnitude of certain variables effective in population dynamics may have been arrived at. But at a certain stage of the development of a scientific subject even such rough estimates may be useful to guide further work. Although population dynamics has been discussed for decades, mainly in connections with evolutionary speculations, few attempts have ever been made to put the discussion on a basis of experimentally determined quantities. (Wright 103, p. 49)

This was indeed a pioneering paper. It pointed toward additional experimental work, some of which was already underway.

While Wright was working on his part of GNP V in the spring and summer of 1941, Dobzhansky was on Mt. San Jacinto gathering the data for GNP

VII. There the idea was to gather data not from isolated populations (as in Death Valley) but from different locations in an apparently continuous terrain and population. The hope was to supply greater experimental data for the problems raised in GNP V.

To be sure, Dobzhansky was not pleased about the time it took Wright to analyze data and fit them into theory. I think Dobzhansky was unaware of the extent of Wright's other responsibilities and also of the conceptual and computational difficulties associated with the analysis of his data. Dobzhansky constantly reminded Wright that the analysis was overdue. On April 18, 1941, he told Wright, "it is sad that other matters deflect you from this most interesting and profitable occupation." And after Wright sent Dobzhansky the draft of the analysis, he wrote to Wright that on August 31, by which time he hoped the manuscript would be submitted for publication (it was submitted August 12), it "will be considerably more than a year since its inception, a period of gestation sufficient, if I am not mistaken, for a baby elefant to be born" (Dobzhansky to Wright, July 6, 1940). The extra time was well worth the wait for Dobzhansky. Not only was the quantitative analysis of the data beyond his capabilities but Wright also greatly stimulated Dobzhansky's ideas for research and helped him with experimental design. Thus on April 22, 1940, Dobzhansky asked Wright for a preliminary analysis of the data for GNP V because Wright's analyses "always spur me on to more work, and, more important still, to devise new projects and experiments." This statement was not mere flattery designed to speed Wright on the analysis of Dobzhansky's data. So far as I can determine, Dobzhansky stated the situation accurately. Another obvious reason why Dobzhansky patiently endured the wait was that he needed Wright for analysis of the new data he was gathering and planned to gather in the future. Just before Wright sent him the analysis for GNP V, Dobzhansky wrote Wright about the progress on the data for GNP VII, and added: "You told me at the Christmas meetings that such data may be of interest, and, unless the world goes to ruin even sooner than expected, you will have these data reasonably soon. In other words, Dobzhansky is planning to annoy Wright also in the future, and he prays God that Wright does not get so much annoyed as to send him to remote places" (Dobzhansky to Wright, June 27, 1940).

Although the analysis of Dobzhansky's data took much of Wright's time, he also found the collaboration rewarding. GNP V represented the opportunity for him to put his quantitative theory into the context of data from natural populations. Not only did Dobzhansky tirelessly supply data, he would almost instantly begin to gather data on the basis of Wright's queries. Sometimes he anticipated Wright's queries. Thus in a letter of June 20, 1940, Wright pointed out a difficulty with Dobzhansky's data on the chance of allelism of lethals within a locality. The estimate of allelism might have been too large because a male carrying a lethal might have fertilized several females in the trap, but this would have been recorded as several different lethals, when actually it was only one. Dobzhansky replied that "this question has worried me, too. . . . Most fortunately, the new data, from experiments wound up just a few days ago, give, it seems to me, an entirely satisfactory answer to this question" (Dobzhansky to Wright, June 27, 1940). So this time Wright got an immediate answer in experimental data; usually, Dobzhansky set to work producing the required data and often had them in short order. Wright was greatly impressed by the sheer quantity of data Dobzhansky produced and by the enormous amount of work required to obtain the data. For example, when Wright suggested that Dobzhansky get more voluminous data on the frequencies of homozygotes and heterozygotes for inversions (used in GNP V), Dobzhansky gathered over twenty times the data he had given Wright earlier. Wright responded: "I was very glad to hear of the quantity of data which you now have on the inversions, though quite overcome by the amount of work it must have been to get it" (Wright to Dobzhansky, July 16, 1940). Finally, Dobzhansky did have ideas about theory, and the data he produced on his hunches about theory were pertinent to Wright's conceptual scheme. Dobzhansky dreamed up experiments—Wright gave help on experimental design. Wright could scarcely have asked for a better collaborator in the field.

The collaboration therefore continued after GNP V. In the spring of 1941, Dobzhansky sent Wright the data for GNP VII. It would take Wright until November to complete his analysis. One reason he took this long was because he augmented the nine population constants he had used earlier, this time expanding them to twelve. The data were more numerous, and comparison with GNP V was possible. Wright therefore spent a great deal of time writing his part. Even with the new data, Wright still could not separate Nm or the relationship of the inbreeding coefficient (F) and selection against heterozygotes (s), where only the joint estimate ($s + F$) was possible. (Both s and F of course could lead to the observed lower frequency of lethals than expected on the basis of the observed mutation rate and random mating.) By assuming certain migration rates, Wright deduced the resulting effective population sizes, which were larger by a factor of ten than those calculated in GNP V for isolated populations in Death Valley.

Less than three years earlier, Wright had advised Dobzhansky that the data on frequency and allelism of lethals provided an inadequate basis for deducing parameters of population structure. Dobzhansky had produced data to answer some of Wright's reservations, and Wright overcame some of his own hesitancy to make estimates on a less-than-perfect data base. Their collaboration showed, although certainly in a tentative case, that population genetics theory and field research on natural populations could interact in a very fruitful way. This was an important contribution to evolutionary theory. Dobzhansky himself was very pleased with GNP VII and wrote to Wright:

Now, although I am congenitally given to overstatement, this is what I believe: this paper is the best of any to which my name has ever been attached, and its quality is due to your analysis. The data with-

out the analysis would be interesting of course, but not too meaningful. Although 99% of geneticists will hardly see the point when they
read it, the above statement represents my conviction. (Dobzhansky
to Wright, November 29, 1941)

Dobzhansky then insisted that Wright's name be placed first on the paper,
and it was in the published version.

To tell the story of GNP V and VII as a separate section is a little
artificial because in the same time period the collaboration of Wright and
Dobzhansky was growing in many other ways, and indeed the research they
would later publish together had already been initiated or envisioned. GNP X
on dispersion was in the research stage and GNP XII on selection in population cages was planned. An excellent measure of their collaboration comes,
however, in a paper on which Wright and Dobzhansky did not formally collaborate, namely GNP VI on *Linanthus parryae,* by Dobzhansky and his
(then) close friend, Carl Epling.

GNP VI: *Linanthus parryae*

Ever since Dobzhansky heard Wright speak at the international congress in
1932, he wanted to find a sophisticated example of Wright's shifting balance
theory of evolution operating in nature. He thought he had found such an example with the Death Valley populations of *Drosophila pseudoobscura,* but
Wright showed without question that the situation was very complex not only
at San Jacinto, in an apparently continuous terrain, but also at the isolated
canyons of Death Valley. To give an example, in the summer of 1939, when
Dobzhansky had first collected data at Mt. San Jacinto, he discovered that in
four of the nine localities the frequencies of inversions changed during the
course of the summer. Later (as I will describe further in this chapter)
Dobzhansky explained these changes of frequency as being caused by natural
selection. But at the time he held this interpretation as expressed in a letter to
Wright:

> I think these changes mean the following thing. The population is
> broken into numerous colonies, each initially with a very small ef-
> fective size. As the season advances the colonies grow. But they *do
> not grow equally*—some of them may at any time increase far more
> than others. Then the "lucky" colonies, so to speak, *overflow* into the
> surrounding territory, producing a real change in the population of
> the latter. Even when the numbers are in general declining, some
> colonies may so "erupt" and invade adjacent places. I suspect the
> competition in nature is not so much between individuals as between
> such colonies. I think you have stated in one of your papers that such
> a competition is evolutionarily more effective? But do not misunder-
> stand me in this: I do not mean to say that at any time a territory even
> so small as the San Jacinto mountain is inhabited by a single popula-
> tion. *This is definitely not so.* Groups of localities ten miles apart are

independent. All the intercolonial migration takes place within a much smaller territory, but the population of such a small territory is in a constant flux! . . . Please, take all this as a very preliminary stuff, subject to change. But such is the picture according to this time's fashion. (Dobzhansky to Wright, November 16, 1939)

Although *D. pseudoobscura* turned out to be less than a simple and obvious case of Wright's shifting balance theory, in the spring of 1941 Dobzhansky found what appeared to be a nearly ideal example.

In April of 1941 Carl Epling (1894–1968), a botanical systematist at UCLA, and an undergraduate senior, Harlan Lewis (who would go on to take his M.A. and Ph.D. in botany at UCLA), were on their way to Arizona to collect and observe a dimorphic species of *Ansonia*. As they were driving through Lucerne Valley of the Mojave Desert, they could see a striking dimorphism of white and blue flowers spread almost like a carpet on the desert floor. The plant was the diminutive (average 1 cm in height) annual, *Linanthus parryae,* which had relatively large white or blue flowers, up to 2 cm across. In rainy years, this plant provided in some places a continuous show of color in late April and early May, earning the popular name "desert snow." Each plant averaged five to ten flowers, and some had over two hundred. In years of little rainfall, however, *Linanthus parryae* grew spottily, if at all.

The spring of 1941 was a bonanza year. *Linanthus parryae* bloomed almost continuously over a whole 200 mile stretch of roads in one district of the Mojave Desert. The white and blue flowers exhibited a curious distribution. Some areas had only white flowers, some only blue, and others were mixed. All this was in an apparently continuous ecological district. Epling was a good friend of Dobzhansky, whom he knew would be interested in *Linanthus parryae*. So Epling contacted Dobzhansky, who, as expected, was greatly interested. Indeed, he encouraged Epling to gather experimental data immediately. Here was the chance to see the shifting balance scenario in action, and with ease. Unlike *Drosophila pseudoobscura, Linanthus parryae* was conspicuous, easy to count and classify, and readily accessible at the roadside. Epling quickly put together a team of students and gathered data at observation spots every half mile along the 200 miles of roadway; they also set up a transect 750 feet long, with observation spots every 25 feet along the transect. Epling had all the data in hand even before Dobzhansky set out from Columbia University for the West Coast in late May 1941.

Dobzhansky and Epling interpreted the data to mean that genetic isolation by distance, even in the ecologically continuous district, had resulted in random drifting apart of the gene frequencies determining the white and blue flower colors. Unfortunately, the mode of inheritance of the flower colors was unknown, so only phenotypic analysis was possible, although various patterns of genetic inheritance could be assumed. Dobzhansky lamented in his October 3 letter to Wright: "Oh, I wish Epling would find out how the flower color is inherited in this plant, so one could express these data as gene frequency and not phenotype frequency curves! He will do his best, but it is very difficult to generate in this material which has seeds that refuse to germinate."

Dobzhansky, characteristically, decided to publish the data right away rather than wait for the genetic analysis of color in the flowers. (This would actually require more than twenty years: Ball to Wright, November 25, 1962; Wright to Ball, December 18, 1962.) He and Epling worked on a draft of the paper at the end of summer 1941. By the beginning of October, Dobzhansky had sent Wright a copy of the draft, asking for his opinion.

Actually, however, Wright had never published a quantitative theory of genetic isolation by distance in a single population distributed over an ecologically continuous area. He had published on two occasions (94, pp. 430–31; 100, pp. 244–48) a qualitative description of the possibility of such genetic isolation by distance, but neither presentation matched Wright's conception of a quantitative theory. Indeed, the problem of isolation by distance was quite difficult to set in a quantitative model. Wright was, however, keenly interested in the issue of isolation by distance because it appeared to occur in nature more frequently than the "island" model he had presented in 1931 and 1932.

Before Wright had the chance to reply to the first draft, on October 30 Dobzhansky sent him a new one:

> Enclosed is a copy of a MS describing the *Linanthus* material which we are planning to submit to *Genetics* eventually, but not before we get your criticism, which both Epling and myself are earnestly asking for. I think this is a very interesting material and much can probably be done with it in the future, especially if a method of germinating seed of this thing will be found.

Dobzhansky did not wait to get Wright's critique but submitted the paper to *Genetics* on November 3.

Wright's response to the Dobzhansky and Epling manuscript began:

> I have gone over the manuscript on *Linanthus* that you sent me. It is certainly a very interesting case. It appears to be a good example of isolation by distance that I discussed in my symposium paper at the Columbus meeting [Wright 100]. I hope that it will be possible to determine the mode of inheritance since if this can be done a rather extensive analysis along this line will be easy to make. It should be possible to get at least a strong indication of the mode of inheritance merely from the progenies of single plants—especially single blues isolated in the midst of whites and single whites isolated in the midst of blues, assuming that there is little or no self fertilization as seems likely from your statement [in the manuscript]. (Wright to Dobzhansky, November, 1941)

The thrust of Wright's response was that the *Linanthus* data were a gold mine of real theoretical interest but that Dobzhansky and Epling had scarcely scratched the surface of analyzing the data.

In this letter, Wright presented in a nutshell a brief (seven handwritten pages) account of the theory of isolation by distance that he was developing and showed how the *Linanthus* data could be analyzed on assumptions that blue was recessive or dominant. Additionally, Wright made several suggestions for improving what data analysis there was in the manuscript. This was an outstanding letter, demonstrating clearly that the *Linanthus* data were worth far more than Dobzhansky and Epling had suspected. Wright did not know that Dobzhansky had already submitted the paper for publication. Neither Dobzhansky nor Epling understood what Wright was talking about with his theory of isolation by distance. Dobzhansky's reply stated:

> Your letter in connection with Epling's paper has been received of course, and it is my fault that I did not thank you for it. Of course, the corrections that you suggested are incorporated. As to the calculations, they have been sent to Epling for inspiration, with a request to return your letter after reading it. Both Epling and myself had doubts whether the data should be published now, since additional information would make them ever so much more useful. We decided to publish the data now, since the necessary additional work is difficult and may take several years. (Dobzhansky to Wright, November 29, 1941)

I interpret this letter as follows. Dobzhansky incorporated Wright's few suggestions about presentation of correlations between frequencies of blue-flowered plants found at different distances. Dobzhansky refers to Wright's theory of isolation by distance as "the calculations" and obviously understood it not at all. He sent Wright's letter to Epling so he could gain inspiration from "the calculations." The "additional information" that Dobzhansky says is desirable (for making the data more useful) has nothing to do with Wright's theory of isolation by distance but refers instead to Epling's experiments to determine the mode of inheritance of flower color. In short, Dobzhansky totally missed the thrust of Wright's letter.

In the published paper, the theoretical backbone of the investigation was drawn entirely from Dobzhansky's interpretation of Wright's theory of evolution. Two conclusions from the data were the most important. First,

> No relation between the compositions of populations of *Linanthus parryae* and the environment in which these populations live is detectable. As pointed out above, the region studied is homogeneous both with respect to physical factors and with respect to plant associations. The boundaries between the variable (with both blue and white flowers) and the predominately white areas do not coincide

with any perceptible migration barriers. An attempt was made to relate the composition and the density of the populations examined.
The outcome of this attempt was negative.

The situation was thus, to Dobzhansky and Epling, ripe for random drift:

> In the samples from the variable areas the distribution curve of the
> frequencies of blue flowered plants resembles Wright's curves for the
> distribution of gene frequencies in effectively small populations.
> This resemblance, as well as the extremely fine subdivision of the
> species in the region studied, furnishes an indication of the agency
> which is primarily responsible for this mosaic-like subdivision. As
> shown by Wright (1931 and later work) the variable genes in small
> populations tend to become stabilized at values of zero and one—that
> is, tend to be lost or to reach fixation. The distribution curve for such
> populations consequently assumes a U-like shape which is symmetri-
> cal in the absence of mutation, selection, and interchange of mi-
> grants among the separate colonies. (Epling and Dobzhansky 1942,
> 331)

Three papers of Wright's were cited, but not the two in which Wright mentioned isolation by distance.

Dobzhansky and Epling totally missed that Wright had yet to publish a
model to account for the situation that they had found in *Linanthus parryae*
and that Wright had suggested to them in a letter just such a model, which
would give their data far greater significance for evolutionary theory.

Wright obviously thought the *Linanthus parryae* data had much more to
offer than Dobzhansky and Epling indicated. Indeed, for Wright the issue
was crucial to his entire shifting balance theory of evolution in nature. The
least likely situation in which random drift could create novel genetic interaction systems (then to be acted upon by natural selection in the shifting balance theory) was in large populations distributed over uniform environments
such as the grasslands of the American plains. If Wright could show that isolation by distance could really work under such circumstances, then he would
have demonstrated the robustness of his shifting balance theory under the
least favorable of circumstances. He was, in other words, trying to have his
theory of evolution work with what Fisher took to be his private domain.
Symmetrically, of course, as I will discuss in the next chapter with the case
of *Panaxia dominula,* Fisher and Ford tried to exhibit the domination of gene
frequencies by natural selection rather than by genetic drift in a population of
rather small size, taken by many to be Wright's private domain. So if *Linanthus parryae* was a good case of isolation by distance, it was indeed of pivotal importance to Wright's theory of evolution in nature. He could not,
therefore, simply allow the *Linanthus* data to pass by with the scant analysis
given it by Dobzhansky and Epling.

Before analyzing the *Linanthus* data, Wright seriously considered the possibility that the differentiation observed in the field resulted from selection rather than primarily random differentiation due to isolation by distance. William Hovanitz, a graduate student at Cal Tech who had earlier helped gather the data for GNP VII (he is listed as the third author), brought this issue forcibly to Wright's attention in a letter dated May 28, 1942:

Yesterday I noticed the paper by Epling and Dobzhansky on the variation in *Linanthus parryae* which was published in *Genetics*. If we take for granted the genetic nature of the differences between the blue and white flowers, I would like very much to have your opinion on the validity of the data. . . .

For the most part it seems to me very difficult to imagine a gene existing in wild populations which has absolutely no different physiologic effect on the individuals carrying it as compared with a type standardized as "wildtype." Your work on the statistics and theory of variation in populations I like very much, but I think that you will agree with me that all statistical reasoning must be based upon a number of assumptions which for mathematical simplicity must be considered as invariable. These assumptions are generally peculiarities in the environment which are too complex to put into a formula and often are but hazily known. Unfortunately, in order to attempt to illustrate the theoretical calculations which you have shown, persons are prone to pan over lightly the assumptions made for them. I feel that this is true in the case of the *Linanthus* paper. The value of that paper as it is written was supposed to be the illustration of field data as agreeing with your calculations. For this purpose it had to be assumed that the blue flowered plants were no different in survival value over the white. Likewise, it had to be assumed that the climate, etc., was uniform over the area. There is definite information to show that the latter is not true and that the climatic changes take place at just exactly the same places where the blue flowered plants are most abundant. There is some evidence that in other plants the same type of color change takes place under the influence of a similar type of environment though I have no data for this particular area. There is fairly good reason to suppose, therefore, that the blue flowered plants are being selected for in these areas and that the assumption of randomness of the gene distribution is unwarranted.

Wright unquestionably seriously considered the issues raised by Hovanitz, and he later cited the letter in his paper on *Linanthus*. His reply is interesting because it states the "balance of forces" aspect of his shifting balance theory rather than emphasizing just the idea of isolation by distance:

With regard to the Linanthus paper, differential selection could as you state be a factor. The irregular variation in frequency illustrated in figure 3 (of the Epling and Dobzhansky paper, referring to variation within a single area) indicates that there must also be a considerable amount of chance variation. I am most frequently quoted in connection with the possibilities of differentiation due to the cumulative effects of accidents of sampling, probably because this is the aspect of the matter that has been most frequently ignored by others. The real purpose of my mathematical theory is, however, to bring all factors, recurrent mutation, crossbreeding and selection, as well as cumulative accidents of sampling under a common viewpoint in a formula upon which the effects of their simultaneous actions may be obtained. (Wright to Hovanitz, June 11, 1942)

Hovanitz would later raise a different objection to Dobzhansky's idea that natural selection was the cause of the shift in inversion frequencies in *Drosophila pseudoobscura* during a single season. Wright always gave Hovanitz's ideas greater credence than did Dobzhansky, who, for example, did not bother to send Hovanitz a copy of the draft of GNP VII before he sent it to press, even though Hovanitz was listed as third author.

The essential problem for Wright in devising a quantitative theory of isolation by distance was the accurate quantification of population structure, especially the measurement of inbreeding in populations and subpopulations. His inbreeding coefficient F that had proved so useful in his inbreeding theory (covered in chapter 5 of this book) and its application to the history of domestic breeds was just too crude a measure for a subdivided population in nature. For example, the index F could not be used to discriminate between the inbreeding that resulted from (1) division of the population into completely isolated small subpopulations within which there was random breeding (here one generation of random mating between the subpopulations would reduce the inbreeding coefficient to zero); (2) frequent mating of close relatives but no permanent separation of subpopulations, in which case one generation of random breeding would also reduce F to zero; and (3) the sires are few in number for many generations. In this case, there would be little close inbreeding as in case (2) and no subdivision as in case (1), yet F (which measures inbreeding relative to the foundation stock) rises each generation and is little reduced by random mating. A more discriminating measure of inbreeding relative to population structure was required.

Wright's response was to invent a series of inbreeding coefficients, each particularly useful for a given kind of population structure, yet all the coefficients were simply related algebraically. Thus Wright developed the separate inbreeding coefficients for an individual relative to its subpopulation (F_{IS}, in Wright's later notation; Wright 140), that for an individual with respect to the total population (F_{IT}), and that for the correlation between random-uniting gametes drawn from the same subpopulation (a measure of current inbreeding, F_{ST}), where $F_{IS} = (F_{IT} - F_{IS})/(1 - F_{IS})$. The same idea could easily be extended to further subdivisions of the subdivisions down to any hierarchical level of subdivision.

These were Wright's famous F-statistics that he thereafter used to great advantage in analyzing the structure of both natural and artificial populations. (For summaries see especially Wright 140; 179; and 188, chapter 12.) The power of the F-statistics became obvious first in the development of the theory of isolation by distance, but later animal and plant breeders found them extremely useful as have evolutionary biologists working on population structure both in theory and in natural populations. The F-statistics are one of Wright's most notable and influential contributions to biology, although their origin and later use by Wright were closely tied to his shifting balance theory of evolution in nature and are in a sense an integral part of that theory.

In order to reach a quantitative model for isolation by distance, Wright first had to derive formulas for relating the primary variables, (1) gene frequencies of subgroups, (2) effective population sizes of subgroups, (3) immigration index, (4) inbreeding coefficient of individuals relative to the total population, and (5) mean gene frequency of the total population. To approximate the effect of isolation by distance in a continuous population in which each individual had only short-range dispersal, he assumed the parents of any individual were taken from a random breeding subgroup (neighborhood) of limited population size. (Later Wright would vary this assumption to include diverse systems of mating in the neighborhood, Wright 121.) He then derived formulas for the inbreeding coefficient of individuals in these neighborhoods relative to the total population and an expression for the variance of gene frequencies in neighborhoods. He worked out the isolation by distance assumption, both for continuity in area (as on a prairie) and for continuity in a linear range (as on a shoreline or in a river). The formulas indicated, as Wright showed in a series of figures of variability in gene frequencies under varied assumptions of neighborhood size, that when N was less than 100, much random differentiation should be expected in both small and large population subdivisions. Values of N over 10,000, on the other hand, were indistinguishable from a purely random breeding population. In the published version, Wright concluded with a paragraph that pointed directly to the centrality of isolation by distance to his shifting balance theory of evolution:

> If different regions are subject to different conditions of selection, the amounts of both adaptive and nonadaptive differentiation depend on the smallness of m (if subdivision into partially isolated "islands"), or of N, size of the random breeding unit (if a continuous distribution). If these are sufficiently large there is no appreciable differentiation of either sort; if sufficiently small there is predominantly adaptive differentiation of the larger subdivisions with predominantly nonadaptive differentiation of smaller subdivisions superimposed on this. Even under uniform environmental conditions, random differentiation tends to create different adaptive trends in different regions and a process of intergroup selection, based on gene systems as wholes, that presents the most favorable conditions for adaptive advance of the species. (Wright 114, p. 137)

It is not difficult to see why Wright considered isolation by distance important for his shifting balance theory or why he has always been particularly proud of his achievement in this paper of providing the first quantitative model of isolation by distance.

In a companion paper, Wright applied the models developed in the isolation by distance paper to the data of Epling and Dobzhansky on *Linanthus parryae*. After patterning the studied range of *Linanthus parryae* into a hierarchy of subdivisions, ranging from within individual collecting stations (ca. 0.02 sq mi) to primary subdivisions (about 140 sq mi), Wright determined that there was much differentiation at all levels. Since the actual mode of inheritance of the flower color was unknown, he used a series of hypotheses from self-fertilization to crossbreeding, with blue dominant, recessive, or determined by multiple factors with a threshold effect. With any of these last three assumptions and crossbreeding (considered highly probable for *Linanthus* at that time), Wright deduced an effective neighborhood population size to be about 25 (of course assuming no mutation, migration, or differential selection). If true, this would guarantee much random differentiation at the local level.

Wright did not, however, try to use random differentiation to account for differentiation at the higher subdivisions. Here he presented Hovanitz's thesis about possible environmental differences in the primary subdivisions range as plausible but added, "it is less plausible for the differences among secondary and tertiary subdivisions of the same primary subdivision." Wright also suggested that mutation between blue and white flower color, or even between very occasional long-range dispersal, might also account for much of the differentiation at the higher subdivisions. Selection could also result from random drift rather than from environmental difference:

> There is a possibility, however, of selective differentiation even in the absence of any environmental differentiation. . . . the random differences in gene frequency, occurring in all series of alleles up to a certain level in the hierarchy, create a unique genetic system in each locality. Slightly different adaptive systems may be arrived at in different localities. If the gene or genes which distinguish blue and white play a role in any such systems, this would give a basis for locally different selection pressures.

In other words, *Linanthus parryae* was an example of the whole shifting balance theory of evolution in nature in action.

> The distribution of blue and white can be accounted for most easily by supposing that most of the differentiation of the smaller categories is random in character and due to the accumulation of sampling accidents in random breeding groups of one or two dozen productive individuals per year but that at the higher levels, processes which tend to pull down random differentiation such as mutation and

especially occasional long range dispersal are counterbalanced by selective differentials between local genetic systems. (Wright 115, p. 155)

The data from *Linanthus parryae* now had the status of being important for Wright's whole theory of evolution in nature, and in a detailed manner.

Wright worked to exhaustion on these two papers. On November 5, 1942, he sent both manuscripts to Dobzhansky, with the suggestion that he pass the "Isolation by Distance" manuscript on to Marcus Rhoades (the editor of *Genetics*). He asked for Dobzhansky's advice on whether or not the *Linanthus parryae* manuscript should be published, and he suggested that Dobzhansky send the *Linanthus* manuscript on to Epling.

Dobzhansky's response was revealing. Now he really understood why Wright's isolation by distance model was so important, not just for *Linanthus parryae* but also for the work on *Drosophila pseudoobscura* at Mt. San Jacinto. Dobzhansky spent a whole week going through Wright's two manuscripts, trying as best he could to understand the quantitative reasoning. Then he sent Wright the longest letter he ever wrote in connection with the GNP work. It began:

> These papers are hellishly difficult going, even for your papers, and in addition I tried this time to follow the mathematical arguments as completely as possible for me. The result is that I had in three days more practice in algebra than at any time since I graduated from the "gymnasium" (high school). Many a time it seemed to me that I discovered a mistake, but in the end it always proved to be that "papa knows best," as it should be. But your mathematics is made so brief and takes so much for granted that in places I am still in the dark. (Dobzhansky to Wright, November 14, 1942)

There followed detailed comments about both papers, most of them to the effect that Wright should try harder to communicate his ideas so that biologists could understand them. Many of Dobzhansky's points were well taken. The handwritten postscript said, "These two papers are what I was praying for several years—transfer of this theory from the island model to the continuum model. It is no flattery to say that these two papers make a milepost in the history of genetics, and just because of this I would like to see them made as clear as humanly possible." Dobzhansky recommended also that the two papers be published together.

Wright, who responded to Dobzhansky's letter with detailed answers to each of his suggestions and criticisms, was obviously tired of working on these two papers. "I was glad to get your favorable comment on the isolation papers though I look at them at the moment with a lack of enthusiasm closely approaching nausea" (Wright to Dobzhansky, November 19, 1942).

Epling, after receiving Wright's manuscript on *Linanthus,* sent him all his available data on pollination, which, however, was insufficient to elucidate the mode of inheritance of blue and white. Hovanitz's suggestion of environmental differences between the areas of white and blue flowers, raised in Wright's manuscript, stimulated Epling to say to Wright:

> As to environmental differentiation, I can say nothing further at present. Familiarity with the region suggested no correlation between the blue areas and any discernible factors. However, I am going to the Mojave this week end, and by a series of transects, intend to get an approximate analysis of the vegetation in different parts of the area concerned, which may permit comparison of the blue areas with each other and with the extensive intermediate white area. I believe that the vegetation can be taken as the best index for comparison of environmental conditions. I shall also get soil samples which will be analyzed for Ph, water retention and salt content. These data I shall send to you as soon as they are available. (Epling to Wright, November 23, 1942)

Epling meticulously analyzed the twenty-one most common shrubs of the region (nearly all that were present) in relation to the areas of all blue and all white *Linanthus parryae* flowers, and on December 3 sent Wright all the data with this conclusion: "Having revisited the area and made these observations, I am still doubtful that there is any significant correlation between the occurrence of the variable areas and environmental factors." It is important to notice here that both Epling and Dobzhansky were trying to persuade Wright that Hovanitz was wrong, and that Wright should extend his idea of isolation by distance to the differentiation of the larger subdivisions where Wright had suggested that mutation, migration, and selection were the controlling factors rather than random drift.

Epling and Wright maintained a strong interest in *Linanthus parryae* after this initial analysis. Beginning in 1944, Epling permanently marked experimental plots and laid out a permanent transect, gathering data each year the plant blossomed from 1944 to 1966. The data from each year he sent to Wright, who spent untold hours analyzing them. The data were generally frustrating because *Linanthus parryae* had a good year only infrequently, and Epling had constructed his permanent plots on the assumption that plants in each plot would be sufficiently numerous for statistical analysis. (Beginning in 1959, the *Linanthus parryae* story has a curious and important development involving both Wright and Epling; I will describe this development in chapter 13.)

Wright of course remained interested in *Linanthus parryae* because it seemed such a good example of isolation by distance. In 1946 he expanded the theoretical basis of the isolation by distance idea by working out the consequences of varied (and more realistic in nature) assumptions about systems of mating within the smallest breeding groups of the entire population, and he also refined the procedure for estimating the effective population size of the small "neighborhoods." Under a variety of systems of mating, he deduced that for effective breeding populations of neighborhoods of 200 or less, much random differentiation should be expected at the neighborhood level and also at the level of larger subgroups. He concluded that "this differentiation, due to the cumulative effects of accidents of sampling, may be expected in actual cases to be complicated by the effects of occasional long range dispersal, mutation, and selection but that in combination with these it gives the foundation for much more significant evolutionary processes than these factors can provide separately" (Wright 121, p. 59).

This initial phase of the *Linanthus parryae* story yields insight into the collaboration of Wright and Dobzhansky on the GNP series, even though they did not formally collaborate on this problem. By itself, the Epling/Dobzhansky paper represented a fine job of data collection and a keen sense about an experimental situation that was pertinent to theory, and to Wright's theory of evolution in particular. But neither Epling nor Dobzhansky could see the concrete and wide theoretical implications of the *Linanthus* data that Wright saw in them, nor could they possibly have developed the quantitative theory that would give the data substantive meaning in Wright's shifting balance theory of evolution as a whole. In a sense, GNP VI (Epling and Dobzhansky) plus Wright's "Isolation by Distance" and "Analysis of Local Variability of Flower Color in *Linanthus parryae*" taken together constitute the real GNP VI, if it were to be properly compared with the other papers in the GNP series upon which Dobzhansky and Wright formally collaborated. The *Linanthus* story gives a glimpse into what the Dobzhansky/Wright collaborative papers would have looked like if Wright had not, in fact, collaborated. Wright's analysis gave the collaborative papers real substantive meaning for evolutionary theory far beyond what they would have exhibited with Dobzhansky alone. No one understood this better than Dobzhansky himself.

GNP X: Dispersion and Population Structure in *Drosophila pseudoobscura*

I have discussed earlier Dobzhansky's desire to find small effective population sizes in natural populations of *Drosophila pseudoobscura*. I have also shown, however, that Wright's analysis of Dobzhansky's data indicated instead that effective population sizes of *D. pseudoobscura* were much larger than Dobzhansky expected, in the neighborhood of 20,000 to 30,000 for the flies at Mt. San Jacinto according to GNP VII. Back in December 1940, before Dobzhansky had sent Wright all the data that would be the basis for GNP VII, Dobzhansky had another idea for proving experimentally the small population size in *Drosophila pseudoobscura*. He got the idea from a series of three papers by N. W. and H. A. Timoféeff-Ressovsky (1940a, 1940b, and 1940c) on dispersion in *Drosophila melanogaster* and *D. funebris* in a field near Berlin. Using easily recognizable mutant characters, the Timoféeff-Ressovskys had released marked flies in the center of a checkerboard pattern of baited traps, a trap being placed every 10 meters. The largest setup was

110 meters square with 121 traps (11 × 11). They found that dispersing flies did not even reach the outside boundaries during the experiment. Dobzhansky wanted to repeat the experiment with *D. pseudoobscura* on the supposition that similarly small dispersal rates would mean that small effective population sizes were inevitable. In other words, this was an experiment that might help to tease apart *Nm,* and yield separate and more accurate estimates of *N* and *m* separately. From the correspondence between Dobzhansky and Wright, there can be no doubt that in beginning this work Dobzhansky hoped and expected to find small effective population sizes. By the end of 1940, Dobzhansky was hesitant to enter upon any experimental program until he had talked it over with Wright. He did not want to waste his time gathering useless data. On December 9, 1940, he wrote to Wright, "It is almost a year since I saw you last, and I feel the necessity of a visit with you." Wright, who planned to attend meetings in Philadelphia later in December, replied that he had no time to come to New York to visit with Dobzhansky, but would be glad to speak with him in Philadelphia during the meetings. Over four months later (the next time they corresponded), Dobzhansky wrote:

> I am planning to leave New York on May 27th for California, to do the marking experiments that we were discussing in Philadelphia. At that time you favored releasing the recaptured marked flies at the place of their recapture, in order to be able to find them again and again on following days. You told me however that you may change your opinion and advise me accordingly. Well, what is the final instruction? I take it that the disposition of the traps on the experimental plot at the ends of equilateral triangles has been your final advice. . . . I hope to keep in contact with you and to ask you for counsel when the work will get under way and I shall feel in doubt— that is bound to happen, and may happen often. (Dobzhansky to Wright, May 4, 1941)

Dobzhansky was right. The dispersion experiments raised many doubts, and Dobzhansky sent Wright a steady stream of letters requesting advice.

When by May 14 Wright had not yet answered the question about where to set the traps, Dobzhansky wrote him again, this time advocating a pattern of concentric circles with a greater concentration of traps near the center than at the periphery. Wright's reply came only a few days before Dobzhansky left for California to initiate the experiments: "Any irregularities in the spacing (such as involved in concentric circles) would raise difficult questions of the weight to be assigned the frequencies at the different traps." Wright recommended instead a distribution in squares like that used by the Timoféeff-Ressovskys.

Dobzhansky's original design consisted of forty-five 10-meter squares with a trap in the center of each. The first experiment began with the release of 2,000 orange-eyed flies. After three days of recaptures, Dobzhansky wrote

Wright: "Enclosed are the results of the first three days of the release-retrap experiment, and your opinion about them is most urgently asked for and supplicated" (Dobzhansky to Wright, June 13, 1941). The problem was that on the very first day the released flies went outside the experimental plot. Dobzhansky had underestimated their dispersion. Thus the basic question became; how best, with available experimental resources and manpower, could the experimental plot be expanded? Wright recommended (in a June 18 letter) going to squares of 50 or 100 meters, but Dobzhansky had insufficient manpower for such expansion, and besides he did not yet think the flies would travel that far. He set up squares of 20 meters on a side for the next release experiment, and on June 21 Dobzhansky reported that the flies were moving toward the outside boundaries. He also detailed many other difficulties, such as varying densities of the wild and released flies because of ecological differences in the experimental plot. But even after telling Wright about all the problems, Dobzhansky had this insight:

> In general, although I am constantly in doubt about many points in these experiments, I am confident that we shall learn something which will give some knowledge of the biology of the flies which will be of use in thinking about the population genetics problems. Thus, I am becoming a bit skeptical about the local inbreeding, at least during the summer season. The beasts seem to be more mobile than at least I was inclined to think.

Here was an important insight. Dobzhansky had at last come to the view that Wright had been advocating for quite a while—that local effective population sizes in *D. pseudoobscura* were too large to permit intense random drift. Dobzhansky's hope of finding random drift was beginning to fade, along with his conception of population structure.

Only two days later this tentative view was reinforced. The marked flies had gone beyond the plot of 20-meter squares. Only then did Dobzhansky, after hurried consultation with Wright, move to the cross pattern so familiar from GNP X. After confirming the design, Wright added, "There seems little doubt that local inbreeding is unimportant under the conditions that you are studying. But it still might be responsible for much excess elimination of lethals among the grandchildren of the first flies to become active after eclipse" (Wright to Dobzhansky, June 26, 1941). It is important to remember here that for Wright, the inbreeding leading to elimination of lethals led to no permanent genetic differentiation because of later migration.

After Dobzhansky and Wright settled on a suitable experimental design, heat waves and a forest fire plagued the dispersion experiments. Dobzhansky, discouraged, told Wright of one heat disaster in a letter of July 18, to which Wright replied that he could not generate missing data but that "after all, it is the job of statistical methods to extract just such regularities [that Dobzhan-

sky was seeking] from data that look hopelessly irregular at first sight"
(Wright to Dobzhansky, July 25, 1941).

The dispersion experiments did better at the end of the summer, and
Dobzhansky was eager to discuss the data with Wright in Chicago on the return trip: "Needless to say, I am extremely anxious to meet you, since there
are more questions than ever which I want to discuss with you, and it is so
much better done personally instead of by letter" (Dobzhansky to Wright,
August 6, 1941). After talking to Wright about the data from the dispersion
experiments, Dobzhansky was unhappy with their completeness and other
problems and decided, despite the difficulties imposed by the war, to gather
more data on dispersion. Again he wanted to keep in close contact with
Wright. In a letter of May 1, 1942, he described the work on dispersion
planned for the summer and said, "If you can think about some more improvements in these experiments or about some other experiments that would
give more information on the population structure which interests us, please
let me know."

The dispersion experiments went well in the summer of 1942, and
Dobzhansky obtained what he described to Wright as "damned good data."
To speed things along, Dobzhansky decided to do as much of the quantitative
data analysis as possible before sending the data to Wright for final analysis.
Thus Dobzhansky tried to follow Wright's formulas (sent to Dobzhansky during the summer) for calculating variances in distance traveled by the flies and
for the relationship of temperature and variance. Unfortunately, Dobzhansky
discovered that he did not know what all the symbols in Wright's equations
stood for, so he had to ask Wright to explain some of the symbols and also to
ask some embarrassing questions as, "Is the square root of variance, i.e., the
standard deviation in meters, the mean distance? Please forgive me if I am
talking nonsense: I am fully aware of my mathematical inferiority"
(Dobzhansky to Wright, October 16, 1942). Wright sent the required information, only to get the reply, "This makes the whole situation clear to me,
and I believe I know how to proceed with the remaining calculations. If nothing intervenes, I hope to have the whole business completed and sent to you
in less than a month—having the calculations ready you will have only to
draw the conclusions" (Dobzhansky to Wright, October 23, 1942).

Dobzhansky's comments in the paragraph above reveal again how little
he really understood of Wright's quantitative analysis of data. In particular,
he did not understand how Wright revised theoretical models in response to
concrete data. Thus during the summer of 1941 Wright had suggested to
Dobzhansky that the variance of distance traveled by the flies in a given time
period was linearly proportional to the total variance along any one axis.
When Wright began serious analysis of the data in November-December
1942, however, the data did not fit the simple model, and Wright had to devise a more complex one, as he wrote to Dobzhansky on December 18:

The distribution of released flies departs so wildly from normal-ity on the first day and changes so much in the later collections that I do not think now that the variance can safely be used as proportional to the total east-west (or north-south) variance. The most satisfactory method of treatment seems to be to treat the frequency surface as that of a solid of revolution about a vertical axis through the point of release, with frequencies along any radius equal to the average of the observed frequencies along the 4 (or 2) radii along which collections were made.

This same letter raised other problems about models of dispersion in relation to the data. The result was that instead of simply "drawing the conclusions" as Dobzhansky hoped and expected, Wright did his usual serious job of data analysis and took several months to do it. Even a cursory reading of the published paper reveals why Wright's analysis took time.

A detailed analysis of GNP X is inappropriate here, but certainly the conclusion of the paper on the issue of effective population size is both clear and instructive. Dobzhansky's long-favored suspicion had been that local populations of *Drosophila pseudoobscura* were small enough for inevitable random drift. He began the experiments on dispersion in an effort to verify this suspicion, but the data revealed otherwise: "The effective size of the panmictic unit in *Drosophila pseudoobscura* turns out to be so large that but little permanent differentiation can be expected in a continuous population of this species owing to the genetic drift alone." Were Dobzhansky and Wright pleased to come to this conclusion? Did they point out that the situation in *Drosophila pseudoobscura* might well be found in many other organisms?

> It must, however, be pointed out that the figures obtained are valid only for the localities in which the experiments have been made and, of course, only for the species under study. Apart from species of *Drosophila* associated with man, *D. pseudoobscura* is the representative of the genus forming by far the densest and most widespread populations in at least the forested regions of the Western United States. Furthermore, Keen Camp and Idyllwild are about as favorable localities for this species as could be found, and they were selected for experiments for just that reason. It is certain that the population densities in less favorable localities are much smaller than indicated above, and consequently the effective size of the panmictic unit arrived at is closer to the largest than to the smallest in this species. The panmictic units are probably smaller still in species that are less common or more restricted to habitats that recur only sporadically in space, especially if these species have lower dispersion rates than *D. pseudoobscura*. As shown above, *D. pseudoobscura* is certainly more mobile than *D. melanogaster* and *D. funebris*, the only two other species of the genus which have been examined in this respect. (Wright 116, pp. 338–39)

It is almost as if Fisher and Ford had moved into the study of *Drosophila pseudoobscura* and found large effective population sizes, only to be answered by Dobzhansky and Wright with all their reasons why the data on *Drosophila pseudoobscura* from the particular locations studied should in no way be generalized to the conclusion of large population sizes even elsewhere in the range of *D. pseudoobscura,* much less generalized to any other organisms.

With results so clear, one might wonder why Dobzhansky decided in 1945–46 to initiate yet another set of dispersion experiments (to appear as GNP XV with Wright). Could he still hope to find effective population sizes small enough for substantial genetic drifting? I will argue shortly that the basic motivation for doing the research behind GNP XV was quite different from that for GNP X, despite the experimental similarity.

The Brazilian Grand Plan

As Wright was finishing the work on GNP X in the spring of 1943, Dobzhansky was arranging to go to Brazil to study some of the many species of *Drosophila* there. On January 26, Dobzhansky first mentioned the plan in a letter to Wright: "This Brazilian venture is, of course, a plan the execution of which would take a number of years. I think there is a possibility of much to be gained by studying the population structure in species living in a climate that changes as little as possible during the year." Precisely what Dobzhansky sought in Brazil is best explained in his own words from his oral memoir of 1962:

> What originally stimulated the Brazilian visits proved to be eventually a mistake, or a misapprehension. It goes back again to Sewall Wright. One of Sewall Wright's contributions which was and is controversial . . . this other fellow, R. A. Fisher, was until his death violently opposed to it . . . is the so-called principle of genetic drift. Genetic drift, which amounts to oscillation of gene frequencies, oscillation which occurs or may occur in small populations and which would be negligible in large populations. Sewall Wright believed genetic drift to be an important agency in evolution, important in interaction or conjunction with natural selection. . . .
>
> But in collaborating with Wright I have, of course, been very excited about this problem of genetic drift. In fact, these experiments on the release and recapture of flies . . . were designed in order to see how small, how limited, the *Drosophila* populations may be. It was, in other words, a test of the possibility of genetic drift.
>
> Now, one of the important variables there, or so it seemed to me at that time, is that in temperate climates, where you have summer and winter seasons, the populations of many animals, including *Drosophila,* pass every year through a series of contractions and expansions. The flies hibernate almost certainly as adults, and during the winter season most of them die out, so that by spring only a few survive, presumably chiefly impregnated females, also some males

are left, and they start the ball rolling from the beginning. As the season progresses, and more fruits and other food is available, the population grows very large.

It is this periodic reduction of the population to small size which seemed important as a possible agency bringing about this genetic drift.

That led to a very simple idea: if the genetic drift is due to seasonal alternations, chiefly winter resulting in destruction of the flies, then what would happen in a tropical climate where winter never comes? There season after season the population should be large enough to eliminate genetic drift. That seemed to be confirmed by a finding which was made as a result of the collection in Mexico and Guatemala, particularly Guatemala, made in 1938. The Guatemalan population had a great number of concealed lethals, much greater than California populations, which I ascribed to this fact of population size being more constant, the population at no season being reduced to a small number of survivors.

So, logically, it seemed that if you get to a real honest to goodness tropical climate, where population is even, genetic drift would not exist. You would have a larger genetic load, etc.

Now, I just as well say at this point that this proved to be wrong. It proved to be wrong because although there are no winter-summer seasons in the tropics, seasonal changes are by no means absent.

I am not going to describe Dobzhansky's Brazilian research in any detail (for this, see Glass 1980a), but it is essential here to understand Dobzhansky's deep dependence upon Wright's conceptualization of the evolutionary process. With Wright's extreme independence of mind, I suspect he did not at the time fully understand Dobzhansky's dependence upon him. Before departing for Brazil, Dobzhansky tried to convince Wright to come to New York City so that Dobzhansky could discuss with him the design of the Brazilian work. Wright was busy and could not come before June. He wrote Dobzhansky:

> I would like very much to make a trip to New York but I doubt whether I would have anything to say in the course of a day or so that would be worth while. I have to work on a problem for a considerable period of time before reaching any conclusion that I have any confidence in. We would like it very much if you could come here though whether I would be able to make any suggestions of value for your Brazilian trip is another matter. (Wright to Dobzhansky, February 20, 1943)

Dobzhansky was teaching four days a week and had a crushing load preparing for the Brazilian trip. He was scheduled to leave the United States on March 18, 1943. In the midst of this furious preparation he wrote to Wright: "I am reminded of the old adage: 'If the mountain does not come to Mahomet, then Mahomet will go to the mountain.' There being no chance that you can come here before June, I am seriously contemplating a trip to Chicago to see you. How I can do it and when, I do not yet know" (Dobzhansky to Wright, February 26, 1943).

It is a measure of Dobzhansky's dependence upon Wright and his ideas that Dobzhansky did indeed jerk his schedule around to such an extent that he went to Chicago on the train and spent two days talking to Wright about the Brazilian research. Although on occasions I have cursed the abundance of research materials on Wright's life and work (though always on reflection I am truly grateful), it would have been nice to play through a videotaped recording of this meeting between Wright and Dobzhansky before writing this section. There can be little doubt that Dobzhansky found the trip rewarding (he would make a similar trip again later, as I will describe in the next section). By visiting Wright, Dobzhansky felt he was warding off the awful possibility that he might go to South America and return with data that were not amenable to analysis or were merely interesting.

GNP XII and XV: Inversions, Selection, and Migration

Even before he began the research for the GNP series, Dobzhansky was convinced that the third-chromosome inversions in *Drosophila pseudoobscura* were selectively neutral. In GNP I he had stated that the differences in inversions between populations from different mountain ranges in Death Valley were caused by random drift. Koller, in GNP III, had argued for genetic drift as the agent for causing observed differences in inversions in populations from adjacent canyons in Death Valley.

As early as the spring of 1939, however, Dobzhansky discovered by monthly sampling of the *D. pseudoobscura* population in Andreas Canyon of Mt. San Jacinto that "in one of the localities there is observed a quite significant and so far uninterrupted change in the composition of the population. What that means, I cannot decide as yet. But one thing is clear, namely that this change is not connected with the passage of the population through the eclipse stage" (Dobzhansky to Wright, June 28, 1939). After the spring of 1939, Dobzhansky kept careful records of monthly sampling of many different populations. On May 25, 1940, he sent Wright the monthly data (from April 1939-May 1940) for Pinyon A (one of the two collecting areas at Pinyon Flats on the slopes of Mt. San Jacinto) with the comment: "Looking at the data as they stand, without any statistics, it seems to me fairly clear that the population has changed at least twice during the period of observation, none of the changes coinciding however with the eclipse periods." Since the changes in frequencies of inversions changed not at population minima, Dobzhansky would have been justified in concluding that random drift could not be the cause of the observed changes.

Almost a year later, however, Dobzhansky persisted in using random drift to explain seasonal frequency changes. In March 1941, he sent to the printer final revisions for the second edition of *Genetics and the Origin of Species*. This edition contained a detailed description, complete with a table, of temporal changes in the frequencies of inversions in a single population. The section concluded: "The character of the chromosomal variability in *Drosophila pseudoobscura,* and especially the temporal changes observed in its populations, seem best accounted for on the assumption that the genetically effective sizes of these populations are small" (Dobzhansky 1941, 184).

Before the second edition of *Genetics and the Origin of Species* was issued, however, and even before Dobzhansky left for the summer fieldwork in California, he discovered by reviewing the data that the changes in frequency of the inversions might be cyclic. He described this discovery and its implications to Wright in a letter of May 4, 1941.

> You may remember that during our conversation at Philadelphia (late December, 1940) I told you that there is a suspicion that the changes in the frequencies of the gene arrangements in San Jacinto populations with time appear to be cyclic. At present there is no longer any doubt that they are cyclic. In two localities on San Jacinto, namely at Andreas and at Pinon, the standard arrangement increases in frequency during the cool and rainy season, and decreases during the warm and dry season. Chiricahua shows the reverse relation; Arrowhead seems to vary in a haphazard fashion, as though its frequency depends just on what the other two are doing.
>
> This is something neither of us expected, and I must confess that when a few years ago Ake Gustafsson suggested that the changes in the old San Gabriel population may be seasonal, both myself and Sturtevant laughed at him. Anyway, such is the unexpected outcome of this work, and now the question is how to interpret it. The simplest interpretation would seem to be that the inversions produce position effects which make the respective chromosomes acquire selective advantages or disadvantages. I regret to say that this simplest assumption does not fit the data any too well. . . . It looks as if the behavior of the Standard and Chiricahua chromosomes is due to their gene contents rather than to the intrinsic properties of the gene arrangements as such.
>
> But when I think about the changes in the frequencies of the gene arrangements themselves, there is another question which arises in my mind. After all, in order to produce such changes within very few generations (I do not think we can assume more than, say, ten generations of *D. pseudoobscura* per year, if that many) the selective advantages and disadvantages of the Standard and Chiricahua chromosomes in these populations must be very appreciable. I am sure, it will take you just a few minutes to estimate how large the selection coefficients may be. If they are large, it may be worth while to undertake experiments in the laboratory to test the hypothesis. I would appreciate it very much if you would let me know what you think about the practicability of such a project.

This letter indicates that already in May of 1941 Dobzhansky had reached the view that cyclic seasonal changes in frequencies of inversions existed in some populations of *D. pseudoobscura,* and that these changes resulted from selective differences in gene contents of the inversions, not from position effects. Furthermore, Dobzhansky set Wright to work calculating the necessary selection rates and suggested the experiments that would eventually be published as GNP XII. In response, Wright made sample calculations of the selection rates (he said they were "rather large") and agreed with Dobzhansky that laboratory experiments on the selective value of different inversions were worthwhile. He added a note of caution about the work Dobzhansky had been doing on "nonadaptive" differentiation: "Differences in selective value of the same arrangement in different localities will seriously affect the value of the variance of arrangement frequencies as an indicator of nonadaptive differentiation" (Wright to Dobzhansky, May 20, 1941).

As I demonstrated in the previous section, by June of 1941 Dobzhansky had, because of the dispersion experiments, reached the view that local inbreeding could not be as important as he had earlier thought. This left either selection or migration as the key to seasonal changes of frequency of inversions, and from the beginning Dobzhansky was drawn to selection as the primary agent. He continued to gather data about the seasonal changes, and on December 2, 1942, he sent to *Genetics* the manuscript of GNP IX ("Temporal Changes in the Composition of Populations of *Drosophila pseudoobscura*"). Here he documented the seasonal changes and advocated natural selection as the primary cause, a view that was "contrary to the surmise which the writer had entertained in the past." He added, "experiments which may possibly shed some light on this problem are being planned" (Dobzhansky 1943, 178–79).

As evident in this letter of December 4 to Wright, Dobzhansky lost no time thinking about the proposed experiments:

> I have sent you yesterday a MS on the temporal changes in the frequencies of the gene arrangements, a topic which has been mentioned several times in our conversations and correspondence. I hardly need say that your criticisms and comments will be greatly appreciated. In a way, this is a very startling thing, because it implies selection coefficients which would seem to me quite unbelievable. As you know, I am planning to check on this matter by means of "box experiments" constructed according to the plan of l'Héritier. . . . In fact, the experiments would be started already if it were not for a mites infection which we are just now fighting. It may, however, be a blessing in disguise if I have an opportunity to discuss the plan of these experiments with you before they are really started.

After reading the manuscript, Wright replied that he was "utterly appalled at the amount of work involved here on top of that represented by the data on lethals, distribution, etc. I imagine that it will be a long time before anyone collects such a body of data on the population genetics of this or any other organism." In the light of Dobzhansky's later emphasis upon the "balance" theory, to which heterozygote advantage was the key (Beatty 1985), Wright's further comment upon selective differences of the inversions is helpful: "With such marked selective differences as seem to be present, it would be remarkable if the opposing trends at different seasons balance each other so perfectly that there is no overall trend. In the long run all but one should probably be eliminated unless there is general selection favoring heterozygosis" (Wright to Dobzhansky, 18 December 1942). Wright did not at this time, however, think this possibility was very likely.

Dobzhansky worked closely with Wright on the box experiments. He had begun the first of these in December 1942, but mites ruined the experiment while he was away in Brazil in July of that year. On December 17, 1942, he sent the data from this first experiment to Wright, asking for advice in setting up the second: "The experiment, if mites do not ruin it again, will last for at least a year and will take a lot of labor, so I would like to avoid making a mistake. Your opinion is solicited!" Wright dutifully responded. Dobzhansky sent Wright data from the box experiments for over a year, beginning early in 1944. A number of difficulties arose in experimental design, and these Dobzhansky was especially eager to discuss with Wright. The box experiments were very laborious, and the streamlining of the experimental process meant the saving of weeks of labor.

In April 1944 Wright decided to attend a meeting of the National Academy of Sciences in Washington. Dobzhansky heard of this and wrote to Wright:

> I am becoming frantic trying to invent a method that would permit me to see you while you are on the Atlantic Coast. I am willing to skip a couple of classes to attain this end. . . .
> I feel that in an hour or two of conversation with you I could get an agreement on how to do the experiments better and more economically as to labor than by writing you many letters and causing you to waste your time on replying to these letters. (Dobzhansky to Wright, April 15, 1944)

Again Wright depreciated his possible contribution: "I think that you are greatly exaggerating the amount that I could contribute to your box problem off hand. On the other hand, I would like very much to discuss it with you if you are able to be in Washington part of the time" (Wright to Dobzhansky, April 19, 1944). Dobzhansky indeed traveled to Washington for the sole purpose of meeting with Wright, but through a series of slapstick-comedy-like events, neither could find the other in Washington and each assumed the other had not come. Very disappointed, Dobzhansky went back to New York and worked again by correspondence. "The most important question,"

Dobzhansky wrote, "is what to do next, and this depends largely on what you think about the data now available and what is preferable for your eventual treatment of these data" (Dobzhansky to Wright, April 24, 1944). Almost every month through the rest of 1944 Dobzhansky sent Wright data from the box experiments. Finally in January 1945 the first set of box experiments was almost done: "It seems to me that the first series of 'box experiments' is just about completed . . . . It seems reasonable to sit down and write a description of these experiments. . . . Then I beg your kind permission to send the draft manuscript to you for your analysis, so that we may have another joint opus, this time evidently on 'artificial' rather than 'natural' populations." Dobzhansky added that he planned to continue and expand the box experiments ("I have half a dozen plans of what to do next") and also to return to California for more fieldwork. "All this brings me back to the same problem, namely how to meet you. I am very anxious to see you, of course before summer in any case" (Dobzhansky to Wright, January 13, 1945).

This letter caught Wright at a difficult time. In part because of the collaboration with Dobzhansky, his own work, especially on guinea pigs (but including also the writing up of his Hitchcock Lectures at Berkeley, a job Wright never completed), was seriously behind schedule. The prospect of being Dobzhansky's never-ending collaborator was hardly happy. This was especially true with the box experiments, which Wright believed were worthy of no major quantitative analysis. I can find only one place that Wright expressed something of this situation in print. As he wrote to Eric Ashby,

> I have been carrying my guinea pig colony on a reduced scale but am badly in arrears in working up such data as I have obtained, partly because of cooperation with Dobzhansky who, with his extraordinary energy, can collect data on *Drosophila* populations in nature and in experimental situations more rapidly than I can make mathematical analysis. (Wright to Ashby, September 6, 1945).

Wright decided to break off the collaboration with Dobzhansky, and so wrote to him about the proposed collaboration on the box experiments:

> I will be very much interested in seeing your draft. I don't think that I had better try to collaborate, however, because (1) there may be very little for me to suggest or at least nothing novel in which case there is no reason why I should be a collaborator or (2) there may be need for working out new methods, comparable to what I did on the wild populations in which case I do not see the possibility of concentrating on it to the required extent. I am a statistical consultant on several war projects. I devoted nearly all of my time outside of teaching and guinea pig routine to those last winter and spring. I didn't do much along this line in summer and fall but problems are now coming up which threaten to use all my time again. I also have a couple of long articles for the Encyclopaedia Britannica (now

owned by the University of Chicago) which have got to be worked in some how in the next few months. My own work is backed up very badly and needs full attention as soon as possible. Altogether I see very little prospect of finding any time for serious collaboration that I would want to put my name on, short of a year or perhaps the end of the war. I will, however, be glad to see your draft and make such suggestions as I can offhand. I will probably go to the Academy meetings but I have at least one other trip that I will have to make in the spring quarter and I will probably not be able to get away any longer time than necessary because of class work. We would, of course, be delighted to have you visit us sometime in February or March. (Wright to Dobzhansky, January 22, 1945)

Dobzhansky was surprised (all along he had supposed that Wright would collaborate on the box experiment) and discouraged by Wright's letter, but he was also a grand master at inducing busy people to collaborate with him. His letter back to Wright skillfully kept the door wide open for future collaboration:

> You will hardly be surprised if I say that your letter of January 22 was very discouraging to me. I certainly could think of nothing better than a continuation of our collaboration, and although I understand that you have lots of your time consumed by war projects, this remains acutely sad, just one more of the sad things which war has brought. Well, there is nothing to be done about it—war has brought even worse things to many. In the meanwhile I shall be hoping that you may be more receptive later in "a year or perhaps the end of the war," and I certainly will use your permission to send you the draft of the paper on the box experiments when it will be ready (which will not be for some months yet) to get the benefit of your suggestions. But, after all, some of my work projects really depend on you—I can, I hope, collect reasonably good data but only you can get the full value from them. The work planned for California for the next two summers is precisely of that kind, and it is, of course, direct extension and development of the work of the last years on which we did collaborate. This is the principal (although only one of the several) subject which I wanted to discuss with you when we meet at long last—here or in Chicago. Of course, I am determined to have such a meeting, probably in March, and shall go to Chicago for this purpose. The work for California was, of course, primarily the new diffusion experiment that I will analyze just ahead. (Dobzhansky to Wright, January 27, 1945)

Dobzhansky did indeed travel to Chicago to meet Wright on April 5, 1945. Before the visit, he sent Wright a copy of the description of the data from the box experiments. During the visit, he induced Wright to contribute a two-page conclusion to the paper and hoped to get it from Wright within a few weeks at most. Wright sent him the section on June 17; it was twenty-three-pages long, not including the tables. Dobzhansky put Wright as the first author.

GNP XII was distinctive for several reasons. The data clearly indicated that at 25° C (and at "room" temperature) a stable equilibrium of the three inversion types established itself, no matter what the starting frequencies happened to be. Thus the inversions were almost certainly subject to selection pressures, as Dobzhansky had surmised in 1941. Wright also investigated the possibility that the inversions were subject to frequency dependent selection as might be expected in nature in a heterogeneous environment. He developed a quantitative model for the frequency dependent selection and found that one rather extreme version of it fit the data reasonably well. Although Wright considered the possibility of heterogeneous environments within a population cage to be improbable, he stated that the possibility of fitnesses depending "on different functions of gene frequencies is not improbable. This could be tested only by special experiments" (Wright 122, p. 147). I think the primary reason that Wright found the data theoretically interesting beyond the bare fact that selection acted on the inversions was the possibility of developing a theory of frequency dependent selection. Dobzhansky was not very keen on this possibility, emphasizing instead his view that heterozygotes were adaptively favored over both corresponding homozygotes.

Seen in the perspective of the rest of the GNP series, GNP XII represents a significant initial attempt to connect evolution in nature to controlled experiments in the laboratory. Indeed, this was unquestionably Dobzhansky's basic motivation for doing the box experiments. The title of the paper laid bare the motivations: "Experimental Reproduction of Some of the Changes Caused by Natural Selection in Certain Populations of *Drosophila pseudoobscura.*" Dobzhansky of course wanted very much to replicate all of the major changes in gene frequency observed in nature. The results of the box experiments were in that way disappointing. The frequency changes observed between June of one year and March of the next were nicely replicated by the laboratory experiments, but frequency changes observed during the spring months were not replicable, despite many attempts.

Yet the spring changes were crucial. If random drift and mutation were eliminated as major causes of observed changes in frequency, then selection and migration were the only two realistic possibilities for such major and rapid shifts in frequency. Dobzhansky strongly believed in selection, but others were not convinced, among them Sturtevant and Hovanitz. They suspected that in the spring, populations from higher altitudes, taking advantage of gravity, migrated to the population under study, causing the shift in inversion frequencies. It was a possibility that Dobzhansky strongly disliked, both because he thought the evidence was against it and because he was feuding with Sturtevant and to some extent Hovanitz.

Wright encountered Hovanitz's thesis on his trip to Berkeley in May-June, 1943. Epling and Hovanitz took him to the Mojave Desert to see *Linan-*

thus parryae, and Hovanitz took him to Mt. San Jacinto to see *Drosophila pseudoobscura*. After Dobzhansky returned from Brazil, Wright wrote: "Hovanitz suggested that the big changes in the frequencies of the inversions with season might be due to mass migration down the mountain. If true, this would affect the interpretation in various ways" (Wright to Dobzhansky, November 30, 1943). Dobzhansky responded:

> You write about Hovanitz's suggestion that changes in the frequencies of the gene arrangements are due to migration rather than selection. He wrote me about this "suggestion," and has inserted it as a flat statement in his paper now in press in *Genetics* (Hovanitz 1944, 55n). I think this suggestion is plain stupid, and propose to ignore it as far as publications are concerned. . . . [There follows a long paragraph detailing Dobzhansky's objections to the Hovanitz suggestion, ending with] Hovanitz is a bright fellow, but as so many bright fellows he got to thinking that he can solve all the problems of the universe much easier than anyone else without working on them, just by sheer power of intellect. I hope time will cure him, and the wisdom of age will give him some humility which a scientist needs as well as a non-scientist. (Dobzhansky to Wright, December 2, 1943)

Wright agreed with Dobzhansky that mass migration was an unlikely solution to the problem of May-June changes in inversion frequency. But he also thought that the problem did exist and that selection was not the obvious answer (which he showed from Dobzhansky's data).

> Summing up, I feel puzzled about the matter. All positive evidence is against the migration theory. There is, however, the fact that the most striking change in frequencies is the shift in both Piñon and Andreas toward Keen in May and especially June, and reversion thereafter, which doesn't seem to make sense from the standpoint of pure selection. There is something here that presents a real problem. (Wright to Dobzhansky, December 15, 1943)

Almost a year and a half later, when Wright was putting together his part of GNP XII, he still maintained that Hovanitz's thesis, although improbable on the evidence, was still a possibility that could not safely be neglected. Dobzhansky wanted to lambast the Hovanitz thesis. When he returned from California in September 1945, he said in a letter to Wright that "the summer experience has strengthened my conviction that Hovanitz's notion is completely wrong" (Dobzhansky to Wright, September 20, 1945). Wright protested:

> I hope that I did not give the impression that I consider the migration hypothesis as very probable. There is certainly strong evidence against it as far as observations on dispersion go. It must, however,

it seems to me, be considered as a possibility, however remote, that there is some sort of mass movement down the mountain at a much greater rate than observed, under certain conditions, until some evidence is found for a form of selection that can reverse the expected direction of change in spring indicated by the experiments. That is, something contrary in a sense to either the dispersion results or the temperature results must be found before either migration or selection can be ruled out or accepted. (Wright to Dobzhansky, October 1, 1945)

In its published form, GNP XII raised Hovanitz's thesis as a possibility but presented evidence indicating that the possibility was remote. Dobzhansky evaluated the final compromise wording as follows: "Hovanitz's notion is having a dignified funeral, but the grave is not yet deep enough, so that if it does become resurrected it can arise gracefully" (Dobzhansky to Wright, October 3, 1945).

Although the effort in GNP XII to relate evolution in natural populations directly to laboratory experiments was to a high degree inconclusive, and although greater sophistication would be required in the future for similar attempts, this paper represents a major pioneering effort in evolutionary biology. Dobzhansky sensed Wright pulling away from collaboration; after placing Wright's name as first author ("I do not think that I should be the senior author," Wright to Dobzhansky, October 1, 1945), he said of the collaboration, "I pray God it shall continue in the future" (Dobzhansky to Wright, October 3, 1945). It would, formally at least, for just one more paper.

GNP XV (as I mentioned earlier) appears to be a simple extension of GNP X, the first paper on dispersion. Dobzhansky initiated the research for GNP X, hoping and expecting to find very small dispersion rates and a correspondingly high amount of random drift. The data contradicted the expectation. In initiating a second round of experiments on dispersion, Dobzhansky had a much different basic motivation. Now he was not trying to prove that dispersion rates were sufficiently small enough to result in random drift but that dispersion rates were sufficiently small enough to contradict the Hovanitz thesis that migration could account for the May-June changes in frequencies of inversions. Thus in both cases he was trying to elucidate the limits of dispersion, but the limit aimed at for GNP XV was much larger than that for GNP X, and for a different purpose.

The experimental novelty of GNP XV was the attempt to recover released flies (or at least their chromosomes from offspring heterozygotes) the following season to furnish a basis for estimating dispersion over the period from late summer to the following late spring. Combined with the other data on dispersion during the summer, year-long dispersion could be estimated. The release and recapture of flies in the summer of 1945 went well and corroborated the results of GNP X. After a massive release of orange-eyed marker stocks (25,134 flies) in late July and early August 1945, Dobzhansky recovered in June of the following year enough orange heterozygotes for Wright to estimate dispersion rates over the period. From the data Wright calculated that about 95% of the progeny of the released flies would, after a year, be found within a circle of radius 1.76 kilometers, and 99% of the progeny within a circle of radius 2.2 kilometers.

There is not one mention of random drift in the paper, nor is there any mention of the Hovanitz migration thesis. But the conclusion—"There can be no doubt that within a year the flies and their progeny have not moved very far from the point of release" (Wright 124, p. 323)—is certainly directed toward the Hovanitz thesis rather than an attempt to resuscitate the hope about random drift that Dobzhansky held going into the work for GNP X. It is perhaps surprising to discover that an issue prominent in the Wright-Dobzhansky correspondence, and central for understanding the motivation for the research on GNP XV, is nowhere visible in the paper.

When Dobzhansky sent his part of the manuscript of GNP XV to Wright on October 18, 1946, he said that "the present data will have to be the last ones on the mobility of *D. pseudoobscura* as far as I am concerned. Although it is fun to make these experiments, there is so much else that can be done during the summers in Sierra Nevada!" Wright, who was just preparing to initiate his least innovative contribution to the GNP series, was glad to hear that the dispersion experiments were ceasing.

Dobzhansky did not permanently lose interest in the problem of dispersion. Near the end of his life he and his student Jeffrey Powell began a new series of experiments on dispersion in *Drosophila pseudoobscura* and its relatives (Dobzhansky and Powell 1974). The very last (forty-third) paper in the GNP series again addressed this topic, although Dobzhansky had died before its publication (Powell et al. 1976). Wright also never lost interest in the quantitative analysis of dispersion, vigorously rebutting one aspect of Wallace's reanalysis in 1966 of the earlier dispersion data (Wallace 1966; Wright 186) and generally keeping up with the expanding literature on dispersion (see Wright 200, pp. 61–71). Dobzhansky and Powell consulted with Wright before launching into their new work on dispersion and sent him the raw data. The letters exchanged between Dobzhansky and Wright over the new dispersion data were a direct continuation of the correspondence concerning the same issue over thirty years earlier. Both seemed to enjoy and benefit from this reincarnation of their earlier collaboration.

End of the Collaboration and Conclusions

When Wright finished the analysis of data for GNP XV, he was already involved with the analysis of Dobzhansky's data from a new box experiment, later to be reported in GNP XVII. But as in the case of GNP XV, Wright felt he was becoming more of Dobzhansky's number cruncher than a contributor to evolutionary theory. *Drosophila pseudoobscura* was obviously not the organism to use for a demonstration of the shifting balance theory of evolution in nature. From these considerations, and also from the desire to devote full time to his own guinea pig research and work in theoretical population genetics, Wright insisted that he not be listed as an author on GNP XVII. The signal to Dobzhansky was unmistakable this time, and the formal collaboration ceased. Dobzhansky and Wright continued to correspond, however, until Dobzhansky's death in 1975. Most of the correspondence was initiated by Dobzhansky, who frequently asked Wright questions about experimental design or analysis of data, just as he had during the period of formal collaboration. Dobzhansky's interests shifted away from population structure, more toward the analysis of selection in nature and in the laboratory. He therefore felt less need of working with Wright, who seems to have felt relief in being saved from the huge amount of data Dobzhansky so quickly generated, especially since the theoretical interest of Dobzhansky's data had begun to seriously decline for Wright.

The collaboration of Wright and Dobzhansky in the GNP series was an influential event in the history of evolutionary biology. During the 1930s, theoretical population genetics was largely independent of the study of populations in nature, which of course had a long and distinguished history (see Mayr 1982, parts 1 and 2). Wright and Dobzhansky demonstrated that the reciprocal interaction of abstract population genetics theory and practical genetic analysis of natural populations in the field (and using the laboratory) yielded highly rewarding results. The data had far greater meaning by being imbedded in a larger theoretical construct, and the abstract theory benefited by the elucidation of constants and hypothetical relations between variables.

The collaboration reveals central aspects of possible relationships between theoretical models in population genetics to field research.

1. Wright's theoretical models identified for Dobzhansky the relevant parameters of the evolutionary process, and of population structure in particular. Dobzhansky's whole experimental program in the GNP series was based directly upon his desire to quantify or clarify aspects of Wright's models.

2. Data gathered had to be relevant to theory. This meant the data must be amenable to quantitative analysis, and, when analyzed, must enable discrimination between distinct hypotheses. Dobzhansky always insisted on these goals, but only with Wright's help could he actually achieve them.

3. Experimental design was tied deeply and equally to theory and specific circumstances in the field. Wright frequently modified Dobzhansky's initial ideas about design for a particular experiment, but with about equal frequency Dobzhansky reported that Wright's ideas about experimental design were unfeasible in the field.

More than anything else, the collaboration of Wright and Dobzhansky revealed that neither could possibly hope to achieve the results that they could attain by working together.

The example set by Wright and Dobzhansky has served as an ideal toward which the study of evolution in nature could move. Many of Dobzhansky's students, for example, have tried to combine in themselves the approaches of both Wright and Dobzhansky. In this regard, two of Dobzhansky's favorite students, Richard Lewontin and Bruce Wallace, come immediately to mind.

Analysis of the collaboration also reveals the hopelessness of trying to reconstruct it solely from the published collaborative papers, or even from the papers as well as interviews of both Dobzhansky and Wright. Only the sustained correspondence yielded the required insight. The published collaborative papers all bear the inscription of the first page, "experimental data by Th. Dobzhansky, mathematical analysis by Sewall Wright." This phrase is wholly inadequate as a guide to understanding the actual roles of Wright and Dobzhansky in the collaboration. Documentation such as the correspondence between Wright and Dobzhansky usually does not exist in collaborative scientific work. I suggest, however, that historical analysis of collaborative research should be considered tentative when based primarily upon published results. Thomas Kuhn, of course, has long emphasized this general point (Kuhn 1962).

Wright unquestionably had a very deep influence upon Dobzhansky, both in *Genetics and the Origin of Species* (all three editions) and in the GNP series. Through these works and his teaching, Dobzhansky became perhaps the most influential evolutionary biologist of his time. Thus, Wright had a considerable influence through the work of Dobzhansky.

There is another reason why Dobzhansky was important in the dissemination of Wright's ideas on evolutionary theory. A basic theme in the Dobzhansky–Wright collaboration on the GNP series, as I have shown in this and the previous chapters, is that Dobzhansky found Wright's quantitative reasoning very difficult or impossible to follow. Dobzhansky's response to Wright's isolation by distance manuscript is typical—Dobzhansky told Wright that the manuscript was so very important that he should really try to make it accessible to biologists.

Other biologists widely agreed with Dobzhansky that Wright was hard to read. Wright's correspondence files are filled with comments about the difficulties encountered in following his quantitative reasoning. One of my favorite examples is a response by Edgar Anderson, director of the Missouri Botanical Garden, to a quantitative letter from Wright: "Thank you for the letter. Your interesting computations I accept as I do the pleasant sunshine this afternoon: something which I appreciate and value even though I do not understand it" (Anderson to Wright, October 14, 1938). A 1940 letter from Wright's longtime dear friend Harrison Hunt, a geneticist specializing in animal breeding at Michigan State University, illustrates the issue:

> I want to thank you very much indeed for the reprints which you recently sent me. I value your writings very highly because I think they contain a great many ideas and methods of value. I am inclined

to believe that they present materials that are likely to be useful in developing new techniques in applied animal and plant breeding. . . .

I have one very serious criticism, however, to offer to all of these papers. I have expended, yes wasted, an enormous amount of time upon them because as a rule too few key equations are given in the mathematical analyses. You have a marked tendency to state your assumptions very briefly and then give the end results of the mathematical analysis without giving a sufficient number of the intermediate steps to make it easy for a non-mathematical person to follow your reasoning. Upon inquiry I have found other geneticists have the same difficulties that I do. I never accept the validity of a formula until I have seen it demonstrated, as to do so may lead me to use it in the wrong place and manner.

I have hesitated somewhat to speak in this rather blunt fashion. Please accept these comments as coming from a person who has a great deal of interest in you and your work, and who wishes to promote your ideas in the realm of practical breeding operations. Once I have succeeded in understanding your concepts I usually find them extremely useful and stimulating. (Hunt to Wright, December 2, 1940)

I am certain that Dobzhansky and many other geneticists and evolutionists would have agreed with Hunt.

Wright's response to this letter from his friend helps in understanding both why Wright wrote his papers the way he did and why persons like Hunt and Dobzhansky had such trouble reading them.

The question which you raise on how much of the mathematics should be published, is a difficult one. I appreciate the importance of giving all of the steps. I sometimes have difficulty myself in seeing how I reached a result. On the other hand mathematical formulae are expensive and not liked by editors or printers. I would probably have had great trouble getting my mathematical papers published (as far as they have been) if there had been no condensation. I have had to pay a good deal for exceeding space limits (in the Proceedings of the National Academy of Sciences) as it is. A minor point is the amount of proof reading. In my 1931 paper hardly a formula was right in the first proof and in the second proof many new errors had been made in correcting old ones. Most of the typographical errors in the printed papers that I have noticed were made by the printer in correcting the last proof. These considerations have perhaps influenced me too much in cutting out formulae but at that I do not think that my papers are anything like as condensed as the papers published by the professional mathematicians (in the P.N.A.S. for example).

The general principle that I have tried to follow is to give enough to show the course of the reasoning but not to go into the details of such processes as algebraic transformations, solutions of sets of equations, integration or complicated expressions or anything whose more or less standard methods are available even though it is expected that the reader may have to go to a text book at times. (Wright to Hunt, simply dated 1940).

Everything Wright said makes perfect sense from his point of view, yet at the same time one can easily appreciate Hunt's frustration. The truth is that, as John A. Moore has emphasized to me on several occasions, evolutionary biologists trained in the 1930s and earlier generally had little sophistication in quantitative reasoning. Wright's mathematics are not highly esoteric by modern standards in theoretical population genetics, but they were certainly beyond the grasp of most geneticists and evolutionists in the early 1940s. I do not think Wright fully appreciated the lack of quantitative sophistication in his audience, certainly not in the way J. B. S. Haldane did in his popular writings on evolutionary theory.

On many occasions evolutionists encouraged Wright to present his evolutionary theory in a popular book for general audiences such as college students. Thus Arthur Banta of Brown University wrote in 1942 to suggest that Wright prepare "an extended treatise on evolution" for a general audience:

> With reference to mathematical treatment in doing a general treatise, many persons are inhibited by encountering mathematical formulae, and I am afraid this inhibition cannot be overridden. Hence if you produce such a book as a general treatise on evolution, I hope you will subordinate the mathematical treatment, or rather, make it supplementary to your general story. The reader who wishes to follow the mathematical concepts will be able to do so. The reader who is appalled by mathematical considerations will also be able to follow. Haldane, I think, has been rather successful in presenting his ideas on evolution, using both the mathematical approach and general statements which do not require mathematical abilities on the part of his readers. I think you would be able to write a extensive treatise which would be perhaps as readily comprehended as Haldane's writings, and which would have even more influence among biologists in general than Haldane's contributions. (Banta to Wright, January 16, 1942)

Wright of course never produced such a treatise. In his famous paper "A Defense of Beanbag Genetics"—a reply to Mayr's challenge to population geneticists to document their contributions to modern evolutionary biology (Mayr 1959; see the section on Wright and Mayr in chapter 13)—J.B.S. Haldane wrote concerning population structure:

> Sewall Wright has been the main mathematical worker in this field, and I do not think Mayr has followed his arguments. Here Wright is

perhaps to blame. So far as I know, he has never given an exposition of his views which did not require some mathematical knowledge to follow. His defense could be that any such exposition would be misleading. (Haldane 1964, 357)

Haldane is correct that, with respect to population structure, Wright was always quantitative in his published work and that he argued that any merely verbal exposition could be misleading. The conclusions of Epling on population structure in *Linanthus parryae* and its genetic consequences as compared to the conclusions of Wright illustrate graphically the danger of intuitive versus quantitative analysis on the question of complicated population structure (see the continuation of the *Linanthus* story in chapter 13). But as Crow pointed out to me early in the research for this book, Wright never at any time was highly quantitative about the shifting balance theory of evolution itself. Wright never had a mathematical theory that explained how a whole population could be selectively improved by transformation of subgroups by the immigration of organisms with superior interaction systems of genes, in the face of the inevitable recombination that would tend to break up the interaction systems. Thus I agree with Banta that Wright could have written a general treatise on his shifting balance theory of evolution that could have appeared, for example, in the Columbia Biological Series along with the major works by Dobzhansky, Mayr, Simpson, Rensch, and Stebbins. It could have been a very influential work.

The point is that many biologists found Wright unreadable or at least difficult enough to make the necessary expenditure of energy unappealing. But in comparison, reading Dobzhansky was vastly easier, so naturally many more biologists read Dobzhansky than Wright himself.

The mediation of Wright's ideas through Dobzhansky had its advantages and disadvantages. The primary advantage was that, because of Dobzhansky's great popularity, many of Wright's ideas and his reputation as a theoretical population geneticist became more widely known. Many budding evolutionary biologists first read about Wright's ideas in *Genetics and the Origin of Species*. The disadvantage was that in the first two editions (1937 and 1941), Wright's shifting balance theory of evolution in nature acquired the distinct flavor of being dominated, at the level of "geographical races" and below, by random genetic drift. No one could read the first two editions or papers of the early GNP series without coming away with the belief that Wright had explained the observable genetic differences between "local races" of snails, *Drosophila, Linanthus parryae*, and other organisms by the mechanism of random genetic drift.

Yet as I have demonstrated in this chapter, the collaboration on the GNP series had changed Wright's views about random drift in the inversions of *Drosophila pseudoobscura*. By 1945 Wright had rejected what he had said in 1940 (in the *New Systematics* essay; Wright 101) about random drift and the "neutrality" of the inversions in *D. pseudoobscura*. Thus arose the irony that working with Dobzhansky had caused Wright to reject random drift as a major factor in *D. pseudoobscura* while at exactly the same time Dobzhansky's *Genetics and the Origin of Species* (in its first two editions) emphasized random drift in the "balance" of evolutionary forces. By the time both Wright and Dobzhansky had moved toward a more selectionist view of the evolutionary process (I will support this assertion in the next chapter), the gloss upon Wright's ideas that biologists had learned through Dobzhansky was by then widespread, and Wright found it nearly impossible to shake the image of a theoretician who emphasized random drift by itself as the mechanism of evolution in nature. Of course, Dobzhansky was not alone in presenting this image of Wright, having been joined by Julian Huxley, Ernst Mayr, George G. Simpson, Ledyard Stebbins, N. W. Timoféeff-Ressovsky, and many others.

Finally, the collaboration between Wright and Dobzhansky illustrates my argument that the tension between the evolutionary viewpoints of Fisher and Wright was a great stimulus to the study of the genetics of natural populations. This is quite clear in Dobzhansky's motivation for doing the GNP series and, of course, in Wright's participation in the series. The issue of isolation by distance is a particularly instructive example. Here Wright chose to analyze precisely the kind of population (large, apparently panmictic) emphasized by Fisher, but he attempted to find in this population random genetic differences induced by isolation by distance.

The extraordinary care and analysis that Wright gave to Dobzhansky's data on *Drosophila pseudoobscura* were unusual only in their duration. Everyone who sent Wright requests for help in data analysis was, in time, rewarded with a similar thoughtful treatment. When unfamiliar with the quantitative techniques that a correspondent had used (a frequent occurrence), Wright would rederive the entire analysis using his own familiar methods and then compare the results. Wright frequently complained that he did not have enough time to put his own research into publishable form. This was true largely because he devoted so much of his time to helping others. Wider recognition came from the very visible collaboration with Dobzhansky, but most of his laborious analyses appeared as only brief acknowledgements if that. Given the huge amount of energy and time that Wright devoted to analyzing the work of others, it is a wonder he was able to do his own research at all.

12
Evolutionary Theory, 1940–1955

During the heart of the "evolutionary synthesis" period (1940–1955), Sewall Wright was involved in many aspects of evolutionary biology. My analysis of the development of his work in this field during 1940–1955 reveals and supports two major themes. First, although Wright and Fisher no longer communicated except through published papers, the tension between their evolutionary views was a highly creative force in evolutionary biology, in both theory and field research. Second, my analysis corroborates Stephen Jay Gould's thesis that the evolutionary synthesis "hardened" during the 1940s and early 1950s, by which he means that natural selection came to be viewed as the highly dominant mechanism of evolution (Gould 1980, 1982, 1983). The consequence was that biologists viewed the evolutionary process as almost exclusively adaptive rather than nonadaptive, even at the lowest taxonomic levels. This represented a significant change in attitude in the field of systematics for Dobzhansky and Wright as well as for many other prominent evolutionists.

Wright and his views were central to a number of public debates during the 1940s and 1950s. Those concerning the moth *Panaxia dominula* and the snail *Cepaea nemoralis* were particularly colorful, in terms of both the organisms studied and human relations. Anecdotes about these episodes still abound among evolutionists, not only because of the powerful personalities involved but also because the issues under discussion were and still are central to evolutionary biology. The debates to be examined in this chapter provide exceptional insight into the development of Wright's evolutionary thought and more generally into the development of the field of evolutionary biology during the 1940s and 1950s.

Wright's Influence upon Systematists and Evolutionists

Unlike such dramatic discoveries as the structure of DNA, ideas in evolutionary biology often take many years to percolate into the evolutionary literature. By the early 1940s Wright's ideas from the early 1930s became entrenched in the works of many major influential evolutionists and systematists. Perhaps because of Dobzhansky's interpretation of Wright's ideas in the first two editions of *Genetics and the Origin of Species,* but certainly also because some evolutionists and systematists interpreted Wright the same way Dobzhansky did, the major works on evolution emphasized random drift as the cause of nonadaptive taxonomic differentiation at the expense of Wright's comprehensive shifting balance theory. Thus at the same time Wright was deemphasizing the role of random drift in the production of observable taxonomic differentiation and emphasizing instead the role of random drift at the level of local populations only, his reputation in the literature was being tied to the concept of random drift as the causal explanation of nonadaptive taxonomic differentiation.

It is well to recall here that the primary reason Wright had extended the idea of random drift to account for differences at the subspecies, species, and sometimes genus level was because systematists had insisted upon the existence of nonadaptive differences at these taxonomic levels. When Wright offered random drift as the explanation of such nonadaptive differences, systematists and evolutionists welcomed the idea. Dobzhansky in 1937 had said that Wright's random drift was "best able" to explain the widely and long-observed nonadaptive differences in races and species. Dobzhansky was just the first of many to make this point. After 1937 Wright's name and the concept of random drift were invoked constantly whenever the evolution of nonadaptive characters came under discussion. Proof for this assertion may be found in almost any book on systematics and evolutionary theory from 1937 through the 1950s. Accordingly, I will give only a few examples here from major works. (For a fuller treatment, see Provine 1983.)

Next to Dobzhansky's *Genetics and the Origin of Species,* Julian Huxley's *Evolution: The Modern Synthesis* (1942) was probably the most influential book of the synthesis period. Huxley had been much taken by Wright's concept of random genetic drift in relation to systematics and cited Wright over and over again in the introduction to *The New Systematics* (1940) and in *Evolution: The Modern Synthesis* (1942).

> The proof given by Wright, that non-adaptive differentiation will occur in small populations owing to "drift," or the chance fixation of some new mutation or recombination, is one of the most important results of mathematical analysis applied to the facts of neomendelism. It gives accident as well as adaptation a place in evolution, and at one stroke explains many facts which puzzled earlier selectionists, notably the much greater degree of divergence shown by island than mainland forms, by forms in isolated lakes than in continuous river systems. (Huxley 1942, 199–200)

> Non-adaptive or accidental differentiation may occur where isolated groups are small. This "drift," which we have also called the Sewall Wright phenomenon, is perhaps the most important of recent taxonomic discoveries. It was deducted mathematically from neomendelian premises, and has been empirically confirmed both in general and in detail. (P. 260)

The major problem, according to Huxley, had not been finding the causes of adaptive differentiation in species with wide geographic distribution; that was a relatively easy task. Instead, "On the contrary, the major biological problem has been that of accounting for that fraction of the divergence which is not adaptive, and this would now appear to have been settled in principle, as due to the Sewall Wright phenomenon of drift" (p. 265). Huxley's name for random genetic drift, the "Sewall Wright effect," accurately reflects the general tendency to associate Wright with it. The index to *Evolution: The Modern Synthesis* contains thirty-seven references to Sewall Wright, and all but a very few of them concerned random drift and nonadaptive taxonomic differentiation in small populations.

David Lack's change of view on the adaptive value of specific and subspecific differences in Darwin's finches (Geospiza) is particularly instructive (Kottler, personal communication; Gould 1983). With Julian Huxley's backing, Lack went to the Galápagos Islands in the fall of 1938, studying the finches there for six months. He spent April through August 1939 at the California Academy of Sciences and several weeks with Ernst Mayr at his home in New Jersey before returning to England in September 1939. There he analyzed his data on Darwin's finches and by June 1940 had completed the final draft of his monograph on them. The California Academy of Sciences, because of the war, was unable to publish the manuscript until May 1945 (Lack 1945). Before the monograph appeared, Lack had completed by 1944 his popular work *Darwin's Finches,* which appeared only after a delay of three years (Lack 1947). In the 1945 monograph he argued that the specific and subspecific differences were mostly nonadaptive, and in the 1947 book that they were mostly adaptive. In autobiographical remarks written several years before his death and published in an obituary, Lack explained his dramatic shift in view:

> Since so knowledgeable a man as Arthur Cain asked me, in the mid–1950s, why I postulated that various subspecific differences in the finches are non-adaptive, it is worth stressing that, before my book, almost all subspecific differences in animals were regarded as nonadaptive (hence the importance of Sewall Wright's theory of genetic drift). The main exceptions were size differences, such as those in accordance with Bergmann's rule, and the coloration of certain desert larks. Still more were the differences between closely related species regarded as non-adaptive, except for the specific recognition characters which reduce hybridisation. *The Variation of Animals in Nature* by G. C. Robson and O. W. Richards in 1936 fairly reflected current opinion on these problems. I reached my conclusion, that most subspecific and specific differences in Darwin's finches are adaptive, and that ecological isolation is essential for the persistence of new species, only when reconsidering my observations five years after I was in the Galapagos. (Lack 1973, 429)

Lack's comment clearly supports my thesis about the way systematists generally understood the significance of Wright's theory of evolution in nature, but one point may remain puzzling. About the weeks Lack spent visiting Mayr in August 1939, Lack said that he "learned much about systematics and speciation from him" (p. 428), and Mayr said, "we discussed again and again the problem of speciation on islands and elsewhere" (Lack 1973, p. 432). Mayr has had since the 1940s a reputation for advocating and defending an adaptationist view of evolution, yet after all the detailed discussion with Mayr, who is a highly persuasive person, Lack immediately thereafter wrote about the nonadaptive quality of most of the specific and subspecific differences in Darwin's finches. Mayr has no specific recollection of having discussed the adaptation question with Lack in 1939 (personal communication, January 10, 1983). The following discussion may help to clarify this apparent puzzle.

Ernst Mayr's influential *Systematics and the Origin of Species* appeared in 1942, the same year as Huxley's *Evolution: The Modern Synthesis*. Although Mayr later developed his reputation as an ardent selectionist, his 1942 book left significant room for nonadaptive differentiation up through the species level:

> It should not be assumed that all the differences between populations and species are purely adaptational and that they owe their existence to their superior selective qualities. We have already pointed out the fallacy of such a point of view in the discussion of neutral polymorphism. Many combinations of color patterns, spots, and bands, as well as extra bristles and wing veins, are probably largely accidental. This is particularly true in regions with many stationary, small, and well-isolated populations, such as we find commonly in tropical and insular species. . . . We must stress the point that not all geographic variation is adaptive. (Mayr 1942, 86)

To what did Mayr attribute the existence of this nonadaptive variation? The answer comes in the long chapter on the biology of speciation (chapter 9), in a section entitled "Population Size and Variability," which began with this statement: "Naturalists have known for a long time that island populations tend to have aberrant characteristics. Wright (1931, 1932 and elsewhere) found the theoretical basis for this by showing that in small populations the accidental elimination of genes may be a more successful process than selection" (Mayr 1942, 234). The entire section was explicitly based upon Wright's theoretical work as well as upon the observations of naturalists, and Mayr's central conclusion, in agreement with Wright, was "that evolution should proceed more rapidly in small populations than in large ones, and this is exactly what we find" (p. 236).

In this same section, Mayr introduced his famous "founder principle,"

> The reduced variability of small populations is not always due to accidental gene loss, but sometimes to the fact that the entire population was started by a single pair or by a single fertilized female. These "founders" of the population carried with them only a very small proportion of the variability of the parent population. This

"founder" principle sometimes explains even the uniformity of rather large populations, particularly if they are well isolated and near the borders of the range of the species. (P. 237).

Mayr claimed no originality for the concept of the founder principle, clearly outlined before him by Wagner, Gulick, Elton, Crampton, and others; but he did invent a highly appropriate term for the principle. Wright, of course, viewed the founder principle simply as a special case of random drift by sampling followed by population expansion (a more careful discussion of this issue will follow in the next chapter).

I do not wish to convey the impression that Mayr in 1942 saw random drift as the primary factor or even a major factor in the evolutionary process; in general, he emphasized the adaptive aspects of evolution. And his book was highly adaptationist in comparison to Robson and Richard's 1936 book, which was the most widely used general book on systematics before being replaced by Mayr's. But he did see isolation and the resulting random drift as explaining those cases of apparently nonadaptive differentiation.

What really happened when Lack visited Mayr in 1939 is that, in my opinion, they had *no* disagreement about the apparently nonadaptive differentiation of Darwin's finches on the Galápagos Islands.

In 1942 Mayr specifically pointed out the tendency for island populations to have "aberrant characteristics" that he thought were explained by Wright's random genetic drift. Only from the perspective of later hindsight is there a puzzle about why Mayr did not disabuse Lack of his belief in nonadaptive differentiation in Darwin's finches. In 1939, Mayr, Lack, Huxley, and Wright could all have agreed on this interpretation of Darwin's finches, with Wright perhaps the most skeptical about the randomness of the differentiation, just as he was more skeptical than Dobzhansky about the application of random drift to explain differences of inversion frequencies in *Drosophila pseudoobscura*.

Wright's ideas also became known in Germany, Russia, Italy, and elsewhere on the European continent through the writings of N. W. Timoféeff-Ressovsky, who was at the Kaiser-Wilhelm Institut in Berlin-Buch from the late 1920s through World War II, when he returned to Russia. Timoféeff-Ressovsky had been one of the great pioneers in the Chetverikov group studying variability in free-living populations of several species of *Drosophila* (Adams 1980). He also worked for many years on the experimental production and analysis of mutations, publishing a book on this topic in 1937 (Timoféeff-Ressovsky 1937). He was highly sophisticated in matters of experimental laboratory populations and natural populations, and worked with a great variety of organisms. His work was highly respected by geneticists and evolutionary biologists, and his role in evolutionary biology in continental Europe was comparable to that of Dobzhansky in North and South America.

Like Dobzhansky, Timoféeff-Ressovsky attended the Sixth International Congress of Genetics in Ithaca, New York, in 1932, heard Wright speak, and was greatly impressed. Indeed, as in Dobzhansky's case, Timoféeff-Ressovsky's conception of the evolutionary process thereafter bore the strong stamp of Wright's conceptions. In 1938–40, Timoféeff-Ressovsky published three major papers on evolution in nature (Buzzati-Traverso, Jucci, and Timoféeff-Ressovsky 1938; Timoféeff-Ressovsky 1939, 1940). All three reflected Wright's ideas.

The first, written with the Italian geneticists Adriano Buzzati-Traverso and his teacher Carlo Jucci, is a little-known but very important paper for understanding this period of the evolutionary synthesis. Timoféef-Ressovsky, again like Dobzhansky, had come to believe that the theoretical population geneticists had elucidated the primary mechanisms of evolution and their interactions; but the theory in itself could not determine the actual sizes of the variables. That was a job for field naturalists. This paper, after a detailed discussion of Wright's theory, including the diagrams from the 1932 paper, outlined an extremely ambitious plan for the study of natural populations all over the Italian peninsula and on nearby islands. Some populations were relatively continuous and others distinctly isolated; some were small and others were large. Basically, what was described was a European GNP series. Had the plans been carried out, the resulting papers would have rivaled Dobzhansky's GNP series in importance and influence.

World War II interrupted the planned research. A year before he died in 1983, Adriano Buzzati-Traverso told me that he was able to conduct only a tiny portion of the research outlined in the paper and that he never published the data. He also said that Timoféeff-Ressovsky was the intellectual leader of the three authors in the writing of the paper.

The other two papers can be treated together because they were both actually written at the same time (1938) and cover the same issues, but one was in German and the other in English. The latter, published in Huxley's *The New Systematics*, was highly visible and influential. Following the mathematical population geneticists Fisher, Haldane, Wright, and Chetverikov, Timoféeff-Ressovsky stated that the three primary determining factors of the evolutionary process were mutation, selection, and "the limitation of panmixy, leading to accidental fluctuations of the concentration of single genotypes, and, in cases of continued isolation, to a statistical divergence of the different parts of a mixed population." The interaction of these three factors was the mechanism of evolution in nature.

> Mathematical analysis shows the quantitative values and limitations of the efficiency of each of these three factors and their interrelations under various arbitrary conditions and values of mutation-pressure, selection pressure, and isolation. This type of mathematical work is of the greatest importance, showing us the relative efficacy of various evolutionary factors under the different conditions possible within the populations (Wright 1932). It does not, however, tell us anything about the real conditions in nature, or the actual empiri-

cal values of the coefficients of mutation, selection, or isolation. It is the task of the immediate future to discover the order of magnitude of these coefficients in free-living populations of different plants and animals; this should form the aim and content of an empirical population-genetics (Buzzati-Traverso, Jucci, and Timoféeff-Ressovsky 1938). (Timoféeff-Ressovsky 1940, 103–104)

According to Timoféeff-Ressovsky, selection and isolation were the only two *directive* factors in evolution, selection leading to adaptation, whereas "Isolation is the main factor of differentiation in space. The latter may be also produced by selection, acting on different parts of the population under different conditions; but here too differentiation is markedly accelerated by isolation" (p. 1940, 122). Although Timoféeff-Ressovsky emphasized the interactive "balance" of evolutionary factors, he clearly followed Wright's 1932 paper in attributing much observable differentiation up through the species level directly to the action of isolation and random drift.

Not one of the works discussed in this section exhibiting Wright's influence discussed or mentioned the shifting balance theory of evolution, in particular the idea that random drift could enhance adaptive evolution. Evolutionists unquestionably viewed Wright's theory of evolution primarily as the concept of random drift, but they left out the levels of "intragroup" and "intergroup" selection that were for Wright the crucial and necessary steps. After the early 1940s, many of the major works of evolutionary biology (with the notable exception of Dobzhansky, beginning particularly with the third edition of *Genetics and the Origin of Species,* 1951) tended to interpret Wright's concept of the primary mechanism of evolution as random drift, pure and simple. Goldschmidt in 1940 (*The Material Basis of Evolution*) said it was Wright's "contention that small isolated populations have the greatest chances from the standpoint of population genetics," and Stebbins in *Variation and Evolution in Plants* (1950) stated:

Population size is important chiefly in connection with the effects of random fluctuations in gene frequency. Numerous workers (cf. Dobzhansky 1941, 161–165) have shown that while in infinitely large populations gene frequencies tend to remain constant except for the effects of mutation and selection pressure, in populations of finite size there is a gradual reduction in variability owing to chance fluctuations in gene frequency and random fixation of individual alleles. This phenomenon of "random fixation," "drift," or the "Sewall Wright effect" is undoubtedly the chief source of differences between populations, races, and species in nonadaptive characteristics. (Stebbins 1950, 145)

Quotations like these from Goldschmidt and Stebbins can be found everywhere in the evolutionary literature from 1940 to 1960 and even up to the present, although a more accurate view of Wright's work began to emerge during the 1960s and certainly in the 1970s. In the 1940s, Wright was believed to have supplied the crucial missing factor for explaining nonadaptive evolution in general and the nonadaptive differences in particular between geographical races and closely allied species. This is of course the view of Wright's evolutionary theory that Fisher and Ford held, and they were intensely opposed to Wright precisely because they thought he held random drift to be an important mechanism of evolution by itself.

None of these quotations, it should be clearly understood, accurately reflected Wright's shifting balance theory of evolution in nature. Wright had repeatedly denied that random drift in small, isolated populations led to anything but extinction. Yet it was true that Wright himself had in the late 1920s and early 1930s clearly suggested the connection of random drift with observed nonadaptive differentiation in nature.

The historical consequences were ironic. Systematists in the 1940s and 1950s, led strongly by Ernst Mayr, E. B. Ford, and David Lack (after he had given up his earlier ideas about nonadaptive differentiation in Darwin's finches), guided systematists to a more adaptationist interpretation of differences between geographical races and allied species. The three most-cited examples of nonadaptive differentiation by random drift—the inversions in *Drosophila pseudoobscura,* banding and coiling patterns in snails, and blood groups—were all found to be subject to substantial or even huge selection pressures. As this shift toward adaptationism occurred, systematists had no further need for their conception of Wright's random drift and minimized the importance of the concept in their interpretation of the evolutionary process. At the same time, of course, Wright had no further reason to bias the balance in his evolutionary theory to fit what systematists had said about nonadaptive racial and species differences in nature. He could now emphasize the view his shifting balance theory had incorporated from the very beginning—that random drift served the important function of providing novel genetic interaction systems upon which natural selection could act to yield a more rapid process of adaptive evolution than could occur under mass selection alone. Thus just as Wright was providing what many prominent evolutionists now judge to be a robust and useful view of the evolutionary process, many evolutionists came to dismiss Wright's evolutionary theory as mere random drift, unimportant in the adaptive world of species in nature.

By the time Wright became involved in the dispute over *Panaxia dominula* with Fisher and Ford in 1947–51, he had redressed the balance in his shifting balance theory of evolution to lessen the role of random drift as the cause of observable taxonomic differences. Wright unfortunately contributed to the confusion by maintaining that he had never emphasized random drift in small populations as an important mechanism of evolution. This was true, but he neglected to say that he really had changed his mind about random drift in relation to distributions of conspicuous polymorphisms and differences between geographical races and allied species.

Reviews of Willis, Goldschmidt, and Simpson

From 1940 to 1948 Wright published no general papers on his shifting balance theory of evolution, with the notable exception of his reviews of *The Course of Evolution by Differentiation or Divergent Mutation Rather Than by Selection* by J. C.Willis (1940), *The Material Basis of Evolution* by Richard Goldschmidt (1940), and *Tempo and Mode in Evolution* by George Gaylord Simpson (1944). These were no ordinary reviews but full and critical essays in which Wright compared each of the theories under review with his own shifting balance theory. The reviews are therefore useful as a record of the development of Wright's own theory during the period when he was becoming more adaptationist in the "balance" of his shifting balance theory.

The first review, of Willis's *Course of Evolution* (Wright 105), presents more clearly than any piece I know the tension between the views Wright expressed in the early 1930s and those he consistently maintained during and after the *Panaxia dominula* debate with Fisher and Ford. Willis was an iconoclastic botanical biogeographer whose views were uninformed by and outside of the mainstream of evolutionary biology; only the barest summary of his views is necessary here. As Robson and Richards (1936) had argued for animals, Willis argued that the structural differences used by taxonomists to distinguish species, genera, and families of plants were nonadaptive.

> There was a great rush into the study of adaptation, especially during the 'eighties and early 'nineties of last century. But in spite of all the work that was put into it, no one ever succeeded in showing that even a small percentage of the structural characters, that were the reason why plants were divided into so many families, genera, and species, had any adaptational meaning or value whatever. No value could be attributed to opposite as against alternate leaves (or vice versa), to dorsal against ventral raphe, to opening of anthers by pores or by slits, and so on. (Willis 1940, 52)

After describing many examples of structural characters that distinguished species, genera, and families, Willis concluded:

> Everywhere one finds that there are plants showing the important characters of classification and distinction, and even showing, in many cases, both members of the contrasting pairs that are given in the list of family characters (Appendix I). These characters show no relation whatever to any of the ecological features that may give the character to the locality. . . . Morphologists have long maintained that structural characters have nothing to do, directly, with the life or functions of the plant, and it would appear that they are right in this contention, which violently contradicts the supposition of selection as a chief cause in evolution. The evolution that has produced more than 12,000 genera and 180,000 species has not been, primarily, an adaptational evolution. (P. 54)

To explain the mechanism of this nonadaptive differentiation, Willis invoked his rather confusing theory of divergent mutation, which need not be discussed here.

The argument Wright made in the review was that Willis was correct in believing that natural selection could not account for the observed taxonomic differences between plants, but his proposed mechanism to account for these differences was unacceptable. Wright's own shifting balance theory, however, provided the required mechanism: "The author's wide acquaintance with plants in nature make his critique of natural selection as the sole factor in evolution of great weight. It does not appear, however, that he has sufficiently considered the alternatives. It does not seem necessary to postulate an unknown sort of mutation" (Wright 105, p. 347). Thus far Wright's argument is no different than that of 1929–32. But the way he states the preferable mechanism, the shifting balance theory, makes it plain that because of intergroup selection the process is in fact highly adaptive, both in a single subdivided population and particularly in a population divided into wholly isolated segments: "If now a portion of the species is completely isolated, the two portions will inevitably drift apart even if conditions are essentially the same. Both groups will always have adaptive combinations of traits but different ones because of the combination of non-adaptive with adaptive processes in their evolution" (p. 346). It is important to understand precisely Wright's statement. Willis had said that the evolutionary process led to many nonadaptive *characters,* by which species, genera, and families differed. Wright offered in response a statement that his shifting balance theory led not to nonadaptive characters by which species differed but instead led to nonadaptive *differences* between species, the characters of which were equally well adapted—on the adaptive landscape, such species were on different peaks of the same selective value, separated by adaptive valleys.

Pulled out of the context of the Willis review, this statement of the shifting balance theory could just as easily be transplanted into the review of a book by a systematist who saw adaptation almost everywhere. Of course, this switch is exactly what Wright would execute as systematists became more and more adaptationist in their outlook. I think what is difficult for evolutionists today to appreciate is that in 1941 Willis was still in the mainstream of botanical systematists in arguing for a nonadaptive interpretation of taxonomic characters (though his mechanism of differentiation was at the same time generally unacceptable). Wright had an efficient mechanism that could produce a high degree of adaptation at the level of a subdivided population, far lower on the taxonomic scale than a race or subspecies. Yet here he was arguing that the shifting balance process could account for observable nonadaptive differences at the species and genus levels, as he had in his 1932 paper.

To put this issue into perspective, I will skip ahead thirteen years to a letter Wright wrote to Dobzhansky after reading his review of *Evolution as a Process* (1954), edited by Huxley, Hardy, and Ford: "The essence of my whole theory is that a tendency toward random differentiation of demes which by itself could never perhaps go so far as to be observable, greatly increases the effectiveness of selection" (Wright to Dobzhansky, October 25, 1954). This is a dramatic change from the Willis review, yet I assert again that Wright's shifting balance process was, from his initial conception, basically an adaptive process that he had to stretch out of shape so that it could account for nonadaptive differences at the species level and above. Thus the change of systematists toward a more adaptive interpretation of taxonomic differences enabled Wright to use his shifting balance theory in the most natural way; it is understandable that Wright believes he did not change his mind substantively during the 1940s, despite the apparent change.

Wright's review of Goldschmidt's *Material Basis of Evolution* was structurally very similar to his review of Willis's book. Here again Wright argued that his shifting balance theory was a better choice of mechanism than that Goldschmidt offered. In his studies particularly of *Lymantria dispar,* the gypsy moth, Goldschmidt had concluded that there was a "bridgeless gap" between species and that speciation was far too rapid to be caused by gradual natural selection. Thus he argued that microevolution (below the species level) proceeded by the usual Darwinian selection but that macroevolution occurred by "systemic mutations." Goldschmidt carefully distinguished himself from the "neo-Darwinians," the category in which he classified Wright. First denying the existence of "bridgeless gaps" ("the data indicate every conceivable intergrade in degrees of morphological and physiological distinction, of chromosomal differentiation and of cross-sterility or hybrid sterility"), Wright then proposed the shifting balance theory as a mechanism of rapid evolutionary change. Here Wright stressed the ad hoc character of Goldschmidt's theory of systemic mutations and the naturalness of his own theory.

Wright's comments upon physiological genetics in relation to Goldschmidt's theory of evolution are particularly interesting given that Wright and Goldschmidt had not only both been leaders in their field but also had agreed upon a basic conception of gene action. Goldschmidt argued in his book that the reorganization of genes by Darwinian selection was inadequate for speciation, which required reorganization of the chromosomes. Goldschmidt, Wright said,

> seems to hold that the conception of the organism as an integrated reaction system requires a corresponding *spatial* integration of the germ plasm and that essential change in the reaction system can thus come about only by repatterning of the chromosomes. To others, a *temporal* integration is all that is necessary, or even possible, with the chain reaction as the simplified model. Within the organism as a more or less integrated reaction system, there is a hierarchy of subordinate reaction systems, each with considerable independence, as shown by capacities for self-differentiation. Thus there must be partially isolated reaction systems for each kind of organ and for each

kind of cell. It is difficult to see how any spatial pattern in the germ plasm can operate in determining these, but there is no theoretic difficulty with a branching hierarchic system of chain reactions in which genes are brought into effective action whenever presented with the proper substrates, irrespective of their locations in the cells. There is no limit to the number of reaction systems that can be based on the same set of genes, and such systems may obviously evolve more or less independently of each other. (Wright 106, pp. 168–69)

After all their mutual compliments about physiological genetics since 1915, it is fascinating to see this disagreement emerge. Wright's criticism was generally considered telling both by physiological geneticists (like G. W. Beadle) and by evolutionists.

Wright left one very pertinent criticism out of the review. Goldschmidt described Wright's theory of evolution in his book as "Wright's calculation (1931) showing that small isolated groups have the greatest chance of accumulating mutants even without favorable selection. . . . It is the contention that small isolated populations have the greatest chances from the standpoint of population mathematics" (Goldschmidt 1940, 137–38). Since Wright had consistently declared that small isolated populations became extinct, Goldschmidt had clearly missed the thrust of the shifting balance theory. Wright did not forget. In his first paper in the *Panaxia dominula* debate with Fisher and Ford, Wright opened with the above quotation from Goldschmidt, followed by a denial that he had ever held the view.

Both Willis and Goldschmidt explicitly set themselves apart from what they considered the rising tide of neo-Darwinism. Simpson's *Tempo and Mode in Evolution* was, on the other hand, a highly influential book of the neo-Darwinian synthesis. Basically, Simpson argued for the consistency of paleontological evidence with known mechanisms of selection, mutation, migration, and isolation as seen in the works of Fisher, Haldane, Wright, Dobzhansky, and Mayr. More than any other work, Simpson's book integrated paleontology into the evolutionary synthesis of the 1930s and 1940s (see essays by Gould and Mayr in Mayr and Provine 1980). Wright therefore had every reason to appreciate the basic aim of the book, and his review praised it highly for this reason.

He had a more specific reason to be pleased with the book. Simpson had explicitly adopted Wright's general conception of the mechanism of the evolutionary process, as had Dobzhansky in *Genetics and the Origin of Species*. From beginning to end, Simpson relied upon Wright's models of the mechanisms of evolution, including the inevitable diagrams of his phenotypic version of the famous fitness surfaces. Although mentioning Fisher and Haldane frequently, Wright was the real model: "Wright's latest work [Wright 112] comes nearest to a generalization in which the most pertinent variables are simultaneously involved, and his models nearly approach the conditions that have been observed in real populations" (Simpson 1944, 65).

As a paleontologist, Simpson was primarily concerned with data from the fossil record. The large-scale evolution visible in the fossil record (generally family level and above) Simpson termed "mega-evolution," and here he invoked Wright's shifting balance theory as the primary mechanism of change:

> The theory here developed is that mega-evolution normally occurs among small populations that become preadaptive and evolve continuously (without saltation, but at exceptionally rapid rates) to radically different ecological positions.
> The typical pattern is probably this: A large population is fragmented into numerous small isolated lines of descent. Within these, inadaptive differentiation and random fixation of mutations occur. Among many such inadaptive lines one or a few are preadaptive, i.e. some of their characters tend to fit them for available ecological stations quite different from those occupied by their immediate ancestors. Such groups are subjected to strong selection pressure and evolve rapidly in the farther direction of adaptation to the new status. The very few lines that successfully achieve this perfected adaptation then become abundant and expand widely, at the same time becoming differentiated and specialized on lower levels within the brand new ecological zone. (P. 123)

Wright obviously could have no fundamental disagreement with Simpson, who was attempting to base his synthesis upon Wright's ideas.

In the review, therefore, Wright simply "corrected" (in a few places) and extended Simpson's analysis and application of the shifting balance theory. Two "corrections" were especially significant. In speaking of the subdivision of a large population, Simpson frequently referred to the subdivisions as "small and completely isolated" (p. 211). Wright emphasized the distinction that his theory required not complete isolation but "almost but not quite complete isolation," and when Simpson declared that intergroup selection "can produce nothing new; it is purely an eliminating, not an originating, force" (p. 31), Wright disagreed strongly. After all, Simpson had invoked intergroup selection as the primary mechanism of megaevolution, and Wright thought this deserved the evaluation of "creative."

To give this last point added force, Wright worked out quantitatively an example in which an individually disadvantageous, but socially advantageous, character could evolve by group selection. This was an old idea expressed by Darwin in reference to the evolution of altruism and neuter castes in social insects, and it was also mentioned by Haldane in *Causes of Evolution* (1932). But Wright was the first to devise a simple quantitative model showing how this mechanism could work. It is upon this starting point that the modern theory of group selection has been based. (See David Sloan Wilson 1980 and 1983 for reviews.) I should emphasize that this example of Wright's "group" selection is a mechanism logically distinct from the "intergroup" selection of the shifting balance theory; the terminology can be confusing for this reason.

Wright closed the review by praising Simpson's "compact marshalling of pertinent paleontological data and the stimulating attempt at interpretation by genetic principles" (Wright 119, p. 419). After reading the review, Simpson wrote the following letter to Wright:

> I have just read your review of my book in the October 1945 issue of *Ecology*. This is an unusually fine review and I want to thank you for both your favorable remarks and your excellent criticisms on some points. It was with great trepidation that I attempted to summarize some of your work and I am glad that I did not go any farther astray than I did. Except for one or two very minor reservations, I quite agree both with your corrections and with the expansion that you have made regarding some of the topics in my book. Your review is a first-rate original contribution in itself. I hope that you are getting reprints and that I may have one for my files and still closer study. (Simpson to Wright, January 8, 1946)

Just as Dobzhansky had brought Wright's ideas to many geneticists and evolutionists, so Simpson brought them to paleontologists.

Even the most careful reader of these three reviews would be hardpressed to decide where Wright stood on the question of adaptation in nature. His shifting balance theory seemed capable of accounting for the whole gamut of possible evolutionary modes, from nonadaptive at high taxonomic levels to adaptive at the lowest observable taxonomic levels. This ambiguity (or flexibility, depending upon one's perspective) would soon begin to diminish. If Dobzhansky and Simpson responded favorably to Wright's ideas, the same was certainly not true of Fisher and Ford. The debates over *Panaxia dominula* and *Cepaea* would bring the issues of adaptation and random drift into much clearer focus.

Conspicuous Polymorphisms

Since the beginnings of evolutionary biology, conspicuous polymorphisms in natural populations had presented evolutionists with a serious problem. How could a single primary mechanism of evolution, whether natural selection, inheritance of acquired characters, or even an orthogenetic force, lead to conspicuous dimorphisms within a single interbreeding population? All naturalists were familiar with at least some cases of such dimorphism.

Debate about what mechanisms of evolution lead to these easily observed conspicuous polymorphisms has persisted from the mid-nineteenth century to the present. An obvious but important reason why the debates over explaining the origin of conspicuous polymorphisms have been so persistent is that, until the rise of molecular biology, conspicuous polymorphisms were the most easily accessible (often, it seemed, the *only*), measurable heritable characteristics. Thus conspicuous polymorphisms have been constantly in the forefront of research on evolution in natural populations, despite the problems associated with explaining their origin.

As pointed out in chapter 7, Darwin concluded in *Origin* that natural selection could not be the explanation of polymorphic species. Indeed, when Darwin defined the concept of natural selection in chapter 4 of *Origin*, he specifically dissociated natural selection from polymorphism: "This preservation of favourable variations and the rejection of injurious variations, I call Natural Selection. Variations neither useful nor injurious would not be affected by natural selection, and would be left a fluctuating element, as perhaps we see in the species called polymorphic" (Darwin 1859, 81).

Darwin made no clear distinction between populational polymorphisms such as conspicuous polymorphisms and racial "polymorphisms" that evolutionists today would no longer (following Ford 1940) call by that name. Perhaps the most prominent case of "racial polymorphism" treated by Darwin concerned the races of man. He found human racial differences very difficult to explain:

> We have now seen that the characteristic differences between the races of man cannot be accounted for in a satisfactory manner by the direct action of the conditions of life, nor by the effects of the continued use of parts, nor through the principle of correlation. We are therefore led to inquire whether slight individual differences, to which man is eminently liable, may not have been preserved and augmented during a long series of generations through natural selection. But here we are at once met by the objection that beneficial variations alone can be thus preserved; and as far as we are enabled to judge (although always liable to error on this head) not one of the external differences between the races of man are of any direct of special service to him. The intellectual and moral or social faculties must of course be excepted from this remark; but differences in these faculties have had little or no influence on external characters. The variability of all the characteristic differences between the races, before referred to, likewise indicates that these differences cannot be of much importance; for, had they been important, they would long ago have been either fixed and preserved, or eliminated. In this respect man resembles those forms, called by naturalists protean or polymorphic, which have remained extremely variable, owing, as it seems, to their variations being of an indifferent nature, and consequently to their having escaped the action of natural selection. (Darwin 1871, 1:248–249)

The explanation for human racial differences upon which Darwin finally settled was sexual selection, the same mechanism he used to explain many cases of the most colorful conspicuous sexual dimorphisms, these being populational conspicuous polymorphisms in the modern sense. Darwin was sure that

such conspicuous polymorphisms could not have been evolved by natural selection because if the character involved were subject to significant selective
forces, it would exhibit a more uniform morph.

I have quoted Darwin at some length here because biologists educated
during or after the early 1950s have strongly interpreted, in accordance with
the neo-Darwinian synthesis, conspicuous polymorphisms as resulting directly from natural selection. I want to emphasize strongly that this consensus
is a phenomenon of the late evolutionary synthesis, from about 1940 to 1955.
Evolutionists now do not generally realize that even in the early 1940s, most
evolutionists still interpreted polymorphisms as selectively neutral variation.

Two prime examples must suffice. In the first edition (1937) of Dobzhansky's *Genetics and the Origin of Species,* the term *polymorphism* does not appear in the index. Chapter 5, "Variation in Natural Populations," contains
many examples of polymorphisms in snails (*Partula, Cepaea*), insects
(*Drosophila, Cynips, Cicada, Coccinella*), fish (*Clupea, Zoarces,* Sal-
monidae), plants (*Iris, Cupressus*), birds (*Ciconia*), and other organisms; in
every case, Dobzhansky attributed the polymorphisms to random genetic
drift. The chapter on selection *followed* this chapter on "nonadaptive polymorphism," which could easily have been the chapter title. The second edition (1941) had the same arrangement of these two chapters, except in the
chapter on nonadaptive polymorphisms (still entitled "Variation in Natural
Populations") Dobzhansky had added more examples.

Almost any evolutionist nowadays knows that Ernst Mayr considers the
vast majority of conspicuous polymorphisms to be the result of natural selection (heterozygote superiority, selective advantage of different genotypes in
different seasons, selective forces of microhabitats, etc.). Yet consider what
Mayr says in *Systematics and the Origin of Species*:

> *Neutral polymorphism* is due to the action of alleles "approximately
> neutral as regards survival value." Ford (following Fisher) believes
> that this kind of polymorphism is relatively rare, because "the bal-
> ance of advantage between a gene and its allelomorph must be ex-
> traordinarily exact in order to be effectively neutral." This reasoning
> may be correct in all the cases in which one of the alternative fea-
> tures has a definite survival value or at least is genetically linked with
> one. There is, however, considerable indirect evidence that most of
> the characters that are involved in polymorphism are completely neu-
> tral, as far as survival value is concerned. There is, for example, no
> reason to believe that the presence or absence of a band on a snail
> shell would be a noticeable selective advantage or disadvantage.
> Among the many species of birds which occur in several clearcut
> color phases (Stresemann 1926 and later papers), there is, with one
> or two exceptions, no evidence for selective mating or any other ad-
> vantage of any of the phases.
>
> Even more convincing proof for the selective neutrality of the al-
> ternating characters is evidenced by the constancy of the proportions

of the different variants in one population. The most striking case is that of the snails *Cepaea nemoralis* and *C. hortensis,* in which Diver (1929) found that the proportions of the various forms from Pleistocene deposits agree closely with those in colonies living today. (Mayr 1942, 75)

In addition Mayr cited examples of nonadaptive polymorphisms from the work of Gordon on the Mexican fish *Platypoecilus,* Gerould on the butterfly *Colias,* and Kinsey on the gall wasp *Cynips.* For additional cases, Mayr referred the reader to the second edition (1941) of Dobzhansky's *Genetics and the Origin of Species.*

Not only did Mayr and Dobzhansky clearly indicate the prevalence of neutral conspicuous polymorphisms, they both (along with Huxley, Timoféeff-Ressovsky, and so many others) attributed to Wright the chief concept by which the distribution of these polymorphisms could be understood—random genetic drift. The close tie that prominent evolutionists in the early 1940s drew between random drift and conspicuous polymorphism has continued in many quarters to this day, much to Wright's chagrin.

It was precisely the view that conspicuous polymorphisms were selectively neutral, with their distributions determined by random drift, that so stimulated Fisher and Ford. They strongly disagreed with this interpretation of polymorphisms. They also clearly considered Wright to be the chief architect of the theory that gave credence to the observations of apparently nonadaptive polymorphisms. Thus Fisher, Ford, Cain, and Sheppard knew their targets when they began research on *Panaxia dominula* and *Cepaea.* The debates over these research projects would exert deep influence upon later evolutionary biology.

The *Panaxia dominula* Debate

Fisher and Ford were particularly aroused by Wright's criticism of their evolutionary views in his article in Huxley's *The New Systematics* (Wright 101, discussed in chapter 10), a book widely read both in England and in America. In it Wright argued that Fisher's near total reliance upon natural selection as the mechanism of evolution was misplaced because random drift was also an important mechanism, especially in combination with selection and migration. Just as Wright had developed and applied his theory of isolation by distance to *Linanthus parryae* in an attempt to find random drift acting in a (for him) "worst case" continuous population, Fisher and Ford wished to work with a small population (their worst case) in hopes of demonstrating that observed fluctuations in gene frequency resulted from natural selection rather than from random genetic drift. For their purposes, an excellent population for this study was available right in the Oxford district.

The moth *Panaxia dominula* existed during the 1940s in only two colonies in the area of Oxford, England. These two colonies were separated by thirteen miles of land inhospitable to *Panaxia,* and migration between colonies was negligible. The moth is a large and conspicuous day flier, easily captured under the proper conditions. One of the two colonies, that of Dry Sanford Marsh, exhibited an easily observed polymorphism controlled by simple Mendelian inheritance. Conveniently, all three Mendelian classes could be distinguished by sight—the typical and most common form, the heterozygote or *medionigra* form, and the rare homozygote or *bimacula* form. Dobzhansky had always wished for such ease of distinction when working with the orange-eye mutant in *Drosophila pseudoobscura,* but it was unfortunately fully recessive and only breeding experiments could distinguish the heterozygotes from the dominant homozygotes. The percentage of heterozygotes in *Panaxia dominula* exhibited marked fluctuations over the years. And finally, the limited area of the population, its relatively small size, and the ease of capture and hardiness of individuals in the marking procedure made possible the use of the method of marking, release, and recapture (Dowdeswell, Fisher, and Ford 1940) to generate data from which the population size could be estimated. This population of *Panaxia dominula* was nearly ideal for the purposes of Fisher and Ford.

The questions thus were How much fluctuation of the *medionigra* gene occurred from year to year, and could this fluctuation be explained by random drift in a population of the size calculated? The marking, release, and recapture experiments began in 1941 (right after Wright's paper in *The New Systematics*) and continued through 1946. Some data from earlier years also existed.

The data from before 1928 indicated a much lower incidence of *medionigra* heterozygotes (2.4%) than existed in 1939 (16.6%) or 1940 (20.5%, but small sample size). The percentage of heterozygotes varied in the 1940s from 20.5% in 1940 to 7.9% in 1946, with corresponding gene frequencies of 11.1% and 4.3%. The marking, release, and recapture data indicated approximate population sizes from about 1,000 in 1943 to 6,000–8,000 in 1946. In a section entitled "The Significance of Changes in Gene-Ratio," Fisher and Ford directly stated what they saw as the fundamental issues:

> Great evolutionary importance has been attached by Sewall Wright (1931, 1932, 1935, 1940) to the fact that small shifts in the gene-ratios of all segregating factors will occur from generation to generation owing to the errors of random sampling in the process by which the gametes available in any one generation are chosen to constitute the next. Such chance deviations will, of course, be greater the smaller the isolated population concerned. Wright believes that such non-adaptive changes in gene-ratio may serve an evolutionary purpose by permitting the occurrence of genotypes harmoniously adapted to their environment in ways not yet explored, and so of opening up new evolutionary possibilities. Consequently, he claims that subdivision into isolated groups of small size is favorable to evo-

lutionary progress, not, as others have thought, through the variety of environmental conditions to which such colonies are exposed, but, even if the environments were the same for all, through the non-adaptive and casual changes favoured by small population size. Those evolutionists who find it difficult to attach any great evolutionary significance to such chance effects, have urged that the normal segregation of all factors in each generation continually supplies new genotypes selected at random from a number usually much greater than the number in a single generation of even a numerous population, and that the selective increase or decrease of any gene is determined by the totality of the life experience of all these, comprising as they do large numbers of harmonious, or successful, and of inharmonious, or unsuccessful, combinations: that the number of genotypes tried will generally be larger in more numerous than in less numerous populations; and that the existence of very small and completely isolated populations, such as Wright seems to postulate, will generally be terminated by extinction in a period which must be thought of as short on an evolutionary scale of time. (Fisher and Ford 1947, 167–68)

Several points in Fisher and Ford's assessment are worthy of careful consideration. First, they accurately state that the inbreeding effect of small population size reveals new genotypes that would not appear without the inbreeding and that these genotypes "open up new evolutionary possibilities." This had been precisely Wright's view since 1925. Fisher and Ford did not, however, go beyond the production of the new genotypes to the ensuing level of what Wright called "intergroup selection," the next step of his shifting balance theory of evolution. Instead, they portrayed Wright's view of the evolutionary process as dominated by random genetic drift alone. Finally, they suggested that, for his theory, Wright postulated "very small and completely isolated populations." This was wholly untrue; Wright had clearly stated in every one of his papers on the subject that such populations would lead to extinction.

I see no reason to believe that Fisher and Ford deliberately distorted Wright's theory of evolution. Instead, they merely reflected the general perception of Wright's theory as being one consisting purely and simply of random drift. Their conception, for example, did not differ substantially from that of Julian Huxley as expressed so many times in *Evolution: The Modern Synthesis.* Furthermore, as I have shown earlier, Wright himself bore a major responsibility for the popularization of the caricature of his evolutionary theory. Even in his 1940 paper in *The New Systematics,* his only examples of evolutionary change were the supposedly neutral inversions in *Drosophila pseudoobscura* and blood groups in man. And he suggested that the nonadaptive differences, observed so frequently by systematists, were the result of random drift. Thus, although it is true that Fisher and Ford had here misstated Wright's theory, they were in good company, for good reason.

In any case, the major points Fisher and Ford wished to make with their

Panaxia dominula research were just as pertinent to Wright's actual theory as to the typical caricature of it. They wanted to "cut at the root of the whole theory" by arguing that all populations, both large and small, were subject to "selective intensities capable of producing greater fluctuations in gene-ratios than could be ascribed to random sampling" (Fisher and Ford 1947, 168). In the case of *Panaxia dominula,* they assumed what they considered a minimum estimate of effective population size of 1,000 each year from 1939 to 1946. Fisher calculated that the observed level of fluctuation in the *medionigra* gene could be accounted for by random drift only on the average of once in a hundred trials; thus random drift was not the cause of the fluctuations. Random drift was an even less likely explanation of changes in gene frequency in populations larger than 1,000.

The conclusions of Fisher and Ford, on the basis of the *Panaxia dominula* work, are worth quoting in full:

> The conclusion that natural populations in general, like that to which this study is devoted, are affected by selective action varying from time to time in direction and intensity, and of sufficient magnitude to cause fluctuating variation in all gene-ratios, is in good accordance with other studies of observable frequencies in wild populations (for instance, considerable seasonal changes affecting gene-ratios or inversion-frequencies have been demonstrated by Dobzhansky and Timoféeff-Ressovsky [1940]). We do not think, however, that it has been sufficiently emphasized that this fact is fatal to the theory which ascribes particular evolutionary importance to such fluctuations in gene-ratio as may occur by chance in very small isolated populations. Evidently, however large a population might be, its gene-ratios will fluctuate in the same manner and to approximately the same extent as those of the smallest isolated population which can be expected to persist in nature.
>
> It is, in fact, found that the observed fluctuations in gene-ratio are much greater than could be ascribed to random survival only. Fluctuations in natural selection (affecting large and small populations equally) must therefore be responsible for them. The possibility that random fluctuations in populations much smaller than 1000 could be of evolutionary importance is improbable in view of the frequency with which such small isolated populations must be terminated by extinction within periods which must be extremely short from an evolutionary point of view.
>
> Thus our analysis, the first in which the relative parts played by random survival and selection in a wild population can be tested, does not support the view that chance fluctuations in gene-ratios, such as may occur in very small isolated populations, can be of any significance in evolution. (Fisher and Ford 1947, 171, 173)

Here again several points are worth noting carefully. Fisher and Ford conclude, on the basis of the *Panaxia dominula* polymorphism (and data from

Dobzhansky on a few inversions and from Timoféeff-Ressovsky on a few gene frequencies), that all gene ratios are subject to selection rates sufficiently high to obliterate any effects of random drift. Second, they show only that random drift cannot explain the observed shifts in the *medionigra* gene and present no data to show that the *medionigra* gene is subject to varying selective intensities from year to year. This conclusion is a deduction by elimination. Third, they assume that any population with an effective size of less than 1,000 is doomed to extinction "within periods which must be extremely short from an evolutionary point of view" and can therefore simply be ignored by the evolutionist. And finally, they claim that random drift cannot be of any significance in evolution.

If the Fisher and Ford paper had merely reported that the effective population size of the *Panaxia dominula* colony at Dry Sanford was always above 1,000 and that random drift could not therefore explain the observed changes in frequency of the *medionigra* gene, I doubt Wright would have responded, or even disagreed, with the conclusion. After all, he had concluded that the effective population sizes in *Drosophila pseudoobscura* even in Death Valley (GNP VII) were around 1,000, and he concluded (to Dobzhansky's keen disappointment) that random drift was of little significance in the distribution of certain inversions in those populations. But of course Fisher and Ford had not stopped there; they had absolutely dismissed the evolutionary significance of any populations of less than 1,000 and any evolutionary significance of random drift.

Anticipating further development of the *Panaxia dominula* debate, I would point out that Fisher and Ford dismissed not only Wright's theory of evolution but also Ernst Mayr's conception of founder populations, an idea central to Mayr's whole scheme of speciation. In one flourish Fisher and Ford had in their view demolished the primary mechanisms of evolution advocated by both Wright and Mayr.

The first Wright heard of the Fisher and Ford paper on *Panaxia* was in a letter from Dobzhansky dated February 8, 1948:

> You may or may not have already seen the second issue of *Heredity*, the journal of which I have the honor to be one of the co-editors. It contains the most vicious, to date, attack on you by Mr. Fisher. The mathematical part of his argument is beyond me but there is no difficulty in seeing that he states your views obviously deliberately and maliciously wrongly, thus making his task a little more easy. Although the thing is, from the ethical standpoint, on a par with my Pasadena ex-friend's [Sturtevant's] attack on me, here we are dealing with a topic on which only so few people can have critical judgement, and yet so many are interested, that I think you should publish a retort, not for your own sake but for the sake of the multitude which Fisher deliberately leads astray. Perhaps a note in *Nature* will do the trick, or is a more extended statement desirable?

This letter was followed seven weeks later by a letter from Stebbins, who also encouraged Wright to reply to Fisher and Ford: "If you have valid criticism, or criticisms, would you do us the favor of publishing them in *Evolution*, so that we non-mathematicians could have the record straight?" Ernst Mayr, editor of the new journal *Evolution* (established in 1947), had from the beginning strongly urged Wright to publish a paper there; he was pleased to receive Wright's reply to Fisher and Ford in mid-July 1948. Mayr's only critical comments as editor concerned several places in which Wright's argument was too terse to be easily followed. Wright made the suggested changes with the comment, "My first draft was considerably more than twice as long as the version which I submitted and I concluded that I had tried to cover too much ground. I suspect that I cut it too much in some places" (Wright to Mayr, August 17, 1948). Obviously Wright put a great deal of thought and care into this paper, which appeared as the lead article in the winter 1948 issue of *Evolution* (Wright 125).

Wright's arguments can be separated into six primary points.

1. He was not, he said, monolithic about the role of random drift. Instead he had always emphasized a "dynamic equilibrium among all factors as the most favorable condition for evolution" (pp. 280–81). Indeed, he had always insisted that random drift in small isolated populations led to degeneration and extinction. In short, he emphasized the "balance" of his shifting balance theory.

2. Fisher and Ford had misinterpreted him by missing that "the point stressed most" in his papers on evolution "was the simultaneous treatment of all factors by the inclusion of coefficients measuring the effects of all of them on gene frequency in a single formula" (p. 281). Fisher and Ford, Wright said,

> insist on an either-or antithesis according to which one must either hold that the fluctuations of *all* gene frequencies that are of any evolutionary significance are due to accidents of sampling (attributed to me) or that they are *all* due to differences in selection, which they adopt. As already noted, I have consistently rejected this antithesis and have consistently accepted both sorts as playing important, complementary and interacting roles. (P. 281)

3. Fisher and Ford were wrong that their data on *Panaxia dominula* were in any way fatal to the evolutionary significance of random drift. Even if Fisher and Ford were totally correct in concluding that the *medionigra* gene was subject to large fluctuations in selective value, it did not follow that all genes varied in frequency for that same reason. "The situation is similar, except for the element of intent, to one that is familiar to livestock breeders. With very intensive selection for particular characters, others must be allowed to vary at random if numbers are to be maintained" (p. 285). Wright's calculations indicated that

Fisher and Ford's contention that variations in the direction and intensity of selection cause much greater fluctuations in the frequencies of *all* genes than does sampling amounts to the contention that all segregating genes in a population such as that studied are subject to fluctuations in selective value with standard deviations much greater than .05. (P. 287)

This value was too high for belief.

4. The deduction by elimination made by Fisher and Ford was unwarranted. It was not true that if sampling drift were incapable of explaining observed fluctuations in gene frequency, then selection must be the cause. There were, Wright argued, some kinds of random drift even in large populations stemming from fluctuations in selection rates and in migration rates. The major theoretical contribution of the paper was Wright's calculation, in the quantitative appendix, of the sizes of random fluctuations in gene frequency caused by fluctuations in selection, in amount of immigration, and in gene frequency of immigrants. Never again to my knowledge would Wright mean by the general term *random drift* only that caused by sampling, which he later (at Arthur Cain's suggestion) called "sampling drift" (Cain and Currey 1963, 59).

5. The actual data from *Panaxia dominula* did not, in fact, lead necessarily to the conclusion that the observed fluctuations were greater than could have resulted from even sampling drift. The only statistically significant changes in gene frequency were between 1929 and 1939, for which time there were no data on population size, and between the summers of 1940 and 1941 where again Fisher and Ford provided no data on population size. Constricted population size and the resultant random drift could have caused the observed statistically significant fluctuations in frequency of the *medionigra* gene for those two periods.

6. Finally, Wright emphasized the ultimate adaptive significance of his shifting balance process:

The significance is in contributing to the material for the selection of genetic systems as wholes, which may be expected to take place through the welling up of population growth and emigration from those centers in which at the moment the most adaptive systems happen to have been arrived at and the modification by immigration of those centers in which population growth has become relatively depressed because of less successful general adaptation.

It must be emphasized again that this contribution of random differentiation of local population with respect to the more neutral sets of alleles is only one aspect of the whole trial and error process. The great importance of the contribution of selective differentiation among such populations with respect to less neutral alleles is obvious. Moreover, the nonadaptive differentiation is obviously significant only as it ultimately creates adaptive differences. (P. 290)

Wright's paper drew many immediate accolades. One of the most significant came from the quantitative geneticist A. J. Bateman, who was at the John Innes Horticultural Institution in London. Bateman had been convinced by the Fisher and Ford argument from the *Panaxia dominula* data, but after reading Wright's reply he wrote: "Your criticism of Fisher's interpretation of the *Panaxia dominula* data seems unanswerable. It just shows how careful one must be when seeking proof of one's own theories. I am afraid that when the expert (Fisher) claims to have proved something the lesser lights like myself are only too liable to take him at his word" (Bateman to Wright, February 1, 1949). The reaction of Fisher and Ford, however, was totally negative. The lines of dispute had been drawn, and the great *Panaxia dominula* debate begun. The effects of the debate were enormous in evolutionary biology and continue even to the present.

The first response of Fisher and Ford came in the form of a letter from Ford to Ernst Mayr as editor of *Evolution*.

> I have just got back, and have seen the article by S. Wright in *Evolution* in which he criticises a paper by Fisher and myself in *Heredity*. It is, I think, not normally worth replying to criticism. The only circumstances in which this should be done are if others are likely seriously to be misled, or in the rare event in which one's views are actually misrepresented. In respect of Wright's criticism, both reasons hold.
>
> Consequently Fisher and I have drafted a note briefly exposing Wright's misrepresentation of our statements and we should be so grateful if you could publish it in *Evolution,* where Wright's article appeared. I send it to you herewith. (Ford to Mayr, February 14, 1949)

The three-page manuscript included with the letter exhibited the same tone as Ford's letter.

> In a recent paper (*Heredity,* 1947) we criticized the theory widely ascribed to Sewall Wright, that the subdivision of a population into small isolated colonies has had important evolutionary effects; and this through the agency of random fluctuation of gene ratios, due to random reproduction in a small population.
>
> We have long felt that there are grave objections to this view to several of which we referred, though briefly as it was to one of them that our new data were directly relevant. This one, however, is completely fatal to the theory in question, namely that it is not only small isolated populations, but also large populations, that experience fluctuations in gene ratio. If this is the case, whatever other results isolation into small communities may have, any effects which flow from fluctuating variability in the gene ratios will not be confined to such subdivided species, but will be experienced also by species having continuous populations.

This central criticism seems to have escaped Wright's attention; consequently he has evidently formed, and in a recent discussion (*Evolution*, 1948) given publicity to a view of our opinions very wide of the mark. In his summary, for example, he says (p. 291):

> They hold that fluctuations of gene frequencies of evolutionary significance must be supposed to be due wholly either to variations in selection (which they accept) or to accidents of sampling. This antithesis is to be rejected.

> There is actually nothing in our paper even to suggest the antithesis which Wright ascribes to us. We presume throughout that accidents of sampling produce their calculable effects (the extent of which we give some care to calculating), in causing fluctuations in the gene ratios. . . .
> [Referring again to the either-or antithesis] Nothing could be further from our actual criticism of the particular contribution to evolutionary theory which is due to Sewall Wright. It is true he now tells us that he attaches importance to it only as one of many factors. This is all to the good. Still, if it has been from the first based on a misrepresentation, and has been accepted in spite of its obvious difficulties on false grounds, it is as well to admit this when the misapprehension is demonstrated and the grounds for its acceptance are shown to be false.

Fisher and Ford concluded the manuscript with the declaration that in light of their criticism, "the claim for ascribing a special evolutionary advantage for small isolated communities, had better be dropped."

At this stage of the story, some critical assessment is necessary. It is obvious that communication between Fisher and Ford on the one hand and Wright on the other had sunk to a low level. Fisher and Ford were convinced that Wright had promulgated and had been widely understood to say that random drift was an important mechanism of evolution leading to nonadaptive taxonomic differentiation. I have already presented abundant evidence to show why Fisher and Ford were justified in this conclusion. Wright had made this argument as recently as 1940. Yet at the same time, Fisher and Ford insisted that Wright emphasized the importance of random drift in very small isolated populations, but Wright had clearly and consistently denied the importance of this scenario.

It was simply untrue that Fisher and Ford's "central criticism" was "completely fatal" to the significance of random drift in evolution and that this point had "escaped Wright's attention." Wright had directly (and I think quite convincingly) answered their central criticism in his points summarized under 3, 4, and 5.

As to the either-or antithesis, Wright was completely justified. Fisher and Ford had denied any evolutionary significance of random drift in changing gene frequencies and had asserted the all-importance of selection. They had indeed presented the "calculable" effects of random drift but denied its "evolutionary significance." Wright had asserted all along the evolutionary significance of both random drift and selection.

Fisher and Ford were particularly indignant that Wright was, to their minds, changing his views substantially while at the same time denying that any such change had occurred. This issue has two pertinent levels. First, the caricature of Wright's views prevalent in England, and naturally adopted by Fisher and Ford, was certainly different from the always "adaptive" shifting balance process described in Wright's *Evolution* paper. Regarding what Wright had actually said in his previous papers, a development in his view had occurred that was significant but more subtle than Fisher and Ford appreciated. What Wright had done was to retain throughout the metaphor of a "balance" of evolutionary factors, but he had in fact begun to change the relative roles of random drift and selection within the balance. By emphasizing the balance issue, Wright was able to convey the impression that his views had not changed in any significant way. If Fisher and Ford had missed the second level here, they nevertheless were fully justified in raising the issue of an apparently significant change in Wright's views. Perhaps the most surprising aspect of this Fisher and Ford manuscript was that it simply ignored Wright's substantive arguments (summarized above in points 3, 4, and 5) that cast doubt on the conclusions Fisher and Ford had drawn from the *Panaxia dominula* data.

To the present-day evolutionist, almost certainly familiar with the public disagreements of Mayr and Wright since 1959, Mayr's response to the Fisher and Ford manuscript (addressed to Ford) may come as a surprise.

> I entirely agree with you that misrepresentations of fact as well as misquotations should be corrected. The pages of *Evolution* will always be open to such objective correction. Consistent with this policy, I would be glad to publish your note if, as you state, it would prevent that "others are . . . seriously . . . misled." However, it seems to me that in its present form your note fails to achieve this object. I have read your note three times and find myself thoroughly confused. I am unable to see what Sewall Wright has misrepresented or of what your correction consists. There seem to be a number of contradictions which I have been unable to dissolve. With the same breath you say that random fixation occurs but that it is of no evolutionary significance. I have always considered it axiomatic that anything that leads to a deviation from the present generation is automatically of evolutionary significance. I fail to see any refutation of this point in your note.
>
> I am also puzzled by some statements in your paper. In point 15 of your summary (Fisher-Ford, 1947) you state, "Fluctuations in natural selection must therefore be responsible for them (fluctuations in gene ratio)." This assertion to me justifies Sewall Wright's statement

on page 291. Incidentally, point 15 in the summary is merely a rephrasing of an earlier statement in your paper (p. 171) which also ascribes no evolutionary importance "to such fluctuations in gene-ratio as may occur by chance in very small isolated populations." This same point is again repeated in your present reply. In each instance you pose an unequivocal alternative between *an effect* or *no effect* of random fixation. Since you state that you are being misrepresented by Sewall Wright, whose concept it is that random sampling plays a major or minor role in addition to selection, it is obvious that I fail to understand the argument. Other readers of *Evolution* might have equal difficulties. If there is any misrepresentation by Sewall Wright, it seems to me that it should be stated more clearly by you what it is. This is not apparent from your present draft.

What the role is that is played in evolution by random fixation is open for discussion. Sewall Wright believes that it depends to a large extent on the numerical value of other factors, such as selection intensity, gene flow, etc. I am inclined to agree with those of modern authors who ascribe to random fixation a rather minor role. However, it seems to me that S. Wright has shown mathematically that it cannot be ruled out altogether. It would seem to me that any constructive criticism of S. Wright would have to start by showing that either the basis or the process of his calculations is wrong.

It also seems to me that the wording of your criticism is occasionally somewhat unfortunate. I would not like to see you exposed to the criticism of being ungenerous or unfair. I am sure that the sentence, "It is true he *now* tells us that he attaches importance to it only as one of many factors. This is all to the good," would cause a considerable raising of eyebrows. Any reader can do what I have done and check back in what year S. Wright first regarded the "Sewall Wright effect" "only as one of many factors." He will find that he has *always done so!* If you are inclined to disbelieve me I suggest that you go back to S. Wright's earlier publications and see for yourself.

It might be helpful to the reader if you would back up by a quotation your statement that Sewall Wright ascribes "a special evolutionary advantage for small isolated communities." As I remember his work he has always pointed out that small isolated communities are at a great evolutionary disadvantage owing to the rapid homogenization caused by random fixation.

As editor, I am interested in any discussion that leads to a clarification of a scientific problem. On the other hand, I am sure that the Editorial Board would never endorse a communication that might lead to a polemic. My personal impression is that the note, submitted by you, was written in haste and perhaps under some emotional stress. I would do you a great disservice by publishing it in its present form. I am therefore returning it to you for consideration. Under the circumstances, I have refrained from the customary editorial procedure of submitting it to two qualified readers. (Mayr to Ford, February 23, 1949)

If witnessing a spirited defense of Wright by Mayr is surprising to the modern reader, then two points should be kept in mind. First, Mayr expressed the belief in this letter that Wright had been emphasizing the same "balance" of evolutionary factors since the early 1930s. Second, the Fisher and Ford conclusions from *Panaxia dominula,* if true, would completely undercut Mayr's central conception of the founder effect in speciation, as pointed out earlier. These two points, combined with the other rather obvious problems of the Fisher and Ford manuscript, make Mayr's response appear wholly reasonable under the circumstances (except for his uncharacteristic avoidance of any hint of a polemic).

The reactions of Ford and Fisher to Mayr's letter are hardly surprising, though no less interesting.

> I of course forwarded your letter to Professor Fisher, the paper in question being a joint one. I now enclose his reply, with which I am in full accord. You will note too that in his recent paper, to which the answer we sent you was directed, Sewall Wright in an important matter attributed to us views we do not hold and have never expressed (we both quoted his statements on the point and gave exact references to them). (Ford to Mayr, March 3, 1949)

> I am writing to withdraw from *Evolution* my joint note, which with Dr. Ford I recently offered to it: I do so on the presumption that it is our opinions to which you take exception, and that only alterations in opinion would satisfy you as Editor.
> I wonder, reading your letter, whether it is possible that you have not read all that Sewall Wright has written. As you say, it might be ungenerous or unfair to suggest, if it were untrue, that Wright had changed or weakened his opinions. Evidently, this makes you challenge us to "go back to S. Wright's earlier publications and see for yourself."
> The first offprint I took out of my case, "Statistical Theory of Evolution" (dated 1931), contained the sentence:—"In short this seems from statistical considerations to be the only mechanism which offers an adequate basis for a continuous and progressive evolutionary process." The reference is to "non-adaptive radiation."
> Compared with this state of mind, in 1931, I do think the 1948 attitude ("one of many factors") shows a real advance.
> The same paper ends with a sentence relevant to your statement that "he has always pointed out that small isolated communities are at a great evolutionary disadvantage."

> > In particular, a state of sub-division of a sexually reproducing population into small, incompletely isolated groups provides the most favorable condition, not merely for branching of the species, but also for its evolution as a single group.

Perhaps you will stress the difference between complete and incomplete isolation; but this does not affect random sampling variation of gene ratios, which is the supposed evolutionary agency in dispute, and the only one through which the *size* of the sub-groups has been supposed to act.

Are you not also perhaps a little captious in taking us to task for inconsistency because we do not accept a proposition, new to us, which you tell us you regard as axiomatic? To be of evolutionary significance I submit that a factor should fulfill two further requirements, beyond what you ask of it

(a) that the deviation it produces shall be cumulative, and not annulled by other deviations.

(b) that the net cumulative change shall not be altogether trifling in magnitude compared with other changes in progress. (Fisher to Mayr, March 1, 1949)

Fisher had obviously accepted Mayr's challenge to go back to Wright's earlier publications to see if he had indeed emphasized random drift more in the earlier days. In Fisher's search, an almost bizarre thing happened. In the discussion of Wright's evolutionary papers written in 1929–32, I have provided many examples where Wright gave a much stronger emphasis upon random drift (mostly in connection with "nonadaptive" taxonomic differences) than he was willing to give in 1948. Fisher could have chosen almost any of these and made his point with an irrefutable flourish. Instead, he chose just the wrong quote and totally misinterpreted it by paying insufficient attention to Wright's language.

The reference in the first sentence quoted from Wright is not "non-adaptive radiation" as Fisher asserts, but "intergroup selection" instead. Thus Fisher had quoted Wright arguing for adaptive evolution. And Fisher's argument that the difference between complete and incomplete isolation made no significant difference to Wright's theory had been countered by Wright explicitly in the 1948 *Evolution* paper and many times before. Fisher did, I think, demolish Mayr's "axiomatic" definition of "evolutionary significance," and Fisher's query about whether Mayr had actually read "all that Sewall Wright had written" would have curious consequences.

The last Fisher and Ford letters arrived just as Mayr left New York City to spend a term at the University of Minnesota. He did not have ready access to any of Wright's papers except for the 1931 "Evolution in Mendelian Populations." He wrote to Fisher:

I do not have Wright's "Statistical Theory of Evolution" available here but I have read once more his 1931 paper in *Genetics* and to me the argument there seems to be essentially the same as that in his 1948 paper. For example, he begins his summary with the sentence: "The frequency of a given gene in a population may be modified by a number of conditions, including recurrent mutation to and from it,

migration, selection of various sorts and, far from least in importance, mere chance variation." (Nothing is said that this is the only
or even the most important evolutionary factor.)

Finally he states: "The type and rate of evolution in such a system depends on the balance among the evolutionary pressures considered here."

It seems to me that, except for adding chance variations as an
evolutionary factor, S. Wright's concept of evolution is not nearly as
far removed from yours as it may appear on first sight. (Mayr to
Fisher, April 4, 1949)

Mayr also replied to Ford saying that "the whole matter distresses me
greatly" and that the reason he had returned the paper was "certainly not because the opinions expressed in it were unacceptable per se" (Mayr to Ford,
April 4, 1949).

Ford's reply to this letter, the last one sent in this series, emphasized
again the supposed deliberate distortions Wright had introduced:

Thank you for your letter and the copy of the one you have sent
to Fisher. It does seem to me that the quotations from S. Wright
which Fisher included in his letter to you were perfectly accurate,
and therefore justified. However, you do not refer in your letter to
the much more important matter to which our reply to Wright was directed: that in essential particulars his paper in *Evolution* attributed to
us views which we do not hold and which were the reverse of those
which our own paper stated (in the draft we sent you, we both quoted
and gave exact references to Wright's statements in this matter).
These statements must of course be contradicted, and we shall have
to do this in another Journal. (Ford to Mayr, April 12, 1949)

As I have shown, the quotes from Wright cited in Fisher's letter did not support Fisher's contention, nor had Wright introduced any distortion when attributing to Fisher and Ford an either-or antithesis regarding the evolutionary
significance of random drift and selection.

Fisher and Ford revised the manuscript returned by Mayr and submitted it
to *Heredity,* where it appeared with the title "The Sewall Wright Effect" in
April 1950. It had a new introduction and conclusion, some of the wording
was made more inflammatory (now Wright had perpetrated "a direct mis-
statement of their views"), and they carefully referred to Wright's theory as
pertaining to "subdivision of a population into small isolated or *semi-isolated*
colonies" [emphasis added]. Otherwise the published paper was little changed
from the draft submitted to Mayr. The new introduction and conclusion were,
however, important for placing the *Panaxia dominula* debate into a larger
framework of evolutionary biology.

The current state of understanding of the Theory of Natural Selection, and the degree of appreciation which it now enjoys as a pre-sumptive agency of evolutionary change, constitute in effect a reversal of the opinions held by the majority of geneticists during the early years of the century. This reversal followed, we believe inevitably, from the better understanding afforded by the Mendelian system of the genetic structure of natural populations, and of selection within them. It is natural enough that progress in such understanding has not always been easy, and that workers with different preconceptions have not always given equal weight to the same circumstances. The widest disparity, however, which has so far developed in the field of Population Genetics is that which separates those who accept from those who reject the theory of "drift" or "non-adaptive radiation," as it has been called by its author, Professor Sewall Wright of Chicago.

Wright, and others who have supported his views, have repeat-edly attempted to produce examples illustrating the spread of non-adaptive qualities. Yet the extreme difficulty of deciding what char-acters are of neutral survival value should be apparent: still more, the difficulty of deciding whether the total effects of the genes, or ge-netic situations, responsible for them are so. The fate of such specu-lations is well illustrated by advancing knowledge respecting the chromosome inversions found in wild populations of *Drosophila pseudoobscura* quoted by Wright [101, p. 178], and by Sturtevant and Dobzhansky [1936b] as selectively neutral. Yet more recent work shows the very reverse [Dobzhansky 1943, Dobzhansky and Levene 1948, Wright 122], and that these chromosome inversions are in fact subject to a delicate balance of selective intensity. (Fisher and Ford 1950, 117, 119)

Fisher and Ford were correct in asserting that a majority of geneticists in the early years were not ardent selectionists. The same could have been said of systematists. And they were right that the synthesis of Mendelism and selec-tion theory had been crucial in laying the framework for a theory of evolution dominated by natural selection. Most important, they saw that the basic cleavage in the field of population genetics was between the Wrightians and Fisherians. The significance of this cleavage can be seen in the personal an-tagonisms it generated, but more important in the field research and theoreti-cal investigations it stimulated.

Fisher and Ford were also undeniably correct that Wright and others who had supported his views (many without really understanding them) had pro-duced examples purporting to illustrate the spread of nonadaptive characters. And Wright, Dobzhansky, and Sturtevant had certainly changed their minds about the intensity of natural selection to which the inversions in *Drosophila pseudoobscura* were subjected. One pertinent sidelight on this issue was a let-ter from Goldschmidt to Wright concerning the *Panaxia dominula* contro-versy, written in response to Wright's 1948 *Evolution* paper. Goldschmidt apologized for having attributed to Wright in *Material Basis of Evolution* the inaccurate view that evolution proceeded primarily by random drift in small isolated populations. Goldschmidt said, "I am sorry to see that I misquoted you, probably under the influence of Dobzhansky's book" (Goldschmidt to Wright, August 1, 1949). I suspect that *Genetics and the Origin of Species* gave the same impression of Wright's ideas to many besides Goldschmidt.

Wright never intended to write a reply to "The Sewall Wright Effect," but an invitation from George Baitsell to publish a reply in the *American Scientist* (circulation then over 50,000) proved irresistible. Wright's seven-page article, "Fisher and Ford on 'The Sewall Wright Effect,'" was basically a restatement in more accessible language of the 1948 *Evolution* paper. He again asserted that the role of random drift was merely to generate new variability upon which natural selection could act through intergroup selection: "Let me emphasize that it is only as stochastic differentiation gives a basis for selection in the form of differential population growth and migration that it contributes significantly to evolution" (p. 456). Wright easily defended himself against the charge that he had directly misstated Fisher and Ford and showed how Fisher had misinterpreted the 1931 quotations. In the last paragraph Wright did admit changes in his view over the years:

> I do not, of course, wish to maintain that my views on evolution have stood entirely still since 1931. . . . Qualifications and additions have been made, beginning in 1932, and have continued up to the present time. The interplay of directed and random processes in populations of suitable structure has, however, continued to be the central theme. (Wright 141, p. 458)

Wright did not, however, detail what were his changes of view, and the overwhelming impression given by the paper was that he had always held about the same conception of the balance of evolutionary factors. To my mind, however (and anyone can easily make the comparison), Wright's views were decidedly more selectionist and adaptationist in 1951 than they were in 1931–32. Fisher's sense that Wright had changed his views substantively was accurate.

Significance of the *Panaxia dominula* Debate

The *Panaxia dominula* debate during the years 1947–51 was highly visible to evolutionary biologists everywhere; both the issues and the personalities were important. The debate had significant consequences for experimental work on natural populations. It also highlights certain important facets of the state of evolutionary biology in the late 1940s and early 1950s. More than anything else, the *Panaxia dominula* debate crystallized the differences between the Fisherian and Wrightian ways of thinking about evolution in nature. Each side was now even more strongly motivated to produce supporting field research. Although the data on *Panaxia dominula* for gene frequency and effective population size were among the very best for any natural population (re-call that Fisher and Ford had used for one of the first times the important marking, release, and recapture technique), the data still were not good enough to discriminate clearly between the Fisherian and Wrightian views. To most evolutionary biologists, the intense disagreement of Wright and Fisher over interpretation of the *Panaxia dominula* data indicated that at the level of microevolution, the mechanisms of evolution were unsettled.

The aftermath of the *Panaxia dominula* debate had specific effects upon each participant. After having strongly defended Wright against Fisher's charge that he had changed his mind significantly (without admitting it), Mayr finally took Fisher's advice and carefully read Wright's papers on evolutionary biology for the years 1929–1932. He became convinced that Wright really had changed his mind about the role of random drift and was chagrined that this agreement with Fisher had not surfaced sooner. Mayr's refusal to publish the Fisher and Ford paper in 1949 totally alienated Fisher, who there-after turned a cold shoulder to Mayr on their few subsequent encounters (Mayr, personal communication).

Ford and his student/associate Philip Sheppard continued the field work on *Panaxia dominula* and also initiated controlled breeding experiments. In part, this work was directed toward the criticisms Wright had raised in his 1948 *Evolution* paper. Wright had pointed out that Fisher and Ford in their 1947 paper had not calculated the selective value of the *medionigra* gene or estimated the effects of this selective value on gene frequencies; he had also suggested the possibility that the effective population size was far less than Fisher and Ford calculated because whole broods could be wiped out by viral diseases. Late in 1951 Sheppard published a long paper detailing the contin-ued research on *Panaxia dominula*. Here he presented evidence that the *medionigra* heterozygotes had been subject to a 10% selective disadvantage since 1939 and that the present females spread their eggs very widely over many days, thus countering Wright's two criticisms. Another of Wright's criticisms, that the effective population size in 1940 might have been very low, Sheppard answered with evidence from collectors who thought the pop-ulation size had been about normal (Sheppard 1951). Ford, Sheppard, and others continued to monitor the few populations of *Panaxia dominula* (For a summary, see the 4th ed. of Ford's *Ecological Genetics,* 1975, chapter 7.)

In later years, Ford described the *Panaxia dominula* debate in these terms:

> It was very curious to what lengths Wright was led to minimize the importance of selection. When Fisher and I produced our first *Panaxia dominula* paper (1947), in which for the first time a decision could be taken between selection and random drift in a wild popula-tion, Wright was shocked to find that the change we demonstrated was selective. He promptly (1948) criticized our work on three grounds so [weak] that we could, and did, reply to them in a way

that, in all the years that have succeeded, he has never attempted to refute. (Ford to Mayr, September 22, 1973)

Wright, of course, recalls the debate wholly differently. His contention is that Fisher and Ford never answered his primary criticism: even if the predominant force changing the frequency of the *medionigra* gene was selection, it did not follow that the frequencies of all genes were subject to similar levels of selective forces. In the fourth volume of *Evolution and the Genetics of Populations,* (Wright 200, pp. 171–77), Wright again argued that the yearly fluctuations in the frequency of the *medionigra* gene probably resulted from "largely sampling effects in a colony of small effective size," and that the rise in frequency of the *medionigra* gene before 1939 was most plausibly "a consequence of near-extinction and accidental presence of *medionigra* in one of the surviving individuals" (p. 177). Although Wright agreed that the observed thirty-year (1939–68) decline in the frequency of the *medionigra* gene was best interpreted as a strong selective disadvantage, he still argued that the *Panaxia dominula* data indicated that sampling drift was occurring (p. 177).

Although the personal antagonism between Wright and Fisher was heightened by the *Panaxia dominula* debate, the consequences for the study of natural populations were extremely positive. Those who agreed with Fisher and Ford worked on natural populations hoping to prove the importance of selection and the unimportance of random drift. All so-called nonadaptive characters were fair game for attack. This experimental work, one example of which follows, formed the backbone of the evidence presented by Ford in the first edition of *Ecological Genetics* (1964), a book that must in my opinion be understood only in the context of the Fisher/Wright disagreements. Those following Wright's line of reasoning tried to find random drift operating in some local populations. Thus for experimental work on the genetics of natural populations, the *Panaxia dominula* debate was a great stimulus.

Cepaea and the Great Snail Debate

From the time of Gulick's work on Hawaiian snails of the genus *Achatinella* in the late nineteenth century, the most frequently cited examples of nonadaptive taxonomic differentiation came from land snails. Slow moving, easily isolated by geographic boundaries, and highly polymorphic, land snails seemed the perfect candidates for small effective population sizes and consequent random drift and nonadaptive differentiation. As described in chapter 9, Crampton's well-known work on *Partula* in the Hawaiian Islands, Tahiti, and Moorea strongly supported the conclusion that most of the observed taxonomic differences between populations of the same species, or between species, were nonadaptive. The most-often-studied land snails in pre-1930 Europe were in the genus *Cepaea*. Beautifully colored and with great variability in their banding, *C. nemoralis* and *C. hortensis* attracted much attention in

England. In the late 1920s, Captain Cyril Diver began publishing the results of his studies on fossil and living *Cepaea*. (For an autobiographical account of Diver's early work on *Cepaea*, see Clarke, Diver, and Murray 1968, appendix, and Diver 1929.) Fisher became interested in the linkage groups in *Cepaea*, and he and Diver together published a short paper on crossing over in *Cepaea nemoralis* (Fisher and Diver 1934). Working with Fisher, Diver was naturally exposed to a primarily selectionist outlook, although he was certainly aware of the views of Robson and Richards.

In 1936 the Royal Society of London sponsored a "discussion on the present state of the theory of natural selection" with such luminaries as D. M. J. Watson, N. W. Timoféeff-Ressovsky, E. J. Salisbury, W. B. Turrill, G. D. H. Carpenter, Haldane, and, of course, Fisher and Diver. Fisher's comment predictably was wholly selectionist, but Diver's raised seeds of doubt. The geographical distribution of forms within and between species of *Cepaea* did not appear to him to be the result of just natural selection. Diver's comment ended with the statement:

> With regard to the origin of species, it has been shown that a type of evolution could be brought about by Natural Selection; but the more deeply the facts relating to closely allied species are investigated, the less likely does it appear to become that the evolution that has taken place is of this particular type; and the more likely is it that species formation may be brought about by a number of different causes of which selection might possibly be one. (Diver 1936, 64–65)

Diver did not elaborate what these "different causes" might be. He did refer to Wright, but only to say that Wright believed that subdivision of a large population into small discontinuous colonies was "the most favorable distribution for the operation of selection" (p. 63). Diver did not mention random drift, but he appeared quite ready to appreciate its possible importance. When Wright sent his manuscript for Huxley's *The New Systematics* in July 1938, one of the persons to whom Huxley showed the manuscript was Diver. Huxley soon reported to Wright: "Captain Diver, who is also writing an article in the book, has read your manuscript and found it most helpful in preparing his article" (Huxley to Wright, August 5, 1938). Indeed, Diver invoked random drift many times in his article. In treating the differences between three species of hover flies (diptera, genus *Syrphus*) he concluded:

> None of the morphological differences appear to have any adaptive significance, and the most likely cause of divergence would seem to be small inherited differences leading to barriers the growth of which may well have been fostered by a distribution in small discontinuous populations such as Wright has shown leads to random differentiation. (Diver 1940, 315)

To Diver, the most certain case of random drift was in differentiation in *Cepaea*:

> It seems that selective forces and adaptive values have played little direct part in these specific differentiations, nor is there any evidence to suggest that geographical isolation, which obviously plays a large part in different circumstances has been operative here, though the possibility is by no means excluded. The most probable general cause is random differentiation in small partially isolated populations which Wright shows . . . to be statistically possible. (P. 327)

In both quotations, Diver referred to Wright's paper in *The New Systematics*. Like Dobzhansky, Huxley, and others, Diver thought random drift was the preferred explanation for hitherto unexplained nonadaptive taxonomic differences.

Fisher and Ford were disappointed that Diver had fallen for the idea of random drift. They had, since the mid-1910s, fought to rehabilitate Darwinian natural selection as the only significant mechanism of evolution. The greatest and most seductive obstacle was random drift. In 1973 Ford described this obstacle as "the great harm done by Wright, with the concept of neutral survival values and genetic drift. No one, as you know, was more conscious of that harm than Fisher" (Ford to Mayr, September 22, 1973). The seduction of Diver was a prime example of the harm caused by Wright. And the harm snowballed—Dobzhansky in the second edition of *Genetics and the Origin of Species* (1941), Huxley in *Evolution: The Modern Synthesis* (1942), and Mayr in *Systematics and the Origin of Species* (1942) all used Diver's analysis of *Cepaea* in 1940 to argue for nonadaptive differentiation in *Cepaea*.

When the selectionist study of conspicuous polymorphisms in *Cepaea* came a decade later, it was an outstanding piece of field research conducted by Arthur J. Cain and Philip M. Sheppard. Cain at this time was a lecturer in Zoological Taxonomy at Oxford, where he had taken his doctorate with John R. Baker. I am distressed to have never had the opportunity to talk with Sheppard, but Cain has generously spoken with me by telephone and has provided helpful biographical information and commentary on the origins of the *Cepaea* research. Cain himself had intense interests in natural history as a boy:

> As a boy I had already been a keen naturalist; my biology master at school encouraged me in local fieldwork, and it was a great revelation to me when we went to the seaside for family holidays. I became fairly expert in knowing exactly where to find what, and in keeping things in home-made vivaria and aquaria, which taught me a good deal about their particular requirements. My parents bought me a (very old fashioned) microscope and I explored every ditch,

rivulet, stream and pond for algae, diatoms, desmids, rotifers and what-have-you all around Rugby. When at about age 16 I actually had a bicycle . . . I extended my searches, especially for plants, all over the Midlands. I was particularly lucky in my school friends— Gerald Thompson, one of the founders of Oxford Scientific Films, Lionel Clowes, a botanist at Oxford, and John Harper, now F. R. S. (Cain to Provine, 22 February 1985)

As Ernst Mayr has pointed out frequently, naturalists who know biological organisms in their ecological settings are generally adaptationists, whether neo-Lamarckians or neo-Darwinians. This was certainly true for Cain, who was keenly interested in the problem of adaptation before going up to Oxford in 1939. There he found the nonadaptive tradition in taxonomy that I have described in chapter 9 alive and well.

I read Robson and Richards, a book (as you say correctly) of much importance in those days, and I was simply appalled at its unscientific arguments. All the time, they were building a specific conclusion, that a given set of characters was natural or non-adaptive, on total ignorance about it. They didn't even dream of doing any actual work or directed observation on the subject—because *they* could see no sense in it, therefore the phenomenon was random, or non-adaptive, or neutral. . . . It was with inexpressible relief that I discovered from Charles Elton's lectures that he really did know something about live animals in the wild and you can imagine the shock when he too started to preach the doctrine of non-adaptive characters, and fluctuations in numbers to account for them. Nevertheless, he was not at all bigoted in conversation, readily admitting that most characters were probably adaptive, and I admired his modes of thought.

Early in his student career, Cain heard Ford give a series of eight lectures on genetics and a few others on taxonomy, but he does not recall being significantly influenced by them, perhaps because, so far as the problem of adaptation was concerned, he was already "preadapted to appreciate Ford's lectures."

Back at Oxford after the war, Cain and Sheppard soon struck up a strong friendship. Cain says that they soon

realized that we thought alike on what to me was one of the most important topics in evolution [adaptation]. He soon began his D.Phil. with Ford on *Panaxia dominula,* work which I followed with great interest—Philip taught me in the course of it some of the maths, and we talked over the various opinions in the literature, and I first heard of the opposition between Fisher and Ford on the one hand and Wright on the other. When Philip was in my room one day complaining bitterly about the poverty of variation in *Panaxia,* I poured out on the table in front of him a sample of *Cepaea nemoralis* shells (he was a good lepidopterist but not versed in other groups) and we decided then and there (a) that it was impossible that such striking variation could be wholly neutral, and (b) that we would work on it. That is the origin of our work on *Cepaea*. I was indeed preadapted to agree with Fisher and Ford, but not as a result of their influence. (Cain to Provine, February 22, 1985)

Early in 1950, A. C. Hardy arranged for Cain to spend several months in New York City working with Mayr at the American Museum of Natural History and with Dobzhansky at Columbia University. For many reasons, Cain was extremely interested in Wright's work. The research on *Cepaea nemoralis* was a direct attempt to discriminate between random drift and selection in the maintenance of polymorphism. Also, Cain had been present at the famous lecture Wright gave at Oxford in 1949 and at the discussion heightened by intense disagreements between Wright, Fisher, and Ford.

While at the American Museum, Cain wrote to Wright asking for reprints. He complimented Wright on the Oxford lecture, but then stated:

Being primarily an ecologist, I tend to question whether any gene can have so low a selection coefficient (for either its direct or indirect effects) that it can drift to any great extent, or at all events to have a coefficient which remains low under normal—that is, changing— conditions. The study of polymorphic land-snails may well prove me wrong, but that is why I have taken it up. (Cain to Wright, March 26, 1950)

Wright, who did not return from England until the beginning of October, sent reprints to Cain in the middle of November with this comment:

If we classify pairs of alleles into two categories-pairs that determine conspicuous differences and ones which do not but which act as modifiers or which contribute to quantitative variability—I suspect that variation in selection would in most cases dominate the gene frequencies of the former practically to the exclusion of drift while selection and sampling would either be more or less equal or there would be predominately more drift in most of the latter. (Wright to Cain, November 14, 1950)

This letter demonstrates that even before Wright read the Cain and Sheppard paper on polymorphism in *Cepaea*, he was prepared to believe that the gene frequencies of "pairs that determine conspicuous differences" (by which I am sure Wright meant conspicuous polymorphisms) were governed primarily by natural selection rather than by random genetic drift. In other words, Wright was arguing that one should expect the color and banding polymorphisms such as found in *Cepaea* to be very much under the control of selection—precisely the conclusion to which Cain and Sheppard were coming in their analysis. So far as I can determine, Fisher, Ford, Huxley, Sheppard, Cain, and their associates and students all expected Wright to argue that the gene frequencies of conspicuous polymorphisms were determined as much by random drift as were those of any other genes. Only on this assumption do all the selection experiments on conspicuous polymorphisms conducted by Ford and his associates and students make sense—the basic purpose was to refute Wright's theory of random drift. Yet in this letter Wright agreed with their experimental results but argued that the importance of random drift lay elsewhere.

Cain responded the next April, enclosing a reprint of the joint paper entitled "Selection in the Polymorphic Land Snail *Cepaea nemoralis* (L.)" (Cain and Sheppard 1950). His letter replied directly to Wright's last one:

> Your remarks on the possibility that drift is of importance in determining the gene-frequencies of modifiers or similar genes are most interesting. As far as our observations on *Cepaea* go, the amount of melanin in the animals' body, the exact shade of the ground color of the shell, and certain other minor features seem to be dominated by selection. I have been trying to work out a suitable experiment to *demonstrate* drift, but it seems very difficult. (Cain to Wright, April 4, 1951)

Cain obviously did not accept Wright's thesis that the frequency of genes of little effect upon the phenotype were determined largely by random drift.

The last sentence of Cain's letter raised a concern Wright had thought about for many years. There was indeed little experimental evidence of the action of random drift. Soon, in the collaboration with Warwick Kerr, Wright would demonstrate random drift with direct experimental evidence.

The stated object of the Cain and Sheppard paper was to assess "the relative importance of selection and drift in determining the distribution of different color and banding patterns in *Cepaea nemoralis*" (Cain and Sheppard 1950, 275). They assumed that distribution of the conspicuous polymorphisms in *Cepaea* was a perfect situation in which to see the operation of Wright's concept of random drift (certainly this had been Diver's interpretation in 1940), and that if the distributions were shown to have resulted from selection, then Wright's theory would have been dealt a severe blow. By 1950, however (as seen in the letter to Cain), Wright specifically disavowed the thesis that conspicuous polymorphisms were primarily influenced by drift.

In their paper, Cain and Sheppard carefully analyzed ecological background in relation to shell color and banding in *Cepaea nemoralis*. They discovered significant and consistent correlations between color and banding patterns and the ecological background, and concluded that Diver had been wrong in 1940: "there is good evidence that the general appearance of any colony is determined by natural selection" (p. 292). Not only did Cain and Sheppard undercut Diver's assertions, they also argued that Crampton's data on the genus *Partula* probably indicated the influence of natural selection rather than random drift and that Welch's extension of Gulick's work on achatinellids (Welch 1938) could be similarly interpreted. They concluded:

> It seems that in view of the results presented above, all such cases of polymorphic species showing apparently random variation should be reinvestigated. No useful conclusions on this matter can be drawn merely from a knowledge of the distribution of varieties without a study of their habitats. . . . all situations supposedly caused by drift should be reinvestigated. (Cain and Sheppard 1950, 292)

This was a hard-hitting paper. Both Fisher and Ford had read and commented upon the paper in manuscript. The challenge to supposed observable instances of the effects of random drift was unmistakable.

Wright was fully convinced by the Cain and Sheppard paper regarding *Cepaea*. In his last paper of the *Panaxia dominula* debate (issued July 1951), Wright argued:

> It is probable that conspicuous polymorphism is usually a device for adaptation to diverse conditions encountered by the species. It may relate to adaptation to different seasonal conditions as shown by Dobzhansky in the case of chromosome patterns in *Drosophila pseudoobscura*. In other instances it may give a basis for adaptation to diverse microenvironments. The observations of Cain and Sheppard on the frequencies of different color patterns in the land snail *Cepaea nemoralis* under different ecological conditions suggest this interpretation. (Wright 141, p. 455)

In this statement Wright admitted that two of the prime examples of random drift he had frequently cited were actually cases of adaptations resulting primarily from selection.

Of course Wright did not know that, at the time he wrote the above statement, the young French biologist Maxime Lamotte had been energetically studying *Cepaea nemoralis* for many years. Wholly unaware of the work of Cain and Sheppard, Lamotte wrote the most substantial analysis then available of evolution in *Cepaea nemoralis*. Published in 1951, his 239-page monograph, *Recherches sur la structure génétique des populations naturelles de Cepaea nemoralis L.*, carefully reviewed the extensive literature on *Cepaea* published after 1800 and reported his studies of over 900 populations of *Cepaea nemoralis* in France. Lamotte had conducted a large number of breeding experiments and examined effective population sizes, ecological settings, patterns of breeding, and migration between populations. He also placed his experimental work into a broad range of ideas about mechanisms of evolution in nature, including those of Dobzhansky, Fisher, Ford, Haldane, Huxley, L'Héritier, Rensch, Teissier, Timoféeff-Ressovsky, Vavilov, and Wright. This was truly a monumental study.

·   Like earlier researchers, Lamotte found *Cepaea nemoralis* highly poly-morphic in both color and banding patterns and spatially distributed into well-isolated colonies that often differed significantly in the frequencies of the genes controlling the polymorphisms. Modeling his procedure on that of Dowdeswell, Fisher, and Ford (1940), Lamotte used mark and recapture studies to estimate that population sizes were generally in the range of 1,000 to 2,000 adults. His experimental studies on mating behavior between differ-ent morphs indicated that within a single population mating was wholly ran-dom, not assortative with respect to morphology in any way. Thus far, Lam-otte's research yielded results consonant with those reported in 1950 by Cain and Sheppard.

Lamotte, however, reached very different conclusions from those of Cain and Sheppard. Most of his monograph was devoted to the analysis of the character "bandless" ($b+$, a dominant), which yielded a striking polymor-phism. "Bandless" exhibited no clines, revealed no frequency differences as-sociated with climactic zones, and showed no discernible influence of ecolog-ical factors. Lamotte thus concluded that not only was the subdivision of *Cepaea nemoralis* into relatively small colonies suggestive of Wright's ideal scenario for the shifting balance theory, but that all the evidence pointed to-ward random drift (*fluctuations fortuit*) as the primary mechanism involved in the distribution of "bandless" between colonies. Variation in gene frequency of $b+$ was more pronounced in small than in large colonies, and migration was minimal. (Lamotte measured migration over 40 kilometers, but found it unmeasurably small over distances greater than 3 kilometers.) The final clinching piece of evidence was that he could find little evidence of selective predation:

Especially favorable circumstances permitted the experimental study of a factor in evolution which is often stressed by the protago-nists of certain theories of evolution: selection by predators. This fa-vorable situation is the persistence of the shells of eaten snails. It was found that the presence of bands confers a selective advantage which is apparently of considerable importance in certain cases. These cases are, however, exceptional. In reality, *Cepaea* is eaten by thrushes and blackbirds often in winter at a time when they are buried in the soil, and, consequently, when the banding of the shell is invisible. If one takes into consideration the proportion of snails which are killed by birds, and the frequency of the colonies where a selective choice seems to have occurred, it appears that, on the whole, the selection which can be attributed to predators is exceed-ingly small. (Lamotte 1951, 234. This paragraph is from the English summary, and is a direct translation of the corresponding paragraph in the "Résumé", p. 229.)

The sharp contrast with the conclusions of Cain and Sheppard is obvious.

Lamotte sent Wright a copy of the monograph, asking in return for some of his reprints, which Wright sent him. Lamotte replied with a revealing letter:

> I have just received your reprints, for which I thank you very much. The aspects of the genetics of populations that you study interest me especially. You have been able to see that my work on natural populations of *Cepaea nemoralis* were wholly oriented towards the interpretation of actual gene distributions in the light of schematic theories in which the role of the greatest number of factors as possible is analyzed quantitatively. Initially, the goal of my work was only to interpret the distribution of the gene *b+* of *Cepaea* by comparison with the families of the "curves of Wright" corresponding to the diverse parameters of mutation, selection and migration. But to be entitled to make a legitimate comparison, I had to analyse very precisely all of the characteristics of the populations of the species, notably the effective size of the populations, their relationships with the terrain, their ecological preferences, their genetic structure— homogamy, consanguinity—and all that led me much farther than I had expected. . . .
>
> I am going to send to *Heredity* a study of the role of "random drift" [*fluctuations fortuit*] in response to the article where Cain and Sheppard denied the role of "genetic drift" upon the diversity of the genic composition of populations of *Cepaea*. It appears that the effective size of colonies plays a fundamental role in the diversity of composition of the colonies, and this is only possible through the mediation of "genetic drift," quite natural for a living species divided into well isolated groups of 500 to 1000 individuals. This naturally does not exclude the possibility that different biotopes [ecological settings] can have an important effect, as appears very clearly in the case of the gene "yellow," which I have actually studied. (Lamotte to Wright, June 3, 1952. I have modified in several places Wright's translation of the original French.)

Wholly unbeknown to Wright, his ideas had strongly influenced Lamotte's work on *Cepaea*. Unlike the typical interpretation of Wright's shifting balance theory, reducing it to mere random drift, Lamotte thoroughly appreciated Wright's attempt to take into account simultaneously all the known factors of evolutionary change. Certainly aware that natural selection could be an important force in the evolution of *Cepaea* (as in the case of the "yellow" gene), Lamotte concluded that his data on the "bandless" gene indicated an important role for random drift.

With the publication of Lamotte's monograph and the summary of it sent to *Heredity* (Lamotte 1952), the beginning lines for the modern version of what I call the Great Snail Debate were drawn. The selectionists, led by Cain and Sheppard and supported by Ford and Fisher, argued for selection as the primary determinant of the frequencies of all genes affecting polymorphisms in *Cepaea*. Lamotte, citing Wright's theoretical work as support, argued for an important role for random genetic drift. The whole debate, and the great stimulus for empirical investigation in the field, rested firmly upon the tension between the theoretical views of Fisher and Wright.

Although much impressed with the volume and care of Lamotte's work on *Cepaea*, Cain and Sheppard disagreed not only with his conclusions but also with some of his experimental designs. They continued their work on *Cepaea* in England, finding more and more convincing evidence of natural selection as the primary determinant of differences between populations. Soon they published (Cain and Sheppard 1954, in the United States in *Genetics*) a summary of the work on *Cepaea* since 1950 and a detailed critique of Lamotte's work, which they considered to be not only important but influential, since they believed most American evolutionists supported Lamotte's interpretations rather than theirs:

> By far the most important single paper on *Cepaea* that has yet appeared is Lamotte's study of *C. nemoralis* (1951). This author's work on population size, panmictic units, homogamy, and migration is the best yet produced. He is also the first to apply quantitatively Sewall Wright's equations for the interaction of selection, mutation, migration, and population size to an ensemble of actual populations. His conclusions, based on a careful study of over 800 French colonies, are diametrically opposed to ours, since he believes that mutation assisted by migration is the prime agent maintaining the variation, that differences between colonies are due to genetic drift, and that selection, although present, plays an almost negligible part. In the colonies investigated by us, selection is the primary factor both in maintaining variation within a colony and in determining the differences between colonies. This apparent contradiction must be investigated. (Cain and Sheppard 1954, 106)

Their examination of the apparent contradiction in the paper included careful critiques of Lamotte's methodology in assessing the effects of visual selection and of his theoretical arguments for random drift. They concluded from their own evidence that natural selection was the primary determinant of the distributions of color and banding polymorphisms.

Accepting some of these criticisms, Lamotte rejected others. In 1959 he attended the Cold Spring Harbor Symposium on Quantitative Biology, in that centennial year naturally devoted to the topic "Genetics and Twentieth Century Darwinism," where he delivered a paper (in French) summarizing his views on *Cepaea;* Sheppard, Wright, Dobzhansky, Rensch, and many other notables were in the audience. Although admitting a significant role for predatory and climactic selection, Lamotte strongly defended his earlier view that random genetic drift was an important factor in the differentiation of colonies of *Cepaea* in France (Lamotte 1959; Dobzhansky translated the paper into English). Sheppard insisted at the meeting that Lamotte's results with *Cepaea* in France were inapplicable to *Cepaea* in England. Although several comments by Rensch, Van Valen, Dobzhansky, and Wright were published with the paper, Sheppard's was left out, an omission he resented (Arthur Cain, personal conversation).

Wright's comment on Lamotte's paper is illuminating. As I pointed out earlier, in 1951, before he had heard about Lamotte's work, Wright was fully prepared to have all the conspicuous polymorphisms in *Cepaea* be subject to strong selection pressures. Lamotte's work convinced him, however, that random drift was a significant factor affecting the distribution of polymorphisms in *Cepaea:* "It seems to me that Dr. Lamotte's results indicate strongly that random drift is playing an important role in variations of gene frequency in the populations of *Cepaea* species, whatever may be the mechanisms of equilibrium under various conditions" (Lamotte 1959, 86). One might think that Wright would be particularly pleased with Lamotte's work because it decisively kept alive the possibility that random drift was an important factor in the distribution of polymorphisms in *Cepaea*. What Wright actually said is this:

> It should be emphasized, however, that the importance of random drift in evolution does not come from cases of this sort in which there is merely local fluctuation about equilibrium for a single pair of alleles. It comes from cases in which there is simultaneous fluctuation in the frequencies at a great many loci in interaction systems with multiple selective peaks. Such fluctuations in conjunction with the selection specified by the surface of selective values, permit the establishment of populations centering about different selective peaks among which interdemic selection may lead to evolutionary advance. (Wright comment in Lamotte 1959, 86)

In other words, just as the general dominance of selection in the evolutionary process cannot be proved by demonstrating its efficacy in determining the gene frequency of a single pair of alleles controlling a conspicuous polymorphism, so the general importance of random drift in evolution cannot be proved by demonstrating its efficacy for a single pair of alleles that might happen to be selectively neutral. I think Wright was attempting to disassociate himself from the view that had been so common in the 1930s and early 1940s—that the general importance of the concept of random drift was indicated by its application to a handful of examples, mostly conspicuous polymorphisms. In short, the key to understanding evolution in nature was not to be found in the study of conspicuous polymorphisms.

The Great Snail Debate, as I have termed it, was lively during the years 1950–59, but it had really only begun in earnest and soon became far more complex. In 1963 Cain and Currey published a major paper detailing what they called "area effects" in *Cepaea*—in some places the morph frequencies changed dramatically over distances as little as 100 to 300 meters, yet were uncorrelated with visual predation or habitat change and occurred in populations of such large size (maintained over long periods) that random drift appeared an unlikely explanation. By the 1970s the Great Snail Debate included arguments over precise forms of selection (visual, climatic, frequency dependent, disruptive, density dependent, stabilizing, heterozygote advantage) in relation to random processes (founder effects, sampling drift, random drift resulting from fluctuations in selection pressures), migration, and reproductive biology (hermaphroditism, sperm storage, multiple matings). A brief but helpful review of the literature by Jones, Leith, and Rawlings (1977), is appropriately entitled, "Polymorphism in *Cepaea:* A Problem with Too Many Solutions?" Another useful review is that of Cain (1977). The entire issue of *Biological Journal of the Linnean Society* 14 (nos. 3 and 4, November/December 1980) is devoted entirely to the Great Snail Debate. One highlight of the issue is found in the article by Woodruff and Gould, who argued that in the snail genus *Cerion* (West Indian land snails), Wright's "interdemic selection based on local peak-shifts among multilocus interaction systems" was the most likely explanation for patterns of variation (Woodruff and Gould 1980, 413–14).

Although obviously much involved as a theorist in the debate, Wright himself had never written anything substantive about the snails until the final volume of his magnum opus that appeared in 1978. There he provided a quantitative reanalysis of the data of Cain and Currey (1963) and a review of the literature. (Wright's copy of the reprint of the Cain and Currey paper is heavily annotated and almost worn out.) His final assessment is worth quoting here:

> The most plausible primary explanation of the polymorphisms seems to be a rather weak frequency-dependent selection for diversity. Secondarily, colony sizes are in general sufficiently small on allowing for the various reasons for reduction of effective size, especially occasional extinction of colonies and reestablishment by small groups of immigrants from other colonies, to permit wide sampling drift and the building up of differentiation among areas of considerable size (of the order of square kilometers). In addition there is much differentiation according to habitat in some places based on appropriate differences in concealing coloration. There is also differentiation among large areas due to climatic selection, principally near the limits of tolerance of the species. In many cases, however, such large-scale differences may be due rather to interaction with different multifactorial background heredities, the development of which is the most important evolutionary consequence of sampling drift. (Wright 200, pp. 239–40)

Thus Wright gave most weight to various kinds of selective forces in determining the reasons for the distributions of conspicuous polymorphisms in *Cepaea*. But he was unwilling to give up an important role for random drift.

After the early 1950s, Wright argued consistently that conspicuous polymorphisms, although particularly useful for the study of genetic variation in natural populations because of their accessibility, were uncommon and unrepresentative of the evolutionary process as a whole. He saw conspicuous polymorphisms as evolutionary adaptations to predation and heterogeneous environments, not as variability upon which evolutionary advance was based. Thus proving that a conspicuous polymorphism was subject to large selection rates was proving the theoretically obvious, and furthermore it was not possible to generalize beyond the specific case to role of the whole genome in the evolutionary process.

But conspicuous polymorphisms nevertheless remained important for Wright. To argue for the shifting balance process, it was pointless to give ironclad cases of truly neutral characters. Such examples gave no insight whatsoever into what was for Wright the most crucial issue—population structure. What Wright wanted to show was that in cases where characters were subject to significant selection pressures, there were populational differences unrelated to the ecological background. The conclusion was that effective population sizes were, or, more likely, had been sufficiently small for sampling drift to play a prominent role. Because of the significant selection rates, different populations would not be expected to differ in the adaptedness of their different conspicuous polymorphisms, but the differences themselves would be caused by random drift (nonadaptive differences, not nonadaptive characters). Thus if Wright could show that in some cases of conspicuous polymorphisms differences between populations were caused by sampling drift, then he could conclude that the rest of the genome, most of which was exposed to far lower selection pressures, was even more subject to sampling drift, leading to the creation of new interactive systems of genes according to the requirements of his shifting balance theory.

To anticipate one of the conclusions of the next chapter, one can easily see that Wright would not be particularly enthusiastic about Kimura's neutral theory of molecular evolution because it did not depend upon population structure at all. Proving the existence of sampling drift in a character subject to selection told a great deal about population structure and in turn about the shifting balance theory of evolution. Proving the existence of wholly neutral (or even almost neutral) genes told nothing about either population structure or the shifting balance theory. Something that Fisher and many others never seemed to understand was that Wright's allegiance was to the shifting balance theory, not to the mechanism of random genetic drift.

Conclusion

The evidence presented in the last four chapters demonstrates clearly that the basic tension between the evolutionary views of Wright and Fisher had an enormous influence upon evolutionary thought in the period after 1930. In particular, the tension between their views stimulated a large number of experiments on evolution in nature. In this sense, the tension between Wrightian and Fisherian views was a highly creative force in evolutionary biology.

By the late 1940s, the antagonism between Wright and Fisher had reached legendary proportions among evolutionary biologists. Their few meetings were carefully staged and observed, any fireworks noted, and conflicting reports issued. Almost any evolutionary biologist active in the 1950s knew many Wright/Fisher stories. During his year at Edinburgh (1949–50), Wright lectured widely in England. Fisher and Ford were both present for his lecture during the fall of 1949 at Oxford University (as were Cain, Sheppard, Darlington, and many other notables). Pitting Fisher and Ford against Wright in one lecture hall was a fight to be savored and talked about. Dobzhansky was eager to hear about it, and of course questioned Cain when he came to visit in 1950. After Wright's return from England, Dobzhansky wrote to him:

> There have reached us some echoes of your fight with Fisher in Oxford. Since the echoes were reflected by a British source, they alleged that Fisher has completely and utterly annihilated you. Well, nobody here has any tendency to be fooled so easily, but I would have been very glad to hear from you what has happened. (Dobzhansky to Wright, October 31, 1950)

Wright wrote back a detailed letter reviewing his Oxford lecture and the reactions of Fisher and Ford to it. He concluded:

> I felt no difficulty in answering all of [Fisher's] points as the views which he attributed to me had, as usual, little relation to what I had actually said or written at any previous time. I certainly did not feel "annihilated" and did not sense that any of those who talked to me afterward thought that I had suffered this fate. I would be glad to know who it was that got this impression and why.
>
> The only point raised in the discussion that I remember feeling any concern about afterward was Ford's complaint that I had seriously misinterpreted Fisher and him in my discussion in *Evolution* of their *Panaxia* paper (Ford, I may say, was extremely cordial throughout this visit to Oxford and later in spite of this complaint).
>
> I felt sufficiently satisfied with my Oxford lecture and its reception to give it unchanged at the University of Durham, before the Royal Society of Edinburgh and before the Zoology Department at Cambridge the next spring. Fisher did not attend the latter. (Wright to Dobzhansky, February 7, 1950)

Dobzhansky replied:

That you and your works have been annihilated in Oxford is well established by testimony of several high authorities. You seem to be the only person who has any doubts about it. (1) E. B. Ford announced it in a triumphant letter to Cain, a very capable (really capable) young Oxford instructor (or whatever they call it in Oxford), who works on snails and who has spent the last winter with Mayr and myself. Talking to Cain, who is very intelligent, I have for the first time appreciated the depth of prejudice in the Fisher-Ford circles. (2) Julian Huxley in the first draft of his speech which he delivered in Columbus last September has stated that genetic drift has recently been shown to be an imaginary phenomenon. I have called Dunn's attention to this statement (Dunn is the editor of the forthcoming volume) and suggested that he try to persuade Huxley that his opinion is at least premature. Since I have departed from Columbus before hearing Huxley I do not know whether he has or has not eliminated this statement [see the pertinent Huxley quote below]. (3) Darlington has spent a few days with us living in our house and he enquired whether or not I am satisfied that all your stuff is dead. He had no fixed opinion himself, but this is clearly the consensus of those who know in England or at least in the Cambridge-Oxford circles. (Dobzhansky to Wright, November 9, 1950)

Here for comparison is Cain's recollection of Wright's Oxford lecture:

I most certainly never told Dobzhansky that Fisher had *annihilated* Sewall Wright at the Oxford meeting. I don't know how far Dobzhansky was pulling Wright's leg to get a detailed response out of him. . . . What happened was that Wright gave a lecture fascinating in what one could understand of its content, but atrociously presented. It went on for an hour and three quarters (and I really believed that he had done so deliberately to prevent time being left for discussion, until I heard from the Edinburgh people some years later that he always lectured like that). He spoke in much too monotonous a voice with little paragraphing or emphasis, and illustrated the lecture with slides of tables of numerous white figures crowded together on a black background—the lecture room was not large, but one could not read them at the back even when in focus. One hour is the maximum for any audience, and when it became evident that he was largely repeating his general views, I must admit that my attention wandered a little. At the end there was some decorous applause. Then Alister Hardy called on Ford to open the discussion. He announced his complete agreement with Sewall Wright (a gasp from the audience!) with a few unimportant exceptions, and then proceeded to take him to pieces for a quarter of an hour, ending up with words something like "With these trivial exceptions, it is a pleasure to express our complete agreement with Mr. Wright." Another gasp. Very little discussion followed, but Wright made at least two answers to questions, both times at too great length; then he actually said something to Fisher, who had sat in the front row with his arms folded and beard bristling, staring out in front of him the whole time. This was unfortunate. Fisher barked out "I only went so far as to disagree with you completely. Could I do more than that?"

. . . . Whether one agreed with him or not, it was clear that Wright had a really great mind. . . . Although Wright's speech confirmed my opinion that neither he nor Dobzhansky (nor anyone else promoting the idea of genetic drift as an evolutionary explanation) were bothering to find out anything about the basic biology of the animals they were talking about (which Ford certainly did) it was clear that he was a theoretician of a very high class indeed. While I thought Fisher and Ford were more likely to be right about most characters of animals, they had done nothing to annihilate Wright. (Cain to Provine, February 22, 1985)

I think that the excitement and interest generated by the meetings between Wright and Fisher is amply illustrated by the letters of Dobzhansky and Cain (who has also nicely expressed the usual assessment of Wright's lecturing style). Such intense interest occurred because so many evolutionary biologists thought they were observing no mere squabble between two evolutionary theorists but the confrontation of the two most basic views in modern evolutionary biology.

This chapter has also provided abundant evidence for a second major theme—that there was indeed a rising tide of adaptationism in the period 1940–55. In 1940–42, Wright, Dobzhansky, Huxley, and Mayr all expressed the view that conspicuous polymorphisms were mostly nonadaptive and their genetic distribution was determined primarily by random drift. By the early 1950s, only one decade later, all four viewed conspicuous polymorphisms as primarily adaptive. The third edition of Dobzhansky's *Genetics and the Origin of Species* (1951) had a whole chapter, following the chapter on selection, devoted to the subject of adaptive polymorphism. The prime example of adaptive polymorphism in this chapter was that of inversions in *Drosophila pseudoobscura;* in the first and second editions, the same inversions had been used as the prime example of nonadaptive differentiation. I challenge anyone who compares the second and third editions to conclude anything other than that Dobzhansky had become more selectionist in the decade 1941–51. (He became even more selectionist between 1951 and 1970, when a much revised edition of *Genetics and the Origin of Species* appeared under the title *Genetics of the Evolutionary Process.*)

Huxley, speaking at the Golden Jubilee of Genetics meeting in Columbus, Ohio, in September of 1950, had greatly changed his opinion about nonadaptive differentiation since the appearance in 1942 of *Evolution: The Modern Synthesis:*

I return to the all-pervading influence of natural selection, and the consequent omnipresence of adaptation, both of characters and gene combinations. The establishment of non-adaptive (that is, selectively neutral or even disadvantageous) characters by "drift" in small populations seems, as Fisher has demonstrated, to happen much less often than its discoverer, Sewall Wright, imagined. As Ford, Muller, and others have shown, truly non-adaptive genes and gene combinations are exceedingly rare as part of the normal genetic outfit of organisms. (Huxley 1950, 597)

Mayr, who argued in 1942 that most conspicuous color polymorphisms in birds and other animals were nonadaptive, soon changed his mind (Mayr, personal communication) and in a 1950 paper with Stresemann stated: "The frequently made assertion that the color characters of species of birds are accidents of variation and without selective significance appears dubious in view of the fact that so many of these characters are either involved in balanced polymorphism or subject to geographical variation or both" (Mayr and Stresemann 1950, 299). The trend toward adaptationism is clear. Furthermore, and this is where Gould, Eldredge, and others may disagree with me, the change toward adaptationism documented here was based upon some vincing experimental and observational evidence and the reasonable expectation that additional evidence would point in the same direction.

Even as late as "Organic Evolution," which Wright wrote for the 1949 *Encyclopaedia Britannica*, he said, speaking of subspecies:

> The boundary between forms exhibiting what seem to be clearly adaptive differences may not coincide with any natural boundary, suggesting that differential population pressures may take precedence over close adaptation. This may be due in part merely to lack of knowledge. Conspicuous but apparently neutral differences may possibly be physiological correlates of intangible but adaptively important characters. It is difficult, however, to avoid the impression that a large portion of the differences merely happen to have arisen and have no adaptive significance. . . .
>
> The differences between species of the same genus are for the most part similar to those between subspecies, though usually greater in degree. . . . It is indeed difficult to cite strikingly adaptive morphological differences at the species level though not at the level of genus or family. Differences in protective coloration are, however, very common. (Wright 126, p. 918)

With a little more hedging, this was the same view Wright had expressed consistently in print since the late 1920s. This was, however, the last time Wright ever expressed the view that taxonomic differences, conspicuous polymorphisms in particular, were largely nonadaptive.

In the period 1949–56, Wright published six major summary papers of his shifting balance theory of evolution in nature. These were "Adaptation and Selection" (128), "Population Structure in Evolution" (132), "Population

Structure as a Factor in Evolution" (139), "The Genetical Structure of Populations" (140), "Classification of the Factors of Evolution" (152), and "Modes of Selection" (151). Not one of these papers mentioned that observable taxonomic differences were largely or even often nonadaptive, and after 1950 Wright regarded conspicuous polymorphisms as generally highly adaptive. The whole tone of Wright's papers had become selectionist and adaptationist. He strongly defended the view that his shifting balance theory had always been selectionist:

> Why was there this emphasis on balance? A partial but very incomplete answer is that in so far as my position could be classified as advocacy of any single principle, this principle was selection which seemed to me then, as now, the only long run guiding principle. (Wright 152, p. 16)

> If my position in 1931 is to be labelled with a single word it would obviously have to be selectionist. (Wright 151, p. 18)

These quotes should of course be contrasted with what Wright actually said in the early 1930s (see chapter 9), as for instance with the summary statement in Wright's famous 1932 paper:

> That evolution involves nonadaptive differentiation to a large extent at the subspecies and even the species level is indicated by the kinds of differences by which such groups are actually distinguished by systematists. It is only at the subfamily and family levels that clearcut adaptive differences become the rule. The principal evolutionary mechanism in the origin of species must thus be an essentially nonadaptive one. (Wright 70, p. 364)

The contrast between Wright's published statements from the early 1930s and the mid-1950s is quite unmistakable.

Yet in his recent published papers, in our detailed interviews, and in his comments upon these chapters in draft, Wright vigorously resisted the notion that he changed his shifting balance theory toward a more adaptationist view:

> Although I touched on the possibility of nonadaptive differentiation of species by sampling drift in a paper published in 1932, I did not consider this phenomenon to contribute more than some unimportant "noise" in the process of adaptive evolution, which was the main subject of this and earlier papers. . . . I emphasize here that while I have attributed great importance to random drift in small local populations as providing material for natural selection among interaction systems, I have never attributed importance to nonadaptive differentiation of species. (Wright 207, p. 12)

Wright also has claimed on many occasions since the early 1950s that his shifting balance theory of evolution was misinterpreted by Fisher, Ford, Huxley, and many other prominent evolutionists, who saw random drift as Wright's entire theory of evolution in nature.

On the basis of the evidence supplied in this and earlier chapters, I hope to explain both the charge of misinterpretation and the apparent changes in Wright's views on adaptation. From the beginning, Wright saw his shifting balance theory as ultimately adaptive, as of course its analogy in domestic breeding was by necessity (i.e., the desired form of the breed was "adaptive"). More specifically, the process was adaptive at the level above the effective action of random drift. But what systematists said in the late 1920s and early 1930s told Wright that differences between species were nonadaptive. Thus Wright adjusted the "balance" in his shifting balance theory to favor the controlling action of random drift at the species level and below, the selective/adaptive levels being primarily those above the species level. The adjustment of Wright's shifting balance theory to meet the observations of systematists was indeed a minor aspect because in theory the action of random drift might easily cease to be effective well below even the subspecies level, yet the shifting balance process remain robust.

As I have shown, however, it was precisely the linkage Wright drew between random drift and observable nonadaptive differences that made the greatest impression upon contemporary evolutionists, who do not seem to have understood the adaptive aspect of Wright's theory. As systematists moved toward the view that differences at the species and subspecies levels were adaptive rather than nonadaptive, they no longer thought random drift could cause those differences. They concluded that Wright's theory of evolution in nature had been seriously undermined. Wright, on the other hand, thought that his theory was enhanced, and not in any way hurt, by the discovery that some differences between species were adaptive rather than nonadaptive. All Wright had to do was move downward on the taxonomic scale the level at which random drift was effective, which was very easy to do. After the early 1950s, Wright's view on this question remained as he explained it to Dobzhansky in 1954: "The essence of my whole theory is that a tendency toward random differentiation of demes which by itself could never perhaps go so far as to be observable, greatly increases the effectiveness of selection" (Wright to Dobzhansky, October 25, 1954).

In this same year Wright and Warwick E. Kerr published the first clearcut evidence of random drift in a controlled experiment. Kerr was a Brazilian geneticist (Universidade de São Paulo) who had come to the University of Wisconsin on a Rockefeller postdoctoral fellowship in 1950. He was primarily interested in random drift in honeybees, but working with Crow (who in turn consulted with Sturtevant), Kerr decided to use *Drosophila melanogaster* as a laboratory model. Using successive generations of small populations of *Drosophila melanogaster* (four males and four

females for each generation) and sex-linked characters, Kerr attempted experimentally to produce and measure the action of random drift. Kerr later
worked with Dobzhansky, who sent the data to Wright. The result was three
very important Wright/Kerr papers (Wright 146–48) in which the theoretical
and observed extent of random drift were in agreement.

Kerr's data indicated that the effective population number was about
three-fourths the number of adults. This interesting result stimulated Crow
and Newton Morton to examine quantitatively the concept of effective population number, the result being a series of papers that laid the theoretical
groundwork for calculation of effective population number in a variety of
breeding systems in nature (Crow 1954; Crow and Morton 1955; Crow and
Kimura 1970, 345–65). Since effective population number was a central variable in Wright's shifting balance theory and his statistical distribution of
genes, the results of this research were of great interest to Wright who followed them closely. Thus in no way did Wright lose interest in random drift
or consider it any less important a factor in the evolutionary process. I think
Wright is justified in arguing that he held to the same basic shifting balance
theory after 1932 and that his theory was badly misinterpreted. But it is also
true that his conception of the "balance" in the shifting balance theory
changed with the attitudes of systematists toward a more adaptive and selectionist stance and that Wright did not clarify this change, thus inviting misinterpretation of his theory. A major factor, of course, was that most of the major textbooks of the 1930s and 1940s described Wright's evolutionary theory
as mere random drift. Since most budding evolutionists learned evolutionary
theory from the major textbooks, and they did not generally read Wright directly for the reasons discussed earlier, they consequently learned a view of
Wright's ideas that often fit poorly with what Wright actually thought from
the early 1950s on. No wonder Wright felt so frustrated at being misinterpreted: he could not seem to stem the tide of the view that had become attached to him, although this situation would change in the 1960s and 1970s.

After his flurry of papers on evolutionary theory in the early to mid-
1950s, Wright knew he had to write up the unpublished results of his many
years of research on physiological genetics of guinea pigs. Thus when he retired at age sixty-five from the University of Chicago in December 1954, he
concentrated first upon the guinea pig work and only then turned his attention
back to the work on evolution that he had been planning since 1925.

13
The Madison Years

As a result of Chancellor Hutchins's antipathy to emeritus professors and the deterioration of the Hyde Park neighborhood that the Wrights had liked so much, Sewall and Louise decided soon after their return from Edinburgh in 1950 to leave Chicago after Sewall's retirement. Word of this decision had spread among the Wrights' closest friends so by early 1953 Sewall had received several offers from other universities. Sewall was keenly aware of the extreme importance of the decision. At stake was not only the completion of his forty-years work on the physiological genetics of guinea pigs but also the writing of his big book on evolutionary theory, which he had been planning almost since he came to the University of Chicago. It was paramount that he find a place congenial for this culmination of his life's work. The wrong decision or lack of possible choices could easily have been a disaster for his ambitious plans.

The opportunity offered to Wright by the University of Wisconsin was almost ideal. Madison was a quiet and congenial place for Sewall and Louise to live. Both of them had old friends there and they made many more. The Department of Genetics was highly supportive in every way and undemanding of his time. For the first time in his professional career Wright had ready access to population geneticists, particularly James F. Crow, and graduate students whose primary interests were in population genetics. After forty years of caring for a large experimental population of guinea pigs and thirty years of full-time teaching, he was at last freed of such responsibilities. Also for the first time in his professional life, Sewall was able to work continuously on his writing. The result that he first squeezed everything of importance from the mountains of data left from the guinea pig experiments at the University of Chicago, and then he wrote an extraordinary four-volume treatise summarizing his thoughts on evolutionary biology. After that, he continued to publish important papers on speciation and macroevolution, topics that he had not addressed to his satisfaction in the four volumes. These accomplishments exemplify Wright's determination and abilities at an advanced age, and they also reveal the wonderful support he received from Louise, his close friends, and the Department of Genetics during his "retirement" years.

The University of Wisconsin

The job market for retired geneticists, even those of Wright's stature, was poor in the mid-1950s. Only two attractive offers appeared. The first came from Wright's old friend Charles W. Metz, then chairman of the zoology de-

Figure 13.1. Wright in 1954, the year of his retirement from the University of Chicago. This is the most widely reproduced photograph of Wright. He did not erase the blackboard with the guinea pig after this session. Wright denies that he ever erased a blackboard with one, despite the many anecdotes to that effect.

Figure 13.2. Four generations of physiological genetics at the twenty-fifth anniversary celebration of the Jackson Laboratory in Bar Harbor, Maine. Castle with his three successive generations: Sewall Wright, Elizabeth S. Russell, and Willys K. Silvers.

partment at the University of Pennsylvania. He wrote on New Year's Day 1953 to say that his department would like to offer Wright a position until the retirement age of seventy. The offer sounded attractive in many ways, especially the opportunity to be with the Metzes and Phineas and Anna Rachel Whiting, all friends since graduate school days. But the neighborhood around the university had many of the same problems as Hyde Park, and the thought of unsettled urban life at advanced age did not appeal to either Sewall or Louise.

Shortly after a trip to Madison in July of 1953, Sewall received a letter from M. R. Irwin, chairman of the Department of Genetics at the University of Wisconsin. It began:

> It has been in the minds of our departmental staff for some time that we would like to be able to discuss with you the possibility that you might be persuaded to join our staff in some mutually agreeable capacity following your retirement from the University of Chicago . . . . It has been our hope that we might be able to work out with you some plan for all or part of the period until you reach the age of 70 which would allow you freedom to carry on your professional interests with a reasonable minimum of requirements from the University. (Irwin to Wright, July 16, 1953)

This was indeed an attractive offer. Wright had enjoyed many ties with the Department of Genetics over the years. L. J. Cole, who had founded the department there in 1910 (it was then called the Department of Experimental Breeding), and who had been Wright's superior as chief of the Animal Husbandry Division of the Bureau of Animal Industry of the USDA in 1923–24, had frequently invited Wright to visit over the years. (Cole was chairman until 1939.) Wright also knew well Irwin, R. A. Brink, A. B. Chapman (who had been a postdoc with Wright), and Crow. Madison appeared to be a congenial place to live.

A year later, after carefully considering both offers, Wright accepted the Wisconsin offer and became the Leon J. Cole Professor of Genetics. The appointment was half time, with the understanding that Wright would give a seminar for graduate students once each year for the five years of the appointment and be available for occasional consultation with faculty and students.

Louise spent many days looking for the right house. From Sewall's perspective the primary requirement was that there be adequate room for his enormous reprint collection (some 15,000 reprints by conservative estimate) and his large collection of journals (among others *American Naturalist, Genetics, Journal of Heredity,* and *Evolution,* the latter three beginning with first issues). They found a lovely modest home with a large basement that suited Sewall's needs well. He walked vigorously several times a week from home to school and back again, a total distance of five-and-one-half miles. I think this walking is related to his longevity.

Sewall and Louise spent much time with their old friends the Brinks, Irwins, and Chapmans. Wright and Irwin ate lunch together at the same restaurant every Friday from 1955 to 1978 and always ordered the same dessert—gooseberry pie. Both Wright and Irwin told me the story of how one Friday the restaurant had no gooseberry pie, much to the consternation of the management. The next Friday, the two were presented with a whole pie for free. The Crows also became close friends of the Wrights. Sewall and Louise found a whole new circle of highly supportive friends by joining the Unitarian church. They made many other friends. Socially, the Madison years were very rewarding.

During the first fifteen years in Madison, Sewall and Louise traveled a great deal as they always had by automobile. These trips were a great joy to them, as were their travels abroad. Sewall did not cease driving until he was eighty-eight.

The great sadness of the Madison years was the death of Louise in 1975. Sewall wrote about this in his autobiographical notes:

Louise developed high blood pressure in her later years. Her doctor got this fairly well under control but her heart became weakened and she often had coughing spells. A breast tumor was removed in November 1974 and while she recovered quickly from the operation, she came out weakened. Nevertheless, she organized a very nice birthday party (December 21) for me at the Ivy Inn. . . .

Shortly after New Years Day, she came down with pneumonia and went to the hospital. She recovered from this and I was waiting for a good day to take her home when it became evident that her heart was getting weaker and her cough worse. Her appetite failed and she died on February 17, aged 80, leaving me very lonely indeed. . . .

She was an unusually sympathetic person. Her long continued activities in connection with housing reflected her compassion for the less fortunate. Her election to Phi Beta Kappa testified to her high intelligence. I always relied much on her strong common sense. She had great courage in facing difficulties and she always did the best that she could under all conditions. She had, moreover, a delightful sense of humor. . . .

Louise was exceedingly fond of travel. She was a wonderful companion on our numerous automobile trips around the country, including ones in all of the 48 states, most of them many times, supplemented by plane trips to Alaska and Hawaii after they became states. She was able to strike up sympathetic conversations with strangers that we met, something that I was not good at. . . .

From a few remarks that she made, I think that Louise was much disappointed that we didn't do more foreign travel than we did after my retirement. I am afraid that I let her down in this respect more than in any other. I became too much concerned with finishing my treatise, and perhaps my lifelong dislike of bus travel (from my sensitivity to tobacco smoke) played a role. After she died I found that she had a whole shelf of travel books, among the many back in her study. I thought at first that these were bought in preparation for trips that we never made, but was somewhat relieved to find that most of them (39 out of 46) were of places that we had visited. . . . Most of them were evidently obtained for trips that we made or for interpretation of what we had seen. I regret greatly, however, that we didn't get to Greece, Spain, Egypt, and perhaps Yugoslavia, and more in Latin America than three brief excursions across the border into Mexico.

Louise's uncomplaining and loyal support, her constant attention to Sewall's needs as he devoted his efforts primarily to his scientific work, were essential to his accomplishments, in particular the writing of the treatise on evolution. She lived long enough to see him through the writing of all four volumes, reporting in the annual Christmastime "Wright Family News" for 1974, six weeks before her death, that "Sewall's volume three has turned out to be volumes three and four. Both are written but, due to shortage of stenographers, a few chapters have not been typed. Volume three should be ready for press before long."

Sewall's formal teaching duties turned out to be minimal. The first year he gave his evolution course for graduate students, Motoo Kimura prominent among them. Crow and many other faculty audited, although the audience dwindled over the semester. The second year, however, only two students took the course, and after that Wright gave no other formal courses, though he often gave guest lectures in courses taught by Crow and others. He was always available for consultation, students and faculty alike receiving long and detailed answers to their questions. He continued in this role well into his nineties.

As I have described in chapter 6, Wright devoted primary attention in the first five years at Madison to the analysis and writing up of his data from the guinea pig experiments conducted at the University of Chicago. He anticipated spending only a year or two at most on this task, but the flood of correspondence, travel to meetings and invited lectures, and complexity of the data all contributed to what he found to be an exasperating delay. When the opportunity to devote years of full-time attention to his treatise on evolutionary biology finally arrived, Wright was seventy years old, but he was healthy and vigorous, and he was eager to begin.

Magnum Opus

The story of the genesis of Wright's treatise on evolutionary biology should be enough to raise the hopes of any struggling scholar under the age of seventy. Wright's primary contact at the University of Chicago Press during his entire tenure at Chicago was Rollin D. Hemens, who started there as a young editor the year before Wright arrived. Probably informed by Frank Lillie that

Wright was planning to write a book on genetics and evolution, Hemens prepared a file on him six months before his arrival, as he told Wright in a letter dated February 12, 1957:

> I have just been going through the file related to a possible book on genetics by you. The first item is a clipping from the May 29, 1925 issue of *Science*. It announces your appointment as Associate Professor of Zoology at Chicago dating from January 1, 1926. In the margin, I pencilled a note to Mr. Bean [director of the press]: "This man has been mentioned as one who has a great deal of good material which should be published."

To this introduction Hemens added: "It isn't often that a youngster in publishing, for that is what I was then, can be so right. I hope that the past year you have had time to work on the manuscript and that it will be completed in the near future." What happened in the years 1926–57 can easily be guessed from this letter—Hemens constantly asked Wright for his manuscript on genetics and evolution, and sometimes Wright appeared almost ready to produce it.

The closest that Wright came to producing a manuscript during the Chicago years was shortly after he gave the Hitchcock Lectures at Berkeley in May 1943. Hemens alerted John T. McNeill, acting editor of the press that year. When McNeill inquired about the manuscript, Wright gave an encouraging reply:

> I have been able to put most of my time for the last month on the manuscript of my lectures "Gene and Organism" and have two of the six chapters completed more or less to my satisfaction. The others are in rough draft only. I should be able to finish it some time this fall provided that it does not become necessary to devote most of my time outside of classes to other matters. (Wright to McNeill, August 31, 1943)

A similar inquiry from McNeill in November, however, got this reply: "I have been making progress on my manuscript but very much more slowly than during the summer quarter. I regret that it is obvious now that I will not be able to finish it this fall" (Wright to McNeill, November 12, 1943). In response to further inquiries from both McNeill and Hemens, Wright reported on January 4, 1944, "I have become involved in statistical work in one of the war research projects to such an extent that I have not been able to make much progress on 'Gene and Organism' recently. I see little prospect of any change in this situation in the near future. I will, however, continue to work on it as time permits." And by November 1945 Wright was holding out little hope: "I regret to say that I have not been able to get back to work on 'Gene and Organism' and see little prospect for some time."

Wright's only other opportunity to finish the manuscript before retirement was during his Fulbright year in Edinburgh (1949–50). After Wright's return without the manuscript and for the following nine years, Hemens wrote frequently to encourage Wright and to ask for the manuscript, all to no avail. Finally in August 1959 Hemens gave up book acquisitions to take charge of a new journal the press was publishing. He wrote to Wright:

> My major regret in making the change is that I will not be working with you and your manuscript when it comes in. You see, after many years, I am still hopeful that you will finish the book and I would have been proud to have been associated with its publication. However, it will be in good hands and I will cheer from the sidelines. (Hemens to Wright, August 5, 1959)

Only a few months later, Wright would begin the intense work on his treatise that would occupy most of his time for the next seventeen years. Neither he nor Hemens had the slightest idea that the treatise would become a set of four hefty volumes. The first was published in 1968, the second in 1969, the third in 1977, and the fourth in 1978, when Wright was eighty-eight years old.

Before turning attention directly to the treatise, some other developments in Wright's thinking about evolutionary theory must be examined. Between 1955 and 1978 much happened in evolutionary biology that would involve Wright's interaction with other evolutionists. The interactions of Wright with Kimura, Mayr, Epling, and Fisher during this period yield important insights into the state of evolutionary biology.

Wright, Kimura, and the Role of Random Drift in Evolution

Thus far in this biography I have utilized two major sets of correspondence. That between Wright and Fisher greatly aided the understanding of evolutionary biology in the 1930s, and that between Wright and Dobzhansky illuminated the 1940s. For the 1950s and 1960s, the correspondence and interaction between Wright and Kimura yields similar helpful insights. In particular, their interaction is important for understanding the most recent version of the long-standing debate, alive since Darwin's day, over whether the prevailing mechanisms of evolution are adaptive or nonadaptive, selectionist or neutralist.

Kimura has written a delightful biographical essay that recounts in some detail his early life and work (Kimura 1985), and because I have no other biographical sources about him, there is no need to simply repeat his account here except for the most essential details. Born in 1924, Kimura was educated primarily in botanical cytology and cytogenetics. At an early age he was fascinated by the application of mathematics to problems in nature. He became interested in genetics during his student days at Kyoto University (1944–47), which was at that time a center of research in genetics with such figures as H. Kihara and T. Komai. During this time he read Dobzhansky's *Genetics and the Origin of Species* and Waddington's *Introduction to Modern Genetics,* both pirated from their first editions. These books led him to Sewall Wright's work, which he found attractive but difficult to read and understand in detail at this time.

Upon graduation Kimura took for two years a position as assistant in the Laboratory of Genetics, Department of Agriculture at Kyoto University, under Kihara's direction. There he continued work in botanical cytogenetics, but he also had considerable time for exploring further his growing interest in mathematical population genetics. He read many of the papers of Wright, Fisher, and Haldane. He was also learning more mathematics on his own, making these papers on population genetics much easier to read and understand. Most of all, he became interested in the concept of random genetic drift.

In 1949, through Kihara's recommendation, Kimura became a research member at the newly established National Institute of Genetics in Mishima, Japan. His efforts, although at first directed to chromosome analysis in plants, became more and more focused upon population genetics. Reading as widely as possible in Japan at this time (he even went to the main library of the University of Tokoyo and copied out papers by hand), Kimura wrote and published in Japanese a comprehensive review of the field of mathematical population genetics (Kimura 1950).

One controversy in particular interested Kimura—the disagreement over evolution in the moth *Panaxia dominula*. Fascinated by the battle of the giants, Wright and Fisher, Kimura decided to join the fray with a quantitative analysis of the effects of random fluctuation in selection pressure upon the distribution of gene frequencies. Wright had carefully pointed to this factor in his 1948 rebuttal to the original report of Fisher and Ford (1947), but he had not developed the idea very far mathematically. Kimura's goal was to push forward this mathematical treatment. To this end, he wrote a preliminary report in the first volume of the newly established *Annual Report of the National Institute of Genetics* (Kimura 1951). The paper began:

> Recently, in the field of population genetics, there is a sharp antagonism between Wright on one side and Fisher and Ford on the other. Wright maintains the evolutionary significance of the "Wright effect," while Fisher and Ford reject it, and emphasize the extreme importance of the role played by natural selection. The present study is to examine this antagonism by means of precise mathematical analyses. (P. 45)

This introduction is revealing because it shows no hint that the controversy was anything other than random drift versus natural selection, precisely the dichotomy emphasized by Fisher and Ford and vigorously denied by Wright.

From his initial analysis in this brief note, Kimura concluded that random fluctuations in selection pressures were relatively unimportant for small populations, where sampling drift was important, and instead produced significant effects in large populations under certain conditions. The next year in the same journal, Kimura published an extension of the analysis, this time using the Markov process for the first time. He also published three other brief papers on population genetics in the same *Report,* including one on effective size of populations and another on the decay of variability due to random extinction of alleles in populations that had not reached equilibrium, a case not analyzed in detail earlier by Wright or Fisher (Kimura 1952a, 1952b, 1952c and 1952d). For someone in Japan who was largely self-taught in mathematics and wholly self-taught in quantitative population genetics, these four papers were nothing less than remarkable. The problem was that Kimura had no way to study professionally what was becoming his favorite field of research. His fondest dream was to study with Wright.

Some members of the Atomic Bomb Casualty Commission, sent to Hiroshima to study the biological effects of radiation, heard about this outstanding, young self-trained population geneticist and decided to help his dream come true. Those who helped most directly were Duncan J. McDonald, Harold H. Plough, and Newton Morton. Wright first heard of Kimura when he received from McDonald a set of Kimura's reprints, his curriculum vitae, a list of publications to date (twenty-eight of them by late 1952), and a letter that said in part:

> In our work here in Japan, we meet most of the Japanese geneticists, either at meetings of the Genetics Society of Japan, or during visits to the various Universities. Amongst the younger geneticists, one, Moto KIMURA, stands out, would be exceptional anywhere. The enclosed series of reprints indicate the nature and calibre of his work, and the list of publications the volume of his work, rather phenomenal for a man only 28 years old. I am not qualified to follow the details of his mathematical developments; but you will be able to do that from the reprints.
>
> More remarkable still is that his mathematics is largely self-taught. He graduated from Kyoto University in botany, wrote his degree thesis on the nucleolus in plant cells. Although he took the full course in botany, he attended courses in the mathematics department, and continued his studies alone. Dr. Komai, under whose direction he works at the new National Institute of Genetics at Mishima writes: "He is quite unusual in mathematical ability. I can also assure you that his knowledge of cytology and genetics fairly compares with that of average professors of cytology or genetics in this country."
>
> But now his development in mathematical genetics is limited by the fact that there is no-one in Japan to teach him. Obviously he should get either to the States or to England. (McDonald to Wright, December 3, 1952)

McDonald suggested that Wright take Kimura on as a graduate student. Wright replied:

> I am very much impressed with Mr. Kimura's capacity for applying mathematics to problems of population genetics. His work is sound in all respects that I have had time to check.
>
> We do not seem to have any scholarship that would meet his needs at the University of Chicago. I also doubt whether I would have much to teach him as far as the pure mathematics is concerned. I am primarily a biologist and have never assigned a thesis problem in purely mathematical genetics. For preparation for an attack on the problems in this field at a higher mathematical level than at present he should go to a department of mathematics. I should also add that I have ceased to take on new graduate students in any field of genetics as I am to retire in less than two years. I am scheduled to give my course in evolution (largely population genetics now) next spring for the last time. (Wright to McDonald, January 20, 1953)

Wright knew that McDonald had written a similar letter to Jay Lush at Iowa State, and he suggested that Kimura study there where he could work in population genetics with Lush and in statistics with Oscar Kempthorne. A scholarship was arranged at Iowa State, and Kimura arrived there in mid-August 1953.

During the voyage from Japan, Kimura wrote up his latest results on the problem of random drift caused by random fluctuations in selection intensities. This time he derived an exact solution for the process under reasonable assumptions, and he also carefully compared the different effects upon the distribution of gene frequencies of random drift due to sampling and that due to fluctuations of selection intensities. He of course wanted to have the paper published, which was easier than he expected. Only three weeks after his arrival at Ames, Kimura attended the Genetics Society of America Annual Meeting in Madison on September 6–10. One reason he wanted to go was to meet Crow, with whom he had already corresponded and exchanged publications and who had expressed a strong interest in his work. Crow met Kimura early at the meeting and invited him to stay at his house. Wright was in Madison at the same time, and Crow invited him over for an evening so that he and Kimura could meet. Kimura wrote Crow after the meeting:

> I shall never forget the beautiful campus close by the lake, the cheerful atmosphere of your home, the refreshing drive way leading to the University, etc. . . . . I had been dreaming for a long time to see and to talk with Professor Wright—once he had been a half-god for me. Therefore it was a great pleasure that I could realize my dream through your arrangement. (Kimura to Crow, September 13, 1953)

Kimura had brought the draft of his paper with him to Madison, where he gave it to Crow who was much impressed and thought it should be published in *Genetics*. (Crow was at this time assistant editor of *Genetics*.) He naturally sent the paper to Wright as referee with the following comment:

> We had hoped not to burden you with so many manuscripts, but in this case you are uniquely able to read it critically. Please don't bother with correcting the English; we can do that. Also we can find someone to check the mathematical details (one of the members of our math department is interested in stochastic processes). But we would like from you a general opinion of the manuscript. (Crow to Wright, September 19, 1953)

True to form, Wright rederived most formulas in the paper. The draft copy of his reply to Crow has twenty densely packed pages of calculations. Wright said:

> I have been over Kimura's paper. It makes a big advance in the mathematical treatment of random processes in genetics. It is a brilliant achievement to reach a complete solution in the case of fluctuations in selection . . . by transformations which normalize the distribution and thus reduce it to the simplest sort of random walk problem. His solution in the case of the inbreeding effect [sampling drift] . . . is also a big advance even though in this case the result is a series applicable only to large t [where t represents the t-th generation]. (Wright to Crow, October 16, 1953)

I do not recall comments this effusive in any of the large number of letters Wright wrote evaluating manuscripts on population genetics. And Kimura wrote this one before he had any formal training in mathematical population genetics.

In our interviews, Wright recalled almost nothing of his first meeting with Kimura who, however, was thrilled to meet Wright. At first Kimura had difficulty speaking and understanding English, especially Wright's quiet, rapid conversation. After receiving Wright's comments on his manuscript Kimura wrote to thank Wright for the comments and also to say:

> Last summer I was very glad to see you in Madison. I had long dreamt to see you and to talk with you. I was sorry, however, that my English, especially my hearing of English, was very poor. Still, it was one of the most important and happy time I ever had in this country.
>
> I would like to see and to talk with you again since I have many questions on population genetics. (Kimura to Wright, January 28, 1954)

In this letter Kimura also told Wright of his plans to return to Japan at the end of the summer. He did make the trip to Chicago and had a very rewarding visit with Wright.

Kimura was not highly pleased with the education in population genetics at Ames. He began in February 1954 to correspond with Crow about the possibility of coming to study with him at the University of Wisconsin. Kimura wrote to Crow:

> Dr. Lush and Prof. Kempthorne have been disputing each other about their works on correlation between relatives and so on. Each seems to be confident that his teacher (Wright or Fisher) is greater than the other. It is quite interesting to listen to their opinions, but I am keeping a neutral position. Frankly speaking, Prof. Kempthorne's opinion is far more easily understandable than that of Dr. Lush, though I am inclined to Dr. Wright's opinion rather than that of Dr. Fisher. . . . Neither Dr. Lush nor Prof. Kempthorne seem to pay any attention to my work, but I am certainly enjoying the life in Ames. New experiences are all valuable unless they are harmful. (Kimura to Crow, May 9,1954)

Kimura won a grant from the Japan Society and, with other money from the graduate school at the University of Wisconsin, was able to go and work with Crow instead of returning to Japan. Kimura wrote from Madison to happily tell Wright of this change of plans, but he seemed very excited by the most recent news: "Recently we received a new with great joy, that you will come this department. It might be one of the most significant matter in my academic life that I shall have you as a teacher" (Kimura to Wright, July 1, 1954). Not only did Kimura have Crow as his understanding and knowledgeable advisor, but now he would be able to take Wright's population genetics course and to talk to him at length on frequent occasions.

Free at last to work full time on population genetics and in an almost ideally supportive setting, Kimura quickly published in 1955 a series of three outstanding papers on random drift and selection (1955a, 1955b, and 1955c), the first of which Wright submitted for publication in *Proceedings of the National Academy of Sciences* and the last of which was delivered at the prestigious Cold Spring Harbor Symposium for 1955. Kimura was the recognized world expert on the mathematical analysis of random genetic drift.

But during the first year at Madison Kimura learned well something that he had not anticipated. He had always thought of random drift as an important process in evolution. He was surprised to learn that Wright was every bit as much a selectionist as Fisher, once above the level of small semi-isolated populations. By the time that Kimura came to Madison, Wright was adamantly arguing, I think largely in response to the criticisms of Fisher and Ford, that evolution was an adaptive process at the level of observable differences at the lowest taxonomic levels. Wright responded so negatively when others attributed to him the idea that random drift could cause observable differences that Kimura believed no other position was reasonable. Wright absolutely dismissed the importance of random drift in large random breeding populations. From the beginning, Kimura was skeptical that natural populations were very often subdivided in the way necessary for Wright's shifting balance theory to work efficiently. Thus Kimura was in the curious position of being the world's expert in a process that, in the minds of most evolutionists, was of rather little importance in the evolutionary process as a whole. He began to focus his attention upon other aspects of population genetics, including selection and genetic load, often collaborating with Crow. Kimura had no way of knowing that advances in molecular biology would make his painstaking work on random drift relate once again to evolution in natural populations, only this time in a new way.

Crow introduced Kimura to Muller in 1955. When Kimura returned to Japan in 1956, he and Muller corresponded and exchanged reprints. Muller became very interested in the earliest results of molecular genetics, and at the end of his life kept insisting that these results must be incorporated into all aspects of biology, especially evolutionary theory. Kimura took Muller's suggestion seriously. Two publications in particular set him to thinking again about the importance of random drift in natural populations. These were *Evolving Genes and Proteins* edited by Bryson and Vogel (1965) and the influential papers by Lewontin and Hubby on allozyme variability in *Drosophila* (Hubby and Lewontin 1966; Lewontin and Hubby 1966). The Bryson and Vogel volume had four papers calculating the rate of amino acid substitution in α and β chains of hemoglobin in the evolutionary history of animals, led by the one by Zuckerkandl and Pauling. The Lewontin and Hubby work revealed substantial levels of genetic variability (about 15% heterozygosis).

Kimura's initial calculations indicated that an apparently low rate of only one amino acid substitution every 28 million years, when extrapolated over the whole genome, gave a phenomenally high rate of nucleotide pair substitutions of something like one every two years in a lineage. This was particularly surprising to Kimura who was very familiar with Haldane's concept of the "cost of selection" (Haldane 1957). The basic idea was that selection for any character meant elimination of individuals, and Haldane quantified the cost in such a way that he estimated that evolution in a lineage would lead to extinction if the average rate of substitution of alleles were greater than one every 300 generations. Haldane of course assumed that selection was the dominant force behind the substitutions.

Faced with this apparent dilemma, Kimura deduced that if the substitutions were taking place at anything like the rate indicated by the new molecular and electrophoretic studies, then the vast majority of the substitutions had to be selectively neutral or almost so. This process of substitution would require a high mutation rate, a requirement that he thought was in harmony with the high frequency of heterozygous loci observed by Lewontin and

Hubby. By November 10, 1967, Kimura had written the draft of a short paper announcing his neutral theory of molecular evolution, which he sent to Wright for comments before sending it on to *Nature*.

Kimura was acutely aware that he was bucking the tide of accepted opinion in the realm of evolutionary biology. Indeed, his thesis was hardly less than heresy in a scientific community in which heresy did not exist by definition. (Scientists, so they continually tell us, do not hold to their views like religious fundamentalists but are always prepared to change them in response to new evidence.) The orthodoxy that Kimura's neutral theory was directed against was clearly expressed by Mayr in his highly influential *Animal Species and Evolution* (1963):

> The order of magnitude of selective difference between alleles co-existing in a population is difficult to determine. It is quite evident that a difference in selective value between two alleles amounting to, let us say, 5 percent or less will be difficult to demonstrate in view of nongenetic variation and experimental errors. However, attempts made in recent years to determine differences in the selective values of alleles found in wild populations often yielded values as high as 10 percent, 30 percent, or even 70 percent. Entirely neutral genes are improbable for physiological reasons. Every gene elaborates a "gene product," a chemical that enters the developmental stream. It seems unrealistic to me to assume that the nature of the particular chemical (enzyme or other product) should be without any effect whatsoever on the fitness of the ultimate phenotype. A gene may be selectively neutral when placed on a particular genetic background in a particular temporary physical and biotic environment. However, genetic background as well as environment change continually in natural populations and I consider it therefore exceedingly unlikely that any gene will remain selectively neutral for any length of time. (Mayr 1963, 207)

Wright, Dobzhansky, Fisher, Ford, Huxley, Stebbins, and almost every other notable evolutionary biologist (including those of the generation younger than those mentioned) would have agreed wholeheartedly with Mayr's position on neutral alleles. Mayr and the others were not, of course, thinking in terms of molecular genetics but in terms of genes that affected measurable elements of phenotype or viability.

Kimura sent the draft of his paper on the neutral theory to Wright with a certain amount of trepidation. Although Kimura traced his work on random drift directly to Wright, he knew that Wright strongly disavowed the role of sampling drift in populations of any appreciable size, and did not see random drift as an alternative to selection but as a mechanism that generated variability upon which selection acted. Kimura's conclusion in the paper was that Wright's concept of random drift was indeed a real alternative to selection even in large populations:

Finally, if the main conclusion of the present paper turns out to be correct, and, if the neutral or nearly neutral mutation is being produced each generation at much higher rate than has been considered before, we will be led to recognize the great importance of random genetic drift due to finite population number (Wright 1931) in forming the genetic architecture of biological populations. The significance of random genetic drift has been deprecated during the last decade by the opinion that almost no mutations are neutral and also the number of individuals forming a species is usually so large that random sampling of gametes should be negligible in determining the course of evolution, except possibly through "founder principle" (cf. Mayr 1965). To emphasize the founder principle but deny the importance of random genetic drift is, in my opinion, somewhat similar to assuming a great flood to explain the formation of deep valleys but rejecting gradual but long lasting process of erosion by water as insufficient.

I think Kimura had good reason to be apprehensive about Wright's reaction to this draft. His letter of transmittal showed his concern:

Enclosed is a Xerox copy of my note that I have just written up and is entitled "The evolutionary rate at the molecular level, etc."

It is a short note and my main thesis is that neutral or nearly neutral mutations are occurring at a very high rate (roughly one mutation per gamete every two years) and that random genetic drift due to finite population number is extremely important in forming the genetic structure of biological populations.

I am very serious on my present note.

Would you be so kind to go over the note and let me have your opinion on the content of my note?

Unless you think that my thesis is wrong or meaningless, I would like to publish it fairly soon.

I am thinking of publishing my note in *Nature* if possible.

I would appreciate any comment from you on my note.

Wright's response to this first statement of the neutral theory came in a detailed letter fully deserving inclusion here:

I was much interested in the Xerox copy of your manuscript and have been trying to collect my ideas on it.

I agree with your "main thesis that neutral or nearly neutral mutations are occurring at a very high rate and that random drift, due to finite population numbers, is extremely important in forming the genetic structure of biologic populations."

This is on the understanding that "mutation" refers to nucleotide replacement. I have often taken "allele" in a physiological and statistical sense. "There may be dozens of distinct albino factors, alike only in that some change in the factor C in each case prevents its

normal activity. Such allelomorphs could not be separated by breeding tests" [Wright 43, p. 238]. "The selection coefficient, s, relating
to a gene A cannot be expected to be constant if the alternative term,
a, includes more than one gene" [Wright 64, p. 106]. In general I
have treated recurrence rates, u and v, of mutations as statistical, referring to physiological classes rather than to single structural alleles.
On the other hand, I was thinking of alleles in an absolute sense in
my recent paper on "polyallelic random drift" [Wright 181] and in
my 1949 *Encyclopedia Britannica* reference to the same subject
[Wright 127].

In a paper [Wright 56] in which I questioned the evolutionary
hypothesis proposed by Fisher to account for the prevailing dominance of "type" alleles over recurrent *deleterious* mutations, I did so
on that basis of a principle, that I later [Wright 151, p. 7] called that
of the pleiotropic threshold, which may seem to be opposed to the
view that neutral or nearly neutral mutations are occurring at a very
high rate. I was here, however, concerned with mutations with dis-
tinguishable physiological effects, not that which are indistinguishable isoalleles included in the "type" allele as a statistical category.
Fisher and I agreed that the selection intensity of his specific
modifiers of the heterozygote was at best of the order of the mutation
rate of the mutation which we took to be about 10^{-6}. My point was
that a mutation with a conspicuous effect (modifying a heterozygote)
would almost certainly have pleiotropic effects, favorable or unfavorable, that would take precedence over the minute selection in-
sity, due to modification of the rare heterozygote. I have observed
many modifiers of dominance in the guinea pig (two before 1929)
but all have had important effects in the homozygotes also and so
were not of the kind required by Fisher's theory. . . .

The point here is that the rarity of extremely low net selection intensities implied by the principle of the pleiotropic threshold applies
only to mutations that have some conspicuous physiological effect,
and has nothing to do with nucleotide replacements or amino acid replacements that have no physiological consequences.

Your figure for the rate of neutral mutation in mammals, two per
gamete, 5×10^{-10} per nucleotide site, seems reasonable enough, assuming that the loci whose protein products are known, are represen-
tative of all loci. This seems reasonable at the nucleotide level. A
rate of 5×10^{-7} per locus (assuming 103 nucleotides per locus) is
less than the usually stated 10^{-5} to 10^{-6} for conspicuous mutations. It
would imply that something like 10% of all nucleotide mutations are
neutral.

With regard to the evolutionary importance of sampling drift,
there is the dilemma that the condition that gives the maximum
amount of such drift is that of complete neutrality and hence of no
evolutionary significance (at least if the neutrality is due to absence
of physiologic effect). The presence of any physiologic (and presumably selective) effect reduces the chance of fixation of a single
new deleterious mutation. . . .

The amount of sampling drift in a reasonably large panmictic species is so slight for mutations of appreciable evolutionary significance that I have always avoided attributing any importance to it in this case. The only role that I have attributed to sampling drift (apart from leading to extinction under sufficiently close inbreeding) is in species that include many demes of small effective size, and here only because such drift, occurring simultaneously at all loci in all demes, tends to lead to local changes in the controlling selective peaks, followed by mass selection toward the system of gene frequencies at the new peak and differential population growth and emigration from the better adapted demes. Thus I have always treated random drift (of all sorts) as a component of a 3-phase system—local random drift, intrademe selection, and interdeme selection. It is possible that I have been too conservative, but this secondary role still seems to me the most significant.

I have never liked Haldane's term "cost of evolution" since it can be taken so easily to imply that the species suffers more than normal losses in mortality or fecundity because of evolution which is obviously the opposite of the actual situation. The cost is to other selective processes, the principle being merely the one painfully familiar to all livestock breeders that there is a limit to the total amount of selection that is compatible with maintaining numbers. The reproductive surplus that is used in replacing one locus by selection is not available for replacements at other loci.

Haldane based his calculations on selective replacement in a panmictic population. It would also undoubtedly apply in some sense to interdeme selection, although differently because of the correlated selection of genes of the same interaction system. There is, on the other hand, no limitation on changes due to sampling drift by itself. It seems to me that the limitation of progress in my 3-phase process, involving considerable change by random drift as well as by intra- and interdeme selection would be much less than under pure mass selection, but I have not developed a mathematical theory. It may well be compatible with the rates of substitution that you estimate, especially as in the 3-phase scheme, a great many substitutions of controlling interaction systems (selective peaks) may occur while frequencies of the many molecular genes at each locus are shifting within rather small ranges instead of from 10^{-4} to near fixation as in Haldane's theory. (Wright to Kimura, November, 30, 1967)

In other words, Kimura was perhaps right that the neutral mutations were occurring at the postulated rate, but if they were, they were evolutionarily insignificant. The most essential thing to understand in Wright's letter is his long-standing emphasis upon population structure as a fundamental factor in the evolutionary process. No theory of evolution would be of serious interest to Wright unless it took into account population structure. Yet Kimura's neutral theory had nothing whatever to say about population structure. Even the most accurate measurements of the pace of neutral molecular evolution had no implications to offer in determining population structure. Thus I think that Wright could not be enthusiastic about it. In his reply, Kimura did not refer to Wright's comments about evolutionary insignificance, but simply said: "I am glad to note from your letter that you agree with my main thesis that neutral or nearly neutral mutations are occurring at a very high rate and that random drift due to finite population numbers is extremely important in forming the genetic structure of biologic populations" (Kimura to Wright, December 15, 1967). I am sure, however, that Kimura did not miss Wright's real criticism.

The published version of Kimura's paper in *Nature* stirred up the expected intense controversy. Many who had admired Kimura's earlier work were dismayed, Dobzhansky being a good example. But soon King and Jukes came roaring to Kimura's defense (1969), and the selectionist-neutralist debate went into full swing. The essentials of this debate are too well known for inclusion here. In this biography, the pertinent issue is the interaction between Wright and the selectionist-neutralist debate.

If Wright were to accept the importance of Kimura's neutralist theory, he would have to repudiate what he had been saying about random drift for decades. But he did think seriously about the issue, thinking up scenarios in which he could accept at least some aspects of the theory. Kimura and his colleague Tomoko Ohta constantly reminded Wright that he, after all, was the one who had invented and mathematically developed the concept of random genetic drift and that he should share the credit for this new application of it.

One example of Wright's attempts to give credit to the neutralist theory can be found in a letter of which only the undated rough handwritten copy survives in Wright's files, but it was written in response to "Protein Polymorphism as a Phase of Molecular Evolution" by Kimura and Ohta, appearing in *Nature* in the issue of February 12, 1971. Wright posed this question:

> Changes in wholly nonfunctional parts of the molecule would be the most frequent ones but would be unimportant, unless they occasionally give a basis for later changes which improve function in the species in question which would then become established by selection. Is there evidence that a succession of neutral changes has in some cases given such a basis?

Many years later, Kimura did indeed begin to use the idea that environmental change might well create selective differences between alleles that had been functionally equivalent for long periods (see Kimura 1983, 325–26), but I have no way to judge whether Wright's suggestion had a significant influence on this issue.

Kimura and Ohta never made much headway convincing Wright that he should welcome the neutral theory as his child, even when Ohta argued that the neutral theory worked best if the mutations were really very slightly dele-

Figure 13.3. *Left to right:* Maruyama, Louise Wright, Sewall Wright, Kimura, Ohta, and unidentified woman at the Mishima Experimental Station at the time of the International Congress of Genetics in Japan in 1968.

terious, almost but not completely neutral. Early in 1972 Ohta sent Wright a draft of a paper entitled "A Model of Evolution by the Substitution of Nearly Neutral Mutations," with a cover letter saying:

> Since the basic idea comes from your original hypothesis on the importance of small population size for the evolution of the species, your comments will be most highly appreciated. I regard the selection coefficient of mutant genes with very small effects as random variables and have tried to relate their advantage or disadvantage with the environmental diversity. I think the model can explain many observed facts which otherwise appear contradictory. (Ohta to Wright, February 2, 1972)

In his reply, Wright disassociated himself from the (almost) neutralist theory and argued again that the truly important role of random drift was to create novel interaction systems upon which selection could act. Ohta's response shows a hint of the frustration that must have come from her failed attempt to change Wright's mind:

> Thank you very much for your comments on my manuscript. . . . I also understand the essential difference of our viewpoints.

> Recent comparative studies on amino acid sequences have shown that many amino acid substitutions are constantly occurring. It seems to me that the gene substitution plays a major role for differentiation of populations. The spreads of favorable interaction systems appears not to be sufficient to account for substantial differentiation. (Ohta to Wright, March 18, 1972)

Wright's final position on the neutralist theory was that first, the theory was not about phenotypic evolution at all and might be irrelevant to it, and second, the data were as consistent with his shifting balance theory as they were with the neutralist theory.

> The probable near-neutrality of substitutions among the leading alleles at thousands of loci provides a favorable condition for the shifting balance process on the basis of any favorable interaction effects that happen to occur among the millions of locus pairs. Moreover there is much less cost per substitution than in the case of genic mass selection. (Wright 200, p. 482)

Thus Wright's position on the neutralist theory did not change substantially from his first reaction to it in 1967.

The interaction between Wright and Kimura is interesting at many levels. Very intriguing is the extent to which Kimura and Wright could agree and commune on the mathematical treatment of random drift but disagree strongly on the integration of random drift into a comprehensive view of the evolutionary process. One is reminded again of the complete agreement in quantitative analysis that Wright and Fisher reached in 1930, while at the same time their evolutionary theories were deeply divergent.

I think also much can be learned about the state of evolutionary biology in the late 1960s and early 1970s by examining Wright's interaction with Kimura and Ohta on the neutralist theory. A particularly revealing passage is found in a paper by Kimura and Ohta that Jack Lester King read for them at a symposium in Berkeley on April 11, 1971:

> In the field of evolutionary genetics, consensus appears to have been reached among leading evolutionists of the world (except possibly Sewall Wright) that natural selection is omnipotent and is the most prevailing factor for evolutionary change. This orthodox view (formed under the dominating influence of R. A. Fisher) also asserts that neutral mutant genes are very rare if they ever exist, and random genetic drift is negligible in determining the genetic structure of biological populations, except possibly for the case of the colonization of a new habitat by a small number of individuals—the founder effect (see Mayr 1963). (Kimura and Ohta 1971b, 63)

This general theme would of course play a prominent role in Kimura's *The*

Neutral Theory of Molecular Evolution (1983; see especially chapter 2). But what is most interesting and perhaps surprising is that Kimura and Ohta list Wright as only "possibly" not in the camp of those who saw natural selection as omnipotent. From their point of view, Wright had relegated random genetic drift to such a limited role in evolution in nature that they thought he was practically in the camp of the ultra-selectionists. When I read this passage, and then found the letter that Wright wrote to Kimura and Ohta after they sent him a copy of the manuscript, the response was delightfully predictable:

> In your reference to my position on the omnipotence of natural selection, I think that you can leave out "possibly." What I (and everyone else until recently) were considering was of course phenotypic evolution. I certainly have never subscribed to the omnipotence of natural selection in the sense of Fisher and his followers, according to which phenotypic evolution is determined by selective differences of alleles in the species as a whole. I have held that random processes (sampling drift, unique events, fluctuations in the coefficients of selection and immigration and even systematic local differences in selection) bring about differences among local populations which are random with respect to the general course of selection and that the genetic composition of the species continually changes because of differences in the composition of population sources and sinks. In a sense this postulates omnipotence of natural selection with respect to phenotypic changes in large species, but it is natural selection which depends on random processes in subpopulations. (Wright to Kimura, June 2, 1971)

Wright was nowhere near letting himself be described as possibly a total selectionist.

The agreement of Wright and Kimura that all the other prominent evolutionists were extreme selectionists is also worth noting. In 1971 this assessment was not far off base. A decade later, leading evolutionary biologists like Gould and Lewontin were roundly criticizing the extreme selectionist interpretation of the evolutionary process. Their reasons for doing so were, however, unrelated to Kimura's neutral theory, but this is not the place to dig into that story (Gould and Lewontin 1979, but see the reply of Mayr 1983).

Mayr, Wright, and Beanbag Genetics

Ernst Mayr and Sewall Wright were an unlikely pair for a serious disagreement. We have seen in the previous chapter that Mayr came strongly to Wright's defense in his dispute with Fisher and Ford over *Panaxia dominula* in the late 1940s. Their amicable relations go back further than that, with Mayr favorably citing Wright in *Systematics and the Origin of Species*. Indeed, in an important paper entitled "Speciation Phenomena in Birds" that

Mayr published in *American Naturalist* in 1940, he argued that of all the definitions of species he had seen in recent years, the one published by Sewall Wright in 1940 showed "the fewest flaws" (Wright 101, p. 162; Mayr 1940, 255). In the 1940s Wright considered Mayr to be the leading systematist in the world on the basis of his 1942 book, and Mayr in turn considered Wright to be the leading population geneticist. In Mayr's important 1954 paper where he introduced the concept of genetic revolutions ("Change of Genetic Environment and Evolution"), he strongly emphasized that "The selective value or viability of a gene is thus not an intrinsic property but is the sum-total of the viabilities on all the genetic backgrounds that occur in a population" (Mayr 1954, 165). As support for this statement, Mayr continued:

> The concept that the viability of a given allele depends on its genetic background is not new. It has been emphasized by several students of this problem. Sewall Wright [64, p. 155], for instance, stated: "The selection coefficient of a particular gene is really a function not only of the relative frequencies and momentary selection coefficients of its different allelomorphs, but also of the entire system of frequencies and selection coefficients of non-allelomorphs."

There was not the slightest hint of rancor in the correspondence between Mayr and Wright from the early 1940s through the mid–1950s. So what caused them to disagree?

The answer begins with the end of the early phase of the *Panaxia dominula* debate. When challenged by Fisher to do so, Mayr finally read Wright's early papers carefully. He found the quotes about nonadaptive evolution that I cited in chapter 9. It appeared to Mayr that Wright really had changed his mind about the role of random drift in the evolutionary process, and he concluded that Wright was a poor historian of evolutionary theory in the early 1930s, particularly concerning Wright's own role. Looking ahead toward the great Darwin centennial of 1959, and keenly aware that evolutionary theory in the 1930s and 1940s had undergone an important synthesis, Mayr began turning some of his attention to the history of evolutionary biology.

By the early 1950s most evolutionary biologists had come to the view that the recent synthesis in evolutionary biology had emanated directly from the theoretical work of Fisher, Haldane, and Wright. Thus Philip Sheppard argued in his contribution to the important 1954 volume *Evolution as a Process* (edited by Huxley, Hardy, and Ford) that

> The great advances in understanding the process of evolution, made during the last thirty years, have been a direct result of the mathematical approach to the problem adopted by R. A. Fisher, J. B. S. Haldane, Sewall Wright, and others. . . . The hypotheses derived by mathematicians have given a great impetus to experimental work on the genetics of populations. (Sheppard 1954, 201–02)

A year later Dobzhansky published a similar assessment in his introductory address to the 1955 Cold Spring Harbor Symposium on population genetics:

> The foundations of population genetics were laid chiefly by mathematical deduction from basic premises contained in the works of Mendel and Morgan and their followers. Haldane, Wright, and Fisher are the pioneers of population genetics whose main research equipment was paper and ink rather than microscopes, experimental fields, *Drosophila* bottles, or mouse cages. This is theoretical biology at its best, and it has provided a guiding light for rigorous titative experiment and observation. (Dobzhansky 1955, 14)

The view offered by Sheppard and Dobzhansky dominated the 1950s literature treating the relation of genetics to evolution. Almost every textbook or paper on the subject cited Fisher, Haldane, and Wright as the cofounders of modern evolutionary theory.

Mayr found this interpretation wrong and distasteful. He had always seen speciation as the primary problem in the evolutionary process and the work of systematists as central to the evolutionary synthesis. The population geneticists, on the other hand, focused upon evolution within populations as their primary, almost exclusive, province. From Mayr's perspective, there was a real question of whether theoretical population geneticists had contributed anything of great significance to the evolutionary synthesis, much less deserved credit for being the primary thinkers.

Mayr was not, however, the first to openly challenge the interpretation that placed Fisher, Haldane, and Wright in the very heart of the evolutionary synthesis. At the Symposium of the Society of Experimental Biology at Oxford in 1952, C. H. Waddington suggested that the "undoubtedly immense importance and prestige" of the mathematical theory of evolution might not deserve such high distinction. The mathematical theories of Fisher, Haldane and Wright

> did not achieve either of the two results which one normally expects from a mathematical theory. It has not, in the first place, led to any noteworthy quantitative statements about evolution. The formulae involve parameters of selective advantage, effective population size, migration and mutation rates, etc., most of which are still too inaccurately known to enable quantitative predictions to be made or verified. But even when this is not possible, a mathematical treatment may reveal new types of relation and of process, and thus provide a more flexible theory, capable of explaining phenomena which were previously obscure. It is doubtful how far the mathematical theory of evolution can be said to have done this. Very few qualitatively new ideas have emerged from it. (Waddington 1953, 186)

Waddington republished these remarks in his widely read *The Strategy of the Genes*, (1957), where Mayr saw and agreed with them.

Already in his address at the 1955 Cold Spring Harbor Symposium, where he was given the task of synthesizing the results of the session entitled "Integration of Genotypes," Mayr in his first paragraph drew a sharp contrast between the contributions of the theoretical population geneticists and those of the field naturalists. He argued that the population geneticists dealt in their mathematical models only with single genes with constant and additive fitnesses, and that the field naturalists had confined themselves to purely evolutionary questions involving selective value or adaptedness. Neither of these approaches, according to Mayr, was a successful way to explain the evolutionary process. A full understanding of evolutionary phenomena was impossible "unless they are interpreted in terms of gene action, that is in terms of physiological genetics" (Mayr 1955, 327). Moreover, he argued, gene interaction was ubiquitous in nature; consequently, what was important in evolution at the genetic level was the evolution of interactive gene complexes rather than the evolution of single genes. Mayr indicated that this approach was new and that it came not at all from the population geneticists but from physiological geneticists and naturalists. What was needed was for population geneticists to work this more sophisticated view of the evolutionary process into their mathematical models. Mayr concluded: "The study of the integration of genotypes has shown that population genetics can no longer operate with the simplified concepts it started out with" (p. 333). Mayr gave no hint that Wright had been the strongest advocate in the world of this "new" approach to evolutionary biology for more than twenty-five years; indeed, Mayr did not even mention Wright in this address. Except for a brief comment by Haldane in the foreword to the 1953 volume in which Waddington's paper had appeared, Fisher, Haldane, and Wright did not reply to Waddington's challenge or to Mayr's characterization of their work on population genetics.

Invited to deliver the introductory address of the 1959 Cold Spring Harbor Symposium entitled "Genetics and Twentieth Century Darwinism" (who could ask for a more prestigious forum?), Mayr decided to renew Waddington's challenge and place it into a historical framework. This was the celebrated "Where Are We?" speech. Mayr divided the history of genetics into three periods: the Mendelian period (1900 to 1920), dominated by the mutation theory of evolution; the period of classical population genetics (1920 to late 1930s), characterized by a resurgence of belief in gradual Darwinian natural selection but dominated by the simplistic view that "evolutionary change was essentially . . . an input or output of genes, as the adding of certain beans to a beanbag and the withdrawing of others" ("beanbag genetics"); and the period (late 1930s to 1959) of "newer population genetics . . . characterized by an increasing emphasis on the interaction of genes." In this period, Mayr contrasted the rich contributions of naturalist-population geneticists like Dobzhansky, Wallace, and Lerner with the arid calculations of the mathematical population geneticists.

The work of Fisher, Haldane, and Wright, Mayr asserted, belonged to and dominated the period of "beanbag genetics." Citing Waddington, Mayr renewed his challenge specifically by asking, "What, precisely, has been the contribution of this mathematical school, if I may be permitted to ask such a provocative question?" He answered his own question:

> It seems to me that the main importance of the mathematical theory was that it gave mathematical rigor to qualitative statements long previously made. It was important to realize and to demonstrate mathematically how slight a selective advantage could lead to the spread of a gene in a population. Perhaps the main service of the mathematical theory was that in a subtle way it changed the mode of thinking about genetic factors and genetic events in evolution without necessarily making any startlingly novel contributions. (Mayr 1959, 2)

This assessment clearly differed from that of Dobzhansky and Sheppard. Added Mayr, "I should perhaps leave it to Fisher, Wright, and Haldane to point out themselves what they consider their major contributions."

Mayr's challenge must be understood both in terms of his own personality and his perception of his role in the community of evolutionary biologists as well as in terms of the substantive issues involved. Mayr has always thought that the quickest and most stimulating way to proceed in science or the writing of history is to make clear and unmistakable assertions based on the available evidence. The dialectical process that ensues then can modify and improve the original assertions. Mayr stated his approach in the preface to *Animal Species and Evolution* (1963): "Where the issue is controversial I have not hesitated to choose the interpretation that seems most consistent with the picture of the evolutionary process as it now emerges. To take an unequivocal stand, it seems to me, is of greater heuristic value and far more likely to stimulate constructive criticism than to evade the issue." And in the first chapter of his *Growth of Biological Thought* (1982), Mayr stated:

> My tactic is to make sweeping categorical statements. Whether or not this is a fault, in the free world of the interchange of scientific ideas, is debatable. My own feeling is that it leads more quickly to the ultimate solution of scientific problems than a cautious sitting on the fence. . . . histories should even be polemical. Such histories will arouse contradiction and they will challenge the reader to come up with a refutation. By a dialectical process this will speed up a synthesis of perspective. The unambiguous adoption of a definite viewpoint should not be confused with subjectivity. (Mayr 1982, 9–10)

Mayr delivered his "Where Are We?" address in this same spirit.

Wright was shocked and appalled by the address. It was bad enough that Mayr had characterized the period of his student days at Harvard as dominated by simple Mendelism and the mutation theory of evolution when actually he had been taught by the two greatest Darwinians of the day, Castle and East. What really rankled was Mayr's lack of appreciation of what Wright had always aimed for in his shifting balance theory of evolution, namely the synthesis of physiological genetics with the factors affecting gene frequencies in nature. Mayr obviously did not appear to understand Wright's great and substantively successful efforts to incorporate genic interaction, multiple factors, frequency dependent selection, degrees of dominance, and many other factors into his quantitative models, nor did Mayr appear to think that Wright's theory of evolution was anything more than random drift. And to characterize Wright's contribution as arid and Dobzhansky's as robust was the last straw.

Wright was asked by *American Journal of Human Genetics* to review the 1959 volume of the *Cold Spring Harbor Symposia on Quantitative Biology*. Wright simply listed the twenty-four papers in the volume, declining to comment upon those by such persons as Lamotte, Carson, Sheppard, and Wallace, electing instead to spend the entire review criticizing Mayr's address. He examined Mayr's three periods, challenging his interpretation in each. First he pointed out the many Darwinians among geneticists in the period 1910–20, and suggested that population genetics already had a significant start in those years. Then he vigorously argued that his own views in Mayr's "classical population genetics" period were not those Mayr had attributed to him but were in fact just those Mayr said were needed for a robust population genetics, and more. To Mayr's accusation that the mathematical population geneticists had only used constant fitnesses in their models, Wright replied that his symbol $\overline{W}$ for selective value in his theory "has always been defined as applying to a *total* genotype in the system under consideration, thus involving whatever interaction effects there may be among the component genes" (Wright 167, p. 369). As for Dobzhansky's work in population genetics, especially his "balance" theory, Wright said: "I fully concur in Mayr's tribute to these tremendous achievements. Nobody, however, has understood better than Dobzhansky the reciprocal relation between concrete facts and abstract theory, as exemplified in his continual collaboration with mathematical population geneticists" (p. 371). No truer words were ever spoken, as the two chapters in this book on the collaboration of Wright and Dobzhansky testify.

Wright obviously felt stung by Mayr's misrepresentation of his life's work. He thought the space allotted to his reply to Mayr in the review was insufficient, and thereafter he never to my knowledge lost the opportunity to criticize Mayr in conversation or print. For his part, Mayr did not reply to Wright's reactions, at least in print, and also minimized the importance of Wright's role in the evolutionary synthesis period. When Mayr organized his conference on the evolutionary synthesis in 1974, he did not invite Wright because he thought Wright's role in the synthesis period was not pivotal and because most of the geneticists to whom he spoke were negative about Wright's participation. (They had all heard him give long-winded and difficult lectures.) No one could possibly guess that Wright was a major figure in twentieth-century evolutionary biology from Mayr's *Growth of Biological Thought,* where Wright rates a very few, all positive but limited remarks.

What was truly at issue in this altercation between Mayr and Wright? Did they really disagree fundamentally about the mechanisms of evolution in nature? Has their disagreement, whatever the personal cost, had salutary effects upon evolutionary biology in the way, let us say, that the disagreements between Wright and Fisher did?

The real bone of contention between Mayr and Wright was the historical interpretation of the evolutionary synthesis rather than substantive questions of evolutionary biology. Mayr had excellent reason to believe that the field of systematics, the problem of speciation, and his own work in particular were being slighted in the interpretation of the evolutionary synthesis provided by Dobzhansky, Sheppard, and so many others in the 1950s. Mayr was fighting for his rightful place and honor for bringing, more than any other single person, systematics and speciation into the evolutionary synthesis. (Julian Huxley also deserves much credit here.)

Mayr's attempt in 1959 to set the record straight, however, went overboard in deprecating the contributions of the mathematical population geneticists and was historically inaccurate in its assessment of Wright's work in particular. In response, Wright was in turn fighting for his own rightful place in the evolutionary synthesis, and also for that of Fisher and Haldane. The vision of Wright defending Fisher from Mayr coming only ten years after Mayr's defense of Wright from Fisher is delightfully ironic.

The disagreement between Wright and Mayr over the interpretation of the evolutionary synthesis has had an enormous impact upon historians and philosophers of science who work on this subject and period. Nothing in my own background has been so essential as the constant interaction with both Mayr and Wright and feeling the tension between their historical interpretations. My first attempt to study the period of the evolutionary synthesis was in "The Role of Mathematical Population Geneticists in the Evolutionary Synthesis of the 1930s and 1940s" (1978), in which I answered Mayr's challenge to Wright, Fisher, and Haldane in 1959. Nearly all of us who work on the period of the evolutionary synthesis have been much influenced.

As for substantive issues in evolutionary biology, when Wright's actual views rather than the ones Mayr imputed to him in 1959 are taken into consideration, Mayr and Wright had little about which to disagree. Of course it is true that they have approached the problem of speciation from opposite ends of the taxonomic hierarchy, Mayr from the species level downward and Wright from the level of populations upward. Mayr has devoted a great deal more thinking and writing to the problem of speciation than has Wright, who has worked a great deal more on microevolution (below the species level) than Mayr. But Wright's conception of speciation from the late 1920s on was that speciation most likely begins with differentiation in small peripheral populations that are geographically separated. (In plants Wright has also emphasized the importance of chromosomal events in speciation.) And both Wright and Mayr have repeatedly emphasized that the speciation events were likely to occur in only a tiny fraction of these peripheral populations across time and space.

Furthermore, Wright thought that the situation most conducive to speciation came when a population crossed from one peak on $\overline{W}$ to another, thus resulting in novel interaction systems of genes and new selective values for great numbers of genes. Indeed, Wright presented this view in his paper "Breeding Structure of Populations in Relation to Speciation," which he delivered in a session organized by Dobzhansky at the 1939 AAAS symposium in Columbus, Ohio; Mayr was the chairman of the session (Wright 100). Mayr on the other hand has emphasized since 1954 that the crucial event leading toward speciation in a very few small peripheral populations was a "genetic revolution" in which the selective values of a great many genes were altered simultaneously, yielding a new state of genetic equilibrium (Mayr 1954, 169–70). I am not arguing that Mayr derived his concept of genetic revolutions from Wright, although there may be some unconscious connection, but simply that their ideas about speciation had converged in important ways despite their different perspectives. One way to gauge this convergence is to try and separate the influences of Mayr and Wright on the founder principle or, for example, on the wonderful analysis of speciation found in the Hawaiian *Drosophila* project by Hampton Carson and his many collaborators. My assessment is that Carson et al. have incorporated the ideas of both Wright and Mayr in their sometimes-changing interpretations of speciation in the nearly 700 species of Hawaiian *Drosophila*. I think that the ideas of Mayr and Wright on the mechanisms of speciation are mostly complementary rather than contradictory.

In 1982 both Wright and Mayr were nominated for the prestigious Balzan Prize in Italy. The selection committee must have had a difficult time choosing. Mayr won his richly deserved Balzan Prize in 1983. He then immediately supported Wright for the next year, and indeed Wright won the Balzan Prize in 1984. Picture for yourself Mayr standing proudly by in Rome as Wright, almost ninety-five, received the prize on November 15, 1984. This vision is not fantasy but fact. Both Wright and Mayr have also won the prestigious Darwin Medal of the Royal Society of London, Wright in 1980 and Mayr in 1984.

Linanthus parryae Revisited

The *Linanthus parryae* story told in chapter 11 yielded much insight into the state of evolutionary biology in general (Dobzhansky and Epling did not hesitate to explain the distribution of flower colors in *Linanthus* by random genetic drift even though they had no theory of isolation by distance), and into the elaboration of Wright's shifting balance theory of evolution (the *thus* data led directly to his quantitative theory of isolation by distance). This continuation of the *Linanthus* story provides insight into the state of Wright's shifting balance theory and and the state of evolutionary biology in general during the heyday of modern neo-Darwinism.

Neither Wright nor Epling ever lost interest in the work on *Linanthus parryae*. Each year Epling gathered more data, which he periodically sent to Wright, who in turn analyzed the data in accordance with the latest report on the mode of inheritance of the blue and white flower color polymorphism. *Linanthus* had great significance for Wright as his first and best example of isolation by distance, a theory that Wright considered to be one of his most important contributions to evolutionary biology.

After seventeen years of gathering data on *Linanthus parryae* in the field, artificially transplanting seeds from blue-flowered plants to plots where only whites had been found earlier, and conducting many controlled experiments on the mode of inheritance, Epling decided with his collaborators Harlan Lewis and Francis Ball to publish the results. By 1959 the experimental evidence was strong that blue was generally dominant to white, but sometimes apparently recessive whites would yield seeds that had blue flowers, so they announced at the beginning of the paper that "the exact mode of inheritance of flower color is still uncertain" (Epling, Lewis, and Ball 1960, 238). The most striking result of the many years of collecting data from the natural populations was that the frequencies of the blue and white phenotypes had remained remarkably stable over the whole transect, almost down to the square meter. Moreover, the stability of the polymorphism over seventeen years seemed to indicate that random drift could not cause the differentiation of the populations, since yearly random drift could be expected to cause changes in the frequencies of the alleles determining flower color. To Epling, that left natural selection as the logical explanation; but that alternative also had problems, the primary one being that the morphs were correlated with absolutely no evidence of ecological differentiation. Another problem was that when the seed from the blue morph was introduced into an area formerly occupied only by the white morph, the blues did quite well, although not perhaps as well as the white morphs; the data were not decisive. Epling, Lewis, and Ball also found that seeds from *Linanthus parryae* could lie dormant for many years, so the population that flowered in any given year was only a sample of a population built up for as much as a decade. Thus effective population size was greater than was apparent from the observable plants. And finally, seed dispersion appeared very limited.

The manuscript was written in 1959, the year of the great Darwin nial (the hundredthth anniversary of the *Origin* and hundred and fiftieth of Darwin's birth). The adaptive spirit of the times that I have described in the last chapter was at its height. Epling, Lewis, and Ball first concluded that "if genetic drift has played a role, it has been of only local consequence and not persistent in its effects," and further that

> The conclusion seems warranted, therefore, that the frequencies of blue and white flowered plants are in the long run the product of selection operating at an intensity we have been unable to measure; and that the large size of the effective population, and the localized dispersion of pollen and seeds, has precluded significant changes in pattern during 15 seasons. (Epling, Lewis, and Ball 1960, 254)

In 1960, incorporating the very same data as in 1942 (but with crucial additions, of course), Epling believed that random drift was ruled out and that selection must be the answer even though he had no direct evidence for that conclusion. Epling sent a copy of the manuscript to Wright and also to Michael Lerner, then editor of *Evolution,* who naturally asked Wright to referee the paper.

Wright had invested hundreds of hours analyzing the *Linanthus parryae* data because he saw them as a prime example of his theory of isolation by distance. Now the researchers who knew *Linanthus* best had declared that random drift played no role at all in the pattern of dispersal of blue and white flower color. Wright was dismayed. First Dobzhansky and Epling had published the view that the pattern of distribution of flower color in *Linanthus* was due entirely to random drift (but with no quantitative analysis), in response to which Wright had developed his theory of isolation by distance and applied it to the data. Now Epling, Lewis, and Ball had published the view that the pattern of distribution was due entirely to selection, without even applying his theory of isolation by distance, in the total absence of direct evidence of selection, and with good evidence of very limited dispersal of pollen and seed.

Wright responded by immediately recommending the paper for publication: "They have given a very detailed account of an extremely interesting and puzzling situation. I think a good deal more can be extracted from the data than in the paper as submitted, though not easily because of the great irregularity in the number of flowers recorded at the different stations" (Wright to Lerner, March 20, 1959). A few weeks later, after a preliminary reanalysis of the data, Wright wrote Epling a dense nine-page letter complete with detailed tables. Wright's conclusion was that the pattern of differentiation was not as stable as Epling, Lewis, and Ball had thought; furthermore:

> There seem to be only two possible explanations of the pattern of differentiation: (1) control by extremely local (as well as long range) selective differences, sufficiently strong to maintain differences against diffusion and (2) local historic accident (i.e. random drift due to accidents of sampling and dispersion) sufficiently rapid to permit the building up of marked differences in spite of diffusion or, of course, some combination of these. In any case there must be extremely slow diffusion. The crucial question is whether persistent major changes can occur naturally over longer periods than the

decade of observation or can be induced. Your evidence of persistence for at least 4 years of one color transferred to a region in which it had been absent seems to be the best evidence on this at present. It seems to show that selection is not over-whelmingly strong. On the whole, it seems to me that the comparison of the correlations at varying distances in determining the three periods indicate that random drift is an important factor in determining the pattern but that it is not a cumulative random drift from year to year but rather one based on periods of sparsity at intervals to be measured in terms of decades rather than years. (Wright to Epling, April 8, 1959)

In response, Epling invited Wright to contribute a technical appendix to their paper. Wright refused, however, and instead by 1962 had written his own analysis of the accumulated *Linanthus* data in a twenty-nine-page paper with fifteen tables and six graphs appended. His conclusion was that random drift had been of much significance in the occasional founding of new colonies and in the repopulation of ones that had become extinct.

Wright had already said many times that conspicuous polymorphisms were relatively unimportant in the evolutionary process. Why, then, did he expend so much time and energy analyzing the case of one conspicuous polymorphism? Even Wright in the first sentences of the concluding paragraph of his manuscript said: "In presenting this interpretation of the complicated pattern of dimorphism in *Linanthus parryae* there is no intention of implying that this case is in itself of more than trivial evolutionary significance. Random drift can rarely be of evolutionary significance with respect to character differences due to only one or two loci." This is almost the same comment that Wright had made about Lamotte's 1959 presentation of his results with *Cepaea*. The catch is that cases of random drift in a conspicuous polymorphism really were important for Wright, as he went on to say:

The type of pattern exemplified here becomes of great evolutionary importance, however, if it applies simultaneously to all of the genes concerned with a system of interlocking multifactorial characters. In such genetic systems there are inevitably numerous disjunct selective peaks, because of intermediate optima and pleiotropy, and selection coefficients for individual loci must in general be very small. Distribution patterns like that of *Linanthus,* occurring more or less independently at many loci would provide the basis for selection among the genetic systems of different localities by the slow but effective processes of differential population growth and dispersion. It is to be noted that this process depends on the joint action of random drift, dispersion and interdemic selection of genetic systems as wholes, and fails if any one of these factors is absent.

Thus Wright's argument really was that conspicuous polymorphisms are relatively unimportant in the evolutionary process and were useless material for selectionists like Fisher and Ford to generalize from. After all, proving that a conspicuous polymorphism was subject to even huge selection rates was no guarantee that other loci were subject to similar selection rates; indeed, others would generally be subject to far lesser selection rates. But if a conspicuous polymorphism that one could almost assume was in some kind of selective equilibrium was subject to random drift, then almost every other independent locus would be undergoing similar random drift. From Wright's perspective, therefore, conspicuous polymorphisms were crucially important cases from which to generalize to population structure and from there to his shifting balance theory. This explains why Wright devoted so much attention to cases of conspicuous polymorphisms while at the very same time proclaiming their unimportance in evolution. The case of *Linanthus parryae* was especially important to Wright because of its relevance to his theory of isolation by distance.

Wright did not, however, publish his manuscript on *Linanthus*. Ball was still working on the problem of the aberrant cases of color inheritance, and Wright did not want to publish again on *Linanthus* without knowing at least the pattern of inheritance. (Ball sent the decisive information on inheritance of flower color to Wright in late December, 1962.) Also, Wright was already deep into his treatise on evolutionary biology, and he did not want anything to interfere with that. Meanwhile, Epling continued to produce more data on *Linanthus* through the summer of 1966, all of which he sent to Wright. In the end, Wright reanalyzed all of the *Linanthus* data when he came to the fourth volume of his treatise devoted to natural populations. This helps explain the twenty-nine-page, wholly original analysis of *Linanthus* that appears in that volume (Wright 200, pp. 194–223), a longer section than Wright devoted to any other organism except humans.

Wright and Fisher: The Final Hurrah

Unlike the tension between Wright and Mayr, that between Wright and Fisher around the time of the Darwin centennial of 1959 continued to raise and fuel important questions about the mechanisms of the evolutionary process. A brief analysis of their last public disagreement before Fisher's death in 1962 illustrates this thesis.

The biological subject of this disagreement was not a center of controversy until Wright and Fisher made it so. It goes back to a study of a rare species of Hugo de Vries's favorite organism, *Oenothera,* which continued to be the object of much genetic research in the 1930s (and later) by geneticists like Ralph Cleland and Sterling Emerson, son of R. A. Emerson. The rare species *Oenothera organensis* was found only in the Organ Mountains of southern New Mexico, and in the 1930s only in a few canyons, with a total population estimated to be not more than 1,000 plants and probably closer to 500. Sterling Emerson began to work on this species because it appeared to have an interesting system of self-incompatibility alleles. Breeding experiments revealed that in *Oenothera organensis* the rigid self-incompatibility pattern was controlled by a series of multiple alleles such as East and Manglesdorf had discovered many years earlier in *Nicotiana* (East and Manglesdorf 1925). Emerson also discovered that the mechanism of the incompatibility was failure of the pollen tubes to develop in any style carrying the same allele as the fertilizing pollen. These results were interesting but hardly of major importance (Emerson 1938).

Of greater evolutionary interest was Emerson's discovery that the small population supported no less than 37 different self-incompatibility alleles in the series, and perhaps more, with a sample of 135 gametes revealing no less than 34 alleles (Emerson 1939). Any new such allele would of course tend to spread in the population, but the small population size would also tend to lose alleles through inbreeding effect—elimination of some alleles and fixation of others through random genetic drift. The question was, how could a population so small carry so many of the alleles? Emerson called the problem to the attention of Wright, whom both he and his father admired.

Wright extended the stochastic distribution he had published in 1937 (Wright 91, 92) to include the situation found with self-incompatibility alleles (Wright 97). Wright's model showed that if the population were random breeding, the necessary mutation rate to maintain such a high number of self-incompatibility alleles was an enormous one in a thousand, unlikely on comparative grounds and from Emerson's data indicating no mutations in 45,000 pollen grains. One possible explanation that Wright gave for the high rate of occurrence of self-incompatibility alleles was "the size of the population is much greater than estimated from the data now at hand or, if not greater now, that it has recently been much greater, with loss of alleles at too slow a rate to have reached equilibrium. . . . a population of some 4000 to 5000 is required to account for the probable number of alleles" (Wright 97, p. 546). Wright did not, however, explore this possibility further; instead he turned to his favorite mechanism for explaining evolutionary questions—population structure under equilibrium.

Subdivision into small wholly isolated populations could explain the large number of alleles (a wholly isolated population of 50 could support an average of 5 or 6), but Emerson's data showed conclusively that such complete isolation was impossible. Wright's suggestion was that the population was actually subdivided into much smaller populations with some gene exchange, in other words the situation in which his shifting balance theory worked most efficiently. Assuming 50 local groups of 10 individuals with 2% foreign pollen by migration, and mutation rate a reasonable 9×10^{-5}, the total population of 500 could support 50 alleles at equilibrium.

Wright found the assumption of fine subdivision with some migration to be the most reasonable explanation of Emerson's data. This appeared to be a nice case of the shifting balance theory of evolution. Wright did not mention

Fisher in this paper, but the implication was clear that Fisherian mass selection could not explain the observations, whereas the shifting balance theory could.

Fisher did not answer this paper by Wright until the appearance of the second edition of *Genetical Theory of Natural Selection* in 1958. Here he added to chapter 4 a section entitled "Self-Sterility Allelomorphs," in which he reanalyzed the stochastic distribution of genes in a population with a series of self-sterility alleles, using the methods he had developed in the 1930 edition. He described his quantitative results as "very different" from those Wright had obtained in 1939, attributing the differences to Wright's "failure to develop any explicit formulae" and his reliance upon "extensive numerical calculations based on trial values of numerous constants he introduces" (Fisher 1958, 109).

I am sure, however, that Fisher's primary motivation in going back to this example then almost twenty-years old was not to deprecate Wright's quantitative reasoning but to counter his interpretation of evolution in *Oenothera organensis* as an example of the shifting balance theory of evolution. Fisher's interpretation was that "species having populations less than 10,000 must, of course, be presumed to have fallen greatly in population during their recent evolutionary history"; as the population size fell, the proportion of self-incompatibility alleles rose accordingly, leaving the population in a state of disequilibrium with the mutation rate. "Many of the rarer alleles have doubtless dropped out, leaving predominantly those well represented in the population, and consequently in little danger of immediate extinction." Isolation of the sort hypothesized by Wright, even if it had occurred, resulted from a "recent large reduction in population numbers" that by itself could explain the high rate of self-incompatibility alleles (p. 109).

As usual, Wright defended himself vigorously. In a paper published in *Biometrics* in 1960, he carefully compared the practical consequences of Fisher's quantitative derivations with his own and discovered that they were virtually identical. Thus Wright and Fisher once again did not really have a quantitative disagreement but a disagreement over the qualitative interpretation of the mechanisms of evolution in nature. Wright did, however, back off some from his hypothesis of fine-scaled subdivision of the population as the most probable primary explanation of the observed frequency of self-incompatibility alleles.

> The hypothesis of fine scaled subdivision of a population in equilibrium at the present estimated size becomes even more improbable than before as the sole explanation of the number of alleles. A combination hypothesis, recent reduction from a population several times as large as at present, aided by partial isolation of colonies, seems the most plausible conclusion. (Wright 160, p. 63)

Like many of the other disagreements between Wright and Fisher over interpretation of mechanisms of evolution, this one refused to die away. Before his death in 1962, Fisher published three more papers on self-incompatibility alleles in *Oenothera organensis* (Fisher 1961a, 1961b, 1962), the first arguing that observed differentiation in frequencies of self-incompatible alleles in the wild populations of *Oenothera organensis* could be due to chance rather than the effect of isolation (in other words the whole population could be random breeding), and the second offering a model for the generation of self-incompatiblity alleles by recombination rather than by new mutation. Others joined the fray, including Moran (1962), Ewens (1964), and Kimura (1965), and of course Wright continued to participate (Wright 178; 193, pp. 402–16). In addition to the published papers, there also exists a substantial set of interesting correspondence primarily between Wright, Ewens, Kimura, and Cornish on the distribution of self-incompatibility alleles.

Here again, Wright and Fisher took a relatively obscure research project and made it into an interesting issue for evolutionary biology. They were able to do this without either of them ever having laid eyes upon the organism, much less having studied its habitat. The tension between Wright and Fisher was highly creative for field research. (Reading the papers and correspondence on self-sterility in *Oenothera organensis* made me, an armchair historian, want to get out into the Organ Mountains and make a meticulous resurvey of the number and distribution of self-sterility alleles.) The tension was also highly creative for mathematical population geneticists, witness the participation of Ewens, Moran, Kimura, and others in this example.

The creative tension and enormous influence of the disagreements between Fisher and Wright did not end with Fisher's death. For proof of this statement one need only go to Wright's four-volume *Evolution and the Genetics of Populations* and check how many of the issues Wright and Fisher had disagreed upon were still alive and flourishing, not just in Wright's eyes but in the mainstream of evolutionary biology. Or one can go to the successive editions of E. B. Ford's *Ecological Genetics* (beginning with the first edition in 1964), which summarizes much of the field work on the genetics of natural populations, and see that this extraordinary book is built squarely upon the exploration in nature of the tension between Fisher and Wright. I would also emphasize yet again that the productive tension between Fisher and Wright emanated from their different views of evolution in nature and not from their differences in quantitative modeling, which were (once the smoke cleared) relatively unimportant so far as evolutionary biology was concerned.

Evolution and the Genetics of Populations

Wright's magnum opus speaks for itself. He included in the four volumes almost everything he had ever wanted to say about population genetics, and I will attempt no substantive summary here. A few comments about *Evolution and the Genetics of Populations* and the relation of this biography to it are, however, in order.

One can look at this book as an introduction to *Evolution and the Genetics of Populations*. I can almost guarantee that the four rather formidable looking volumes will be more accessible to the reader who understands the historical context of the issues that Wright addresses in the work. I cannot emphasize too strongly, however, that this book must not for many reasons be viewed as a volume to be read instead of *Evolution and the Genetics of Populations,* if the object is to understand Wright's views on genetics in relation to evolution. For obvious reasons this book does not discuss or even mention many aspects of Wright's work or his relations with others, and only begins to address the technical richness of his work. Perhaps more important, even when I have discussed an issue in detail, my analysis has ended mostly before Wright's move to the University of Wisconsin in 1955.

Wright's wide-ranging grasp and use of the literature in genetics and evolutionary biology did not end in 1955. Indeed, Wright read voraciously in these years, went to seminars and meetings, and generally sought out the literature he needed to remain current on whatever issue he happened to address in *Evolution and the Genetics of Populations.* The following graph, (fig. 13.4), made by Bruce Wallace for a class he was teaching on Wright, dramatically illustrates the almost incredible range of Wright's citations in the four volumes. Wright not only cited but carefully analyzed much of the literature between 1960 and 1976 when he was writing the volumes.

What the table cannot show is the richness of Wright's reworking of old problems or the great extent to which he incorporated the frontiers of research in his analyses. Thus an issue like the evolution of dominance received a thorough analysis, but mostly based upon recent rather than earlier publications. The same could be said for the problems of dispersion, effective population size, isolation by distance, and a host of others. Similarly, Wright reanalyzed the cases of individual organisms such as *Drosophila pseudoobscura, Panaxia dominula, Linanthus parryae, Cepaea nemoralis* and a great many others. In each case he reviewed most of the pertinent literature from the time he had earlier written about the organism to the time of actually writing the section of *Evolution and the Genetics of Populations.* Thus Wright's analyses are new for the most part, and when recapitulation of one of his earlier writings was called for, Wright referenced rather than repeated it.

Originally I had intended to track each of the major controversies in evolutionary biology up through Wright's treatment of them in *Evolution and the Genetics of Populations.* I thought this plan would give the biography an attractive temporal symmetry, complete with recapitulation and coda. But as I worked through each of the central issues and controversies, it became apparent that another large monograph was required to bring the plan to fruition. The amount of new material in the four volumes was staggering. Thus my hearty recommendation to any reader of this biography who has not already read *Evolution and the Genetics of Populations* is to settle down with it soon. Starting with the fourth volume and working back toward the first is a reason-

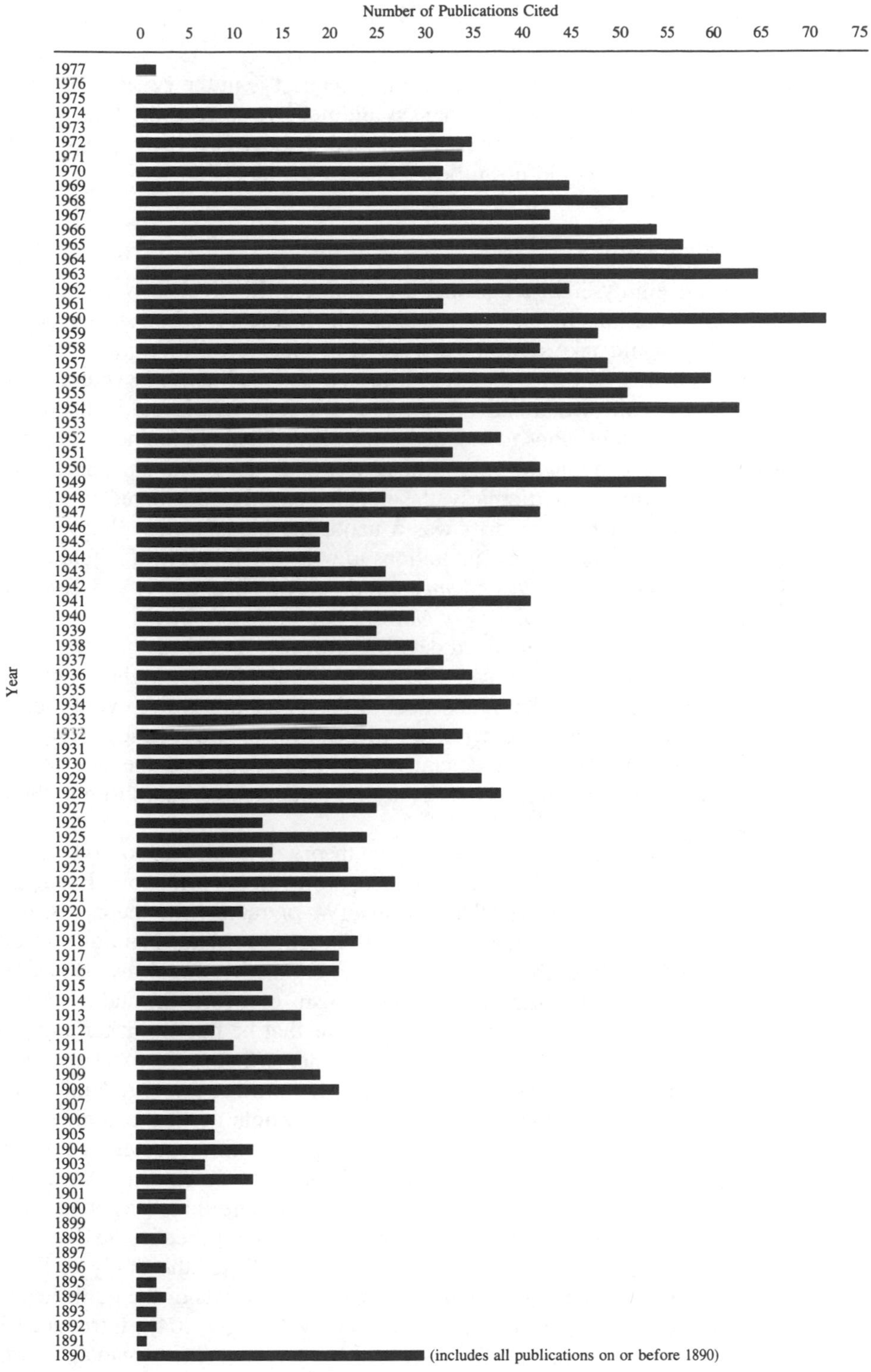

Figure 13.4. Publications cited by Wright in *Evolution and the Genetics of Populations* by year of original publication.

able procedure for anyone whose primary interest is in the genetics of natural populations. It is sad that Dobzhansky did not live to read the final volume, since it was the one he knew he could read without becoming bogged down in mathematics he could not follow.

After the Magnum Opus

Wright was eighty-eight when the last volume of his treatise was published. One might think that a person of this age who had just finished such a gargantuan effort would take a well-deserved rest. Not Sewall Wright. While the last volume was in press, he was disturbed about all the new literature on evolutionary biology that had appeared too late for analysis and inclusion. He again accepted invitations to lecture at universities and conferences, in part to renew the attack upon the many misconceptions about his shifting balance theory of evolution, particularly the popular one that attributed to him the view that straight random drift was a major mechanism of evolution in nature. He even accepted a few invitations to write book reviews, including one on M. J. D. White's *Modes of Speciation* (1978; Wright 203) and Lewontin et al.'s *Dobzhansky's Genetics of Natural Populations* (Wright 206).

As always, Wright remained deeply involved with the major questions of the day in evolutionary biology. I have already discussed Wright's participation in the debate over Kimura's neutral theory of molecular evolution. He also wrote substantial papers related to Dawkins's selfish gene theory, the growing controversy over mechanisms of speciation, and the debate over macroevolution stimulated primarily by the "punctuated equilibrium" theory of Eldredge and Gould (1972).

Wright found Dawkins's selfish gene theory (Dawkins 1976) to be a simple extension of Fisher's genetical theory of natural selection. Population structure played no role at all in the theory—or rather, the theory assumed only one population structure, large and random breeding. Natural selection of the smallest recombinable pieces of genome ("genes" by Dawkins's definition) was indeed the primary mechanism of evolution under such assumptions. Wright raised the same objections that he had to Fisher's theory. If random drift created novel interaction systems that were fixed in a local population by a combination of random drift followed by intrademic selection, then selective diffusion could spread the whole genetic combination to other local populations, in time transforming the entire species. To Wright this was organismic selection rather than merely genic selection as emphasized by Dawkins. Of course this argument was one that Wright had been making since the 1920s, but his presentation at a joint meeting in Tucson of the American Society of Naturalists and the Society for the Study of Evolution was particularly forceful and the demand for reprints of the published paper soon exhausted Wright's generous supply (Wright 204). Introduced by Simpson at the Tucson meeting, pleased by a very positive response to the lecture, and surprised by the overwhelming demand for the reprint, Wright began to think that the tide had shifted away from the situation that had endured for so many years during which he felt his work was misunderstood and unrecognized.

Mechanisms of speciation and macroevolution had never been major interests for Wright, although he had occasionally written briefly about them. In the 1960s and 1970s, however, both became the object of intense interest and controversy among evolutionary biologists. Mayr played a central role in focusing the attention of evolutionists upon speciation, which (beyond an initial superficial level) occurs by a bewildering variety of mechanisms and which naturally led to much controversy among evolutionary biologists. (For a review, see Templeton 1981.) Eldredge and Gould proposed their theory of punctuated equilibria in 1972, challenging the neo-Darwinian view of macroevolution as the natural extension of gradual microevolution (Eldredge and Gould 1972). The geological record with its sharp discontinuities between species found in different layers was not, according to Eldredge and Gould, the imperfect record Darwin imagined it to be. Instead, new species really did appear rather suddenly after long periods of stasis, and such changes were reflected in the geological record. Adherents of the neo-Darwinian view countered that gradual microevolution is perfectly consistent with the known geological record. (See, for example, Stebbins and Ayala 1981.) The intensity of the debate between these two camps was evident at the Macroevolution Conference held at Chicago's Field Museum of Natural History late in 1980.

Wright attended the conference and applied his shifting balance theory of evolution in nature to speciation and macroevolution in a paper entitled "Character Change, Speciation, and the Higher Taxa," as well as in another paper that he wrote for the *Annual Review of Genetics* entitled "The Shifting Balance Theory and Macroevolution"; both were published in 1982 when Wright was ninety-two (Wright 205, 207). The major thrust of Wright's arguments in these two papers was that nothing in the raging controversies about speciation or macroevolution disturbed the consistency of his shifting balance theory with the available evidence. According to Wright, the most crucial variable in both speciation and macroevolution was the appearance or availability of a new ecological niche into which a species could then radiate. Although agreeing that speciation could and perhaps did occur by a large number of mechanisms, the movement of a species into a new niche was to his way of thinking the ideal circumstance for the maximum efficiency of his shifting balance process. The rapid progress possible under the shifting balance process likewise could explain the discontinuities in the geological record emphasized by Gould, Eldredge, and Stanley. Because his shifting balance theory fit the paleontological data so well, Wright disagreed strongly with Gould's suggestion that something like Goldschmidt's hopeful monsters was required for explaining the observed geological record.

I should also add that a great environmental change such as the appearance of a new ecological niche was precisely the circumstance under which Fisher's genetical theory of natural selection operated most efficiently, so the assumptions underlying Wright's argument favored the neo-Darwinian argument as well. Wright was careful to say that his argument was one of consistency rather than contingency.

There is something basically heartening about the thought of Wright in his early nineties traveling around the United States vigorously defending his shifting balance theory, describing what it was and what it was not, and showing how it applied to the burning questions of the day in evolutionary biology.

Wright's Influence upon Modern Evolutionary Biology

If Wright had died in 1960, the year that he began writing his treatise, my assessment would be that his influence upon modern evolutionary biology was immense. As evidence we have his publications that directly influenced an important minority of evolutionists; his correspondence; his great indirect influence through the writings of Dobzhansky, Timoféeff-Ressovsky, Simpson, Mayr, and others; his impact upon animal breeding theory; his application of animal breeding theory, physiological genetics, and path coefficents to the quantitative theory of evolution in nature; his tension with Fisher that so strongly influenced both mathematical population genetics and field research in population genetics; his theory of isolation by distance; his invention of F-statistics; and all the other work that has been detailed in this biography. But in 1960 Wright himself saw his own work as misrepresented and unappreciated as typified by Mayr's attitude in his "Where Are We?" address at Cold Spring Harbor in 1959. One effect of the movement of the evolutionary synthesis toward a more thoroughly adaptationist stance in the 1940s and 1950s was indeed that many evolutionists tended to dismiss any significant evolutionary effect of random genetic drift, and Wright's work was emphatically dismissed by those who mistakenly viewed his theory of evolution as pure random drift.

The 1970s and 1980s, however, have witnessed a dramatic shift in the status of Wright's work among evolutionary biologists. First and foremost, the four volumes of *Evolution and the Genetics of Populations* have had and continue to have a significant impact. Although the first three volumes are difficult to read, evolutionary geneticists now are vastly better prepared in quantitative reasoning than their counterparts who read, for example, Wright's big 1931 paper or Fisher's *Genetical Theory of Natural Selection* at the time of publication. It would be difficult for a reader to misinterpret Wright's extremely careful presentation of his shifting balance theory along with his exposure of previous misinterpretations. Reviewers of the treatise generally understood the shifting balance theory, and one particularly thoughtful and accurate review by Daniel Hartl appeared in *BioScience*, where it reached a very large audience (Hartl 1979). A whole generation of graduate students are reading Wright's four volumes as a general education in twentieth-century evolutionary biology, a strategy that I heartily endorse. Every university that I visit now has a growing number of graduate students who have read or are reading Wright.

Another reason for Wright's growing recognition is the decline of the more extreme selectionist attitude of the 1950s. Kimura, Gould, Eldredge, Lewontin, Stanley, and a great many others have argued for a more balanced view of the evolutionary process, with random drift alloted a significant role. In the last decade evolutionary biology has exhibited great intellectual ferment, and Wright's ideas have benefited greatly from this reconsideration of the factors of the evolutionary process.

Recognition came in other ways. He won the National Medal of Science in 1967, and was awarded the Darwin Medal of the Royal Society of London in 1980 and the Balzan Prize in 1984. He also won the Laing Prize for the best book of the year from the University of Chicago Press in 1979. His work is known and studied all over the world, wherever there is interest in evolutionary biology. It would be an enormous task to detail Wright's current influence in evolutionary biology.

Later Years

After Louise Wright died in 1975, Sewall continued to live in the house on Council Crest in Madison. He enjoyed wonderful support from his friends the Crows, Chapmans, Brinks, his minister Max Gabler and members of the Unitarian church, and many others. The Department of Genetics instituted a policy of calling him on the telephone each morning to ascertain that he was well. I witnessed a measure of the affection for him in Madison when Crow asked me to speak at a ninetieth birthday party (one of four that were thrown for him that year) that the genetics department was sponsoring. My lecture was scheduled for late on a cold Friday afternoon in February 1980. I expected to see an audience of perhaps twenty hardy souls. Instead I was greeted by a large lecture hall jammed with Wright's admirers, eagerly waiting to hear yet something more about *their* Sewall Wright.

Beginning in the early 1980s, Wright began to lose his vision to irreversible macular degeneration. Especially affected was his ability to read, since this utilized the central portions of the retina most affected by the degeneration. His peripheral vision remained good enough for him to continue taking his customary long walks. When I visited him in 1982 he was both amused and fuming about an experience he had had the week before. He had been walking and was several miles away from home when a young woman saw him and asked if he might be lost. When Wright told her that he was fine and where he had already walked, she did not believe him and insisted upon driving him home; she also had trouble believing that he lived alone.

Living at home alone was becoming a growing burden, and in 1982

Figure 13.5. Wright at age ninety, taken at one of his many birthday parties. Photograph by Hildegard Adler.

Wright sold his house and moved to a modest and supportive retirement community. With bright lighting, a powerful magnifying glass, and enlarged text he could read at a very slow pace. In this manner he read the entire manuscript of this book in less than six weeks beginning in January 1985.

The term *declining years* is not applicable to Wright. On my last visit to Madison in February 1985 to record his comments on the manuscript of the biography, he handed me reprints of two new papers published in 1984, his ninety-fifth year, one a rebuttal to Samuel Karlin's critique of his method of path analysis (Wright 209) and the other a summary of his work in physiological genetics (Wright 210). In the early summer 1985 Wright purchased a closed-circuit television apparatus for magnifying reading materials. With it he has been able to read again at reasonable speeds, much to his enjoyment.

Wright continues to take long walks, and he looks and sounds much as he did twenty years ago. His influence as an evolutionary biologist has never been greater and is growing substantially. I predict that historians and biologists in the twenty-first century will look upon Wright as perhaps the single most influential evolutionary theorist of this century.

References

References within year for a single author are listed in chronological rather than alphabetical order.

Adams, M. B. 1980. Sergei Chetverikov, the Kol'tsov Institute, and the evolutionary synthesis. In *The evolutionary synthesis*, ed. E. Mayr and W. B. Provine, 242–78. Cambridge: Harvard University Press.

Allard, R. W. and C. Wehrhahn. 1964. A theory which predicts stable equilibrium for inversion polymorphisms in the grasshopper, *Moraba scurra*. *Evolution* 18:129–30.

Allen, G. E. 1968. Thomas Hunt Morgan and the problem of natural selection. *Journal of the History of Biology* 1:113–39.

————. 1975. The introduction of *Drosophila* into the study of heredity and evolution, 1900–1910. *Isis* 66:322–33.

————. 1978. *Thomas Hunt Morgan: The man and his science*. Princeton: Princeton University Press.

Allen, S. L. 1955. Linkage relations of the genes histocompatibility brachury and kinky tail in the mouse as determined by tumor transplantation. *Genetics* 40:627–50.

Arnold, S. J., M. J. Wade, and R. Lande. 1985. Measuring selection in natural populations. Typescript.

Avery, O. T., C. M. MacLeod, and M. McCarty. 1944. Studies on the chemical nature of the substance inducing transformation of pneumococcal types. *Journal of Experimental Medicine* 79:137–57.

Babcock, E. B. 1933. John Belling: Pioneer in the study of cell mechanics. *Journal of Heredity* 24:296–300.

Barlow, N., ed. 1958. *The autobiography of Charles Darwin*. London: Collins.

Bateson, W. 1894. *Materials for the study of variation*. London: Macmillan.

————. 1902. *Mendel's principles of heredity: A defence*. Cambridge: Cambridge University Press.

————. 1903. The present state of knowledge of colour-heredity in mice and rats. *Proceedings of the Zoological Society of London* 2:71–99. Reprinted in *The scientific papers of William Bateson*, ed. R. C. Punnett, 2:76108. Cambridge: Cambridge University Press.

————. 1909. *Mendel's principles of heredity*. Cambridge: Cambridge University Press. 3d impression with additions, 1913.

Bateson, W., and E. R. Saunders. 1902. *Report I to the Evolution Committee of the Royal Society of London*. London: Harrison & Sons.

Baur, E. 1907. Untersuchungen über die Erblichkeitsverhältnisse einer nur in Bastardform lebensfähigen Sippe von *Antirrhinum majus*. *Berichte der Deutschen Botanischen Gesellschaft* 25:442.

————. 1911. *Einführung in die experimentelle Vererbungslehre*. Berlin: Borntraeger.

Beadle, G.W. 1951. Chemical genetics. In *Genetics in the 20th Century,* ed. L. C. Dunn, 221–39. New York: Macmillan.

Beatty, J. 1985. Dobzhansky and drift: Facts, values, and chance in evolutionary biology. Typescript.

Berg, L. 1926. *Nomogenesis.* London: Constable.

Bennett, J. H., ed. 1983. *Natural selection, heredity, and eugenics: Including selected correspondence of R. A. Fisher with Leonard Darwin and others.* Oxford: Oxford University Press.

Bowler, P. J. 1978. Hugo de Vries and Thomas Hunt Morgan: The mutation theory and the spirit of Darwinism. *Annals of Science* 35:55–73.

———. 1983. *The eclipse of Darwinism: Anti-Darwinian evolution theories in the decades around 1900.* Baltimore: Johns Hopkins University Press.

Box, J. F. 1978. *R. A. Fisher: The life of a scientist.* New York: Wiley.

Brandon, R. N. and R. M. Burian. 1984. *Genes, organisms, populations: Controversies over the units of selection.* Cambridge: MIT Press.

Bryson, V., and H. J. Vogel, eds. 1965. *Evolving genes and proteins.* New York: Academic Press.

Bumpus, H.C. 1899. The elimination of the unfit as illustrated by the introduced sparrow, *Passer domesticus.* In *Biological lectures from the marine laboratory, 1898,* 209–26. Boston: Ginn.

Buri, P. 1956. Gene frequency in small populations of mutant *Drosophila. Evolution* 10:367–402.

Burks, B. S. 1928. The relative influence of nature and nurture upon mental development. In *Nature and nurture: Their influence on intelligence,* 1:219–316. National Society for the Study of Education Yearbook, No.27.

Buzzati-Traverso, A., C. Jucci, and N. W. Timoféeff-Ressovsky. 1938. Genetica di popolazioni. Consiglio Nazionale delle Ricerche. *La ricerva scientifica.* series 2, vol. 1:3–30.

Cain, A. J. 1977. The efficacy of natural selection in wild populations. In *The Changing Scenes in Natural Sciences,* special publication 12, 111–33. Academy of Natural Sciences.

Cain, A. J., and J. D. Currey. 1963. Area effects in *Cepaea. Philosophical Transactions of the Royal Society of London* B 246:1–81.

Cain, A. J., and P. M. Sheppard. 1950. Selection in the polymorphic land snail *Cepaea nemoralis* (L.). *Heredity* 4:275–94.

———. 1954. Natural selection in *Cepaea. Genetics* 39:89–116.

Carlson, E. A. 1966. *The gene: A critical history.* Philadelphia: Saunders.

———. 1971. An unacknowledged founding of molecular biology: H. J. Muller's contribution to gene theory, 1910–1936. *Journal of the History of Biology* 4:149–70.

Castle, W. E. 1903a. The heredity of sex. *Bulletin of the Museum of Comparative Zoology* 40:187–218.

———. 1903b. Mendel's law of heredity. *Science,* 18:396–406.

———. 1903c. The laws of heredity of Galton and Mendel, and some laws governing race improvement by selection. *Proceedings of the American Academy of Arts and Sciences* 39: 223–42.

———. 1905a. *Heredity of coat colors in guinea pigs and rabbits.* Carnegie Institution of Washington Publication No. 23.

———. 1905b. Mutation theory of organic evolution. *Science* 21:521–25.

————. 1908. A new color variety of the guinea-pig. *Science* 28:250–52.

————. 1911. *Heredity in relation to evolution and animal breeding*. New York: Appleton.

————. 1912a. The inconstancy of unit characters. *American Naturalist* 46: 352–62.

————. 1912b. On the origin of a pink-eyed guinea-pig with colored coat. *Science* 35:508–10.

————. 1912c. On the inheritance of tricolor coat in guinea-pigs, and its relation to Galton's law of ancestral heredity. *American Naturalist* 46:437–40.

————. 1914a. Variation and selection: A reply. *Zeitschrift für induktive Abstammungs- und Vererbungslehre* 12:257–64.

————. 1914b. Some new varieties of rats and guinea-pigs and their relation to problems of color inheritance. *American Naturalist* 48:65–73.

————. 1915. Mr. Muller on the constancy of Mendelian factors. *American Naturalist* 49:37–42.

————. 1916. Further studies of piebald rats and selection, with observations on gametic coupling. In Carnegie Institution of Washington Publication No. 241, pp. 163–92.

————. 1917. Piebald rats and multiple factors. *American Naturalist* 51: 102–14.

————. 1919a. Piebald rats and selection: A correction. *American Naturalist* 53:370–75.

————. 1919b. Piebald rats and the theory of genes. *Proceedings of the National Academy of Sciences* 5:126–30.

————. 1930. *The genetics of domestic rabbits*. Cambridge: Harvard University Press.

————. 1951. The beginnings of Mendelism in America. In *Genetics in the 20th century*, ed. L. C. Dunn, 59–76. New York: Macmillan.

Castle, W. E., and G. M. Allen. 1903. The heredity of albinism. *Proceedings of the American Academy of Arts and Sciences* 38:603–22.

Castle, W. E., F. W. Carpenter, A. H. Clark, S. O. Mast, and W. M. Barrows. 1906. The effects of inbreeding, cross-breeding, and selection upon the fertility and variability of *Drosophila*. *Proceedings of the American Academy of Arts and Sciences* 41:729–86.

Castle, W. E., and A. Forbes. 1906. Heredity of hair-length in guinea-pigs and its bearing on the theory of pure gametes. Carnegie Institution of Washington Publication No. 49.

Castle, W. E., and C. C. Little. 1909. The peculiar inheritance of pink eyes among colored mice. *Science* 30:313–15.

————. 1910. On a modified Mendelian ratio among yellow mice. *Science* 32: 868–70.

Castle, W. E., and J. C. Phillips. 1914. Piebald rats and selection: An experimental test of the effectiveness of selection and of the theory of gametic parity in Mendelian crosses. In Carnegie Institution of Washington Publication No. 195.

Castle, W. E., H. E. Walter, R. C. Mullenix, and S. Cobb. 1909. Studies of inheritance in rabbits. In Carnegie Institution of Washington Publication No. 114.

Chase, H.B. 1939a. Studies on the tricolor pattern of the guinea pig, I: The relations between different areas of the coat in respect to the presence of color. *Genetics* 24:610–21.

———. 1939b. Studies on the tricolor pattern of the guinea pig, II: The distribution of black and yellow as affected by white spotting and by imperfect dominance in the tortoise shell series of alleles. *Genetics* 24:622–43.

Child, C. M. 1921. *The origin and development of the nervous system, from a physiological viewpoint.* Chicago: University of Chicago Press.

Clarke, B., C. Diver, and J. Murray. 1968. Studies in *Cepaea*, VI: The spacial and temporal distributions of phenotypes in a colony of *Cepaea nemoralis* L. *Philosophical Transactions of the Royal Society of London* B 253: 519–48.

Cock, A. G. 1973. William Bateson, Mendelism, and biometry. *Journal of the History of Biology* 6:1–36.

Crampton, H. E. 1904. Experimental and statistical studies upon Lepidoptera, I: Variation and elimination in *Philosamia cynthia. Biometrika* 3:115–30.

———. 1905. On a general theory of adaptation and selection. *Journal of Experimental Zoology* 2:425–30.

———. 1916. Studies on the variation, distribution, and evolution of the genus Partula: The species inhabiting Tahiti. Carnegie Institution of Washington Publication No. 228.

———. 1925. Contemporaneous organic differentiation in the species of *Partula* living in Moorea, Society Islands. *American Naturalist* 59:5–35.

Crane, J. H. C. 1975. Carl Sandburg, Philip Green Wright, and the Asgard Press, 1900–1910. Charlottesville: University of Virginia Press.

Crow, J. F. 1954. Breeding structure of populations, II: Effective population number. In *Statistics and mathematics in biology,* ed. O. Kempthorne, T. A. Bancroft, J. W. Gowen, and J. L. Lush, 543–60. Ames: Iowa State College Press.

Crow, J. F., and M. Kimura. 1970. An introduction to population genetics theory. New York: Harper and Row. Crow and Kimura dedicated this book to Wright.

Crow, J. F., and N. E. Morton. 1955. Measurement of gene frequency drift in small populations. *Evolution* 9:202–14.

Cuénot, Lucien. 1902. La loi de Mendel et l'hérédité de la pigmentation chez les souris. 2d note. *Archives de Zoologie expérimentale et générale,* Notes et revues, 33–41.

———. 1904. 3d note. *Archives de Zoologie expérimentale et générale,* Notes et revues, 45–56.

———. 1907. 5th note. *Archives de Zoologie expérimentale et générale,* Notes et revues, 123–32.

Curtsinger, J. W. 1984a. Evolutionary landscapes for complex selection. *Evolution* 38:359–67.

———. 1984b. Evolutionary principles for polynomial models of frequency-dependent selection. *Proceedings of the National Academy of Sciences* USA, 2840–42.

Darwin, C. 1859. *On the origin of species.* London: Murray. 6th ed., 1872. Title changed to *The origin of species.*

————. 1868. *The variation of animals and plants under domestication.* 2 vols. London: John Murray.

————. 1873. *The expression of the emotions in man and animals.* New York: D. Appleton.

————. 1876. *The effects of cross and self fertilization in the vegetable kingdom.* London: Murray.

Darwin, F. ed. 1887. *The life and letters of Charles Darwin.* 3 vols. London: Murray.

Darwin, F., and A. C. Seward, eds. 1903. *More letters of Charles Darwin.* 2 vols. New York: Appleton.

Davenport, C.B. 1897–99. *Experimental morphology.* New York: Macmillan.

————. 1899. *Statistical methods with special reference to biological variation.* New York: John Wiley.

————. 1903. The animal ecology of the Cold Spring Sand Spit, with remarks on the theory of adaptation. *Chicago Decennial Publications.* 1st series, vol. 10:155–76.

————. 1917. Inheritance of stature. *Genetics* 2:313–89.

Dawkins, R. 1976. *The selfish gene.* New York: Oxford University Press.

Detlefsen, J. A. 1914. Genetic studies on a cavy species cross. Carnegie Institution of Washington Publication No. 205.

————. 1925. The inheritance of acquired characters. *Physiological Reviews* 5:244–78.

de Vries, H. 1905. *Species and varieties: Their origin by mutation.* Chicago: Open Court.

Diver, C. 1929. Fossil records of Mendelian mutants. *Nature* 124:183.

————. 1936. The problem of closely related species and the distribution of their populations. *Proceedings of the Royal Society of London* B 121:62–65.

————. 1940. The problem of closely related species living in the same area. In *The new systematics,* ed. J. S. Huxley, 303-28. Oxford: Oxford University Press.

Dobzhansky, T. 1927. Studies on the manifold effect of certain genes in *Drosophila melanogaster. Zeitschrift für induktive Abstammungs- und Vererbungslehre* 43:330–88.

————. 1933. Geographical variation in lady-beetles. *American Naturalist* 67:97–126.

————. 1934. Are racial and specific characters non-Mendelian? *Journal of Mammalogy* 15:1–3.

————. 1935. A critique of the species concept in biology. *Philosophy of Science* 2:344–55.

————. 1937. *Genetics and the origin of species.* New York: Columbia University Press. 2d ed., 1941; 3d ed., 1951; 1st edition reprinted, with a new introduction by S. J. Gould, 1982.

————. 1943. Genetics of natural populations, IX: Temporal changes in the composition of populations of *Drosophila pseudoobscura. Genetics* 28:162–86.

————. 1955. A review of some fundamental concepts and problems of population genetics. *Cold Spring Harbor Symposia on Quantitative Biology* 20:1–15.

———. 1962. Oral history memoir. Columbia University, Oral History Research Office, New York.

———. 1970. *Genetics of the evolutionary process*. New York: Columbia University Press.

Dobzhansky, T., and C. Epling. 1944. Contributions to the genetics, taxonomy, and ecology of *Drosophila pseudoobscura* and its relatives. In Carnegie Institution of Washington Publication No. 554, 1–183.

Dobzhansky, T., and H. Levine. 1948. Genetics of natural populations, XVII: Proof of operation of natural selection in wild populations of *Drosophila pseudoobscura*. *Genetics* 33:537–47.

Dobzhansky, T., and J. R. Powell. 1974. Rates of dispersal of *Drosophila pseudoobscura* and its relatives. *Proceedings of the Royal Society of London* B. 187:281–98.

Dobzhansky, T., and M. L. Queal. 1938a. Chromosome variation in populations of *Drosophila pseudoobscura* inhabiting isolated mountain ranges. *Genetics* 23:239–51.

———. 1938b. Genic variation in populations of *Drosophila pseudoobscura* inhabiting isolated mountain ranges. *Genetics* 23:463–84.

Dobzhansky, T., and A. H. Sturtevant. 1938. Inversions in the chromosomes of *Drosophila pseudoobscura*. *Genetics* 23:28–64.

Dowdeswell, W. H., R. A. Fisher, and E. B. Ford. 1940. The quantitative study of populations in the Lepidoptera. *Annals of Eugenics* 10:123–36.

Dunn, L. C., ed. 1951. *Genetics in the 20th century: Essays on the progress of genetics during its first 50 years*. New York: Macmillan.

———. 1961. The reminiscences of L. C. Dunn. Columbia University, Oral History Research Office, New York.

———. 1965. William Ernest Castle. In *Biographical memoirs of the National Academy of Sciences* 38:33–80.

East, E. M. 1907. The relation of certain biological principles to plant breeding. Connecticut Agricultural Experiment Station Bulletin No. 158.

———. 1908. A study of the factors influencing the improvement of the potato. University of Illinois Agricultural Experiment Station Bulletin No. 127.

———. 1910. A Mendelian interpretation of variation that is apparently continuous. *American Naturalist* 44:65–82.

———. 1912. The Mendelian notation as a description of physiological facts. *American Naturalist* 46:633–55.

———. 1923. *Mankind at the crossroads*. New York: Scribners.

East, E. M., and D. F. Jones. 1919. *Inbreeding and outbreeding: Their genetic and sociological significance*. Philadelphia: Lippincott.

East, E. M., and A. J. Manglesdorf. 1925. A new interpretation of the hereditary behavior of self-sterile plants. *Proceedings of the National Academy of Science* 11:161–71.

Eldredge, N., and S. J. Gould. 1972. Punctuated equilibria: An alternative to phyletic gradualism. In *Models in Paleobiology,* ed. T. J. M. Schopf, 82–115. San Francisco: W. H. Freeman.

Elton, C. 1927. *Animal ecology*. London: Macmillan.

Emerson, R.A. 1910. The inheritance of size and shapes in plants: A preliminary note. *American Naturalist* 44:739–46.

Emerson, S. 1938. The genetics of self-incompatibility in *Oenothera organensis*. *Genetics* 23:190–202.

———. 1939. A preliminary survey of the *Oenothera organensis* population. *Genetics* 24:524–37.

Epling, C., and T. Dobzhansky. 1942. Genetics of natural populations, VI: Microgeographic races in *Linanthus parryae*. *Genetics* 27:317–32.

Epling, C., H. Lewis, and F. M. Ball. 1960. The breeding group and seed storage: A study in population dynamics. *Evolution* 14:238–55.

Evans, H. E., and M. A. Evans. 1970. *William Morton Wheeler, biologist*. Cambridge: Harvard University Press.

Ewens, W. J. 1964. On the problem of self-sterility alleles. *Genetics* 50: 1433–38.

Fisher, R. A. 1918. The correlation between relatives on the supposition of Mendelian inheritance. *Transactions of the Royal Society of Edinburgh* 52:399–433.

———. 1922. On the dominance ratio. *Proceedings of the Royal Society of Edinburgh* 42:321–41.

———. 1925. *Statistical methods for research workers*. Edinburgh: Oliver and Boyd.

———. 1927. On some objections to mimicry theory: Statistical and genetic. *Transactions of the Royal Entomological Society of London* 75:269–78.

———. 1928a. The possible modification of the response of the wild type to recurrent mutation. *American Naturalist* 62:115–26.

———. 1928b. Two further notes on the origin of dominance. *American Naturalist* 62:571–74.

———. 1929. The evolution of dominance: A reply to Professor Sewall Wright. *American Naturalist* 63:553–56.

———. 1930a. *The genetical theory of natural selection*. Oxford: Oxford University Press. 2d., 1958. New York: Dover.

———. 1930b. The evolution of dominance in certain polymorphic species. *American Naturalist* 64:385–406.

———. 1930c. The distribution of gene ratios for rare mutations. *Proceedings of the Royal Society of Edinburgh* 50:205-20.

———. 1934. Professor Wright on the theory of dominance. *American Naturalist* 68:370–74.

———. 1961a. Possible differentiation in the wild population of *Oenothera organensis*. *Australian Journal of Biological Sciences* 14:76–78.

———. 1961b. A model for the generation of self-sterility alleles. *Journal of Theoretical Biology* 1:411–14.

———. 1962. Self-sterility alleles: A reply to Professor D. Lewis. *Journal of Theoretical Biology* 3:146–47.

Fisher, R. A., and C. Diver. 1934. Crossing-over in the land snail *Cepaea nemoralis* L. *Nature* 133:834–35.

Fisher, R. A., and E. B. Ford. 1926. Variability of species. *Nature* 118:515–16.

————. 1928. The variability of species in the Lepidoptera, with reference to abundance and sex. *Transactions of the Royal Entomological Society of London* 76:367–79.

————. 1947. The spread of a gene in natural conditions in a colony of the moth *Panaxia dominula. Heredity* 1:143-74.

————. 1950. The "Sewall Wright effect." *Heredity* 4:117-19.

Ford, E. B. 1931. *Mendelism and evolution.* London: Methuen. 5th ed., 1949.

————. 1940. Polymorphism and taxonomy. In *The new systematics,* ed. J.S. Huxley, 493–513. Oxford: Oxford University Press.

————. 1964. *Ecological genetics.* London: Methuen. 2d ed., 1965 (London: Chapman and Hall); 3d ed., 1971; 4th ed., 1975.

————. 1980. Some recollections pertaining to the evolutionary synthesis. In *The evolutionary synthesis,* ed. E. Mayr and W. Provine, 334–42. Cambridge: Harvard University Press.

Fox, A. 1949a. Immunogenetic studies of *Drosophila melanogaster,* I: The development of techniques and the detection of strain differences. *Journal of Immunology* 62:13–27.

————. 1949b. Immunogenetic studies of *Drosophila melanogaster,* II: Interaction between the *Rb* and *V* loci in production of antigens. *Genetics* 34:647–64.

Freeman, R. B. 1977. *The works of Charles Darwin.* 2d ed. London: Dawson.

Froggatt, P., and N. C. Nevin. 1971. The "law of ancestral heredity" and the Mendelian-ancestrian controversy in England. *Journal of Medical Genetics* 8:1–36.

Galton, F. 1889. *Natural inheritance.* London: Macmillan.

Garrod, A. E. The incidence of alcaptonuria, a study in chemical individuality. *Lancet* 2:1616–20.

Ginsburg, B. 1944. The effects of the major genes controlling coat color in the guinea pig on the dopa oxidase activity of skin extracts. *Genetics* 29:176–98.

Glass, H. B., ed. 1980a. *Roving naturalist: Travel letters of Theodosius Dobzhansky.* Memoirs of the American Philosophical Society, vol. 139.

————. 1980b. The strange encounter of Luther Burbank and George Harrison Shull. *Proceedings of the American Philosophical Society* 124:133–53.

Goldschmidt, R. B. 1927. *Physiologische Theorie der Vererburg.* Berlin: Springer.

————. 1934. *Lymantria. Biblioteca Genetica* 11:1–189.

————. 1937. *Cynips* and *Lymantria. American Naturalist* 71:508–14.

————. 1938. *Physiological genetics.* New York: McGraw-Hill.

————. 1940. *The material basis of evolution.* New Haven: Yale University Press. Reprinted 1982, with introduction by S. Rachootin.

————. 1960. *In and out of the ivory tower.* Seattle: University of Washington Press.

Gortner, R. A. 1911. Studies on melanin, IV. The origin of the pigment and color pattern in the elytra of the Colorado potato beetle. *American Naturalist* 45:743–55.

Gould, S. J. 1980. G. G. Simpson, paleontology, and the modern synthesis. In *The evolutionary synthesis,* ed. E. Mayr and W. Provine, 153–71. Cambridge: Harvard University Press.

———. 1982. Introduction. In T. Dobzhansky, *Genetics and the origin of species,* xvii-xli. Reprint of 1st ed. New York: Columbia University Press.

———. 1983. The hardening of the modern synthesis. In *Dimensions of Darwinism,* ed. M. Grene, 71–93. New York: Cambridge University Press.

Gould, S. J., and R. C. Lewontin. 1979. The spandrels of San Marco and the Panglossian paradigm: A critique of the adaptationist programme. *Proceedings of the Royal Society of London* B 205:581–98.

Gregory, R. P. 1914. On the genetics of tetraploid plants in *Primula sinensis. Proceedings of the Royal Society of London* B 87.

Gulick, A. 1932. *John Thomas Gulick: Evolutionist and missionary.* Chicago: University of Chicago Press.

Gulick, J. T. 1872. On diversity of evolution under one set of external conditions. *Journal of the Linnaean Society of London* (Zoology) 11:496–505.

———. 1888. Divergent evolution through cumulative segregation. *Journal of the Linnaean Society of London* (Zoology) 20:189-274.

Hagedoorn, A. L., and A. C. Hagedoorn. 1914. Studies on variation and selection. *Zeitschrift für Induktive Abstammungs- und Vererbungslehre* 11:145–83.

———. 1921. *On the relative value of the processes causing evolution.* The Hague: Martinus Nijhoff.

Haldane, J. B. S. 1924. A mathematical theory of natural and artificial selection, I. *Transactions of the Cambridge Philosophical Society* 23:19–41.

———. 1930. A note on Fisher's theory of the origin of dominance, and on a correlation between dominance and linkage. *American Naturalist* 64:87–90.

———. 1931. A mathematical theory of natural selection, part VIII: Metastable populations. *Proceedings of the Cambridge Philosophical Society* 27:137–42.

———. 1932. *The causes of evolution.* London: Longmans.

———. 1957. The cost of natural selection. *Journal of Genetics* 55:511–24.

———. 1964. A defense of beanbag genetics. *Perspectives in Biology and Medicine* 7:343–30.

Haldane, J. B. S., N. M. Haldane, and A. D. Sprunt. 1915. Reduplication in mice. *Journal of Genetics* 5:133–35.

Haller, M.H. 1963. *Eugenics: Hereditarian attitudes in American thought.* New Brunswick, N. J.: Rutgers University Press.

Hamilton, W. D. 1964. The genetical evolution of social behavior, I and II. *Journal of Theoretical Biology* 17:1–54.

Hartl, D. L. 1979. Four volume treatise on population biology. *BioScience* 29:179–80.

Harwood, J. 1984. The reception of Morgan's chromosome theory in Germany: Inter-war debate over cytoplasmic inheritance. *Medizin historiisches Journal* 19:3–32.

Heidenthal, G. 1939. A colorimetric study of genic effect on guinea pig coat color. *Genetics* 25:197–214.

Heincke, E. 1898. Naturgeschichte der Herings. *Abhandlungen der Deutschen Leerfischereivereins* 2:1–223.

Henderson, L. J. 1913. *The fitness of the environment: An enquiry into the biological significance of the properties of matter.* New York: Macmillan.

Hubby, J. L., and R. C. Lewontin. 1966. A molecular approach to the study of genic heterozygosity in natural populations, I: The number of alleles at different loci in *Drosophila pseudoobscura. Genetics* 54:577–94.

Huxley, J. S. 1932. *Problems of relative growth.* London: Methuen.

———. 1942. *Evolution: The modern synthesis.* London: Allen and Unwin.

———. 1950. Genetics, evolution and human destiny. In *Genetics in the 20th century,* ed. L. C. Dunn, 591–621. New York: Macmillan

———, ed. 1940. *The new systematics.* Oxford: Oxford University Press.

Huxley, J. S., A. C. Hardy, and E. B. Ford, eds. 1954. *Evolution as a process.* London: Allen and Unwin.

Jacot, A. P. 1932. The status of the species and the genus. *American Naturalist* 66:346–64.

Jennings, H. S. 1931. *The biological basis of human nature.* New York: Norton.

Jones, J. S., B. H. Leith, and P. Rawlings. 1977. Polymorphism in *Cepaea*: A problem with too many solutions? *Annual Review of Ecology and Systematics* 8:109–43.

Jordan, D. S. 1898. *Footnotes to evolution.* New York: D. Appleton.

———. 1923. John Thomas Gulick, missionary and Darwinian. *Science* 44:509.

———. 1925. Isolation with segregation as a factor in organic evolution. In *Smithsonian Report for 1925,* 321–26. Washington, D.C.

Jordan, D. S., and V. L. Kellogg. 1907. *Evolution and animal life.* New York: Appleton.

Kellogg, V. L. 1907. *Darwinism To-Day.* New York: Holt.

Kimura, M. 1950. Mathematical population genetics (in Japanese). *Modern Biology* 2:289–341.

———. 1951. Effect of random fluctuation of selective value on the distribution of gene frequencies in natural populations. *Annual Report of the National Institute of Genetics* 1:45-47.

———. 1952a. Fluctuations of adaptive values and the frequency distribution of heterotic genes in natural populations. *Annual Report of the National Institute of Genetics* 2:53–56.

———. 1952b. Process of irregular change of gene frequencies due to the random fluctuation of selection intensities. *Annual Report of the National Institute of Genetics* 2:56–57.

———. 1952c. On "effective size of populations." *Annual Report of the National Institute of Genetics* 2:57–60.

———. 1952d. On the process of decay of variability due to random extinction of alleles. *Annual Report of the National Institute of Genetics* 2:60–61.

———. 1954. Process leading to quasi-fixation of genes in natural populations due to random fluctuation of selection intensities. *Genetics* 39:280–95.

———. 1955a. Solution of a process of random genetic drift with a continuous model. *Proceedings of the National Academy of Sciences* USA 41:144–50.

———. 1955b. Random genetic drift in a multi-allelic locus. *Evolution* 9: 419–35.

———. 1955c. Stochastic processes and distribution of gene frequencies under natural selection. *Cold Spring Harbor Symposia* 20:33–53.

———. 1965. Simulation studies on the number of self-sterility alleles maintained in a small population. *Annual Reports of the National Institute of Genetics* 16:86–88.

———. 1968. Evolutionary rate at the molecular level. *Nature* 217:624–26.

———. 1983. *The neutral theory of molecular evolution.* Cambridge: Cambridge University Press.

———. 1985. Blooming of diffusion models of population genetics in the age of molecular biology. In *The craft of probabilistic modelling,* ed. J. Gani. New York: Springer Verlag.

Kimura, M., and J. F. Crow. 1963. The measurement of effective population number. *Evolution* 17:279–88.

Kimura, M., and T. Ohta. 1971a. Protein polymorphism as a phase of molecular evolution. *Nature* 229:467–69.

———. 1971b. Population genetics, molecular biometry, and evolution. *Proceedings of the Sixth Berkeley Symposium on Mathematical Statistics and Probability.* Berkeley: University of California Press, 43–68.

King, J. L., and T. H. Jukes. 1969. Non-Darwinian evolution: Random fixation of selectively neutral mutations. *Science* 164:788–98.

Kinsey, A. C. 1930. *The gall wasp genus Cynips: A study in the origin of species.* Indiana University Studies 16, nos. 84, 85, and 86.

Koller, P. C. 1932. The relation of fertility factors to crossing-over in the *Drosophila obscura* hybrid. *Zeitschrift für induktive Abstammungs- und Vererbungslehre* 60:137–51.

———. 1939. Genetics of natural populations, III: Gene arrangements in populations of *Drosophila pseudoobscura* from contiguous localities. *Genetics* 24:22–33.

Kottler, M. J. 1974. Alfred Russel Wallace, the origin of man, and spiritualism. *Isis* 65:144–92.

———. 1976. Isolation and speciation, 1837–1900. Ph.D. diss., Yale University.

Kruskal, W. 1980. The significance of Fisher: A review of *R. A. Fisher: The life of a scientist. Journal of the American Statistical Association* 75: 1019–30.

Kuhn, T. S. 1962. *The structure of scientific revolutions.* Chicago: University of Chicago Press.

Lack, D. L. 1945. *The Galapagos finches (Geospizinae): a study in variation.* Occasional Papers, no. 21. San Francisco: California Academy of Sciences.

———. 1947. *Darwin's finches.* Cambridge: Cambridge University Press. Reprinted 1983.

———. 1973. David L. Lack: Obituary. *Ibis* 115:421–41.

Lamotte, M. 1951. *Rescherches sur la structure génétique des populations naturelles de Cepaea nemoralis L.* Supplement to *Bulletin Biologique de France et de Belgique,* no. 35.

———. 1952. Le rôle des fluctations fortuites dans la diversité des populations naturelles de *Cepaea nemoralis L. Heredity* 6:333–43.

———. 1959. Polymorphism of natural populations of *Cepaea nemoralis. Cold Spring Harbor Symposia on Quantitative Biology* 24:65–86.

Lancefield, D. E. 1929. A genetic study of crosses of two races or physiological species of *Drosophila obscura. Zeitschrift für induktive Abstammungs- und Vererbungslehre* 52:287–317.

Lande, R. 1976. Natural selection and random genetic drift in phenotypic evolution. *Evolution* 30:314–34.

———. 1979. Quantitative genetic analysis of multivariate evolution, applied to brain: Body size allometry. *Evolution* 33:402–16.

Lande, R., and S. J. Arnold. 1983. The measurement of selection on correlated characters. *Evolution* 37:1210–26.

Lang, Arnold. 1910. Die Erblichkeitsverhältnisse der Ohrenlänge der Kaninchen nach Castle und das Problem der intermediären Vererbung und Bildung konstanter Bastard-rassen. *Zeitschrift für induktive Abstammungs- und Vererbungslehre* 4:1–23.

Lesch, J. E. 1975. The role of isolation in evolution: George J. Romanes and John T. Gulick. *Isis* 66:483–503.

Lewontin, R. C. 1970. The units of selection. *Annual Review of Ecology and Systematics* 1:1–18.

Lewontin, R. C., and J. L. Hubby. 1966. A molecular approach to the study of genic heterozygosity in natural populations, II: Amount of variation and degree of heterozygosity in natural populations of *Drosophila pseudoobscura. Genetics* 54:595–609.

Lewontin, R. C., J. A. Moore, W. B. Provine, and B. Wallace. 1981. *Dobzhansky's genetics of natural populations.* New York: Columbia University Press.

Lewontin, R. C., and M. J. D. White. 1960. Interaction between inversion polymorphisms and two character pairs in the grasshopper, *Moraba scurra. Evolution* 14:116–29.

Little, C. C. 1911. The "dilute" forms of yellow mice. *Science* 33:896–97.

———. 1913. Experimental studies of the inheritance of color in mice. Carnegie Institution of Washington Publication No. 179.

———. 1957. *The inheritance of coat color in dogs.* Ithaca, N. Y.: Cornell University Press.

Lock, R. H. 1906. *Recent progress in the study of variation, heredity, and evolution.* London: Murray.

Lush, J. L. 1925. The possibility of sex control by artificial insemination with centrifuged spermatozoa. *Journal of Agricultural Research* 30:893–913.

———. 1937. *Animal breeding plans.* Ames, Iowa: Collegiate Press.

MacCurdy, H. and W.E. Castle. 1907. Selection and cross breeding in relation to the inheritance of coat-pigments and coat-patterns in rats and guinea pigs. Carnegie Institution of Washington Publication No. 70.

MacDowell, E. C. 1914. Size inheritance in rabbits. Carnegie Institution of Washington Publication No. 196.

———. 1915. Bristle inheritance in *Drosophila*. *Journal of Experimental Zoology* 19:61–97.

———. 1916. Piebald rats and multiple factors. *American Naturalist* 50: 719–42.

———. 1917. Bristle inheritance in *Drosophila*, II: Selection. *Journal of Experimental Zoology* 23:109–46.

Mayo, O. 1966. On the problem of self-incompatibility alleles. *Biometrics* 22:111–20.

Mayr, E. 1940. Speciation phenomena in birds. *American Naturalist* 74:249–78.

———. 1942. *Systematics and the origin of species*. New York: Columbia University Press.

———. 1954. Change of genetic environment and evolution. In *Evolution as a process,* ed. J. Huxley, A. C. Hardy, and E. B. Ford, 157–80. London: Allen and Unwin.

———. 1955. Integration of genotypes: Synthesis. *Cold Spring Harbor Symposia on Quantitative Biology* 20:327–33.

———. 1959. Where are we? *Cold Spring Harbor Symposia on Quantitative Biology* 24:1–14.

———. 1963. *Animal species and evolution*. Cambridge: Harvard University Press.

———. 1964. Introduction. In Charles Darwin, *On the Origin of Species,* a facsimile of the first edition. Cambridge: Harvard University Press.

———. 1969. Introduction: The role of systematics in biology. In *Systematic biology,* 4–15. Publication no. 1692 of the National Academy of Sciences-National Research Council, Washington, D. C. Reprinted as "The role of systematics in biology." In E. Mayr, *Evolution and the diversity of life: Selected essays,* 416–424. Cambridge: Harvard University Press, 1976.

———. 1982. *The growth of biological thought*. Cambridge: Harvard University Press.

———. 1983. How to carry out the adaptationist program? *American Naturalist* 121:324–34.

Mayr, E., and Provine, W. B., eds. 1980. *The evolutionary synthesis*. Cambridge: Harvard University Press.

Mayr, E., and E. Stresemann. 1950. Polymorphism in the chat genus *Oenanthe* (Aves). *Evolution* 4:291–300.

Michod, R.E. 1982. The theory of kin selection. *Annual Review of Ecology and Systematics* 13:23–55.

Moran, P. A. P. 1962. *The statistical processes of evolutionary theory*. Oxford: Oxford University Press.

Morgan, T. H. 1903. *Evolution and adaptation*. New York: Macmillan.

———. 1907. *Experimental zoology*. New York: Macmillan.

———. 1927. *Experimental embryology*. New York: Columbia University Press.

———. 1934. *Embryology and genetics*. New York: Columbia University Press.

Morgan, T. H., H. J. Muller, A. H. Sturtevant, and C. B. Bridges. 1915. *The mechanism of Mendelian heredity*. New York: Holt.

Muller, H. J. 1914a. The bearing of the selection experiments of Castle and Phillips on the variability of genes. *American Naturalist* 48:567–76.

———. 1914b. A new mode of segregation of Gregory's tetraploid *Primulas*. *American Naturalist* 48:508–12.

———. 1932. Further studies on the nature and causes of gene mutations. *Proceedings of the Sixth International Congress of Genetics* 1:213–55.

Newman, H. H. 1921. *Readings in evolution, genetics, and eugenics*. Chicago: University of Chicago Press.

———. 1948. History of the Department of Zoology in the University of Chicago. *Bios* 19:215–39.

Niles, H. E. 1922. Correlation, causation, and Wright's theory of "path coefficients." *Genetics* 7:258–73.

———. 1923. The method of path coefficients: An answer to Wright. *Genetics* 8:256–60.

Nilsson-Ehle, H. 1909. Kreuzungsuntersuchungen an Hafer und Weizen. *Lunds Universitets Årsskrift*, n.s., series 2, vol. 5, no. 2.

Norton, B. J. 1973. The biometric defense of Darwinism. *Journal of the History of Biology* 6:283–316.

Ohta, T. 1973. Slightly deleterious mutant substitutions in evolution. *Nature* 246:96–98.

Onslow, H. 1915. A contribution to our knowledge of the chemistry of coat colors in animals and of dominant and recessive whiteness. *Proceedings of the Royal Society of London* B 89:36–58.

Osborn, H. F. 1910. *The age of mammals*. New York: Scribners.

———. 1927. The origin of species, V: Speciation and mutation. *American Naturalist* 61:5–42.

Osgood, W. 1909. *Revision of mice of the American genus Peromyscus*. Biological survey, U.S. Department of Agriculture.

Payne, F. 1918a. The effect of artificial selection on bristle number in *Drosophila ampelophila* and its interpretation. *Proceedings of the National Academy of Sciences* 4:55-58.

———. 1918b. *An experiment to test the nature of variation on which selection acts*. Indiana University Studies No. 36.

———. 1920. Selection for high and low bristle number in the mutant strain "reduced." *Genetics* 5:501–42.

Pearl, R. 1913. A contribution towards an analysis of the problem of inbreeding. *American Naturalist* 47:577–614.

———. 1914. On the results of inbreeding a Mendelian population: A correction and extension of previous conclusions. *American Naturalist* 48:57–62.

———. 1917. The probable error of a Mendelian class frequency. *American Naturalist*. 51:144–56.

Pearson, K. 1900. *The grammar of science*. 2d ed., revised and enlarged. London: A.& C. Black.

———. 1903. Mathematical contributions to the theory of evolution, XI: On the influence of natural selection on the variability and correlation of organs. *Philosophical Transactions of the Royal Society of London* A 200:1–66.

―――. 1904. A Mendelian's view of the law of ancestral inheritance. *Biometrika* 3:109–12.

―――. 1910. Darwinism, biometry, and some recent biology, I. *Biometrika* 7:368–85.

Phillips, J. C. 1912. Size inheritance in ducks. *Journal of Experimental Zoology* 12:369–80.

Pictet, A., and A. Ferrero. 1934. Sur des cas d'apparentes anomalies Mendéliennes et sur deux mutations dominantes du cobaye domestique. *Genetica* 16:77–110.

Plunkett, C. R. 1932a. A contribution to the theory of dominance. *American Naturalist* 67:84–85.

―――. 1932b. Temperature as a tool of research in phenogenetics: Methods and results. *Proceedings of the Sixth International Congress of Genetics* 2:158–60.

Popenoe, P., and R. H. Johnson. 1918. *Applied eugenics*. New York: Macmillan.

Poulton, E. B. 1908. *Essays on evolution*. Oxford: Oxford University Press.

Powell, J. R., T. Dobzhansky, J. E. Hook, and H. E. Wistrand. 1976. Genetics of natural populations, XLIII: Further studies on rates of dispersal of *Drosophila pseudoobscura* and its relatives. *Genetics* 82:493–506.

Provine, W. B. 1971. *Origins of theoretical population genetics*. Chicago: University of Chicago Press.

―――. 1978. The role of mathematical population geneticists in the evolutionary synthesis of the 1930s and 1940s. *Studies in the History of Biology* 2:167–92.

―――. 1979. Francis B. Sumner and the evolutionary synthesis. *Studies in the History of Biology* 3:211–40.

―――. 1981. Origins of the genetics of natural populations series. In *Dobzhansky's genetics of natural populations,* ed. Lewontin, Moore, Provine, and Wallace, 1–76. New York: Columbia University Press.

―――. 1983. The development of Wright's theory of evolution: Systematics, adaptation, and drift. In *Dimensions of Darwinism,* ed. M. Grene, 43–70. Cambridge: Cambridge University Press.

―――. 1985. Adaptation and mechanisms of evolution after Darwin: A study in persistent controversies. In *The Darwinian heritage,* ed. D. Kohn, 825–866. Princeton: Princeton University Press.

Punnett, R. C. 1905. *Mendelism*. Cambridge: Macmillan & Bowes.

―――. 1911. Mendelism. In *Encyclopaedia Britannica,* 11th ed., vol. 18, 118–20.

―――. 1912. Inheritance of coat-colour in rabbits. *Journal of Genetics* 2:221–38.

―――. 1915. *Mimicry in butterflies*. Cambridge: Cambridge University Press.

―――. 1941. Narrative. *Proceedings of the seventh international congress of genetics*. Supplementary volume, *Journal of Genetics*. Cambridge: Cambridge University Press.

Ravin, A. W. 1977. The gene as catalyst: The gene as organism. *Studies in the History of Biology* 1:1–45.

Reid, G. A. 1905. *The principles of heredity*. New York: Dutton.

Rensch, B. 1929. *Das Prinzip geographischer Rassenkneise und das Problem der Artbildung*. Berlin: Borntraeger.

Rhodes, F. H. T. 1983. Gradualism, punctuated equilibrium, and the *Origin of species*. *Nature* 305:269–72.

Richards, O. W., and G. C. Robson. 1926. The species problem and evolution. *Nature* 117:345–47, 382–84.

Riddle, Oscar. 1909. Our knowledge of melanin color formation and its bearing on the Mendelian description of heredity. *Biological Bulletin* 16:316–51.

Robson, G. C. 1928. *The species problem*. London: Oliver and Boyd.

Robson, G. C., and O. W. Richards. 1936. *The variation of animals in nature*. London: Longmans.

Romanes, G. J. 1886. Physiological selection: An additional suggestion on the origin of species. *Journal of the Linnaean Society of London* 19:337–411.

———. 1894, 1895, 1897. *Darwin and After Darwin*. 3 vols. London: Longmans, Green, & Co.

Ruse, M. 1980. Charles Darwin and group selection. *Annals of Science* 37:615- 30.

Russell, E. S. 1939. A quantitative study of genic effects on guinea pig coat color. *Genetics* 24:332–55.

Russell, L. B. 1950. X-ray induced developmental abnormalities in the mouse and their use in the analysis of embryological pattern. *Journal of Experimental Zoology* 114:545–96.

Russell, W. L. 1939. Investigation of the physiological genetics of hair and skin color in the guinea pig by means of the dopa reaction. *Genetics* 24:645–67.

Ruthven, A. G. 1908. *Variations and genetic relationships of the garter snakes*. U.S. National Museum, bulletin 61. Washington, D. C.: Smithsonian Institution.

Savage, L. J. 1976. On rereading R. A. Fisher. *Annals of Statistics* 4:441–500.

Schmidt, J. 1918. Racial studies in fishes, I: Statistical investigations with *Zoarces viviparous* L. *Journal of Genetics* 7:107–18.

Schwab, J. J. 1940. A further study of the effects of a random group of genes on shape of spermatheca in *Drosophila melanogaster*. *Genetics* 25:157–77.

Scott, J. P. 1937a. The embryology of the guinea pig, I: A table of normal development. *American Journal of Anatomy* 60:397-432.

———. 1937b. The embryology of the guinea pig, III: The development of the polydactylous monster. A case of growth acceleration at a particular period by a semidominant gene. *Journal of Experimental Zoology* 77:123–57.

———. 1938. The embryology of the guinea pig, II: The development of a polydactylous monster. A new teras produced by the genes *PxPx*. *Journal of Morphology* 62:299–321.

Sheppard, P. M. 1951. A quantitative study of two populations of the moth, *Panaxia dominula*. *Heredity* 5:349–78.

———. 1954. Evolution in bisexually reproducing organisms. In *Evolution as a process*, ed. J. Huxley, A. C. Hardy, and E. B. Ford, 201–18. London: Allen and Unwin.

Shine, I. B., and S. Wroble. 1976. *Thomas Hunt Morgan: Pioneer of genetics*. Lexington: University of Kentucky Press.

Shull, A. F. 1936. *Evolution*. New York: McGraw-Hill.

Silvers, W. K. 1956. Pigment cell migration following transplantation. *Journal of Experimental Zoology* 132:539-56.

———. 1979. *The coat colors of mice: A model for mammalian gene action and interaction*. New York: Springer Verlag.

Simpson, G. G. 1944. *Tempo and mode in evolution*. New York: Columbia University Press.

Slatis, H. M. 1955. Position effects of the brown locus in *Drosophila melanogaster*. *Genetics* 40:5–23.

Sober, E. 1984. *The nature of selection: A philosophical inquiry*. Cambridge: MIT Press.

Spofford, J. B. 1956. The relation between expressivity and selection against eyeless in *Drosophila melanogaster*. *Genetics* 41:938–59.

Stebbins, G. L. 1950. *Variation and evolution in plants*. New York: Columbia University Press.

Stebbins, G. L., and F. G. Ayala. 1981. Is a new evolutionary synthesis necessary? *Science* 213:967–71.

Stent, G. S. 1970. DNA. *Daedalus* 99:909–37.

Sturtevant, A. H. 1913. The Himalayan rabbit case, with some considerations on multiple allelomorphs. *American Naturalist* 47:234–38.

———. 1918. An analysis of the effects of selection. Carnegie Institution of Washington Publication No. 264.

———. 1920. Genetic studies on *Drosophila simulans*, I: Introduction. Hybrids with *Drosophila melanogaster*. *Genetics* 5:488–500.

———. 1937. Autosomal lethals in wild populations of *Drosophila pseudoobscura*. *Biological Bulletin* 73:542–51.

———. 1944. *Drosophila pseudoobscura*. *Ecology* 25:476-77.

Sturtevant, A. H., and T. Dobzhansky. 1936a. Geographical distribution and cytology of "sex ratio" in *Drosophila pseudoobscura* and related species. *Genetics* 21:473–90.

———. 1936b. Inversions in the third chromosome of wild races of *Drosophila pseudoobscura* and their use in the study of the history of the species. *Proceedings of the National Academy of Science* 22:448–50.

Sulloway, F. J. 1979. Geographic isolation in Darwin's thinking: The vicissitudes of a crucial idea. *Studies in the History of Biology* 3:23–65.

Tait, P. G. 1890. *Elementary treatise on quaternions*. 3d ed. Cambridge: Cambridge University Press.

Tan, C. C. 1935. Salivary gland chromosomes in the two races of *Drosophila pseudoobscura*. *Genetics* 20:392–402.

Templeton, A. R. 1981. Mechanisms of speciation: A population genetic approach. *Annual Review of Ecology and Systematics* 12:23–48.

Thayer, H. S., ed. 1953. *Newton's philosophy of nature*. New York: Harper.

Thompson, D. H. 1931. Variation in fishes as a function of distance. *Transactions of the Illinois State Academy of Science* 23:276–81.

Thomson, J. A. 1910. *Darwinism and human life*. London: Andrew Melrose.

Timoféeff-Ressovsky, N. W. 1937. *Experimentelle Mutationsforschung in der Vererbungslehre*. Dresden: Steinkopff.

————. 1939. Genetik and Evolution. *Zeitschrift für induktive Abstammungs- und Vererbungslehre* 76:158–218.

————. 1940. Mutations and geographical variation. In *The new systematics*, ed. J. S. Huxley, 73–136. Oxford: Oxford University Press.

Timoféeff-Ressovsky, N. W., and H. A. Timoféeff-Ressovsky 1940a. Populationsgenetische Versuche an *Drosophila*, I: Zeitliche und räumliche Verteilung der Individuen einiger *Drosophila*-Arten über das Gelände. *Zeitschrift für induktive Abstammungs- und Vererbungslehre* 79:28–34.

————. 1940b. Populationsgenetische Versuche an *Drosophila*, II: Aktionsbereiche von *Drosophila funebris* und *Drosophila melanogaster*. *Zeitschrift für induktive Abstammungs- und Vererbungslehre* 79:35–43.

————. 1940c. Populationsgenetische Versuch an *Drosophila*, III: Quantitative Untersuchung an einigen *Drosophila* populationen. *Zeitschrift für induktive Abstammungs- und Vererbungslehre* 79:44–49.

Tower, W. L. 1903. The development of the colors and color patterns of Coleoptera, with observations upon the development of color in other orders of insects. *Chicago Decennial Publications*, 1st series, vol. 10:31–70.

————. 1906. An investigation of evolution in Chrysomelid beetles of the genus Leptinotarsa. Carnegie Institution of Washington Publication no. 48.

Turner, J. R. G. 1972. Selection and stability in the complex polymorphism of *Moraba scurra*. *Evolution* 26:334–43.

Waddington, C. H. 1939. *Introduction to modern genetics*. London: Macmillan.

————. 1953. Epigenetics and evolution. *Symposia of the Society for Experimental Biology* 7:186.

————. 1957. *The strategy of the genes*. London: Allen and Unwin.

Wagner, M. 1868. *Die Darwin'sche Theorie und das Migrationsgesetz der Organismen*. Leipzig: Duncker and Humblot.

————. 1870. *Naturwissenschaftliche Reisen in tropischen Amerika*. Stuttgart: Verlag der J. G. Gotta'schen Buchhandlung.

Wallace. A. R. 1889. *Darwinism*. London: Macmillan.

Wallace, B. 1966. On the dispersal of *Drosophila*. *American Naturalist* 100:551–63.

Weinstein, A. 1980. A note on W. L. Tower's *Leptinotarsa* work. In *The Evolutionary Synthesis*, ed. E. Mayr and W. Provine, 352–53. Cambridge: Harvard University Press.

Welch, D'A. A. 1938. Distribution and variation of *Achatinella mustelina* Mighels in the Waianae mountains, Oahu. *Bishop Museum Bulletin* 151:1–164.

Weldon, W. F. R. 1895. An attempt to measure the death-rate due to the selective destruction of *Carcinus moenas* with respect to a particular dimension. *Proceedings of the Royal Society of London* 58:557–61.

White, M. J. D. 1978. *Modes of Speciation*. San Francisco: W. H. Freeman.

Whiting, P. W. 1914. Heredity of the bristles in the common greenbottle fly, *Lucilia sericata*: A study of factors governing distribution. *American Naturalist* 48:339–55.

Williams, G. C. 1966. *Adaptation and natural selection*. Princeton: Princeton University Press.

Willis, J. C. 1940. *The course of evolution by differentiation or divergent mutation rather than by selection*. Cambridge: Cambridge University Press.

Wilson, D. S. 1980. *The natural selection of populations and communities*. Menlo Park, Calif.: Benjamin/Cummings.

———. 1983. The theory of kin selection. *Annual Review of Ecology and Systematics* 13:23–55.

Winters, L. M. 1930. *Animal breeding*. 2d ed. New York: Wiley.

Wolff, G. L. 1954. A sex difference in the color change of a specific guinea pig genotype. *American Naturalist* 87:381-85.

———. 1955. The effects of environmental temperature on coat color in diverse genotypes of the guinea pig. *Genetics* 40:90–106.

Woodruff, D. S., and S. J. Gould. 1980. Geographic differentiation and speciation in *Cerion*: A preliminary discussion of patterns and processes. *Biological Journal of the Linnean Society* 14:389–416.

Wright, P. G. 1906. *The dreamer*. Galesburg, Ill.: Asgard Press.

Wright, P. G., and E. Q. Wright. 1937. *Elizur Wright: The father of life insurance*. Chicago: University of Chicago Press.

Publications by Sewall Wright

1. 1912 Notes on the anatomy of the trematode, *Microphallus opacus*. *Transactions of the American Microscopical Society* 31:167–75.
2. 1914 Duplicate genes. *American Naturalist* 48:638–39.
3. 1915 The albino series of allelomorphs in guinea pigs. *American Naturalist* 49:140–48.
4. Two color mutations of rats which show partial coupling. With W. E. Castle. *Science* 42:193–95.
5. 1916 An intensive study of the inheritance of color and of other coat characters in guinea pigs with especial reference to graded variation. In Carnegie Institution of Washington Publication No. 241,59–160.
6. 1917 On the probable error of Mendelian class frequencies. *American Naturalist* 51:373–75.
7. The average correlation within subgroups of a population. *Journal of the Washington Academy of Science* 7:532–35.
8. Color inheritance in mammals, I. *Journal of Heredity* 8:224–35.
9. II: The mouse. *Journal of Heredity* 8:373–78.
10. III: The rat. *Journal of Heredity* 8:426–30.
11. IV: The rabbit. *Journal of Heredity* 8:473–75.
12. V: The guinea pig. *Journal of Heredity* 8:476–80.
13. VI: Cattle. *Journal of Heredity* 8:521–27.
14. VII: The horse. *Journal of Heredity* 8:561–64.
15. 1918 VIII: Swine. *Journal of Heredity* 9:33–38.
16. IX: The dog. *Journal of Heredity* 9:89–90.
17. X: The cat. *Journal of Heredity* 9:139–44.
18. XI: Man. *Journal of Heredity* 9:227–40.
19. Pigmentation in guinea pig hair. With H. R. Hunt. *Journal of Heredity* 9:178–81.
20. On the nature of size factors. *Genetics* 3:367–74.
21. 1919 Scientific principles applied to breeding. *Breeders Gazette* 75:401–2.
22. 1920 The relative importance of heredity and environment in determining the piebald pattern of guinea pigs. *Proceedings of the National Academy of Science* 6:320–32.
23. *Principles of livestock breeding*. Bulletin No. 905. U.S. Department of Agriculture.
24. 1921 Correlation and causation. *Journal of Agricultural Research* 20:557–85.
25. Factors in the resistance of guinea pigs to tuberculosis with especial regard to inbreeding and heredity. With Paul A. Lewis. *American Naturalist* 55:20–50.
26. Review of *The origin and development of the nervous system*, by C. M. Child. *Journal of Heredity* 12:72–75.
27. Systems of mating, I: The biometric relation between parent and offspring. *Genetics* 6:111–23.

28.   II: The effects of inbreeding on the genetic composition of a population. *Genetics* 6:124–43.

29.   III: Assortative mating based on somatic resemblance. *Genetics* 6:144–61.

30.   IV: The effects of selection. *Genetics* 6:162–66.

31.   V: General considerations. *Genetics* 6:168–78.

32. 1922 Coefficients of inbreeding and relationship. *American Naturalist* 56:330–38.

33.   The effects of inbreeding and crossbreeding on guinea pigs, I: Decline in vigor. In Bulletin No. 1090, 1–36. U.S. Department of Agriculture.

34.   II: Differentiation among inbred families. Bulletin No. 1090, 37–63. U.S. Department of Agriculture.

35.   III: Crosses between highly inbred families. Bulletin No. 1121. U.S. Department of Agriculture.

36. 1923 Two new color factors of the guinea pig. *American Naturalist* 57:42–51.

37.   The theory of path coefficients: A reply to Niles' criticism. *Genetics* 8:239–55.

38.   Mendelian analysis of the pure breeds of livestock, I: The measurement of inbreeding and relationship. *Journal of Heredity* 14:339–48.

39.   II: The Duchess family of shorthorns as bred by Thomas Bates. *Journal of Heredity* 14:405–22.

40.   The relation between piebald and tortoise shell color pattern in guinea pigs. *Anatomical Record* 23:393.

41.   Factors which determine otocephaly in guinea pigs. With O. N. Eaton. *Journal of Agricultural Research* 26:161–82.

42. 1925 Corn and hog correlations. Bulletin No. 1300. U.S. Department of Agriculture.

43.   The factors of the albino series of guinea pigs and their effects on black and yellow pigmentation. *Genetics* 10:223–60.

44.   Mendelian analysis of the pure breeds of livestock, III: The shorthorns. With H. C. McPhee. *Journal of Heredity* 16:205–15.

45.   An approximate method of calculating coefficients of inbreeding and relationship from livestock pedigrees. With H. C. McPhee. *Journal of Agricultural Research* 31:377–83.

46. 1926 A frequency curve adapted to variation in percentage occurrence. *Journal of the American Statistical Association* 21: 161–78.

47.   Mutational mosaic coat patterns of the guinea pig. With O. N. Eaton. *Genetics* 11:333–51.

48.   Effects of age of parents on characteristics of the guinea pig. *American Naturalist* 60:552–59.

49.   Mendelian analysis of the pure breeds of livestock, IV: The British dairy shorthorns. With H. C. McPhee. *Journal of Heredity* 17:397–401.

50.   Review of *The biology of population growth,* by Raymond Pearl, and *The Natural Increase of Mankind,* by J. Shirley Sweeney. *Journal of the American Statistical Association* 21:493–97.

51. 1927 Transplantation and individuality differentials in inbred families of guinea pigs. With Leo Loeb. *American Journal of Pathology* 3:251–85.

52. The effects in combination of the major color-factors of the guinea pig. *Genetics* 12:530–69.

53. 1928 Review of *The rate of living,* by Raymond Pearl. *Journal of the American Statistical Association* 23:336–39.

54. An eight-factor cross in the guinea pig. *Genetics* 13:508–31.

55. 1929 The persistence of differentiation among inbred families of guinea pigs. With O. N. Eaton. Technical Bulletin No. 103. U.S. Department of Agriculture.

56. Fisher's theory of dominance. *American Naturalist* 63:274–79.

57. The evolution of dominance. *American Naturalist* 63:556–61.

58. The dominance of bar over infra bar in *Drosophila. American Naturalist* 63:479–80.

59. Evolution in a Mendelian population. *Anatomical Record* 44: 287.

60. 1930 Review of *The genetical theory of natural selection,* by R. A. Fisher. *Journal of Heredity* 21:349–56.

61. 1931 Statistical theory of evolution. *Journal of the American Statistical Association* 26, suppl., 201–208.

62. Statistical methods in biology. *Journal of the American Statistical Association* 26, suppl., 155–163.

63. Review of *The measurement of man,* by J. A. Harris, C. M. Jackson, D .C Paterson, and R. E. Scammon. *Journal of the American Statistical Association* 26:358–60.

64. Evolution in Mendelian populations. *Genetics* 16:97–159.

65. On the genetics of number of digits of the guinea pig. *Anatomical Record* 51:115.

66. 1932 On the evaluation of dairy sires. *Proceedings of the American Society for Animal Production.*

67. Complementary factors for eye color in Drosophila. *American Naturalist* 66:282–83.

68. 1932 General, group and special size factors. *Genetics* 17:603–19.

69. Hereditary variations of the guinea pig. *Proceedings of the Sixth International Congress of Genetics* 2:247–49.

70. The roles of mutation, inbreeding, crossbreeding and selection in evolution. *Proceedings of the Sixth International Congress of Genetics* 1:356–66.

71. 1933 Inbreeding and homozygosis. *Proceedings of the National Academy of Science* 19:411–20.

72. Inbreeding and recombination. *Proceedings of the National Academy of Science* 19:420–33.

73. Review of *Order of birth, parent-age, and intelligence,* by L. L. Thurstone and R. L. Jenkins. *Journal of Heredity* 24: 193–94.

74. Review of *Some recent researches in the theory of statistics and actuarial science,* by J. F. Steffensen. *Journal of Heredity* 24:364–66.

75. 1934 Physiological and evolutionary theories of dominance. *American Naturalist* 68:25–53.
76.   The genetics of growth. *Proceedings of the American Society for Animal Production* (1933): 233–37.
77.   Types of subnormal development of the head from inbred strains of guinea pigs and their bearing on the classification and interpretation of vertebrate monsters With K. Wagner. *American Journal of Anatomy* 54:383–447.
78.   On the genetics of submornal development of the head (otocephaly) in the guinea pig. *Genetics* 19:471–505.
79.   Polydactylous guinea pigs: Two types respectively heterozygous and homozygous in the same mutant gene. *Journal of Heredity* 25:359–62.
80.   An analysis of variability in number of digits in an inbred strain of guinea pigs. *Genetics* 19:506–36.
81.   The results of crosses between inbred strains of guinea pigs differing in number of digits. *Genetics* 19:537–51.
82.   Genetics of abnormal growth in the guinea pig. *Cold Spring Harbor Symposium on Quantitative Biology* 2:137–47.
83.   Professor Fisher on the theory of dominance. *American Naturalist* 68:562–65.
84.   The method of path coefficients. *Annals of Mathematical Statistics* 5:161–215.
85. 1935 A mutation of the guinea pig, tending to restore the pentadactyl foot when heterozygous, producing a monstrosity when homozygous. *Genetics* 20:84–107.
86.   The analysis of variance and the correlations between relatives with respect to deviations from an optimum. *Journal of Genetics* 30:243–56.
87.   Evolution in populations in approximate equilibrium. *Journal of Genetics* 30:257–66.
88.   The emergency of novelty: A review of Lloyd Morgan's "emergent" theory of evolution. *Journal of Heredity* 26: 369–73.
89.   On the genetics of rosette pattern in guinea pigs. *Genetica* 17:547–60.
90. 1936 On the genetics of the spotted pattern of the guinea pig. With Herman B. Chase. *Genetics* 21:758–87.
91. 1937 The distribution of gene frequencies in populations. *Science* 85:504.
92.   The distribution of gene frequencies in populations. *Proceedings of the National Academy of Science* 23:307–20.
93.   The hereditary factor in abnormal development. *Proceedings of the Institute of Medicine*. vol. 11, November 15, 1937.
94. 1938 Size of population and breeding structure in relation to evolution. *Science* 87:430–31.
95.   The distribution of gene frequencies under irreversible mutation. *Proceedings of the National Academy of Science* 24: 253–59.
96.   The distribution of gene frequencies in populations of polyploids. *Proceedings of the National Academy of Science* 24: 372–77.

97. 1939 The distribution of self-sterility alleles in populations. *Genetics* 24:538–52.

98. Genetic principles governing the rate of progress of livestock breeding. *Proceedings of the American Society for Animal Production* (1939):18–26.

99. *Statistical genetics in relation to evolution: Actualités scientifiques et industrielles.* 802. Exposés de Biometrie et de la statistique biologique, XIII. Paris: Hermann & Cie.

100. 1940 Breeding structure of populations in relation to speciation. *American Naturalist* 74:232–48.

101. The statistical consequences of Mendelian heredity in relation to speciation. In *The new systematics,* ed. Julian S. Huxley, 161–83. Oxford: Oxford University Press.

102. 1941 A quantitative study of the interactions of the major colour factors of the guinea pig. *Proceedings of the Seventh International Genetics Congress* (1939):319–29.

103. Genetics of natural populations, V: Relations between mutation rate and accumulation of lethals in populations of *Drosophila pseudoobscura.* With T. Dobzhansky. *Genetics* 26:23–51.

104. Review of *A philosophy of science,* by W. J. Werkmeister. *The American Biology Teacher* 3:276–78.

105. Review of *The "Age and Area" concept extended,* by J. C. Willis. *Ecology* 22:345–47.

106. Review of *The material basis of evolution,* by R. Goldschmidt. *The Scientific Monthly* 53:165–70.

107. The physiology of the gene. *Physiological Reviews* 21:487–527.

108. Tests for linkage in the guinea pig. *Genetics* 26:650–69.

109. On the probability of fixation of reciprocal translocations. *American Naturalist* 75:513–22.

110. 1942 Genetics of natural populations, VII: The allelism of lethals in the third chromosome of *Drosophila pseudoobscura.* With T. Dob- zhansky and W. Hovanitz. *Genetics* 27:363–94.

111. The physiological genetics of coat color of the guinea pig. *Biological Symposia* 6:337–55.

112. Statistical genetics and evolution. *Bulletin of the American Mathematical Society* 48:223–46.

113. Comment on "Mating customs in North Carolina," by Florence C. Dudley and William Allen. *Journal of Heredity* 33:333–34.

114. 1943 Isolation by distance. *Genetics* 28:114-38.

115. Analysis of local variability of flower color in *Linanthus parryae. Genetics* 28:139–56.

116. Genetics of natural populations, X: Dispersion rates in *Drosophila pseudoobscura.* With T. Dobzhansky. *Genetics* 28:304– 40.

117. 1945 Physiological aspects of genetics. *Annual Reviews of Physiology* 5:75–106.

118. Genes as physiological agents. General considerations. *American Naturalist* 79:289–303.

119. *Tempo and mode in evolution*: A critical review. *Ecology* 26: 415–19.

120.   The differential equation of the distribution of gene frequencies. *Proceedings of the National Academy of Science* 31: 383–89.
121. 1946   Isolation by distance under diverse systems of mating. *Genetics* 31:39–59.
122.   Genetics of natural populations, XII: Experimental reproduction of some of the changes caused by natural selection in certain populations of *Drosophila pseudoobscura*. With T. Dobzhansky. *Genetics* 31:125–156.
123. 1947   On the genetics of several types of silvering in the guinea pig. *Genetics* 32:115–41.
124.   Genetics of natural populations, XV: Rate of diffusion of a mutant gene through a population of *Drosophila pseudoobscura*. With T. Dobzhansky. *Genetics* 32:303–24.
125. 1948   On the roles of directed and random changes in gene frequency in the genetics of populations. *Evolution* 2:279–94.
126.   Evolution, organic. *Encyclopaedia Britannica*. 14th ed. rev. 8:915–29.
127.   Genetics of populations. *Encyclopaedia Britannica*. 10:111–112.
128. 1949   Adaptation and selection. In *Genetics, paleontology, and evolution*, ed. G.L. Jepson, G. G. Simpson, and E. Mayr, 365–89. Princeton: Princeton University Press.
129.   Colorimetric determination of the amounts of melanin in the hair of diverse genotypes of the guinea pig. With Zora I. Braddock. *Genetics* 34:223–44.
130.   Estimates of the amounts of melanin in the hair of diverse genotypes of the guinea pig from transformation of empirical grades. *Genetics* 34:245–71.
131.   Differentiation of strains of guinea pigs under inbreeding. *Proceedings of the First National Cancer Conference*, 13–27.
132.   Population structure in evolution. *Proceedings of the American Philosophical Society* 93:471–78.
133.   On the genetics of hair direction in the guinea pig, I: Variability in the patterns found in combinations of the R and M loci. *Journal of Experimental Zoology* 112:303–24.
134.   II: Evidences for a new dominant gene, Star, and tests for linkage with eleven other loci. *Journal of Experimental Zoology* 112:325–40.
135. 1949   Dogma or opportunism. Commentary on "The Russian purge of Genetics": History of the conflict. *Bulletin of Atomic Scientists* 5:141–42.
136. 1950   III: Interaction between the processes due to loci R and St. *Journal of Experimental Zoology* 113:33–64.
137.   Discussion on population genetics and radiation. *Journal of Cell and Comparative Physiology* 35:187–210.
138.   Genetical structure of populations. *Nature* 166:247–53.
139.   Population structure as a factor in evolution. In *Moderne Biologie: Festschrift für Hans Nachtsheim*, 274–87. Berlin: F. W. Peter.

140. 1951 The genetical structure of populations. *Annals of Eugenics* 15:323–54.

141.   Fisher and Ford on "the Sewall Wright effect." *American Scientist* 39:452–58, 479.

142. 1952 The genetics of quantitative variability. Agricultural Research Council, Quantitative Inheritance. London: Her Majesty's Stationery Office (1) 1952, pp. 5–41. Reprinted in *Yearbook of Physical Anthropology,* 1952, 159–95.

143.   The theoretical variance within and among subdivisions of a population that is in a steady state. *Genetics* 27:312–21.

144. 1953 Gene and organism. *American Naturalist* 87:5–18.

145.   The interpretation of multivariate systems. In *Statistics and mathematics in biology,* ed. O. Kempthorne, T. A. Bancroft, J. W. Gowen and J. L. Lush, 11–33. Ames: Iowa State College Press.

146. 1954 Experimental studies of the distribution of gene frequencies in very small populations of *Drosophila melanogaster.* With W. E. Kerr. I: Forked. *Evolution* 8:172–77.

147.   II: Bar. *Evolution* 8:225–40.

148.   III: Aristapedia and spineless. *Evolution* 8:293–302.

149.   Summary of patterns of mammalian gene action. *Journal of the National Cancer Institute* 15:837–51.

150. 1955 Discussion on responses of populations to radiation. Conference on genetics held at Argonne Nat. Lab., November 19–20, 1954, 59–62. AEC Division of Biology and Medicine Washington, D.C.

151. 1956 Modes of selection. *American Naturalist* 90:5–24.

152.   Classification of the factors of evolution. *Cold Spring Harbor Symposium on Quantitative Biology* (1955) 20:16–24.

153. 1959 Genetics, the gene, and the hierarchy of biological sciences. *Proceedings of the Tenth International Congress of Genetics* 1: 475–89. Also in *Science* 130:959–65.

154.   On the genetics of silvering in the guinea pig with especial reference to interaction and linkage. *Genetics* 44:383.

155.   Silvering (si) and diminution (dm) of coat color of the guinea pig and male sterility of the white or near-white combination of these. *Genetics* 44:563–90.

156.   A quantitative study of variations in intensity of genotypes of the guinea pig at birth. *Genetics* 44:1001–26.

157.   Qualitative differences among the colors of the guinea pig due to diverse genotypes. *Journal of Experimental Zoology* 142:75–114.

158.   Physiological genetics, ecology of populations, and natural selection. *Perspectives in Biology and Medicine* 3:107–31. Also in *Evolution after Darwin,* ed. Sol Tax, 1:429–55. Chicago: University of Chicago Press, 1960.

159. 1960 On the appraisal of genetic effects of radiation in man. In *The Biological Effects of Atomic Radiation: Summary Reports*

1960, 18–24. National Academy of Science, National Research Council.

160.   On the number of self-incompatibility alleles maintained in equilibrium by a given mutation rate in a population of given size: A re-examination. *Biometrics* 16:61–85.

161.   Path coefficients and path regression: Alternative or complementary concepts. *Biometrics* 16:189–202.

162.   The treatment of reciprocal interaction, with or without lag, in path analysis. *Biometrics* 16:423–45.

163.   Residual variability of intensity of coat color in the guinea pig. *Genetics* 45:583–612.

164.   Thomas Park: President elect. *Science* 131:502–3.

165.   Postnatal changes in the intensity of coat color in diverse genotypes of the guinea pig. *Genetics* 45:1503–29.

166.   The genetics of vital characters of the guinea pig. *Journal of Cell and Comparative Physiology* 56, suppl. 1, 123–51.

167.   Genetics and twentieth century Darwinism: A review and discussion. *American Journal of Human Genetics* 12:365–72.

168. 1962 Review of *The statistical processes of evolutionary theory*, by P. A. P. Moran. *American Scientist*, December 1962, p. 460A.

169. 1963 Discussion of "Systems of mating in mammalian genetics," by E. L. Green and D. F. Doolittle. In *Methodology in mammalian genetics*, ed. W. J. Burdette, 42–53. San Francisco: Holden-Day. 170.Genic interaction. In *Methodology in mammalian genetics*, ed. W. J. Burdette, 159–88. San Francisco: Holden-Day.

171.   William Ernest Castle, 1867–1962. *Genetics* 48:1–5.

172.   Plant and animal improvement in the presence of multiple selective peaks. In *Statistical genetics in plant breeding*, ed. W. D. Hanson and H. F. Robinson, 116–22. National Academy of Science, National Research Council.

173.   Selection toward an optimum and linkage disequilibrium. *Proceedings of the Eleventh International Congress of Genetics* (The Hague) 1:147.

174. 1964 Pleiotropy in the evolution of structural reduction and of dominance. *American Naturalist* 98:65–69.

175.   Biology and the philosophy of science. *Monist* 48:265–90. Also in *The Hartshorne festschrift: Process and divinity*, ed. W. R. Freese and E. Freeman, 101–25. La Salle, Ill: Open Court Publishing.

176.   Stochastic processes in evolution. In *Stochastic models in medicine and biology*, ed. John Gurland, 199–242. Madison: University of Wisconsin Press.

177. 1965 Factor interaction and linkage in evolution. *Proceedings of the Royal Society of London* B 162:80–104.

178.   The distribution of self-incompatibility alleles in populations. *Evolution* 18:609–19.

179.   The interpretation of population structure by F-statistics with special regard to systems of mating. *Evolution* 19: 355–420.

180.   Dr. Wilhelmina Key. *Journal of Heredity* 56:195–96.
181. 1966 Polyallelic random drift in relation to evolution. *Proceedings of the National Academy of Science* 55:1074–81.
182.   Mendel's ratios. In *The origin of genetics: A Mendel source book,* ed. Curt Stern and Eva A. Sherwood, 173–79. San Francisco: W. H. Freeman.
183. 1967 Comments on the preliminary working papers of Eden and Waddington. In *Mathematical challenges to the neo-Darwinian interpretation of evolution,* Wistar Symposium monograph no. 5, 117–20.
184.   The foundations of population genetics. In *Heritage from Mendel,* ed. R. Alexander Brink, 245–63. Madison: University of Wisconsin Press.
185.   "Surfaces" of selective value. *Proceedings of the National Academy of Science* 102:81–84.
186. 1968 Dispersion of *Drosophila melanogaster. American Naturalist* 102:81–84.
187.   Contributions to genetics by J. B. S. Haldane. In *Haldane memorial volume,* ed. K. R. Dronamraju. Baltimore: Johns Hopkins University Press.
188.   *Evolution and the genetics of populations, vol. 1: Genetic and biometric foundations.* Chicago: University of Chicago Press.
189. 1969 Haldane's contributions to population and evolutionary genetics. *Proceedings of the Twelfth International Congress of Genetics* 2:445–51.
190.   Deviations from random combination in the optimum model. *Proceedings of the Twelfth International Congress of Genetics* 2:150.
191.   Deviations from random combination in the optimum model. *Japanese Journal of Genetics* 44, suppl. 1,152–59.
192.   The theoretical course of directional selection. *American Naturalist* 103:561–74.
193.   *Evolution and the genetics of populations, vol. 2: The theory of gene frequencies.* Chicago: University of Chicago Press.
194. 1970 Random drift and the shifting balance theory of evolution. In *Mathematical topics in population genetics,* ed. K. Kojima. Heidelberg: Springer-Verlag.
195. 1971 Evolution, organic. Rev. *Encyclopedia Britannica.*
196. 1973 The origin of the F-statistics for describing the genetic aspects of population structure. In *Genetic structure of populations,* ed. N. E. Morton, 3–25. Honolulu: University of Hawaii Press.
197. 1975 Panpsychism and science. In *Mind in nature,* ed. J. E. Cobb, Jr. and D. R. Griffin. Washington, D.C.: University Press of America.
198. 1977 *Evolution and the genetics of populations, vol. 3: Experimental results and evolutionary deductions.* Chicago: University of Chicago Press.
199.   Modes of evolutionary change of characters. In *Proceedings of*

the International Conference on Quantitative Genetics, pp. 681–701. Ames: Iowa State University Press.

200. 1978 *Evolution and the genetics of populations, vol. 4: Variability within and among natural populations.* Chicago: University of Chicago Press.

201.　The relation of livestock breeding to theories of evolution. *Journal of Animal Science* 46:1192–1200.

202.　The application of path analysis to etiology. In *Genetic Epidemiology,* ed. N. E. Morton. New York: Academic Press.

203.　Review of *Modes of speciation,* by Michael J. D. White. *Paleobiology* 4:373–79.

204. 1980 Genic and organismic selection. *Evolution* 34:825–43.

205. 1982 Character change, speciation, and the higher taxa. *Evolution* 36:427–43.

206.　Dobzhansky's genetics of natural populations. *Evolution* 36:1102–6.

207.　The shifting balance theory and macroevolution. *Annual Review of Genetics* 16:1–19.

208. 1983 On "Path analysis in genetic epidemiology: A critique." *American Journal of Human Genetics* 35:757–68.

209. 1984 The first Meckel oration: On the causes of morphological differences in a population of guinea pigs. *American Journal of Medical Genetics* 18:591–616.

210.　Diverse uses of path analysis. In *Human population genetics,* ed. A. Chakravarti, 1–34. New York: Van Nostrand Reinhold.

Index